STRUCTURAL LOADS

2012 IBC® and ASCE/SEI 7-10

David A. Fanella, Ph.D., S.E., P.E., F.ASCE

Structural Loads - 2012 IBC and ASCE/SEI 7-10

ISBN: 978-1-60983-437-1

Cover Art Director:	Dianna Hallmark
Project Editor:	Daniel Mutz
Typesetting:	Sue Brockman
Project Head:	John Henry
Publications Manager:	Mary Lou Luif

COPYRIGHT © 2012

ALL RIGHTS RESERVED. This publication is a copyrighted work owned by the International Code Council, Inc. Without advance written permission from the copyright owner, no part of this book may be reproduced, distributed or transmitted in any form or by any means, including, without limitation, electronic, optical or mechanical means (by way of example, and not limitation, photocopying or recording by or in an information storage retrieval system). For information on permission to copy material exceeding fair use, please contact: Publications, 4051 West Flossmoor Road, Country Club Hills, IL 60478. Phone 1-888-ICC-SAFE (422-7233).

The information contained in this document is believed to be accurate; however, it is being provided for informational purposes only and is intended for use only as a guide. Publication of this document by the ICC should not be construed as the ICC engaging in or rendering engineering, legal or other professional services. Use of the information contained in this book should not be considered by the user to be a substitute for the advice of a registered professional engineer, attorney or other professional. If such advice is required, it should be sought through the services of a registered professional engineer, licensed attorney or other professional.

Trademarks: "International Code Council" and the "International Code Council" logo and the "International Building Code" are trademarks of International Code Council, Inc.

Errata on various ICC publications may be available at www.iccsafe.org/errata.

First Printing: December 2012

PRINTED IN THE U.S.A.

Contents

Preface v

Acknowledgements vii

About the Author ix

Chapter 1: Introduction 1
    1.1   Overview 1
    1.2   Scope 2
    1.3   References 4

Chapter 2: Load Combinations 5
    2.1   Introduction 5
    2.2   Load Effects 5
    2.3   Load Combinations Using Strength Design or Load and
           Resistance Factor Design 6
    2.4   Load Combinations Using Allowable Stress Design 9
    2.5   Load Combinations with Overstrength Factor 13
    2.6   General Structural Integrity 15
    2.7   Extraordinary Loads and Events 17
    2.8   Examples 18
    2.9   References 28
    2.10 Problems 28

Chapter 3: Dead, Live, Rain and Soil Lateral Loads 31
    3.1   Dead Loads 31
    3.2   Live Loads 31
    3.3   Rain Loads 44
    3.4   Soil Lateral Loads 49
    3.5   Flowcharts 50
    3.6   Examples 54
    3.7   References 71
    3.8   Problems 72

Chapter 4: Snow and Ice Loads 73
    4.1   Introduction 73
    4.2   Snow Loads 73
    4.3   Ice Loads 96
    4.4   Flowcharts 101
    4.5   Examples 111
    4.6   References 140
    4.7   Problems 140

Chapter 5: Wind Loads 145
    5.1   Introduction 145
    5.2   General Requirements 153

5.3	Main Windforce-resisting Systems (MWFRSs)	164
5.4	Components and Cladding	186
5.5	Wind Tunnel Procedure	195
5.6	Alternate All-heights Method	196
5.7	Flowcharts	198
5.8	Examples	232
5.9	References	328
5.10	Problems	328

Chapter 6: Earthquake Loads ... 331

6.1	Introduction	331
6.2	Seismic Design Criteria	337
6.3	Seismic Design Requirements for Building Structures	347
6.4	Seismic Design Requirements for Nonstructural Components	387
6.5	Seismic Design Requirements for Nonbuilding Structures	391
6.6	Flowcharts	393
6.7	Examples	426
6.8	References	473
6.9	Problems	474

Chapter 7: Flood Loads ... 477

7.1	Introduction	477
7.2	Flood Hazard Areas	478
7.3	Flood Hazard Zones	480
7.4	Design and Construction	482
7.5	Examples	494
7.6	References	504
7.7	Problems	505

Chapter 8: Load Paths ... 507

8.1	Introduction	507
8.2	Load Paths for Gravity Loads	508
8.3	Load Paths for Lateral Loads	511
8.4	References	533

Preface

This edition updates this publication to the 2012 *International Building Code®* (IBC®) and the 2010 edition of *Minimum Design Loads for Buildings and Other Structures* (ASCE/SEI 7-10).

Readers who have used previous editions of this publication will immediately notice a significant increase in the explanatory material that is provided in each chapter. The main reason for including this basic background information is to help the reader understand the fundamental concepts that are behind the provisions in the IBC and ASCE/SEI 7.

Like the previous editions, this edition is an essential resource for civil and structural engineers, architects, plan check engineers and students who need an efficient and practical approach to load determination under the 2012 IBC and ASCE/SEI 7-10 standard. It illustrates the application of code provisions and methodology for determining structural loads through the use of numerous flowcharts and practical design examples. Included are the following major topics:

- Load combinations for allowable stress design, load and resistance factor (strength) design, seismic load combinations with vertical load effect and special seismic load combinations, and

- Dead loads, live loads (including live load reduction), rain loads, snow loads, ice loads, wind loads, earthquake load effects and flood loads.

New problem sections are included at the ends of most of the chapters. Solutions to these problems, which are available in a companion document to this publication, further illustrate the proper application of the code provisions.

A new section on ice loads has been added in Chapter 4. Also, a new Chapter 8 was added on load paths. Once loads are properly determined, it is important to understand the paths that these loads take through a structure. Gravity and lateral load paths are presented for various types of structures. The role of diaphragms and collectors is discussed for wind and seismic loads. Included are details that illustrate common load paths for a variety of situations.

Structural Loads - 2012 IBC and ASCE/SEI 7-10 is a multipurpose resource for civil and structural engineers, architects and plan check engineers because it can be used as a self-learning guide as well as a reference manual.

Enhance Your Study Experience

The Solutions Manual to Structural Loads is a free bonus learning tool just right for you.

STRUCTURAL LOADS - 2012 IBC® AND ASCE/SEI 7-10 includes a companion Solutions Manual to help enhance your understanding of how to solve structural load problems. The Solutions Manual restates each problem in the book and provides complete solutions to many practical situations. The Solutions Manual covers Chapters 2 through 7 and includes:

- Chapter 2 – Load combinations
- Chapter 3 – Dead, live, rain and soil lateral loads
- Chapter 4 – Snow and ice loads
- Chapter 5 – Wind loads
- Chapter 6 – Earthquake loads
- Chapter 7 – Flood loads

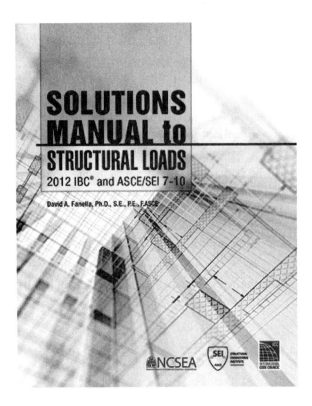

To download your free bonus Solutions Manual to Structural Loads, visit: http://www.iccsafe.org/2012StructuralSolution

Acknowledgements

The writer is deeply grateful to John R. Henry, P.E., Principal Staff Engineer, and Sandra Hyde, P.E., Staff Engineer, both of the International Code Council, Inc., for their thorough review of this publication. Their insightful comments and suggestions for improvement have added significant value to this edition. An additional special thanks goes out to John for suggesting the new chapter on load paths and for making significant contributions to that chapter.

About the Author

David A. Fanella, Ph.D., S.E., P.E., F.ASCE, is Principal and Vice President at Klein and Hoffman Inc., Chicago, Illinois. Dr. Fanella holds a Ph.D. in structural engineering from the University of Illinois at Chicago and is a licensed Structural Engineer in the State of Illinois and a licensed Professional Engineer in numerous states. He is an active member of a number of American Concrete Institute (ACI) Committees and is an Associate Member of the ASCE 7 Committee. Dr. Fanella is past-president and a current board member of the Structural Engineers Association of Illinois. He has authored or coauthored many structural publications, including a recent textbook on reinforced concrete design.

About the International Code Council

The International Code Council (ICC), a membership association dedicated to building safety, fire prevention and energy efficiency, develops the codes and standards used to construct residential and commercial buildings, including homes and schools. The mission of ICC is to provide the highest quality codes, standards, products and services for all concerned with the safety and performance of the built environment. Most United States cities, counties and states choose the International Codes, building safety codes developed by the International Code Council.

The International Codes also serve as the basis for construction of federal properties around the world and as a reference for many nations outside the United States. The International Code Council is also dedicated to innovation and sustainability. ICC Evaluation Service, a subsidiary of ICC, issues Evaluation Reports and Listings for innovative building products as well as environmental documents such as ICC-ES VAR Environmental Reports and ICC-ES Environmental Product Declarations (EPDs).

ICC Headquarters

500 New Jersey Avenue, NW
6th Floor
Washington, DC 20001

District Offices

Birmingham, AL • Chicago, IL • Los Angeles, CA

Telephone

1-888-422-7233 (ICC-SAFE)

www.iccsafe.org

About the Structural Engineering Institute of ASCE

The Structural Engineering Institute (SEI) is a vibrant community of more than 20,000 structural engineers within the American Society of Civil Engineers. SEI started on October 1, 1996 in order to serve the unique needs of the structural engineering community while influencing change on broader issues that shape the entire civil engineering community. Because SEI members are leaders in structural engineering practice and academia, SEI provides great networking opportunities while stimulating coordination and understanding between academia and practicing engineers - driving the practical application of cutting edge research. SEI produces technical publications, journals, conferences, continuing education, codes and standards, and professional practice documents that advance the structural engineering profession. SEI advances our members' careers, stimulates technological advancement, and improves professional practice.

Structural Engineering Institute of ASCE

1801 Alexander Bell Drive
Reston, VA 20191

Telephone

1-800-548-2723

www.asce.org/sei

About the National Council of Structural Engineers Associations

The National Council of Structural Engineers Associations (NCSEA) is comprised of 43 structural engineering associations throughout the United States. NCSEA serves to advance the practice of structural engineering and, as the autonomous national voice for practicing structural engineers, protect the public's right to safe, sustainable and cost effective buildings, bridges, and other structures. NCSEA offers continuing education and communication opportunities for structural engineers by publishing technical books and other publications, including STRUCTURE magazine, providing live monthly webinars by nationally-known speakers, and providing in-person technical programming at the Annual Conference, as well as thought provoking learning and networking opportunities at the Winter Leadership Forum. NCSEA generates and responds to code changes, promotes structural engineering certification and separate licensure, and promotes the practice of structural engineering to students and the general public.

National Council of Structural Engineers Associations

645 North Michigan Avenue
Suite 540
Chicago, IL 60611

Telephone

312-649-4600

www.NCSEA.com

CHAPTER 1

Introduction

1.1 Overview

The purpose of this publication is to assist in the proper determination of structural loads in accordance with the 2012 edition of the *International Building Code®* (IBC®) (Reference 1.1) and the 2010 edition of ASCE/SEI 7 *Minimum Design Loads for Buildings and Other Structures* (Reference 1.2). Chapter 16 of the IBC, Structural Design, prescribes minimum structural loading requirements that are to be used in the design of all buildings and structures. The intent is to subject buildings and structures to loads that are likely to be encountered during their life span, thereby minimizing hazard to life and improving performance during and after a design event.

The snow load provisions in Section 1608, the wind load provisions in Section 1609, the flood load provisions in Section 1612, the earthquake load provisions in Section 1613 and the atmospheric ice loads in Section 1614 of the IBC are based on the provisions of Chapter 7, Chapters 26 through 31, Chapter 5, Chapters 11 through 23 (with some exceptions) and Chapter 10 of ASCE/SEI 7, respectively. These ASCE/SEI 7 chapters are referenced in the aforementioned sections of the IBC. Note that ASCE/SEI 7 is one of a number of codes and standards that is referenced by the IBC. These documents, which can be found in Chapter 35 of the IBC, are considered part of the requirements of the IBC to the prescribed extent of each reference (see Section 101.4 of the IBC).

The seismic requirements of the 2012 IBC and ASCE/SEI 7-10 are based primarily on those in the 2009 edition of *NEHRP Recommended Seismic Provisions for New Buildings and Other Structures* (Reference 1.3). The NEHRP document, which has been updated every three to five years since the first edition in 1985, contains state-of-the-art criteria for the design and construction of buildings anywhere in the United States and its territories that are subject to the effects of earthquake ground motion. Life safety is the primary goal of the provisions. The requirements are also intended to enhance the performance of high-occupancy buildings and to improve the capability of essential facilities to function during and after a design-basis earthquake.

In addition to minimum design load requirements, Chapter 16 contains other important criteria that have a direct impact on the design of buildings and structures, including permitted design methodologies and design load combinations. For example, Section 1615 contains provisions for structural integrity, which are applicable to high-rise buildings that are assigned to Risk Category III or IV and that are bearing wall structures or frame structures. "High-rise buildings" are defined in Section 202 as a building with an occupied floor located more than 75 feet above the lowest level of fire department vehicle access. Definitions of bearing wall structures and frame structures are also given in Section 202.

Risk Categories are defined in IBC Table 1604.5. These categories are used to relate the criteria for maximum environmental loads or distortions that are specified in the code or referenced standards to the consequence that would occur to the structure and its occupants if such loads were exceeded. Prior to the 2012 edition of the IBC, "Occupancy Category" was used. The term "occupancy" relates primarily to issues associated with fire and life safety protection as opposed to the risks associated with structural failure. As such, "Risk Category" was adopted to more clearly identify the nature of the categorization.

Risk Category I buildings and structures are those that are usually unoccupied and, as such, result in negligible risk to the public should they fail. Included are agricultural facilities (such as barns), certain temporary facilities and minor storage facilities.

The vast majority of buildings and structures, including most residential, commercial and industrial facilities, fall under Risk Category II. According to IBC Table 1604.5, any building or structure that is not listed in Risk Category I, III or IV is assigned to Risk Category II.

Included in Risk Category III are buildings and structures that house large numbers of persons, including places of public assembly, educational facilities and institutional facilities. Also included are structures associated with utilities that are required to protect the health and safety of a community, such as power-generating stations and water and sewage treatment facilities. Buildings and other structures that contain certain amounts of toxic or explosive materials also fall within this risk category.

Risk Category IV includes buildings and structures that are essential for a community to cope with emergency situations. Hospitals, fire stations, police stations, rescue facilities and designated emergency shelters are some of the types of structures included in this risk category. Power-generating stations and other public utility facilities required as emergency backup facilities and buildings or structures that house certain quantities of highly toxic materials also fall in this risk category.

It is shown in subsequent chapters of this publication that important factors are directly related to the risk category of a building or structure. In particular, importance factors are referenced in ASCE/SEI 7 Chapter 7 (snow), Chapter 10 (atmospheric ice) and Chapter 11 (seismic) for the four risk categories noted above. The magnitude of an importance factor is different for different risk categories and is based on the statistical characteristics of the environmental loads and the manner in which a building or structure responds to these loads. In general, larger importance factors are assigned in situations where the consequence of failure may be severe.

1.2 Scope

The content of this publication has been substantially increased from previous editions and includes more of the underlying theory and background information for all of the load types that are covered. The load requirements of the IBC and ASCE/SEI 7 are presented in a straightforward manner with emphasis placed on the proper application of the provisions in everyday practice.

Code provisions have been organized in comprehensive flowcharts, which provide a road map that guides the reader through the requirements. Included in the flowcharts are the applicable section numbers and equation numbers from the IBC and ASCE/SEI 7 that pertain to the specific requirements. A basic description of flowchart symbols used in this publication is provided in Table 1.1.

Introduction 3

Table 1.1 Summary of Flowchart Symbols

Symbol		Description
(rounded rectangle)	Terminator	The terminator symbol represents the starting or ending point of a flowchart.
(rectangle)	Process	The process symbol indicates a particular step or action that is taken within a flowchart.
(diamond)	Decision	The decision symbol represents a decision point, which requires a "yes" or "no" response.
(pentagon)	Off-page Connector	The off-page connector symbol is used to indicate continuation of the flowchart on another page.
(circle with cross)	Or	The logical "Or" symbol is used when a process diverges in two or more branches. Any one of the branches attached to this symbol can be followed.
(arrow)	Connector	The connector symbol indicates the sequence and direction of a process.

Numerous completely worked-out design examples are included in the chapters that illustrate the proper application of the code requirements. These examples follow the steps provided in the referenced flowcharts. A section of problems is now available at the end of each chapter. Solutions to these problems are available in a companion document to this publication, which is on ICC's website.

Readers who are interested in the history and design philosophy of the requirements can find detailed discussions in the commentary of *Minimum Design Loads for Buildings and Other Structures* and *Recommended Seismic Provisions for New Buildings and Other Structures* (References 1.2 and 1.3).

Practicing structural engineers, engineers studying for licensing exams, structural plan checkers and others involved in structural engineering, such as advanced undergraduate students and graduate students, will find the flowcharts and the worked-out design examples and problems to be very useful.

Throughout this publication, section numbers from the IBC are referenced as illustrated by the following: Section 1613 of the IBC is denoted as IBC 1613. Similarly, Section 11.4 from ASCE/SEI 7-10 is referenced as ASCE/SEI 11.4 or as 11.4.

Chapter 2 outlines the required load combinations that must be considered when designing a building or its members for a variety of load effects. Load combinations using strength design or load and resistance factor design and load combinations using allowable stress design are both covered. Examples are provided that illustrate the strength design and allowable stress design load combinations for different types of members subject to a variety of load effects.

Dead, live and rain loads are discussed in Chapter 3. The general method and an alternate method of live load reduction are covered, and flowcharts and examples illustrate both methods. The rain load provisions of IBC 1611 are also described, and examples demonstrate the calculation of design rain loads for roofs with scuppers and with a circular drain.

Design provisions for snow loads are given in Chapter 4. A series of flowcharts highlight the requirements, and examples show the determination of flat roof snow loads, sloped roof snow loads, unbalanced roof snow loads and snow drift loads on a variety of flat and sloped roofs, including gable roofs, monoslope roofs, sawtooth roofs and curved roofs. Examples are also given that illustrate design snow loads for parapets and rooftop units. New to Chapter 4 are design provisions for atmospheric ice loads.

Chapter 5 presents the design requirements for wind loads. Flowcharts are provided for the procedures that are allowed to be used when analyzing main windforce-resisting systems and components and cladding. Other flowcharts give step-by-step procedures on how to determine design wind pressures on main windforce-resisting systems and components and cladding of enclosed, partially enclosed and open buildings using the procedures outlined in ASCE/SEI 7. A number of worked-out examples illustrate the design requirements for a variety of buildings and structures.

Earthquake loads are presented in Chapter 6. Information on how to determine design ground accelerations, site class and the seismic design category (SDC) of a building or structure is included, as are the various methods of analysis and their applicability for regular and irregular buildings and structures. Flowcharts and examples are provided that cover seismic design criteria, seismic design requirements for building structures, seismic design requirements for nonstructural components and seismic design requirements for nonbuilding structures.

Chapter 7 contains the requirements for flood loads. Included is information on flood hazard areas and flood hazard zones. Equations are provided for the following types of flood loads: hydrostatic loads, hydrodynamic loads, wave loads (breaking wave loads on vertical pilings and columns, breaking wave loads on vertical and nonvertical walls, and breaking wave loads from obliquely incident waves), and impact loads. Examples illustrate load calculations for a residential building in a Noncoastal A Zone, a Coastal A Zone, and a V Zone.

Load paths are covered in Chapter 8. Gravity and lateral load paths are presented for various types of structures. The role of diaphragms and collectors are discussed for wind and seismic loads. Included are details that illustrate common load paths for a variety of situations.

1.3 References

1.1. International Code Council. 2011. 2012 *International Building Code*. Washington, DC.

1.2. Structural Engineering Institute of the American Society of Civil Engineers (ASCE). 2010. *Minimum Design Loads for Buildings and Other Structures*, ASCE/SEI 7-10. Reston, VA.

1.3. Building Seismic Safety Council. 2009. *NEHRP Recommended Seismic Provisions for New Buildings and Other Structures*, FEMA P-750. Washington, DC.

CHAPTER 2
Load Combinations

2.1 Introduction

In accordance with IBC 1605.1, structural members of buildings and other structures must be designed to resist the load combinations of IBC 1605.2, 1605.3.1 or 1605.3.2. Load combinations that are specified in Chapters 18 through 23 of the IBC, which contain provisions for soils and foundations, concrete, aluminum, masonry, steel and wood, must also be considered. The structural elements identified in ASCE/SEI 12.2.5.2, 12.3.3.3 and 12.10.2.1 must be designed for the load combinations with overstrength factor of ASCE/SEI 12.4.3.2. These load combinations and their applicability are examined in Section 2.5 of this publication.

IBC 1605.2 contains the load combinations that are to be used when strength design or load and resistance factor design is utilized. Load combinations using allowable stress design are given in IBC 1605.3. Both sets of combinations are covered in Sections 2.3 and 2.4 of this publication, respectively. The combinations of IBC 1605.2 or 1605.3 can also be used to check overall structural stability, including stability against overturning, sliding and buoyancy (IBC 1605.1.1).

It is important to understand the difference between permanent loads and variable loads and their role in load combinations. Permanent loads, such as dead loads, do not change or change very slightly over time. Live loads, roof live loads, snow loads, rain loads, wind loads and earthquake loads are all examples of variable loads. These loads are not considered to be permanent because of their inherent degree of variability with respect to time (see the definition of "Loads" in IBC 202).

According to IBC 1605.1, load combinations must be investigated with one or more of the variable loads set equal to zero. It is possible that the most critical load effects on a member occur when one or more variable loads are not present.

ASCE/SEI 2.3 and 2.4 contain load combinations using strength design and allowable stress design, respectively. The load combinations are essentially the same as those in IBC 1605.2 and 1605.3 with some exceptions. Differences in the IBC and ASCE/SEI 7 load combinations are covered in the following sections.

Prior to examining the various load combinations, a brief introduction on load effects is given in Section 2.2.

2.2 Load Effects

The load effects that are included in the IBC and ASCE/SEI 7 load combinations are summarized in Table 2.1. More details on these load effects can be found in those documents, as well as in subsequent chapters of this publication (see the Notes column in Table 2.1, which gives specific locations where more information can be found on the various load effects).

Table 2.1 Summary of Load Effects

Notation	Load Effect	Notes
D	Dead load	See IBC 1606 and Chapter 3 of this publication
D_i	Weight of ice	See IBC 1614, Chapter 10 of ASCE/SEI 7 and Chapter 4 of this publication
E	Combined effect of horizontal and vertical earthquake-induced forces as defined in ASCE/SEI 12.4.2	See IBC 1613, ASCE/SEI 12.4.2 and Chapter 6 of this publication
E_m	Maximum seismic load effect of horizontal and vertical forces as set forth in ASCE/SEI 12.4.3	See IBC 1613, ASCE/SEI 12.4.3 and Chapter 6 of this publication
F	Load due to fluids with well-defined pressures and maximum heights	---
F_a	Flood load	See IBC 1612 and Chapter 7 of this publication
H	Load due to lateral earth pressures, ground water pressure or pressure of bulk materials	See IBC 1610 and Chapter 3 of this publication for soil lateral loads
L	Live load, except roof live load, including any permitted live load reduction	See IBC 1607 and Chapter 3 of this publication
L_r	Roof live load including any permitted live load reduction	See IBC 1607 and Chapter 3 of this publication
R	Rain load	See IBC 1611 and Chapter 3 of this publication
S	Snow load	See IBC 1608 and Chapter 4 of this publication
T	Self-straining force arising from contraction or expansion resulting from temperature change, shrinkage, moisture change, creep in component materials, movement due to differential settlement or combinations thereof	See ASCE/SEI 2.3.5 and 2.4.4
W	Load due to wind pressure	See IBC 1609 and Chapter 5 of this publication
W_i	Wind-on-ice load	See IBC 1614, Chapter 10 of ASCE/SEI 7 and Chapter 4 of this publication

2.3 Load Combinations Using Strength Design or Load and Resistance Factor Design

The basic load combinations where strength design or, equivalently, load and resistance factor design is used are given in IBC 1605.2 and summarized in Table 2.2. These equations establish the minimum required strength that needs to be provided in the members of a building or structure.

Table 2.2 Summary of Load Combinations Using Strength Design or Load and Resistance Factor Design (IBC 1605.2)

IBC Equation No.	Load Combination
16-1	$1.4(D + F)$
16-2	$1.2(D + F) + 1.6(L + H) + 0.5(L_r \text{ or } S \text{ or } R)$
16-3	$1.2(D + F) + 1.6(L_r \text{ or } S \text{ or } R) + 1.6H + (f_1L \text{ or } 0.5W)$
16-4	$1.2(D + F) + 1.0W + f_1L + 1.6H + 0.5(L_r \text{ or } S \text{ or } R)$
16-5	$1.2(D + F) + 1.0E + f_1L + 1.6H + f_2S$
16-6	$0.9D + 1.0W + 1.6H$
16-7	$0.9(D + F) + 1.0E + 1.6H$

f_1 = 1 for places of public assembly live loads in excess of 100 psf, and for parking garages
 = 0.5 for other live loads
f_2 = 0.7 for roof configurations (such as sawtooth) that do not shed snow off the structure
 = 0.2 for other roof configurations

These load combinations apply only to strength limit states; serviceability limit states for deflection, vibration, drift, camber, expansion and contraction and durability are given in Appendix C of ASCE/SEI 7.

The load factors were developed using a first-order probabilistic analysis and a broad survey of the reliabilities inherent in contemporary design practice. The equations in Table 2.2 are meant to be used in the design of any structural member regardless of material in conjunction with the appropriate nominal resistance factors set forth in the individual material specifications. References 2.1 and 2.2 provide information on the development of these load factors along with additional background material.

Factored loads are determined by multiplying nominal loads (that is, loads specified in Chapter 16 of the IBC) by a load factor, which is typically greater than or less than 1.0. Earthquake and wind load effects are an exception to this: a load factor of 1.0 is used to determine the maximum effects from these loads, since they are considered strength-level loads (wind loads are defined for the first time as strength-level loads in the IBC and in ASCE/SEI 7-10; see Chapter 5 of this publication for more information).

Load combinations are constructed by adding to the dead load one or more of the variable loads at its maximum value, which is typically indicated by a load factor of 1.6. Also included in the combinations are other variable loads with load factors less than 1.0; these are companion loads that represent arbitrary point-in-time values for those loads. Certain types of variable loads, such as wind and earthquake loads, act in more than one direction on a building or structure, and the appropriate sign of the variable load must be considered in the load combinations.

Fluid load effects, F, occur in tanks and other storage containers due to stored liquid products. The stored liquid is generally considered to have characteristics of both a dead load and a live load. It is not a purely permanent load since the tank or storage container can go through cycles of being emptied and refilled. The fluid load effect is included in IBC Equations 16-1 through 16-5 where it adds to the effects from the other loads. It is also included in IBC Equation 16-7 where it counteracts the effects from uplift due to seismic load effects, E. Because the wind load effects, W, can be present when the tank is either full or empty, F is not incorporated in IBC Equation 16-6; that is, the maximum effects occur when F is set equal to zero.

Two exceptions are given in IBC 1605.2. According to the first exception, factored load combinations that are specified in other provisions of the IBC take precedence to those listed in IBC 1605.2.

The second exception is applicable where the load, H, resists the primary variable load effect. In cases where H is not permanent, the load factor on H must be taken equal to zero (that is, H is not permitted to resist the primary variable load effect if it is not permanent). The 1.6 load factor on H accounts for the higher degree of uncertainty in lateral forces from bulk materials (which are included in H) compared to that from fluids, F, especially when considering the dynamic effects that are introduced as the bulk material is set in motion by filling operations.

The load combinations given in IBC 1605.2 are the same as those in ASCE/SEI 2.3.2 with the following exceptions:

- The variable f_1 that is present in IBC Equations 16-3, 16-4 and 16-5 is not found in ASCE/SEI combinations 3, 4 and 5. Instead, the load factor on the live load, L, in the ASCE/SEI 7 combinations is equal to 1.0 with the exception that the load factor on L is permitted to equal 0.5 for all occupancies where the live load is less than or equal to 100 psf, except for parking garages or areas occupied as places of public assembly (see exception 1 in ASCE/SEI 2.3.2). This exception makes these load combinations the same in ASCE/SEI 7 and the IBC.

- The variable f_2 that is present in IBC Equation 16-5 is not found in ASCE/SEI combination 5. Instead, a load factor of 0.2 is applied to S in the ASCE/SEI 7 combination. The second exception in ASCE/SEI 2.3.2 states that in ASCE/SEI combinations 2, 4 and 5, S shall be taken as either the flat roof snow load, p_f, or the sloped roof snow load, p_s. This essentially means that the balanced snow load defined in ASCE/SEI 7.3 for flat roofs and in ASCE/SEI 7.4 for sloped roofs can be used in ASCE/SEI combinations 2, 4 and 5. Drift loads and unbalanced snow loads are covered by ASCE/SEI combination 3.

According to IBC 1605.2.1, the load combinations of ASCE/SEI 2.3.3 are to be used where flood loads, F_a, must be considered in design (flood loads are determined by Chapter 5 of ASCE/SEI 7 and are covered in Chapter 7 of this publication). In particular, the following modifications are to be made:

- V Zones or Coastal A Zones

    $1.0W$ in IBC Equations 16-4 and 16-6 shall be replaced by $1.0W + 2.0F_a$

- Noncoastal A Zones

    $1.0W$ in IBC Equations 16-4 and 16-6 shall be replaced by $0.5W + 1.0F_a$

Definitions of Coastal High Hazard Areas (V Zones) and Coastal A Zones are given in ASCE/SEI 5.2 (see Chapter 7 of this publication).

The load factors on F_a are based on a statistical analysis of flood loads associated with hydrostatic pressures, pressures due to steady overland flow, and hydrodynamic pressures due to waves, all of which are specified in ASCE/SEI 5.4.

In cases where self-straining loads, T, must be considered, their effects in combination with other loads are to be determined by ASCE/SEI 2.3.5 (IBC 1605.2.1). Instead of calculating self-straining effects based on upper bound values of this variable like other load effects, the most probable effect expected at any arbitrary point in time is used. More information, including load combinations that should be considered in design, is given in ASCE/SEI C2.3.5.

IBC 1605.2.1 requires that the load combinations of ASCE/SEI 2.3.4 be used where atmospheric ice loads must be considered in design. The following modifications to the load combinations must be made when a structure is subjected to atmospheric ice and wind-on-ice loads (atmospheric and wind-on-ice loads are determined by Chapter 10 of ASCE/SEI 7; see IBC 1614 and Chapter 4 of this publication):

- $0.5(L_r$ or S or $R)$ in ASCE/SEI combination 2 (IBC Equation 16-2) shall be replaced by $0.2D_i + 0.5S$

- $1.0W + 0.5(L_r$ or S or $R)$ in ASCE/SEI combination 4 (IBC Equation 16-4) shall be replaced by $D_i + W_i + 0.5S$

- $1.0W$ in ASCE/SEI combination 6 (IBC Equation 16-6) shall be replaced by $D_i + W_i$

See ASCE/SEI C2.3.4 for more information on the load factors used in these equations.

ASCE/SEI 2.3.6 provides information on how to develop strength design load criteria where no information on loads or load combinations is given in ASCE/SEI 7 or where performance-based design in accordance with ASCE/SEI 1.3.1.3 is being utilized. Detailed information on how to develop such load criteria that is consistent with the methodology used in ASCE/SEI 7 can be found in ASCE/SEI C2.3.6.

2.4 Load Combinations Using Allowable Stress Design

2.4.1 Overview

The basic load combinations where allowable stress design (working stress design) is used are given in IBC 1605.3. A set of basic load combinations is given in IBC 1605.3.1, and a set of alternative basic load combinations is given in IBC 1605.3.2. Both sets are examined below.

2.4.2 Basic Load Combinations

The basic load combinations of IBC 1605.3.1 are summarized in Table 2.3. A factor of 0.75 is applied where these combinations include more than one variable load, since the probability is low that two or more of the variable loads will reach their maximum values at the same time.

Table 2.3 Summary of Basic Load Combinations Using Allowable Stress Design (IBC 1605.3.1)

Equation No.	Load Combination
16-8	$D + F$
16-9	$D + H + F + L$
16-10	$D + H + F + (L_r$ or S or $R)$
16-11	$D + H + F + 0.75L + 0.75(L_r$ or S or $R)$
16-12	$D + H + F + (0.6W$ or $0.7E)$
16-13	$D + H + F + 0.75(0.6W) + 0.75L + 0.75(L_r$ or S or $R)$
16-14	$D + H + F + 0.75(0.7E) + 0.75L + 0.75S$
16-15	$0.6D + 0.6W + H$
16-16	$0.6(D + F) + 0.7E + H$

A factor of 0.6 is applied to the dead load, D, in IBC Equations 16-15 and 16-16, which is meant to limit the dead load that resists horizontal loads to approximately two-thirds of its actual value. Previous editions of the legacy building codes specified that the overturning moment and sliding due to wind load could not exceed two-thirds of the dead load stabilizing moment. This provision was not typically applied to all members in the building. These load combinations apply to the design of all members in a structure and also provide for overall stability of a structure.

As noted in Section 2.3 of this publication, the combined effect of horizontal and vertical earthquake-induced forces, E, is a strength-level load. A factor of 0.7, which is approximately equal to 1/1.4, is applied to E in IBC Equations 16-12, 16-14 and 16-16 to convert the strength-level load to a service-level load. Similarly, a factor of 0.6 is applied to W in IBC Equations 16-12, 16-13 and 16-15.

Five exceptions are given in IBC 1605.3.1. The first exception states that crane hook loads need not be combined with roof live loads or with more than three-fourths of the snow load or one-half of the wind loads. It is important to note this exception does not eliminate the need to combine live loads other than crane live loads with wind and snow loads in the prescribed manner. In other words, the load combinations in IBC Equations 16-11, 16-13 and 16-14 must be investigated without the crane live load and with the crane live load using the criteria in the exception. In particular, the following load combinations must be investigated where crane live loads, L_c, are present:

- IBC Equation 16-11:

  $D + H + F + 0.75L + 0.75(L_r \text{ or } S \text{ or } R)$

  and

  $D + H + F + 0.75(L + L_c) + 0.75(0.75S \text{ or } R)$

- IBC Equation 16-13:

  $D + H + F + 0.75(0.6W) + 0.75L + 0.75(L_r \text{ or } S \text{ or } R)$

  and

  $D + H + F + 0.75[0.6(0.5W)] + 0.75(L + L_c) + 0.75(0.75S \text{ or } R)$

- IBC Equation 16-14:

  $D + H + F + 0.75(0.7E) + 0.75L + 0.75S$

  and

  $D + H + F + 0.75(0.7E) + 0.75(L + L_c) + 0.75(0.75S)$

The second exception in IBC 1605.3.1 states that flat roof snow loads, p_f, that are less than or equal to 30 psf and roof live loads, L_r, that are less than or equal to 30 psf need not be combined with seismic loads. Also, where p_f is greater than 30 psf, 20 percent of the snow load must be combined with seismic loads.

According to the third exception, a load factor of 0.6 shall be included with H when it is permanent and resists the primary variable load effect. For all other conditions, including where H is not permanent, H is to be set equal to zero. In other words, an H effect that is variable is not permitted to help resist the effects from overturning due to wind or earthquake effects; in such cases, H must be set equal to zero in IBC Equations 16-15 and 16-16, respectively.

The fourth exception permits the effects from wind in IBC Equation 16-15 to be reduced in accordance with Exception 2 of ASCE/SEI 2.4.1: W can be replaced with $0.9W$ in that equation for the design of nonbuilding structure foundations and self-anchored, ground-supported tanks provided the conditions set forth in that exception are satisfied. The rationale behind this reduction can be found in ASCE/SEI C2.4.1.

In the fifth exception, $0.6D$ in IBC Equation 16-16 is permitted to be increased to $0.9D$ for the design of special reinforced masonry shear walls that comply with the design requirements in IBC Chapter 21. More information on the reasons behind this permitted increase is given in ASCE/SEI C2.4.1.

Increases in allowable stresses that are given in the materials chapters of the IBC or in referenced standards are not permitted when the load combinations of IBC 1605.3.1 are used (IBC 1605.3.1.1). However, it is permitted to use the duration of load factor when designing wood structures in accordance with Chapter 23 of the IBC, which references the 2012 edition of the *National Design Specification for Wood Construction with 2012 Supplement* (NDS-2012) (Reference 2.3).

According to IBC 1605.3.1.2, the load combinations of ASCE/SEI 2.4.2 are to be used where flood loads, F_a, must be considered in design. In particular, the following modifications are to be made:

- V Zones or Coastal A Zones

  $1.5F_a$ must be added to the other loads in IBC Equations 16-12, 16-13, 16-14 and 16-15, and E is set equal to zero in IBC Equations 16-12 and 16-14.

- Noncoastal A Zones

  $0.75F_a$ must be added to the other loads in IBC Equations 16-12, 16-13, 16-14 and 16-15, and E is set equal to zero in IBC Equations 16-12 and 16-14.

Where self-straining loads, T, must be considered in design, the provisions of ASCE/SEI 2.4.4 are to be used to determine the proper combination of T with other loads (IBC 1605.3.1.2). ASCE/SEI C2.4.4 provides load combinations for typical situations.

IBC 1605.3.1.2 requires that the load combinations of ASCE/SEI 2.4.3 be used where atmospheric ice loads must be considered in design. The following modifications to the load combinations must be made when a structure is subjected to atmospheric ice and wind-on-ice loads:

- $0.7D_i$ shall be added to ASCE/SEI combination 2 (IBC Equation 16-9)

- (L_r or S or R) in ASCE/SEI combination 3 (IBC Equation 16-10) shall be replaced by $0.7D_i + 0.7W_i + S$

- $0.6W$ in ASCE/SEI combination 7 (IBC Equation 16-15) shall be replaced by $0.7D_i + 0.7W_i$

The load combinations of IBC 1605.3.1 and ASCE/SEI 2.4.1 are the same except for the following:

- There is no specific exception for crane loads in ASCE/SEI 2.4.1.

- The first exception in ASCE/SEI 2.4.1 states that in ASCE/SEI combinations 4 and 6, S shall be taken as either the flat roof snow load, p_f, or the sloped roof snow load, p_s. The balanced snow load defined in ASCE/SEI 7.3 for flat roofs and in ASCE/SEI 7.4 for sloped roofs can be used in ASCE/SEI combinations 4 and 6, and drift loads and unbalanced snow loads are covered by ASCE/SEI combination 3.

2.4.3 Alternative Basic Load Combinations

The alternative basic load combinations can be found in IBC 1605.3.2 and are summarized in Table 2.4.

Table 2.4 Summary of Alternative Basic Load Combinations Using Allowable Stress Design (IBC 1605.3.2)

Equation No.	Load Combination
16-17	$D + L + (L_r$ or S or $R)$
16-18	$D + L + 0.6\omega W$
16-19	$D + L + 0.6\omega W + S/2$
16-20	$D + L + S + 0.6\omega W/2$
16-21	$D + L + S + E/1.4$
16-22	$0.9D + E/1.4$

These load combinations are based on the allowable stress load combinations that appeared in the *Uniform Building Code* (Reference 2.4) for many years.

Unlike the basic load combinations of IBC 1605.3.1, allowable stresses are permitted to be increased or load combinations are permitted to be reduced where permitted by the material chapters of the IBC (Chapters 18 through 23) or by referenced standards when the alternative basic load combinations of IBC 1605.3.2 are used. This applies to those load combinations that include wind or earthquake loads.

The alternative allowable stress design load combinations do not include a load combination comparable to IBC Equation 16-15 for dead load counteracting wind load effects. Instead of a specific load combination, IBC 1605.3.2 states that for load combinations that include counteracting effects of dead and wind loads, only two-thirds of the minimum dead load that is likely to be in place during a design wind event is to be used in the load combination.

As noted in the preceding discussion, the combined effect of horizontal and vertical earthquake-induced forces, E, is a strength-level load. This strength-level load is divided by 1.4 in IBC Equations 16-21 and 16-22 to convert it to a service-level load. Similarly, W is multiplied by 0.6 in IBC Equations 16-18, 16-19 and 16-20.

The coefficient ω in IBC Equations 16-18, 16-19 and 16-20 is equal to 1.3 where wind loads are calculated in accordance with ASCE/SEI Chapters 26 through 31 and where allowable stresses have been increased or load combinations have been reduced as permitted by the material chapters of the IBC or the referenced standards in IBC Chapter 35. In all other cases, the coefficient ω is to be taken as 1.0. It is shown in Chapter 5 of this publication that the wind directionality factor, which is equal to 0.85 for building structures, is explicitly included in the velocity pressure equation for wind. In earlier editions of ASCE/SEI 7 and in the legacy codes, the directionality factor was part of the load factor, which was equal to 1.3 for wind. Thus, for allowable stress design, $\omega = 1.3 \times 0.85 \approx 1.0$ and for strength design, $\omega = 1.6 \times 0.85 \approx 1.3$.

ASCE/SEI 12.13.4 permits a reduction of foundation overturning due to earthquake forces, provided that the criteria of that section are satisfied. Such a reduction is not permitted when the alternative basic load combinations are used to evaluate sliding, overturning and soil bearing at the soil-structure interface. Also, the vertical seismic load effect, E_v, in ASCE/SEI Equation 12.4-4 may be taken as zero when proportioning foundations using these load combinations.

The two exceptions in IBC 1605.3.2 for crane hook loads and for combinations of snow loads, roof live loads, and earthquake loads are the same as those in IBC 1605.3.1, which were discussed previously.

IBC 1605.3.2.1 requires that where F, H or T must be considered in design, each applicable load is to be added to the load combinations in IBC Equations 16-17 through 16-22. As noted previously, the effects of T in combination with other loads must be determined in accordance with ASCE/SEI 2.4.4.

ASCE/SEI 7 does not contain provisions for the alternative basic load combinations of IBC 1605.3.2.

2.5 Load Combinations with Overstrength Factor

The following load combinations, which are given in ASCE/SEI 12.4.3.2, must be used where required by ASCE 12.2.5.2, 12.3.3.3 or 12.10.2.1 instead of the corresponding load combinations in IBC 1605.2 and 1605.3 (IBC 1605.1, Item 3):

- Basic Combinations for Strength Design with Overstrength Factor

    IBC Equation 16-5: $(1.2 + 0.2S_{DS})D + \Omega_0 Q_E + L + 0.2S$

    IBC Equation 16-7: $(0.9 - 0.2S_{DS})D + \Omega_0 Q_E$

- Basic Combinations for Allowable Stress Design with Overstrength Factor

    IBC Equation 16-12: $(1.0 + 0.14S_{DS})D + 0.7\Omega_0 Q_E$

    IBC Equation 16-14: $(1.0 + 0.105S_{DS})D + 0.525\Omega_0 Q_E + 0.75L + 0.75S$

    IBC Equation 16-16: $(0.6 - 0.14S_{DS})D + 0.7\Omega_0 Q_E$

- Alternative Basic Combinations for Allowable Stress Design with Overstrength Factor

    IBC Equation 16-21: $\left(1.0 + \dfrac{0.2S_{DS}}{1.4}\right)D + \dfrac{\Omega_0 Q_E}{1.4} + L + S$

    IBC Equation 16-22: $\left(0.9 - \dfrac{0.2S_{DS}}{1.4}\right)D + \dfrac{\Omega_0 Q_E}{1.4}$

where E_m = $E_{mh} + E_v = \Omega_0 Q_E + 0.2S_{DS}D$ for use in IBC Equations 16-5, 16-12, 16-14 and 16-21

= $E_{mh} - E_v = \Omega_0 Q_E - 0.2S_{DS}D$ for use in IBC Equations 16-7, 16-16 and 16-22

Ω_0 = system overstrength factor obtained from ASCE/SEI Table 12.2-1 for a particular seismic-force-resisting system

Q_E = effects of horizontal seismic forces on a building or structure

S_{DS} = design spectral response acceleration parameter at short periods determined by IBC 1613.3.4 or ASCE/SEI 11.4.4

Notes 1 and 2 in ASCE/SEI 12.4.3.2 pertain to the strength design load combinations. Note 1 permits a load factor of 0.5 on L in Equation 16-5 for all occupancies where L_o is less than or equal to 100 psf, except for parking garages and places of public assembly. Note 2 requires that F be included with the same load factor as D in Equations 16-5 and 16-7. Additionally, H must be included with a load factor of 1.6 in situations where it adds to the primary variable load effect. Where H is permanent and resists the primary variable load effect, a load factor of 0.9 is to be used; for all other conditions, including cases where H is variable, a load factor of zero must be used.

When the simplified procedure of ASCE/SEI 12.14 is utilized in determining seismic load effects, the strength design and allowable stress design load combinations with overstrength factor of ASCE/SEI 12.14.3.2 are to be used instead of the equations presented above.

ASCE/SEI 12.4.3.3 permits allowable stresses to be increased by a factor of 1.2 where allowable stress design is used with seismic load effect including overstrength factor. This increase is not to be combined with increases in allowable stresses or reductions in load combinations that are otherwise permitted in ASCE/SEI 7 or in other referenced materials standards. However, the duration of load factor is permitted when designing wood members in accordance with the referenced standard (Reference 2.3).

As noted in the preceding discussion, load combinations with the overstrength factor apply only to specific types of members or systems; they are not applicable in the design of all members subjected to the effects from earthquakes. Provisions for cantilever column systems are given in ASCE/SEI 12.2.5.2. In addition to the design requirements of that section, the members in such systems must be designed to resist the strength design or allowable stress design load combinations of IBC 1605.2 or 1605.3 and the applicable load combinations with overstrength factor specified in ASCE/SEI 12.4.3.

The provisions of ASCE/SEI 12.3.3.3 apply to structural members that support discontinuous frames or shear wall systems where the discontinuity is severe enough to be deemed a structural irregularity. In particular, structural elements that support discontinuous walls or frames having horizontal irregularity Type 4 of ASCE/SEI Table 12.3-1 or vertical irregularity Type 4 of ASCE/SEI Table 12.3-2 must be designed to resist the load combinations with overstrength factor specified in ASCE/SEI 12.4.3 in addition to the strength design or allowable stress design load combinations described previously. Additional information on structural irregularities can be found in ASCE/SEI 12.3 and Chapter 6 of this publication.

An example of columns supporting a shear wall that has been discontinued at the first floor of a multistory building is illustrated in Figure 2.1. The columns in this situation must be designed to resist the load combinations with overstrength factor.

Figure 2.1
Example of Columns Supporting Discontinuous Shear Wall

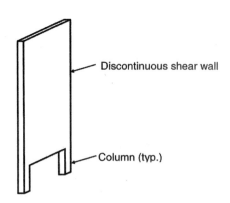

ASCE/SEI 12.10.2.1 applies to collector elements in structures assigned to Seismic Design Category (SDC) C and higher (more information on how to determine the SDC of a building or structure is given in IBC 1613.3.5, ASCE/SEI 11.6 and Chapter 6 of this publication). Collectors, which are also commonly referred to as drag struts, are elements in a structure that are used to transfer the loads from a diaphragm to the elements of the lateral-force-resisting system (LFRS) where the lengths of the vertical elements in the LFRS are less than the length of the diaphragm at that location. An example of collector beams and a shear wall is shown in Figure 2.2. The collector beams collect the force from the diaphragm and distribute it to the shear wall.

Figure 2.2
Example of Collector Beams and Shear Walls

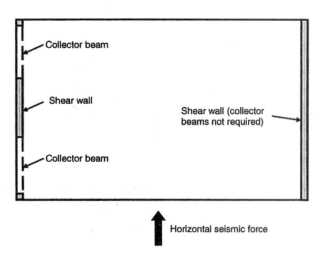

In general, collector elements and their connections to vertical elements must be designed to resist the maximum of the following:

1. Forces calculated using the seismic load effects including the overstrength factor of ASCE/SEI 12.4.3 with the seismic forces determined by the procedures given in ASCE/SEI 12.8 or 12.9.

2. Forces calculated using the seismic load effects including the overstrength factor of ASCE/SEI 12.4.3 with the seismic forces determined by ASCE/SEI Equation 12.10-1 for diaphragms.

3. Forces calculated using the seismic load effects including the overstrength factor of ASCE/SEI 12.4.2.3 with the seismic forces determined by ASCE/SEI Equation 12.10-2 for diaphragms (lower-bound diaphragm force).

Two exceptions to the above requirements are given in this section. The first exception states that the governing force calculated by the three methods above need not exceed the forces calculated using the load combinations of ASCE/SEI 12.4.2.3 where the seismic force is determined by Equation 12.10-3, which is the upper-bound diaphragm design force.

The second exception is applicable to structures utilizing light-frame shear walls. In such cases, collector elements and their connections to the vertical elements need only be designed to resist the forces determined using the load combinations of ASCE/SEI 12.4.2.3 where the seismic forces are determined in accordance with the provisions of diaphragm design forces in ASCE/SEI 12.10.1.1.

2.6 General Structural Integrity

Provisions for structural integrity are contained in IBC 1615, and they are applicable to buildings classified as high-rise buildings in accordance with IBC 403 and assigned to Risk Category III or IV with frame structures or bearing wall structures. A high-rise building is defined in IBC 202 as a building with an occupied floor located more than 75 feet above the lowest level of fire department vehicle access. Risk Categories III and IV are defined in IBC Table 1604.5. Specific load combinations are not included in these prescriptive requirements; rather, the requirements are meant to improve the redundancy and ductility of these types of framing systems in the event of damage due to an abnormal loading event. General design and detailing requirements are provided for frame structures and bearing wall structures.

ASCE/SEI 1.4 also contains general structural integrity requirements that are applicable to all structures, and are as follows:

- **A continuous load path in accordance with ASCE/SEI 1.4.2 must be provided.** A continuous path to the lateral-force-resisting system is essential to ensure that the loads are transmitted properly. The members and connections in that path must be designed to resist the applicable load combinations, and any smaller part of the structure is to be tied to the remainder of the structure with elements that can resist a minimum force equal to 5 percent of its weight.

- **A complete lateral-force-resisting system with adequate strength to resist the forces indicated in ASCE/SEI 1.4.3 must be provided.** In each of two orthogonal directions, a structure must be able to resist lateral forces at each floor level that are equal to 1 percent of the total dead load assigned to that level (see Equation 1.4-1). These forces are to be applied at each floor level simultaneously in the direction of analysis and can be applied independently in the two orthogonal directions. Any structure that has been explicitly designed for stability, which includes the necessary second-order effects, automatically complies with this requirement.

- **Members of the structural system must be connected to their supporting members in accordance with ASCE/SEI 1.4.4.** A positive connection must be provided to resist horizontal forces acting parallel to the member. In particular, each beam, girder, or truss must have adequate connections to its supporting elements or to slabs that act as diaphragms. The connection shall have the strength to resist a minimum force equal to 5 percent of the unfactored dead load plus live load reaction that is imposed by the supported member on the supporting member. In cases where the supported element is connected to a diaphragm, the supporting member also must be connected to the diaphragm.

- **Structural walls must be anchored to diaphragms and supports in accordance with ASCE/SEI 1.4.5.** Load-bearing walls and shear walls (that is, walls that provide lateral shear resistance) must be adequately anchored to the floor and roof members that provide lateral support of the wall or that are supported by the wall. A direct connection must be provided that is capable of resisting a strength level horizontal force perpendicular to the plane of the wall of at least 0.2 times the weight of the wall that is tributary to the connection or 5 psf, whichever is greater.

These minimum strength criteria help ensure that structural integrity is maintained for anticipated and minor unanticipated loading events that have a reasonable chance of occurring during the life of the structure. Guidelines for providing general structural integrity are given in ASCE/SEI C1.4.

The load combinations that include the effects from integrity loads are given in ASCE/SEI 1.4.1 for strength design and allowable stress design:

- Strength design notional load combinations

    a. $1.2D + 1.0N + L + 0.2S$
    b. $0.9D + 1.0N$

- Allowable stress design notional load combinations
  a. $D + 0.7N$
  b. $D + 0.75(0.7N) + 0.75L + 0.75(L_r \text{ or } S \text{ or } R)$
  c. $0.6D + 0.7N$

The effects from the loads specified in ASCE/SEI 1.4.2 through 1.4.5 are defined as notional load effects, N. Note that wind load effects, W, and seismic load effects, E, are not included in these load combinations.

ASCE/SEI C1.4 also contains information on general collapse and limited local collapse and provides case studies of the former. A definition is also given for progressive collapse, which is related to the aforementioned types of collapse. It is important to note that at this time, ASCE/SEI 7 does not provide specific events to be considered during design or specific design criteria to minimize the risk of progressive collapse.

2.7 Extraordinary Loads and Events

Requirements for extraordinary loads and events are given in ASCE/SEI 1.4.6, which references ASCE/SEI 2.5. That section provides minimum requirements for strength and stability of a structure where it has been required by the owner or code having jurisdiction that the structure be able to withstand the effects from extraordinary events.

In general, extraordinary events arise from service or environmental conditions that are not traditionally considered in the design of ordinary buildings because their probability of occurrence is low and their duration is short. Fires, explosions, vehicular impact and tornadoes are all examples of such events. The purpose of these requirements is to help ensure that buildings and structures have sufficient strength and ductility and are adequately tied together so that damage caused by the extraordinary event is relatively small.

A load combination for checking the capacity of a structure or structural element to withstand the effects of an extraordinary event is given in ASCE/SEI Equation 2.5-1:

$$(0.9 \text{ or } 1.2)D + A_k + 0.5L + 0.2S \qquad (2.1)$$

In this equation, A_k is the load or load effect that results from the extraordinary event A. Similar to what is done in the strength design load combinations, the load factor on A_k is set equal to 1.0 because it can be considered a strength-level load.

A factor of 0.9 should be used on the dead load effect, D, if the dead load has a stabilizing effect; otherwise, a load factor of 1.2 should be used. The load factors on L and S correspond approximately to the mean of the yearly maximum live and snow loads. Roof live loads, L_r, and rain loads, R, are not included in this load combination because they have short durations in comparison to S, and thus, the probability of them occurring with A_k is negligible.

ASCE/SEI Equation 2.5-2 is to be used to check the residual load-carrying capacity of a structure or structural element following the occurrence of a damaging event. In particular, selected load-bearing members are to be removed from the structure assuming that they have been critically damaged (that is, they have essentially no load-carrying capacity) and the capacity of the damaged structure is to be evaluated by the following load combination:

$$(0.9 \text{ or } 1.2)D + 0.5L + 0.2(L_r \text{ or } S \text{ or } R) \qquad (2.2)$$

The stability of the entire structure and each of its members must be checked after an extraordinary event as well. ASCE/SEI C2.5 provides a rational method for meeting this requirement.

2.8 Examples

The following examples illustrate the load combinations that were discussed in the previous sections of this publication.

2.8.1 Example 2.1 – Column in Office Building, Strength Design Load Combinations for Axial Loads

Determine the strength design load combinations for a column in a multistory office building using the nominal axial loads in the design data. The live load on the floors is less than 100 psf and the roof is a gable roof.

DESIGN DATA

	Axial Load (kips)
Dead load, D	78
Live load, L	38
Roof live load, L_r	13
Balanced snow load, S	19

SOLUTION

The load combinations using strength design of IBC 1605.2 are summarized in Table 2.5 for this column. The load combinations in the table include only the applicable load effects from the design data.

Table 2.5 Summary of Load Combinations Using Strength Design for Column in Example 2.1

IBC Equation No.	Load Combination
16-1	$1.4D = 1.4 \times 78 = 109$ kips
16-2	$1.2D + 1.6L + 0.5(L_r \text{ or } S) = (1.2 \times 78) + (1.6 \times 38) + (0.5 \times 13) = 161$ kips $= (1.2 \times 78) + (1.6 \times 38) + (0.5 \times 19) = 164$ kips
16-3	$1.2D + 1.6(L_r \text{ or } S) + f_1 L = (1.2 \times 78) + (1.6 \times 13) + (0.5 \times 38) = 133$ kips $= (1.2 \times 78) + (1.6 \times 19) + (0.5 \times 38) = 143$ kips
16-4	$1.2D + f_1 L + 0.5(L_r \text{ or } S) = (1.2 \times 78) + (0.5 \times 38) + (0.5 \times 13) = 119$ kips $= (1.2 \times 78) + (0.5 \times 38) + (0.5 \times 19) = 122$ kips
16-5	$1.2D + f_1 L + f_2 S = (1.2 \times 78) + (0.5 \times 38) + (0.2 \times 19) = 116$ kips
16-6, 16-7	$0.9D = 0.9 \times 78 = 70$ kips

The constant f_1 is taken as 0.5 (live load is less than 100 psf) and f_2 is taken as 0.2 (the gable roof can shed snow). When one or more of the variable loads (live, roof live or snow loads) in IBC Equations 16-1 through 16-5 are taken equal to zero, the resulting factored loads are less than those in Table 2.5.

2.8.2 Example 2.2 – Column in Office Building, Strength Design Load Combinations for Axial Loads and Bending Moments

Determine the strength design load combinations for a column in a multistory office building using the nominal axial loads and maximum bending moments in the design data. The live load on the floors is less than 100 psf and the roof is essentially flat.

Design Data

	Axial Load (kips)	Bending Moment (ft-kips)
Dead load, D	78	15
Live load, L	38	5
Roof live load, L_r	13	0
Balanced snow load, S	19	0
Wind load, W	± 32	± 75

Solution

The load combinations using strength design of IBC 1605.2 are summarized in Table 2.6 for this column and include only the applicable load effects from the design data. The constant f_1 is taken as 0.5, since the live load is less than 100 psf. The constant f_2 is taken as 0.2, since the flat roof can shed snow, unlike a sawtooth roof.

Table 2.6 Summary of Load Combinations Using Strength Design for Column in Example 2.2

IBC Equation No.	Load Combination	Axial Load (kips)	Bending Moment (ft-kips)
16-1	$1.4D$	109	21
16-2	$1.2D + 1.6L + 0.5L_r$	161	26
	$1.2D + 1.6L + 0.5S$	164	26
16-3	$1.2D + 1.6L_r + f_1 L$	133	21
	$1.2D + 1.6L_r + 0.5W$	130	56
	$1.2D + 1.6L_r - 0.5W$	98	−20
	$1.2D + 1.6S + f_1 L$	143	21
	$1.2D + 1.6S + 0.5W$	140	56
	$1.2D + 1.6S - 0.5W$	108	−20
16-4	$1.2D + 1.0W + f_1 L + 0.5L_r$	151	96
	$1.2D + 1.0W + f_1 L + 0.5S$	154	96
	$1.2D - 1.0W + f_1 L + 0.5L_r$	87	−55
	$1.2D - 1.0W + f_1 L + 0.5S$	90	−55
16-5	$1.2D + f_1 L + f_2 S$	116	21
16-6	$0.9D + 1.0W$	102	89
	$0.9D - 1.0W$	38	−62
16-7	$0.9D$	70	14

Since the wind loads cause the structure to sway to the right and to the left, load combinations must be investigated for both cases. This is accomplished by taking both "plus" and "minus" load effects of the wind.

In general, all of the load combinations in Table 2.6 must be investigated when designing the column. It is usually possible to anticipate which of the load combinations is the most critical. Setting one or more of the variable loads (live, roof live, snow or wind loads) in IBC Equations 16-1 through 16-5 equal to zero was not considered, since these load combinations typically do not govern the design of such members.

2.8.3 Example 2.3 – Beam in University Building, Strength Design Load Combinations for Shear Forces and Bending Moments

Determine the strength design load combinations for a beam in a university building using the nominal shear forces and bending moments in the design data. The occupancy of the building is classified as a place of public assembly.

DESIGN DATA

	Shear Force (kips)	Bending Moment (ft-kips)	
		Support	Midspan
Dead load, D	50	−250	170
Live load, L	15	−50	35
Wind load, W	± 16	± 160	---
Seismic, Q_E	± 5	± 50	---

The seismic design data are as follows (more information on seismic design can be found in Chapter 6 of this publication):

ρ = redundancy factor = 1.0

S_{DS} = design spectral response acceleration parameter at short periods = $0.5g$

SOLUTION

The load combinations using strength design of IBC 1605.2 are summarized in Table 2.7 for this beam and include only the applicable load effects from the design data.

The quantity, Q_E, is the effect of code-prescribed horizontal seismic forces on the beam determined from a structural analysis.

In accordance with ASCE/SEI 12.4.2, the seismic load effect, E, is defined as follows:

- For use in load combination 5 of ASCE/SEI 2.3.2 or, equivalently, in IBC Equation 16-5:

$E = E_h + E_v = \rho Q_E + 0.2 S_{DS} D = Q_E + 0.1D$

- For use in load combination 7 of ASCE/SEI 2.3.2 or, equivalently, in IBC Equation 16-7:

$E = E_h - E_v = \rho Q_E - 0.2 S_{DS} D = Q_E - 0.1D$

Substituting for E, IBC Equation 16-5 becomes: $1.2D + 1.0E + f_1 L = 1.3D + L + Q_E$

Similarly, IBC Equation 16-7 becomes: $0.9D + 1.0E = 0.8D + Q_E$

Table 2.7 Summary of Load Combinations Using Strength Design for the Beam in Example 2.3

IBC Equation No.	Load Combination	Location	Bending Moment (ft-kips)	Shear Force (kips)
16-1	$1.4D$	Support	−350	70
		Midspan	238	---
16-2	$1.2D + 1.6L$	Support	−380	84
		Midspan	260	---
16-3	$1.2D + L$	Support	−350	75
		Midspan	239	---
	$1.2D + 0.5W$	Support	−380	68
		Midspan	204	---
16-4	$1.2D + 1.0W + L$	Support	−510	91
		Midspan	239	---
16-5	$1.3D + L + Q_E$	Support	−425	85
		Midspan	256	---
16-6	$0.9D − 1.0W$	Support	−65	29
		Midspan	153	---
16-7	$0.8D − Q_E$	Support	−150	35
		Midspan	136	---

Like wind loads, sidesway to the right and to the left must be investigated for seismic loads. In IBC Equation 16-5, the maximum effect occurs when Q_E is added to the effects of the gravity loads. In IBC Equation 16-7, Q_E is subtracted from the effect of the dead load, since maximum effects occur, in general, when minimum dead load and the effects from lateral loads counteract. The same reasoning is applied in IBC Equations 16-3, 16-4 and 16-6 for the wind effects, W.

The constant f_1 is taken as 1.0, since the occupancy of the building is classified as a place of public assembly.

2.8.4 Example 2.4 – Beam in University Building, Basic Allowable Stress Design Load Combinations for Shear Forces and Bending Moments

Determine the basic allowable stress design load combinations of IBC 1605.3.1 for the beam described in Example 2.3.

SOLUTION

The basic load combinations using allowable stress design of IBC 1605.3.1 are summarized in Table 2.8 for this beam and include only the applicable load effects from the design data.

Table 2.8 Summary of Basic Load Combinations Using Allowable Stress Design for Beam in Example 2.4

IBC Equation No.	Load Combination	Location	Bending Moment (ft-kips)	Shear Force (kips)
16-8 16-10	D	Support	−250	50
		Midspan	170	---
16-9	$D + L$	Support	−300	65
		Midspan	205	---
16-11	$D + 0.75L$	Support	−288	61
		Midspan	196	---
16-12	$D + 0.6W$	Support	−346	60
		Midspan	170	---
	$1.07D + 0.7Q_E$	Support	−303	57
		Midspan	182	---
16-13	$D + 0.75(0.6W) + 0.75L$	Support	−360	69
		Midspan	196	---
16-14	$1.05D + 0.525Q_E + 0.75L$	Support	−326	66
		Midspan	205	---
16-15	$0.6D - 0.6W$	Support	−54	20
		Midspan	102	---
16-16	$0.53D - 0.7Q_E$	Support	−98	23
		Midspan	90	---

In accordance with ASCE/SEI 12.4.2, the seismic load effect, E, is defined as follows:

- For use in load combinations 5 and 6b of ASCE/SEI 2.4.1 or, equivalently, in IBC Equations 16-12 and 16-14:

$$E = E_h + E_v = \rho Q_E + 0.2 S_{DS} D = Q_E + 0.1D$$

- For use in load combination 8 of ASCE/SEI 2.3.2 or, equivalently, in IBC Equation 16-16:

$$E = E_h - E_v = \rho Q_E - 0.2 S_{DS} D = Q_E - 0.1D$$

Thus, substituting for E, IBC Equations 16-12 and 16-14 become, respectively,

$$D + 0.7E = 1.07D + 0.7Q_E$$

$$D + 0.75L + 0.525E = 1.05D + 0.525Q_E + 0.75L$$

Similarly, IBC Equation 16-16 becomes: $0.6D + 0.7E = 0.53D + 0.7Q_E$

Like wind loads, sidesway to the right and to the left must be investigated for seismic loads. In IBC Equations 16-12 and 16-14, the maximum effect occurs when Q_E is added to the effects of the gravity loads. In IBC Equation 16-16, Q_E is subtracted from the effect of the dead load, since maximum effects occur, in general, when minimum dead load and the effects from lateral loads counteract. The same reasoning is applied in IBC Equations 16-12, 16-13 and 16-15 for the wind effects, W.

2.8.5 Example 2.5 – Beam in University Building, Alternative Basic Allowable Stress Design Load Combinations for Shear Forces and Bending Moments

Determine the alternative basic allowable stress design load combinations of IBC 1605.3.2 for the beam described in Example 2.3. Assume that the wind forces have been determined using the provisions of Chapters 26 through 31 of ASCE/SEI 7.

SOLUTION

The alternative basic load combinations using allowable stress design of IBC 1605.3.2 are summarized in Table 2.9 for this beam and include only the applicable load effects from the design data.

The factor ω is taken as 1.3, since the wind forces have been determined by ASCE/SEI 7.

In accordance with ASCE/SEI 12.4.2, the seismic load effect, E, is defined as follows:

- For use in IBC Equation 16-21:

$$E = E_h + E_v = \rho Q_E + 0.2 S_{DS} D = Q_E + 0.1D$$

- For use in IBC Equation 16-22:

$$E = E_h - E_v = \rho Q_E - 0.2 S_{DS} D = Q_E - 0.1D$$

Thus, substituting for E, IBC Equation 16-21 becomes: $D + L + E/1.4 = 1.07D + L + Q_E/1.4$

Similarly, IBC Equation 16-22 becomes: $0.9D + E/1.4 = 0.83D + Q_E/1.4$

Table 2.9 Summary of Alternative Basic Load Combinations Using Allowable Stress Design for Beam in Example 2.5

IBC Equation No.	Load Combination	Location	Bending Moment (ft-kips)	Shear Force (kips)
16-17	$D + L$	Support	−300	65
		Midspan	205	---
16-18 16-19	$D + L + 0.78W$	Support	−425	78
		Midspan	205	---
	$0.67D + L - 0.78W$	Support	−93	36
		Midspan	149	---
16-20	$D + L + 0.39W$	Support	−362	72
		Midspan	205	---
16-21	$1.07D + L + (Q_E/1.4)$	Support	−353	72
		Midspan	217	---
16-22	$0.83D - (Q_E/1.4)$	Support	−172	38
		Midspan	141	---

Like wind loads, sidesway to the right and to the left must be investigated for seismic loads. In IBC Equation 16-21, the maximum effect occurs when Q_E is added to the effects of the gravity loads. In IBC Equation 16-22, Q_E is subtracted from the effect of the dead load, since maximum effects occur, in general, when minimum dead load and the effects from lateral loads counteract. In accordance with IBC 1605.3.2, two-thirds of the dead load is used in IBC Equations 16-18 and 16-19 to counter the maximum effects from the wind pressure.

2.8.6 Example 2.6 – Collector Beam in Residential Building, Load Combinations Using Strength Design and Basic Load Combinations for Strength Design with Overstrength Factor for Axial Forces, Shear Forces, and Bending Moments

Determine the strength design load combinations and the basic combinations for strength design with overstrength factor for a simply supported collector beam in a residential building using the nominal axial loads, shear forces and bending moments in the design data. The live load on the floors is less than 100 psf.

DESIGN DATA

	Axial Force (kips)	Shear Force (kips)	Bending Moment (ft-kips)
Dead load, D	0	56	703
Live load, L	0	19	235
Seismic, Q_E	± 50	0	0

The seismic design data are as follows (more information on seismic design can be found in Chapter 6 of this publication):

Ω_0 = system overstrength factor = 2.5

S_{DS} = design spectral response acceleration parameter at short periods = 1.0g

Seismic design category: D

The axial seismic force, Q_E, corresponds to the portion of the diaphragm design force that is resisted by the collector beam. This force can be tensile or compressive.

SOLUTION

The governing load combination in IBC 1605.2 is as follows:

- IBC Equation 16-2:

    Bending moment: $1.2D + 1.6L = (1.2 \times 703) + (1.6 \times 235) + 1{,}220$ ft-kips

    Shear force: $1.2D + 1.6L + (1.2 \times 56) + (1.6 \times 19) = 98$ kips

Since the beam is a collector beam in a building assigned to SDC D, the beam must be designed for the following basic combinations for strength design with overstrength factor (see IBC 1605.1 and 1605.2; ASCE/SEI 12.4.3.2 and 12.10.2.1):

- IBC Equation 16-5: $(1.2 + 0.2S_{DS})D + \Omega_0 Q_E + 0.5L$

    Axial force: $\Omega_0 Q_E = 2.5 \times 50 = 125$ kips tension or compression

    Bending moment: $(1.2 + 0.2S_{DS})D + 0.5L = (1.4 \times 703) + (0.5 \times 235) = 1{,}102$ ft-kips

    Shear force: $(1.2 + 0.2S_{DS})D + 0.5L = (1.4 \times 56) + (0.5 \times 19) = 88$ kips

    Note that the load factor on L is permitted to equal 0.5 in accordance with Note 1 in ASCE/SEI 12.4.3.2.

- IBC Equation 16-7: $(0.9 - 0.2S_{DS})D + \Omega_0 Q_E$

    Axial force: $\Omega_0 Q_E = 2.5 \times 50 = 125$ kips tension or compression

Bending moment: $(0.9 - 0.2S_{DS})D = 0.7 \times 703 = 492$ ft-kips

Shear force: $(0.9 - 0.2S_{DS})D = 0.7 \times 56 = 39$ kips

The collector beam and its connections must be designed to resist the combined effects of (1) flexure and axial tension, (2) flexure and axial compression and (3) shear as set forth by the above load combinations.

2.8.7 Example 2.7 – Collector Beam in Residential Building, Load Combinations Using Allowable Stress Design (Basic Load Combinations) and Basic Combinations for Allowable Stress Design with Overstrength Factor for Axial Forces, Shear Forces, and Bending Moments

Determine the load combinations using allowable stress design (basic load combinations) and the basic combinations for allowable stress design with overstrength factor for the beam in Example 2.6.

SOLUTION

The governing load combination in IBC 1605.3.1 is as follows:

- IBC Equation 16-9:

    Bending moment: $D + L = 703 + 235 = 938$ ft-kips

    Shear force: $D + L = 56 + 19 = 75$ kips

Since the beam is a collector beam in a building assigned to SDC D, the beam must be designed for the following basic combinations for allowable stress design with overstrength factor (see IBC 1605.1 and 1605.3.1; ASCE/SEI 12.4.3.2):

- IBC Equation 16-12: $(1.0 + 0.14S_{DS})D + 0.7\Omega_0 Q_E$

    Axial force: $0.7\Omega_0 Q_E = 0.7 \times 2.5 \times 50 = 88$ kips tension or compression

    Bending moment: $(1.0 + 0.14S_{DS})D = 1.14 \times 703 = 801$ ft-kips

    Shear force: $(1.0 + 0.14S_{DS})D = 1.14 \times 56 = 64$ kips

- IBC Equation 16-14: $(1.0 + 0.105S_{DS})D + 0.525\Omega_0 Q_E + 0.75L$

    Axial force: $0.525\Omega_0 Q_E = 0.525 \times 2.5 \times 50 = 66$ kips tension or compression

    Bending moment: $1.105D + 0.75L = (1.105 \times 703) + (0.75 \times 235) = 953$ ft-kips

    Shear force: $1.105D + 0.75L = (1.105 \times 56) + (0.75 \times 19) = 76$ kips

- IBC Equation 16-16: $(0.6 - 0.14S_{DS})D + 0.7\Omega_0 Q_E$

    Axial force: $0.7\Omega_0 Q_E = 0.7 \times 2.5 \times 50 = 88$ kips tension or compression

    Bending moment: $(0.6 - 0.14S_{DS})D = 0.46 \times 703 = 323$ ft-kips

    Shear force: $(0.6 - 0.14S_{DS})D = 0.46 \times 56 = 26$ kips

The collector beam and its connections must be designed to resist the combined effects of (1) flexure and axial tension, (2) flexure and axial compression and (3) shear as set forth by the above load combinations (see ASCE/SEI 12.4.3.3 for allowable stress increase for load combinations with overstrength).

2.8.8 Example 2.8 – Collector Beam in Residential Building, Load Combinations Using Allowable Stress Design (Alternative Basic Load Combinations) and Basic Combinations for Allowable Stress Design with Overstrength Factor for Axial Forces, Shear Forces, and Bending Moments

Determine the load combinations using allowable stress design (alternative basic load combinations) and the basic combinations for allowable stress design with overstrength factor for the beam in Example 2.6.

SOLUTION

The governing load combination in IBC 1605.3.2 is as follows:

- IBC Equation 16-17:

    Bending moment: $D + L = 703 + 235 = 938$ ft-kips

    Shear force: $D + L = 56 + 19 = 75$ kips

Since the beam is a collector beam in a building assigned to SDC D, the beam must be designed for the following basic combinations for allowable stress design with overstrength factor (see IBC 1605.1 and 1605.3.2; ASCE/SEI 12.4.3.2):

- IBC Equation 16-21: $\left(1.0 + \dfrac{0.2 S_{DS}}{1.4}\right) D + \dfrac{\Omega_0 Q_E}{1.4} + L$

    Axial force: $\dfrac{\Omega_0 Q_E}{1.4} = \dfrac{2.5 \times 50}{1.4} = 89$ kips tension or compression

    Bending moment: $\left[\left(1.0 + \dfrac{0.2}{1.4}\right) \times 703\right] + 235 = 1{,}038$ ft-kips

    Shear force: $\left[\left(1.0 + \dfrac{0.2}{1.4}\right) \times 56\right] + 19 = 83$ kips

- IBC Equation 16-22: $\left(0.9 - \dfrac{0.2 S_{DS}}{1.4}\right) D + \dfrac{\Omega_0 Q_E}{1.4}$

    Axial force: $\dfrac{\Omega_0 Q_E}{1.4} = \dfrac{2.5 \times 50}{1.4} = 89$ kips tension or compression

    Bending moment: $\left(0.9 - \dfrac{0.2}{1.4}\right) \times 703 = 532$ ft-kips

    Shear force: $\left(0.9 - \dfrac{0.2}{1.4}\right) \times 56 = 42$ kips

The collector beam and its connections must be designed to resist the combined effects of (1) flexure and axial tension, (2) flexure and axial compression and (3) shear as set forth by the above load combinations (see ASCE/SEI 12.4.3.3 for allowable stress increase for load combinations with overstrength).

2.8.9 Example 2.9 – Timber Pile in Residential Building, Basic Allowable Stress Design Load Combinations for Axial Forces

Determine the basic allowable stress design load combinations for a timber pile supporting a residential building using the nominal axial loads in the design data. The residential building is located in a Coastal A Zone.

DESIGN DATA

	Axial Force (kips)
Dead load, D	8
Live load, L	6
Roof live load, L_r	4
Wind, W	± 26
Flood, F_a	± 2

SOLUTION

The basic load combinations using allowable stress design of IBC 1605.3.1 are summarized in Table 2.10 for this pile. The load combinations in the table include only the applicable load effects from the design data. Also, since flood loads, F_a, must be considered, the load combinations of ASCE/SEI 2.4.2 are used (IBC 1605.3.1.2). In particular, $1.5F_a$ is added to the other applicable loads in IBC Equations 16-12, 16-13 and 16-15.

Table 2.10 Summary of Basic Load Combinations Using Allowable Stress Design (IBC 1605.3.1) for Timber Pile in Example 2.9

IBC Equation No.	Load Combination	Axial Load (kips)
16-8	D	8
16-9	$D + L$	14
16-10	$D + L_r$	12
16-11	$D + 0.75L + 0.75L_r$	16
16-12	$D + 0.6W + 1.5F_a$	27
	$D - 0.6W - 1.5F_a$	−11
16-13	$D + 0.75(0.6W) + 0.75L + 0.75L_r + 1.5F_a$	30
	$D - 0.75(0.6W) + 0.75L + 0.75L_r - 1.5F_a$	1
16-16	$0.6D + 0.6W$	20
	$0.6D - 0.6W$	−11

The pile must be designed for the axial compression and tension forces in Table 2.10 in combination with bending moments caused by the wind and flood loads. Shear forces and deflection at the tip of the pile must also be checked. Finally, the embedment length of the pile must be sufficient to resist the maximum net tension force.

More information on flood loads can be found in Chapter 7 of this publication.

2.9 References

2.1. Ellingwood, B. 1981. "Wind and Snow Load Statistics for Probabilistic Design." *Journal of the Structural Division*, 107(7): 1345–1350.

2.2. Galambos, T.V., Ellingwood, B., MacGregor, J.G., and Cornell, C.A. 1982. "Probability-based Load Criteria: Assessment of Current Design Practice." *Journal of the Structural Division*, 108(5): 959–977.

2.3. American Forest and Paper Association. 2012. *National Design Specification for Wood Construction with 2012 Supplement* (NDS-2012). Washington, DC.

2.4. International Conference of Building Officials. 1997. *Uniform Building Code*. Whittier, CA.

2.10 Problems

2.1. Determine the strength design load combinations for a reinforced concrete beam on a typical floor of a multistory residential building using the nominal bending moments in Table 2.11. All bending moments are in foot-kips. Assume the live load on the floor is less than 100 psf.

Table 2.11 Design Data for Problem 2.1

	External Negative	Positive	Interior Negative
Dead load, D	−13.3	43.9	−53.2
Live load, L	−12.9	42.5	−51.6

2.2. Determine the strength design load combinations for a steel beam that is part of an ordinary moment frame in an office building using the nominal bending moments and shear forces in Table 2.12. All bending moments are in foot-kips and all shear forces are in kips. Assume the live load on the floor is less than 100 psf.

Table 2.12 Design Data for Problem 2.2

	Bending Moment		Shear Force
	Support	Midspan	Support
Dead load, D	−57.6	41.1	11.8
Live load, L	−22.5	16.2	4.6
Wind, W	± 54.0	---	± 4.8

2.3. Given the information in Problem 2.2, determine the basic allowable stress design load combinations.

2.4. Given the information in Problem 2.2, determine the alternative basic allowable stress design load combinations. Assume the wind loads have been determined using ASCE/SEI 7.

2.5. Determine the strength design load combinations for a reinforced concrete column that is part of an intermediate moment frame in an office building using the nominal axial forces, bending moments and shear forces in Table 2.13. All axial forces are in kips, all bending moments are in foot-kips and all shear forces are in kips. Assume the live loads on the floors are equal to 100 psf and $S_{DS} = 0.41g$.

Table 2.13 Design Data for Problem 2.5

	Axial Force	Bending Moment	Shear Force
Dead load, D	167.9	21.3	2.3
Live load, L	41.5	21.0	2.2
Roof live load, L_r	14.9	---	---
Wind, W	± 13.6	± 121.0	± 11.1
Seismic, Q_E	± 36.4	± 432.1	± 42.2

2.6. Determine the strength design load combinations for a reinforced concrete shear wall in a parking garage using the nominal axial forces, bending moments and shear forces in Table 2.14. All axial forces are in kips, all bending moments are in foot-kips and all shear forces are in kips. Assume that $\rho = 1.0$ and $S_{DS} = 1.0g$.

Table 2.14 Design Data for Problem 2.6

	Axial Force	Bending Moment	Shear Force
Dead load, D	645	0	0
Live load, L	149	0	0
Seismic, Q_E	0	± 4,280	± 143

2.7. Determine the strength design load combinations and the basic combinations for strength design with overstrength factor for a simply supported steel collector beam in an assembly building using the nominal axial forces, bending moments and shear forces in Table 2.15. All axial forces are in kips, all bending moments are in foot-kips and all shear forces are in kips. Assume $S_{DS} = 0.9g$ and $\Omega_0 = 2.0$.

Table 2.15 Design Data for Problem 2.7

	Axial Force	Bending Moment		Shear Force
		Negative	Positive	
Dead load, D	0	80.6	53.7	29.7
Live load, L	0	42.1	30.4	19.0
Seismic, Q_E	± 241	0	0	0

2.8. Given the information in Problem 2.7, determine the basic allowable stress design load combinations.

30 Chapter 2

2.9. Determine the strength design load combinations for a wood roof truss in a commercial building with a curved roof using the nominal distributed loads in Table 2.16. All loads are in pounds per linear foot.

Table 2.16 Design Data for Problem 2.9

	Load
Dead load, D	75
Rain load, R	200
Roof live load, L_r	100
Balanced snow load, S	125

2.10. Given the information in Problem 2.9, determine the basic allowable stress design load combinations.

CHAPTER 3
Dead, Live, Rain and Soil Lateral Loads

3.1 Dead Loads

Nominal dead loads, D, are determined in accordance with IBC 1606. In general, design dead loads are the actual weights of construction materials and fixed service equipment that are attached to or supported by the building or structure. Various types of such loads are listed in IBC 202 under "Dead Load."

Dead loads are considered to be permanent loads; that is, loads in which variations over time are rare or of small magnitude. Variable loads, such as live loads and wind loads, are not permanent. It is important to know the distinction between permanent and variable loads when applying the provisions for load combinations (see IBC 1605, ASCE/SEI Chapter 2 and Chapter 2 of this publication for information on load combinations).

The weights of materials and service equipment (such as plumbing stacks and risers, HVAC equipment, elevators and elevator machinery, fire protection systems and similar fixed equipment) are not usually known during the design phase. Estimated material and equipment loads are often used in design. Typically, estimated dead loads are assumed to be greater than the actual dead loads so that the design is conservative. While such practice is acceptable when considering load combinations where the effects of gravity loads and lateral loads are additive, it is not acceptable when considering load combinations where gravity loads and lateral loads counteract. For example, it would be unconservative to design for uplift on a structure using a value of dead load that is overestimated.

ASCE/SEI Table C3-1 provides minimum design dead loads for various types of common construction components, including ceilings, roof and wall coverings, floor fill, floors and floor finishes, frame partitions and frame walls. Minimum densities for common construction materials are given in ASCE/SEI Table C3-2.

The weights in ASCE/SEI Tables C3-1 and C3-2 can be used as a guide when estimating dead loads. Actual weights of construction materials and equipment can be greater than tabulated values, so it is always prudent to verify weights with manufacturers or other similar resources prior to design. In cases where information on dead load is unavailable, values of dead loads used in design must be approved by the building official (IBC 1606.2).

3.2 Live Loads

3.2.1 General

Nominal live loads are determined in accordance with IBC 1607. Live loads are those loads produced by the use and occupancy of a building or structure and do not include construction loads, environmental loads (such as wind loads, snow loads, rain loads, earthquake loads and flood loads) or dead loads (see the definition of "Live Load" in IBC 202).

In general, live loads are transient in nature and vary in magnitude over the life of a structure. Studies have shown that building live loads consist of both a sustained portion and a variable portion. The sustained portion is based on general day-to-day use of the facilities and will generally vary during the life of the structure due to tenant modifications and changes in occupancy, for example. The variable portion of the live load is typically created by events such as remodeling, temporary storage and similar unusual events.

Nominal design values of uniformly distributed and concentrated live loads are given in IBC Table 1607.1 as a function of occupancy or use. The occupancy category listed in the table is not necessarily group specific (occupancy groups are defined in IBC Chapter 3). For example, an office building with a Business Group B classification may also have storage areas that may warrant live loads of 125 psf or 250 psf depending on the type of storage.

The design values in IBC Table 1607.1 are minimum values; actual design loads can be determined to be larger than these loads, but in no case shall the structure be designed for live loads that are less than the tabulated values. For occupancies that are not listed in the table, live loads used in design must be approved by the building official. It is also important to note that the provisions do not require concurrent application of uniform and concentrated loads. Structural members are designed based on the maximum effects due to the application of either a uniform load or a concentrated load, and they need not be designed for the effects of both loads applied at the same time. Unless specified otherwise, concentrated loads are to be applied over an area of 2.5 by 2.5 feet and are to be located so as to produce the maximum load effects in the supporting structural members.

ASCE/SEI Table 4-1 also contains minimum uniform and concentrated live loads, which differ in some cases with the corresponding ones in IBC Table 1607.1. ASCE Tables C4-1 and C4-2 can also be used as a guide in establishing live loads for commonly encountered occupancies.

3.2.2 Partitions

Provisions for partitions are given in IBC 1607.5. Partitions that can be relocated (that is, those types that are not permanently attached to the structure) are considered to be live loads in office and other buildings because they are considered to be variable by nature. A live load equal to at least 15 psf must be included for moveable partitions where the nominal uniform floor load is less than or equal to 80 psf.

The weight of any built-in partitions that cannot be moved is considered a dead load in accordance with IBC 1602.

3.2.3 Helipads

Uniform and concentrated live loads that are to be used in the design of helipads are given in IBC 1607.6 and are summarized in Table 3.1. The concentrated loads that are specified are not required to act concurrently with other uniform or concentrated live loads.

Table 3.1 Summary of Nominal Live Loads for Helipads

Load Type	Minimum Load
Uniform	• 40 psf for a helicopter with a maximum take-off weight of 3,000 pounds • 60 psf for a helicopter with a maximum take-off weight greater than 3,000 pounds
One Single, Concentrated	3,000 pounds applied over an area of 4.5 by 4.5 inches located to produce maximum load effects on the structural elements
Two Single, Concentrated	0.75 times the maximum take-off weight of the helicopter applied over an area of 8 by 8 inches and located 8 feet apart on the landing pad to produce maximum load effects on the structural elements

Landing areas that are designed for a helicopter with a maximum take-off weight of 3,000 pounds must be identified as such by providing a numeral "3" (for 3 kips) that is at least 5 feet in height in the bottom right corner of the landing area.

ASCE/SEI Table 4-1, including footnotes d, e, f and g, contain the same requirements for helipads as those covered in the preceding discussion.

3.2.4 Heavy Vehicle Loads

Live load requirements for heavy vehicle loads (that is, vehicle loads greater than 10,000 pounds) are given in IBC 1607.7. In general, portions of structures where such vehicles can have access must be designed for live loads—including impact and fatigue—determined in accordance with codes and specifications for the design and construction of roadways and bridges (IBC 1607.7.1). The American Association of State Highway and Transportation Officials (AASHTO) design specifications (Reference 3.2) have been adopted by many jurisdictions throughout the United States and contain provisions on how to determine these live loads.

In the case of fire trucks and emergency vehicles, the actual operational loads and the reactions from the vehicle outriggers, where applicable, must also be considered. The supporting structure must be designed for the greater of the operational loads or the loads described in IBC 1607.7.1.

Garages that accommodate vehicle loads exceeding 10,000 pounds must also be designed for the live loads specified in IBC 1607.7.1, except that the design need not include the effects due to impact or fatigue. The exception that is provided to this requirement in IBC 1607.7.3 permits garage floors to be designed for the actual weights of the vehicles that are anticipated to occupy the garage as long as (1) the loads and their placement are based on a rational analysis, (2) the loads are greater than or equal to 50 psf and are not reduced and (3) the loads are approved by the building official.

For forklifts and similar moveable equipment, the supporting structure is to be designed for a minimum live load corresponding to the total vehicle or equipment load and the individual wheel loads. Additionally, impact and fatigue loads must be considered in design. It is permitted to account for these additional loads by increasing the vehicle and wheel loads by 30 percent (IBC 1607.7.4.1).

Similar to heavy live loads for floors, the live load for heavy vehicles must be posted in accordance with IBC 106.1.

ASCE/SEI Table 4-1 makes specific reference to the AASHTO *LRFD Bridge Design Specifications* (Reference 3.2) for the design of garages that can accommodate trucks and buses.

3.2.5 Handrails and Guards

IBC 1607.8.1 requires that handrails and guards for stairs, balconies and similar elements be designed for the live loads specified in ASCE/SEI 4.5.1. In particular, the following live loads are applicable:

1. A single concentrated load of 200 pounds applied in any direction at any point on the handrail or top rail to produce the maximum load effect on the element being considered (load condition 1 in Figure 3.1);

2. A load of 50 pounds per linear foot applied in any direction along the handrail or top rail (load condition 2 in Figure 3.1); and

3. A load of 50 pounds distributed normal to a 12-inch by 12-inch area on intermediate rails (that is, on all elements except the handrail or top rail) located to produce the maximum load effects (see Figure 3.2).

Figure 3.1
Handrail Live Load Conditions

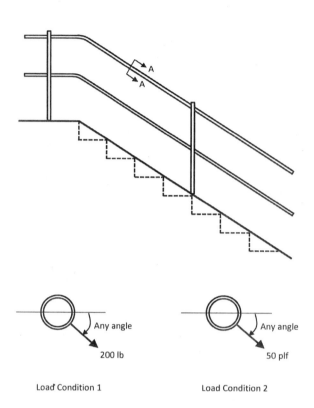

Figure 3.2
Intermediate Rail Load

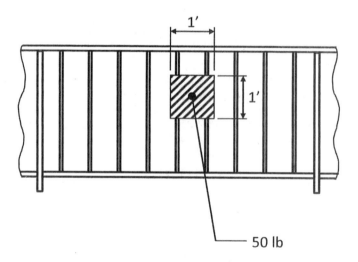

The concentrated load in load condition 1 is meant to simulate the maximum anticipated load from a person grabbing or falling into the handrail or guard while the line load in load condition 2 is the maximum anticipated load on a handrail or guard from use by a crowd of people on a stairway. These loads can occur in any direction, as shown in Figure 3.1, and need not be applied concurrently.

Exceptions to these requirements recognize the circumstances that are applicable in occupancies where the handrail or guard is inaccessible to the public. In particular, load condition 2 need not be considered in one- and two-family dwellings or in factory, industrial and storage occupancies that are not accessible to the public and that have an occupant load of 50 or less.

The second exception in IBC 1607.8.1 differs from that in ASCE/SEI 4.5.1 in three ways. First, Group I-3 institutional occupancies (correctional centers, jails, prisons and the like) are included in addition to factory, industrial and storage occupancies. Second, this exception is applicable to occupant loads less than 50 instead of 50 or less. Finally, a minimum line load of 20 pounds per linear foot is required in such occupancies instead of the line load not being considered (that is, set equal to zero).

The area load specified in item 3 is a localized load for the guard members; as shown in Figure 3.2, the balusters that would resist this load are those within the one-square-foot area in the plane of the guard. This load need not be superimposed with any other loads.

IBC 1607.8.1 also stipulates that glass handrail assemblies and guards comply with IBC 2407; that section includes minimum requirements related to material properties and support conditions.

3.2.6 Grab Bars, Shower Seats and Dressing Room Bench Seats

Live load requirements for grab bars, shower seats and dressing room bench seats are given in IBC 1607.8.2. Such elements are to be designed for a single concentrated load of 250 pounds applied in any direction and at any point of the grab bar or seat so as to produce the maximum load effects. This load is anticipated to be encountered from the use of such elements.

The same requirement is given in ASCE/SEI 4.5.2, but only for the case of grab bars.

3.2.7 Vehicle Barriers

Vehicle barriers are defined in IBC 202 as a component or a system of components that are positioned at open sides of a parking garage floor or ramp or at building walls that act as restraints for vehicles. In other words, these barriers provide a passive restraint system that is located where vehicles could fall to a lower level.

IBC 1607.8.3 refers to ASCE/SEI 4.5.3, which requires that vehicle barrier systems be designed to resist a single load of 6,000 pounds applied horizontally over an area not to exceed 12 by 12 inches in any direction to the barrier system at heights between 18 and 27 inches above the floor or ramp surface (see Figure 3.3). This load is to be located so as to produce the maximum load effects, and it need not act concurrently with any handrail or guard loading discussed in Section 3.2.5 of this publication.

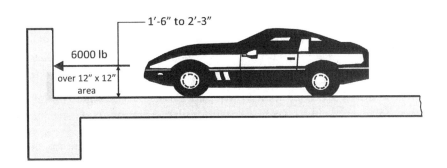

Figure 3.3
Vehicle Barrier Load Requirements

These provisions are applicable to barrier systems in garages with passenger vehicles or light trucks, and the required load includes the effects from impact. In the case of garages and buildings that can accommodate trucks, buses and similar type vehicles, the loads associated with the traffic railing provisions of *AASHTO LRFD Bridge Design Specifications* (Reference 3.2) should be used instead.

3.2.8 Fixed Ladders

ASCE/SEI 4.5.4 contains live load requirements for fixed ladders. A single concentrated load of 300 pounds is to be applied at any point to produce the maximum load effect on the element under consideration. Additional 300-pound loads are required for every 10 feet of ladder height.

A concentrated load of 100 pounds is required to be applied to the top of each rail extension of a fixed ladder that extends above a floor or platform at the top of the ladder. This load, which can act in any direction and at any height, is based on a 250-pound person standing on a rung of the ladder and accounts for reasonable angles of pull on the extension of the rails above the floor or platform level.

Ship ladders that have treads instead of rungs are required to be designed as stairs in accordance with ASCE/SEI Table 4-1.

3.2.9 Impact Loads

According to IBC 1607.9, the live loads that are specified in IBC 1607.3 through 1607.8 include an allowance for impact that is normally attributed to such loads. Impact loads that involve unusual vibration and impact forces, such as those from elevators and machinery, must be accounted for in design because they result in additional forces and deflections in the structural systems that support them.

The static load of an elevator must be increased to account for motion effects. The loads on the members supporting an elevator are significantly higher than the weight of the elevator and its occupants when the elevator comes to a stop. The rate of acceleration and deceleration has a significant impact on this effect. IBC 1607.9.1 and ASCE/SEI 4.6.2 require that elements that are subjected to dynamic loads imparted by elevators be designed for the impact loads and deflection limits specified in ASME A17.1 *Safety Code for Elevators and Escalators*. This document specifies an impact factor equal to 100 percent, which means that the weight must be increased by a factor of 2 to account for impact. A dynamic analysis is generally not required; use of an equivalent static load is usually sufficient.

The weight of machinery and moving loads are to be increased as follows to account for impact:

- Shaft- or motor-driven light machinery: 20 percent
- Reciprocating machinery or power-driven units: 50 percent

These impact factors include the effects due to vibration; thus, a larger impact factor is assigned to reciprocating machinery and power-driven units because this type of equipment vibrates more than shaft- or motor-driven light machinery. It is always good practice to acquire impact factors for specific pieces of equipment from manufacturers since such factors can be larger than those that are specified in IBC 1607.9.2 and ASCE/SEI 4.6.3.

ASCE/SEI C4.6 contains information regarding loads on grandstands, stadiums and other similar assembly structures. Although assembly loads are given in IBC Table 1607.1 and ASCE/SEI Table 4-1, the possibility of impact and/or vibration from crowd activity (such as swaying in unison, jumping to their feet or stomping) should be considered in design.

Dead, Live, Rain and Soil Lateral Loads 37

3.2.10 Reduction in Live Loads

According to IBC 1607.10, the minimum nominal uniformly distributed live loads, L_o, in IBC Table 1607.1 are permitted to be reduced by either the provisions of IBC 1607.10.1 or 1607.10.2. Both methods are discussed below.

Uniform roof live loads are not permitted to be reduced by these provisions; reduction of such roof live loads is covered in IBC 1607.12.2 (see Section 3.2.12 of this publication).

Basic Uniform Live Load Reduction

The method in IBC 1607.10.1 of reducing uniform live loads other than uniform live loads at roofs is based on the provisions in ASCE/SEI 4.7. IBC Equation 16-23 can be used to obtain a reduced live load, L, for members where $K_{LL}A_T \geq 400$ square feet, subject to the limitations of IBC 1607.10.1.1 through 1607.10.1.3:

$$L = L_o\left(0.25 + \frac{15}{\sqrt{K_{LL}A_T}}\right) \qquad (3.1)$$

$\geq 0.50L_o$ for members supporting one floor

$\geq 0.40L_o$ for members supporting two or more floors

In this equation, K_{LL} is the live load element factor given in IBC Table 1607.10.1 and A_T is the tributary area in square feet.

The live load element factor, K_{LL}, converts the tributary area of a structural member, A_T, to an influence area, which is considered to be the adjacent floor area from which the member derives any of its load. In general,

$$K_{LL} = \frac{\text{Influence area}}{\text{Tributary area}} \qquad (3.2)$$

Consider interior column B3 in Figure 3.4. The influence area for this column is equal to the four bays adjacent to the column:

Influence area = $(\ell_A + \ell_B)(\ell_2 + \ell_3)$

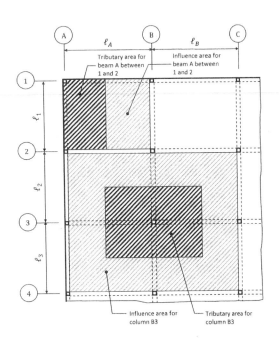

Figure 3.4
Influence Areas and Tributary Areas for Columns and Beams

The tributary area, A_T, supported by this column is equal to the product of the tributary widths in both directions (the tributary width in this case is equal to the sum of one-half of the span lengths on both sides of the column):

$$A_T = \left(\frac{\ell_A}{2} + \frac{\ell_B}{2}\right)\left(\frac{\ell_2}{2} + \frac{\ell_3}{2}\right)$$

Using Equation 3.2 it is evident that the live load element factor, K_{LL}, is equal to 4 for this interior column; this matches the value given in IBC Table 1607.10.1. It can also be shown that K_{LL} is equal to 4 for any exterior column (other than corner columns) without cantilever slabs.

Now, consider the spandrel beam on line A between lines 1 and 2. The influence area for this beam is equal to the area of the bay adjacent to the beam:

$$\text{Influence area} = \ell_A \ell_1$$

The tributary area, A_T, supported by this beam is equal to the tributary width times the length of the beam:

$$A_T = \left(\frac{\ell_A}{2}\right)\ell_1$$

Thus, K_{LL} is equal to 2 for an edge beam without a cantilever slab. Values of K_{LL} for other members can be derived in a similar fashion. Illustrated in ASCE/SEI Figure C4-1 are typical tributary and influence areas for a variety of elements in a structure with regular bay spacing.

Figure 3.5 illustrates how the reduction multiplier $0.25 + 15/(\sqrt{K_{LL}A_T})$ varies with respect to the influence area $K_{LL}A_T$. Included in the figure are the minimum influence area of 400 square feet and the limits of 0.5 and 0.4, which are the maximum permitted reductions for members supporting one floor and two or more floors, respectively.

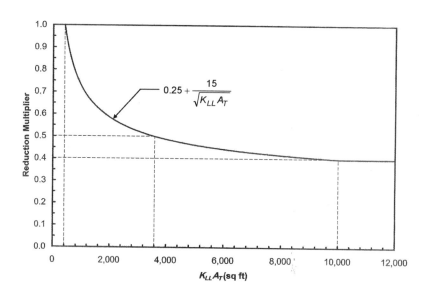

Figure 3.5 Reduction Multiplier for Live Load in Accordance with IBC 1607.10.1

One-way Slabs. Live load reduction on one-way slabs is permitted provided the tributary area, A_T, does not exceed an area equal to the slab span times a width normal to the span of 1.5 times the slab span (that is, an area with an aspect ratio of 1.5). The live load will be somewhat higher for a one-way slab with an aspect ratio of 1.5 than for a two-way slab with the same aspect ratio. This recognizes the benefits of higher redundancy that results from two-way action.

ASCE/SEI 4.7.6 has the same requirements for live load reduction on one-way slabs as that in IBC 1607.10.1.1.

Heavy Live Loads. According to IBC 1607.10.1.2, live loads that are greater than 100 psf shall not be reduced except for the following:

1. Live loads for members supporting two or more floors are permitted to be reduced by a maximum of 20 percent, but L shall not be less than that calculated by IBC 1607.10.1.

2. In occupancies other than storage, additional live load reduction is permitted if it can be shown by a registered design professional that such a reduction is warranted.

In buildings that support relatively large live loads, such as storage buildings, several adjacent bays may be fully loaded; as such, live loads should not be reduced in those situations. Data in actual buildings indicate that the floor in any story is seldom loaded with more than 80 percent of the nominal live load. Thus, a maximum live load reduction of 20 percent is permitted for members that support two or more floors, such as columns and walls.

Passenger Vehicle Garages. The live load in passenger vehicle garages is not permitted to be reduced, except for members supporting two or more floors; in such cases, the maximum reduction is 20 percent, but L shall not be less than that calculated by IBC 1607.10.1 (IBC 1607.10.1.3).

Parking garage live loads are unlike live loads in office and residential occupancies: vehicles are parked in regular patterns and the garages are often full compared to office or residential space where the live loads are most of the time spatially random. Thus, live load reduction is not permitted except for members that support two or more floors.

Group A (Assembly) Occupancies. Due to the nature of assembly occupancies, there is a high probability that the entire floor is subjected to full uniform live load. According to Footnote m in IBC Table 1607.1, live load reduction is not permitted in assembly areas, except for follow spot, projection and control rooms, unless specific exceptions of IBC 1607.10 apply.

Flowchart 3.1 in Section 3.5 of this publication can be used to determine basic uniform live load reduction in accordance with IBC 1607.10.1.

Alternative Uniform Live Load Reduction

An alternative method of uniform live load reduction, which is based on provisions in the 1997 *Uniform Building Code* (Reference 3.1), is given in IBC 1607.10.2. IBC Equation 16-24 can be used to obtain a reduction factor, R, for members that support an area greater than or equal to 150 square feet where the live load is less than or equal to 100 psf:

$$R = 0.08(A - 150) \tag{3.3}$$

$$\leq \text{the smallest of} \begin{cases} 40 \text{ percent for horizontal members} \\ 60 \text{ percent for vertical members} \\ 23.1(1 + D/L_o) \end{cases}$$

In this equation, A is the area of floor supported by a member in square feet, and D is the dead load per square foot of area supported.

The reduced live load, L, is then determined by the following equation:

$$L = L_o\left(1 - \frac{R}{100}\right) \quad (3.4)$$

Similar to the general method of live load reduction, live loads are not permitted to be reduced in the following situations:

1. In Group A (assembly) occupancies;

2. Where the live load exceeds 100 psf except (a) for members supporting two or more floors in which case the live load may be reduced by a maximum of 20 percent or (b) in occupancies other than storage where it can be shown by a registered design professional that such a reduction is warranted; and

3. In passenger vehicle garages except for members supporting two or more floors in which case the live load may be reduced by a maximum of 20 percent.

Reduction of live load on one-way slab systems is permitted by this method when the area, A, is limited to the slab span multiplied by one-half of the slab span in IBC 1607.10.2(4).

Flowchart 3.2 in Section 3.5 of this publication can be used to determine alternative uniform live load reduction in accordance with IBC 1607.10.2.

3.2.11 Distribution of Floor Loads

IBC 1607.11 requires that the effects of partial uniform live loading (or alternate span loading) be investigated when analyzing continuous floor members. Such loading produces greatest effects at different locations along the span. Reduced floor live loads may be used when performing this analysis.

Figure 3.6 illustrates four loading patterns that need to be investigated for a three-span continuous system subject to dead and live loads. The different loading patterns must be considered in order to obtain the maximum effects on the system. Similar load patterns can be derived for systems with other span conditions.

Figure 3.6
Distribution of Floor Loads for a Three-span Continuous System in Accordance with IBC 1607.11

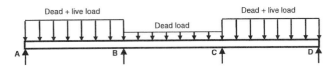
Loading pattern for maximum negative moment at support A or D and maximum positive moment in span AB or CD

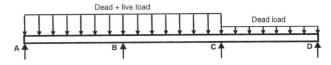
Loading pattern for maximum negative moment at support B

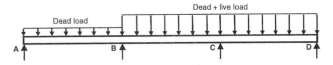
Loading pattern for maximum negative moment at support C

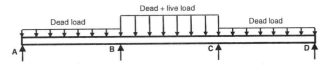
Loading pattern for maximum positive moment in span BC

3.2.12 Roof Loads

In general, roofs are to be designed to resist dead, live, wind, and where applicable, rain, snow and earthquake loads. A minimum roof live load of 20 psf is prescribed in IBC Table 1607.1 for typical roof structures, while larger live loads are required for roofs used as gardens or places of assembly.

IBC 1607.12.2 permits nominal roof live loads on flat, pitched and curved roofs and awnings and canopies other than fabric construction supported by a skeleton frame to be reduced in accordance with IBC Equation 16-26:

$$L_r = L_o R_1 R_2, \quad 12 \leq L_r \leq 20 \tag{3.5}$$

where

L_o = unreduced roof live load per square foot of horizontal roof projection supported by the member

L_r = reduced roof live load per square foot of horizontal roof projection supported by the member

$$R_1 = \begin{cases} 1 \text{ for } A_t \leq 200 \text{ square feet} \\ 1.2 - 0.001 A_t \text{ for } 200 \text{ square feet} < A_t < 600 \text{ square feet} \\ 0.6 \text{ for } A_t \geq 600 \text{ square feet} \end{cases}$$

$$R_2 = \begin{cases} 1 \text{ for } F \leq 4 \\ 1.2 - 0.05 F \text{ for } 4 < F < 12 \\ 0.6 \text{ for } F \geq 12 \end{cases}$$

A_t = tributary area (span length multiplied by effective width) in square feet supported by a member

F = the number of inches of rise per foot for a sloped roof

= the rise-to-span ratio multiplied by 32 for an arch or dome

It is evident from IBC Equation 16-26 that roof live load reduction is based on the tributary area of the member being considered and the slope of the roof. No live load reduction is permitted for members supporting less than or equal to 200 square feet as well as for roof slopes less than or equal to 4:12. In no case is the reduced roof live load to be taken less than 12 psf. The minimum load determined by this equation accounts for occasional loading due to the presence of workers and materials during repair operations.

Live loads are permitted to be reduced on areas of occupiable roofs using the provisions of IBC 1607.10 for floor live loads (IBC 1607.12.3). Live loads that are equal to or greater than 100 psf at areas of roofs that are classified as Group A (assembly) occupancies are not permitted to be reduced unless specific exceptions of IBC 1607.10 apply (see Footnote m in IBC Table 1607.1).

A minimum roof live load of 20 psf is required in unoccupied landscaped areas on roofs (IBC 1607.12.3.1). The weight of landscaping material is considered a dead load and must be determined based on the saturation level of the soil.

A minimum roof live load of 5 psf is required for awnings and canopies in accordance with IBC Table 1607.1 (IBC 1607.12.4). Such elements must also be designed for the combined effects of snow and wind loads in accordance with IBC 1605.

3.2.13 Crane Loads

Design provisions for runway beams that support moving bridge cranes and monorail cranes are given in IBC 1607.13. In general, the support structure of the crane must be designed for the maximum wheel load, vertical impact and horizontal impact as a simultaneous load combination.

A typical top-running bridge crane is depicted in Figure 3.7. The trolley and hoist move along the crane bridge, which is supported by the runway beams and support columns. The entire crane assemblage can also move along the length of the runway beams.

Figure 3.7
Top-running Bridge Crane

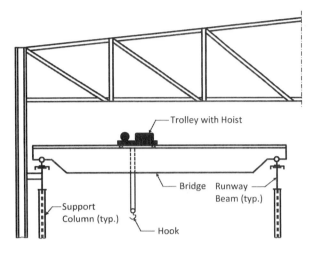

The maximum wheel loads that are to be used in the design of the supporting members are due to the weight of the bridge plus the sum of the rated capacity and the weight of the trolley. The trolley is to be positioned on its runway at the location where the resulting load effect is maximum; generally, this occurs when the trolley is moved as close to the supporting members as possible.

The maximum wheel loads must be increased by the percentages given in IBC 1607.13.2 to account for the vertical impact force that is caused by the starting and stopping movement of the suspended weight from the crane and by the movement of the crane along the rails.

A lateral force that acts perpendicular to the crane runway beams is generated by the transverse movement of the crane, that is, by movement that occurs perpendicular to the runway beam (see Figure 3.8). According to IBC 1607.13.3, the magnitude of this load on crane runway beams with electronically powered trolleys is to be taken equal to 20 percent of the sum of the rated capacity of the crane and the weight of the hoist and trolley. It is assumed that this load acts horizontally at the traction surface of the runway beam and that it is distributed to the runway beam and supporting structure (such as columns) based on the lateral stiffness of the members.

Figure 3.8 Crane Loads on a Runway Beam

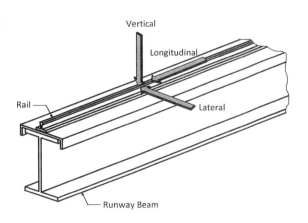

A longitudinal force is generated on a crane runway beam by acceleration, deceleration and braking of the crane bridge beam (see Figure 3.8). This load is taken as 10 percent of the maximum wheel loads of the crane and is assumed to act horizontally at the traction surface of the runway beam in either direction parallel to the beam (IBC 1607.13.4). Bridge cranes with hand-geared bridges are exempt from this provision.

Reference 3.3 contains additional information on the determination of crane loads for industrial buildings.

3.2.14 Interior Walls and Partitions

Interior walls and partitions (including their finishing materials) that are greater than 6 feet in height are required to be designed for a horizontal load of 5 psf (IBC 1607.14). This requirement is intended to provide sufficient strength and durability of the wall framing and its finished construction when subjected to nominal impact loads, such as those from moving furniture or equipment and from HVAC pressurization.

Requirements for fabric partitions that exceed 6 feet in height are given in IBC 1607.14.1. A horizontal load equal to 5 psf is to be applied to the partition framing, which acts over the area of the fabric face between the framing members to which the fabric is attached. The distributed load on these framing members shall be uniform and is based on the length of the member. Additionally, a 40-pound load must be applied over an 8-inch diameter area of the fabric face at a height of 54 inches above the floor. This condition is meant to simulate the load caused by a person leaning against the fabric using their hand as the point of contact.

3.3 Rain Loads

3.3.1 Overview

IBC 1611 and ASCE/SEI Chapter 8 contain requirements for design rain loads. The nominal rain load, R, is determined based on the amount of water that can accumulate on a roof assuming that the primary roof drainage system is blocked. When this occurs, water will rise above the primary roof drain until it reaches the elevation of the roof edge or the secondary drainage system. The depth of water above the primary drain at the design rainfall intensity is based on the flow rate of the secondary system, which varies widely depending on the type of secondary system that is used.

The type and location of the secondary drains and the amount of rainwater above their inlets under design conditions must be known in order to determine R. Coordination among the design team (architectural, structural and plumbing) is very important when establishing rain loads.

Chapter 11 of the *International Plumbing Code*® (IPC®) contains requirements on the design of roof drainage systems, including the required size and number of drains based on the area that is serviced (Reference 3.4). Rainfall rates are given for various cities in the United States in Appendix B of the IPC. These rates are based on the maps in IPC Figure 1106.1, which have the same origin as the maps in the IBC (see the following discussion).

3.3.2 Design Rain Load

IBC Equation 16-36 or ASCE/SEI Equation 8.3-1 is used to determine the rain load, R:

$$R = 5.2(d_s + d_h) \tag{3.6}$$

The constant in this equation is equal to the unit load per inch depth of rainwater, that is, the density of water divided by 12 inches per foot: 62.4/12 = 5.2 psf per inch.

The total depth of rainwater on a roof that is to be used in determining R consists of two parts:

1. The depth of water on the undeflected roof up to the inlet of the secondary drainage system when the primary drainage system is blocked (d_s); and

2. The additional depth of water on the undeflected roof above the inlet of the secondary drainage system at its design flow (d_h).

The static head, d_s, is the distance from the primary drain to the secondary drain and is determined in the design of the combined drainage system.

The rainwater depth, d_h, (which is also referred to as the hydraulic head) is a function of the rainfall intensity, i, at the site, the area of roof, A, that is serviced by that drainage system, and the size of the drainage system.

IBC Figure 1611.1 provides the rainfall rates for a storm of 1-hour duration that has a 100-year return period. These rates have been determined by a statistical analysis of weather records. Both the primary and secondary drainage systems must be designed for the prescribed rainfall rate (see Section 1108 of the IPC on how to size secondary drainage systems). It is always good practice to check with local building authorities to ensure that the proper rainfall rate is used in design.

In general, d_h depends on the type and size of the secondary drainage system and the flow rate, Q, it must handle. The following equation from ASCE/SEI C8.5 can be used to determine the flow rate through a single secondary drainage system:

$$Q = 0.0104 Ai \qquad (3.7)$$

In this equation, Q is in gallons per minute, A is in square feet and i is in inches per hour. The constant, 0.0104, is obtained based on the units associated with the variables in the equation:

$$\text{Constant} = \text{ft}^2 \times \frac{\text{in}}{\text{hr}} \times \frac{1 \text{ ft}}{12 \text{ in}} \times \frac{7.48 \text{ gal}}{\text{ft}^3} \times \frac{1 \text{ hr}}{60 \text{ minutes}} = 0.0104$$

The following equations relate the flow rate, Q, in gallons per minute to the hydraulic head, d_h, in inches for channel- and closed-type scuppers (see Figure 3.9 and Reference 3.5):

Channel-type scuppers: $Q = 2.9 b d_h^{1.5}$ when $h \geq d_h$ \qquad (3.8)

Closed-type scuppers: $Q = 2.9 b [d_h^{1.5} - (d_h - h)^{1.5}]$ when $h < d_h$ \qquad (3.9)

where b and h are the width and depth of the scupper, respectively, in inches. Note that Equation 3.8 is also applicable to closed-type scuppers when $h \geq d_h$.

Figure 3.9 Roof Scuppers

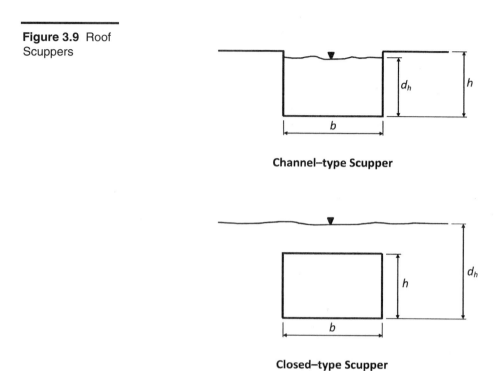

ASCE/SEI Table C8-1 gives flow rates in gallons per minute and corresponding hydraulic heads for circular drain, channel-type scupper and closed-type scupper drainage systems. For example, a 6-inch wide by 4-inch high closed scupper with 3 inches of hydraulic head will discharge 90 gallons per minute. If Equation 3.8 is used to calculate Q, the result is 90.4 gallons per minute, which is essentially the same value as that given in the table. Similarly, a hydraulic head, d_h, can be determined from this table or from Equations 3.8 and 3.9 for a required flow rate, Q.

Figure 3.10 illustrates the rainwater depths, d_s and d_h, that are to be used in determining R for the case of a scupper secondary drainage system while Figure 3.11 illustrates these water depths for a typical interior circular secondary drainage system.

Figure 3.10 Water Depths, d_s and d_h, in Accordance with IBC 1611 for Typical Perimeter Scuppers

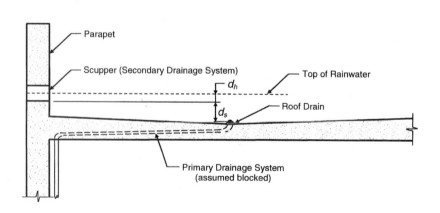

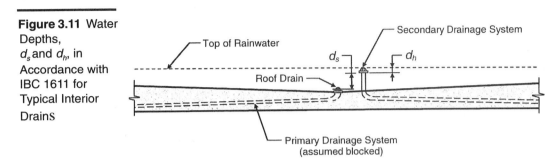

Figure 3.11 Water Depths, d_s and d_h, in Accordance with IBC 1611 for Typical Interior Drains

As noted in the preceding discussion, d_s and d_h are determined assuming an undeflected roof; deflections due to any types of loads, including dead loads, are not considered when determining the amount of rainwater on the roof. This assumption eliminates the complexities of determining the amount of rain that would be present in deflected areas of a roof.

Where buildings are configured such that rainwater will not collect on the roof, no rain load is required in the design of the roof, and a secondary drainage system is not needed. What is important to note is that the provisions of IBC 1611 and ASCE/SEI Chapter 8 must be considered wherever the potential exists that water may accumulate on the roof.

3.3.3 Ponding Instability

In situations where roofs do not have adequate slope or have insufficient and/or blocked drains to remove water due to rain or melting snow, water will tend to pond in low areas, which will cause the roof structure to deflect. These low areas will subsequently attract even more water, leading to additional deflection. Sufficient stiffness must be provided so that deflections will not continually increase until instability occurs, resulting in localized failure.

The provisions in ASCE/SEI 8.4 are provided to help ensure that ponding instability does not occur (IBC 1611.2). Susceptible bays, that is, bays with a roof slope less than $1/4$ inch per foot or those where the primary drain system is blocked and the secondary drain system is functional, must be analyzed for the effects from the larger of rain loads or snow loads (see Chapter 4 of this publication for the determination of snow loads). The roof structure in these bays must be designed with adequate stiffness to preclude ponding instability.

ASCE/SEI Figure C8.3 illustrates susceptible bays in a building where the roof slopes are greater than or equal to $1/4$ inch per foot (see Figure 3.12). Water can accumulate in the identified bays, regardless of the roof slope, and can lead to instability if sufficient stiffness is not provided. Note that the bay on the far right is not susceptible because the slope is at least $1/4$ inch per foot and water can drain freely over the edge (ASCE/SEI 8.4). If the roof slope is less than $1/4$ inch per foot, then all bays are susceptible.

Figure 3.12 Example of Susceptible Bays for Ponding Evaluation for Roof Slopes Greater than $1/4$ inch per foot

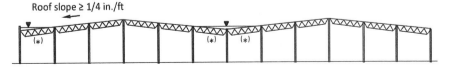

(*)Susceptible bays

ASCE/SEI Figure C8.4 shows a roof with interior primary drains and a secondary drainage system consisting of perimeter scuppers (see Figure 3.13). All bays are susceptible in this situation irrespective of the roof slope because water can be impounded on the roof when the primary drainage system is blocked.

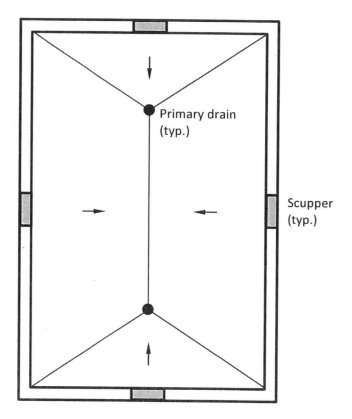

Figure 3.13 Example of Susceptible Bays for Ponding Evaluation Where Rainwater Is Impounded on the Roof Due to Blocked Primary Drains

3.3.4 Controlled Drainage

In some jurisdictions, the rate of rainwater flow from roofs into storm drains is limited. Controlled-flow drains are typically utilized in such cases and are designed in accordance with IPC 1110 (Reference 3.4). The drainage flow rate provided by the primary drainage system (controlled-flow drains) is less than the rainfall rate and water intentionally accumulates on the roof.

According to IBC 1611.3 and ASCE/SEI 8.5, a secondary drainage system at a higher elevation than the primary system must be provided when controlled-flow drains are used; this is meant to limit the accumulation of water on the roof above that elevation.

Roofs must be designed for the load of rainwater that will accumulate on the roof between the primary drainage system and the secondary system plus the amount that will accumulate above the inlet of the secondary drain when the design flow rate is reached. Determining the rain load in this situation is identical to that discussed previously in Section 3.3.2 of this publication.

Such roofs must also be checked for ponding instability in accordance with IBC 1611.2 or ASCE/SEI 8.4.

3.4 Soil Lateral Loads

Foundation walls of a building or structure and retaining walls must be designed to resist the lateral loads caused by adjacent soil. A geotechnical investigation is usually undertaken to determine the magnitude of the soil pressure. In cases where the results of such an investigation are not available, the lateral soil loads in IBC Table 1610.1 are to be used (similar design lateral loads are provided in ASCE/SEI Table 3.2-1).

The design lateral soil load, H, depends on the type of soil and the boundary conditions at the top of the wall. Walls that are restricted to move at the top are to be designed for the at-rest pressures tabulated in IBC Table 1610.1 while walls that are free to deflect and rotate at the top are to be designed for the active pressures in that table. Figure 3.14 illustrates the distribution of active soil pressure over the height of a reinforced concrete foundation wall.

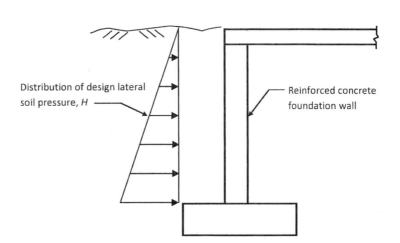

Figure 3.14
Distribution of Active Soil Pressure on a Foundation Wall

Foundation walls that do not extend more than 8 feet below grade and that are laterally supported at the top by flexible diaphragms are permitted to be designed for the active pressure values given in the table.

Lateral soil pressures are not provided for the expansive soils identified by Note b in IBC Table 1610.1 because these soils have unpredictable characteristics. These soils absorb water and tend to shrink and swell to a higher degree than other soils. As these soils swell, they are capable of exerting relatively large forces on the soil-retaining structure. As such, expansive soils are not suitable as backfill (see Note b).

In addition to lateral pressures from soil, walls must be designed to resist the effects of hydrostatic pressure due to undrained backfill (unless a drainage system is installed) and to any surcharge loads that can result from sloping backfills or from drive ways or parking spaces that are close to a wall. Submerged or saturated soil pressures include the weight of the buoyant soil plus the hydrostatic pressure.

ASCE/SEI 3.2.2 contains requirements for the design of any horizontal element supported directly on soil, such as slabs on grade and basement slabs. Full hydrostatic pressure must be applied over the entire area of such elements where applicable. Elements that are supported by expansive soils must be designed to accommodate the upward loads caused by the expansive soil, or the expansive soil is to be removed or stabilized around and beneath the structure.

3.5 Flowcharts

A summary of the flowcharts provided in this chapter is given in Table 3.2.

Table 3.2 Summary of Flowcharts Provided in Chapter 3

Flowchart	Title
Flowchart 3.1	Basic Uniform Live Load Reduction in accordance with IBC 1607.10.1
Flowchart 3.2	Alternative Uniform Live Load Reduction in accordance with IBC 1607.10.2

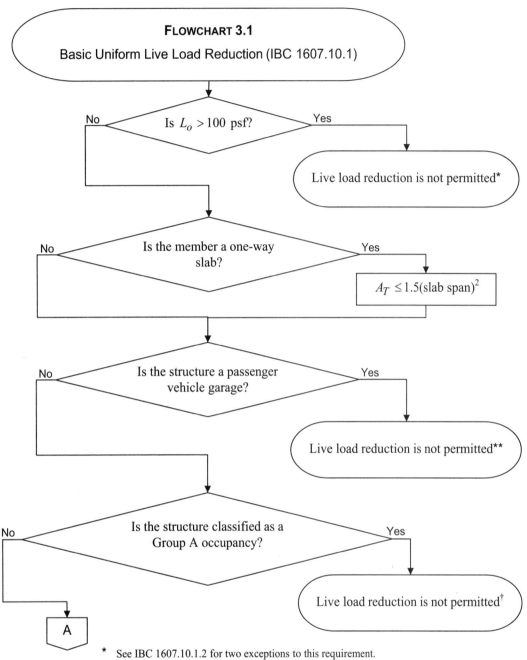

* See IBC 1607.10.1.2 for two exceptions to this requirement.

** Live loads for members supporting two or more floors are permitted to be reduced by a maximum of 20 percent (IBC 1607.10.1.3).

† Live loads for members supporting follow spot, projections and control rooms are permitted to be reduced (see Footnote m in IBC Table 1607.1).

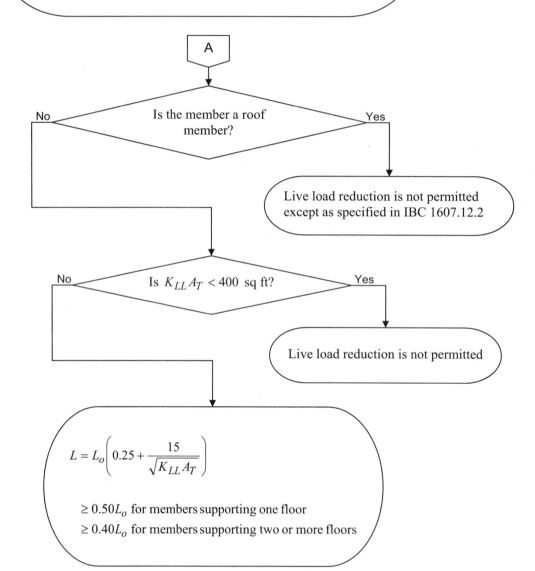

Dead, Live, Rain and Soil Lateral Loads

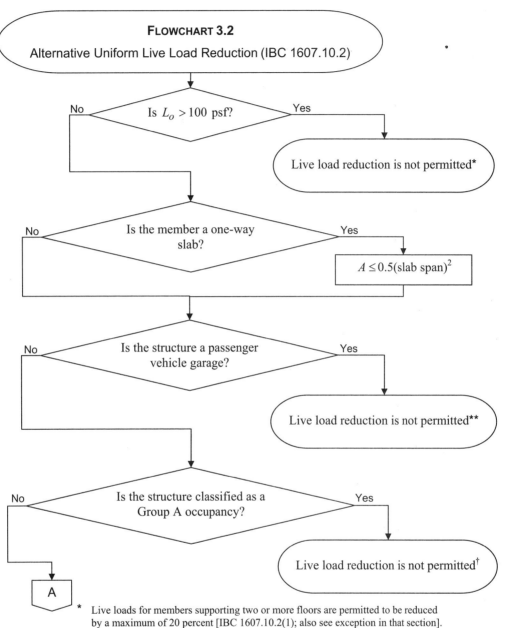

* Live loads for members supporting two or more floors are permitted to be reduced by a maximum of 20 percent [IBC 1607.10.2(1); also see exception in that section].

** Live loads for members supporting two or more floors are permitted to be reduced by a maximum of 20 percent [IBC 1607.10.2(2)].

† Live loads for members supporting follow spot, projections and control rooms are permitted to be reduced (see Footnote m in IBC Table 1607.1).

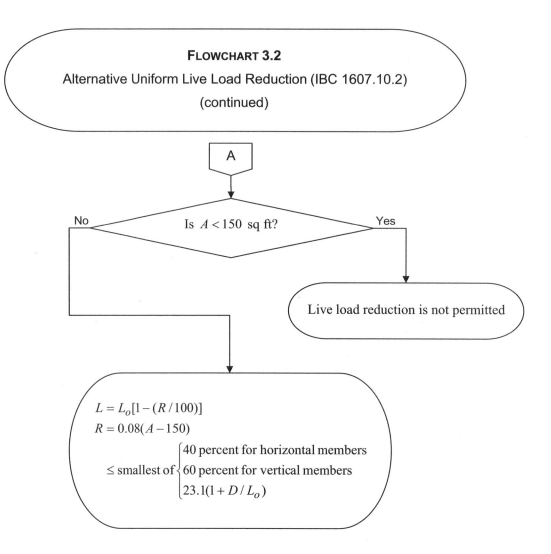

3.6 Examples

The following examples illustrate the IBC requirements for live load reduction and rain loads.

3.6.1 Example 3.1 – Basic Uniform Live Load Reduction, IBC 1607.10.1

The typical floor plan of a 10-story reinforced concrete office building is illustrated in Figure 3.15.

Determine reduced live loads for: (1) column A3, (2) column B3, (3) column C1, (4) column C4, (5) column B6 and (6) two-way slab AB23.

Figure 3.15
Typical Floor Plan of 10-story Office Building, Example 3.1

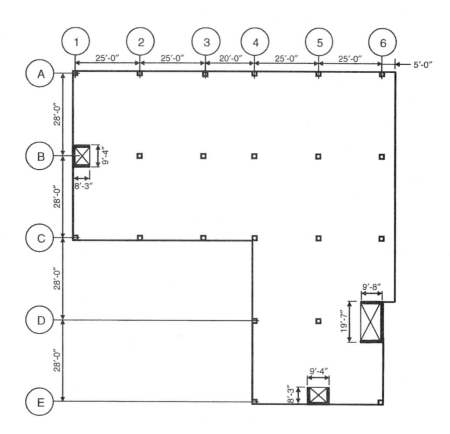

The ninth floor is designated as a storage floor (125 psf) and all other floors are typical office floors with moveable partitions.

The roof is an ordinary flat roof (slope of $1/2$ on 12) that is not occupiable. Assume that rainwater does not collect on the roof, and neglect snow loads.

Neglect lobby/corridor loads on the typical floors for this example.

SOLUTION

Nominal Loads

- Roof: 20 psf in accordance with IBC Table 1607.1, since the roof is an ordinary flat roof that is not occupiable.
- Ninth floor: storage load is given in the design criteria as 125 psf.
- Typical floor: 50 psf for office space in accordance with IBC Table 1607.1 and 15 psf for moveable partitions in accordance with IBC 1607.5, since the live load does not exceed 80 psf. The partition load is not reducible; only the minimum loads in IBC Table 1607.1 are permitted to be reduced (IBC 1607.10).

Part 1: Determine reduced live load for column A3

A summary of the reduced live loads is given in Table 3.3. Detailed calculations for various floor levels follow.

Table 3.3 Summary of Reduced Live Loads for Column A3

Story	Live Load (psf) N	Live Load (psf) R	$K_{LL}A_T$ (sq. ft.)	Reduction Multiplier	Reduced Live Load (psf)	N + R (kips)	Cumulative N + R (kips)
10	---	20	---*	---	17.8	5.6	5.6
9	125	---	---**	---	---	39.4	45.0
8	15	50	1,260	0.67	33.5	15.3	60.3
7	15	50	2,520	0.55	27.5	13.4	73.7
6	15	50	3,780	0.49	24.5	12.4	86.1
5	15	50	5,040	0.46	23.0	12.0	98.1
4	15	50	6,300	0.44	22.0	11.7	109.8
3	15	50	7,560	0.42	21.0	11.3	121.1
2	15	50	8,820	0.41	20.5	11.2	132.3
1	15	50	10,080	0.40	20.0	11.0	143.3

N = nonreducible live load, R = reducible live load
* Roof live load reduced in accordance with IBC 1607.12.2
** Live load > 100 psf is not permitted to be reduced (IBC 1607.10.1.2)

- Roof

    The reduced roof live load, L_r, is determined by IBC Equation 16-26:

    $$L_r = L_o R_1 R_2 = 20 R_1 R_2$$

    The tributary area, A_t, of column A3 = $(28/2) \times 22.5 = 315$ square feet

    Since 200 square feet $< A_t <$ 600 square feet, R_1 is determined by IBC Equation 16-28:

    $$R_1 = 1.2 - 0.001 A_t = 1.2 - (0.001 \times 315) = 0.89$$

    Since $F = 1/2 < 4$, $R_2 = 1$ (IBC Equation 16-30)

    Thus, $L_r = 20 \times 0.89 \times 1 = 17.8$ psf

    Axial load = $17.8 \times 315 / 1,000 = 5.6$ kips

- Ninth floor

    Since the ninth floor is storage with a live load of 125 psf, which exceeds 100 psf, the live load is not permitted to be reduced (IBC 1607.10.1.2).

    Axial load = $125 \times 315 / 1,000 = 39.4$ kips

- Typical floors

    Reducible nominal live load = 50 psf

    Since column A3 is an exterior column without a cantilever slab, the live load element factor $K_{LL} = 4$ (IBC Table 1607.10.1). Equivalently, K_{LL} = influence area/tributary area = $28(25 + 20)/315 = 4$.

    Reduced live load, L, is determined by IBC Equation 16-23:

    $$L = L_o \left(0.25 + \frac{15}{\sqrt{K_{LL} A_T}} \right)$$

Dead, Live, Rain and Soil Lateral Loads

$\geq 0.50 L_o$ for members supporting one floor

$\geq 0.40 L_o$ for members supporting two or more floors

The reduction multiplier is equal to 0.40 where $K_{LL} A_T \geq 10,000$ square feet (see Figure 3.5).

Axial load $= (L + 15) A_T = 315(L + 15)$

Part 2: Determine reduced live load for column B3

A summary of the reduced live loads is given in Table 3.4. Detailed calculations for various floor levels follow.

Table 3.4 Summary of Reduced Live Loads for Column B3

Story	Live Load (psf) N	Live Load (psf) R	$K_{LL}A_T$ (sq. ft.)	Reduction Multiplier	Reduced Live Load (psf)	N + R (kips)	Cumulative N + R (kips)
10	---	20	---*	---	12.0	7.6	7.6
9	125	---	---**	---	---	78.8	86.4
8	15	50	2,520	0.55	27.5	26.8	113.2
7	15	50	5,040	0.46	23.0	23.9	137.1
6	15	50	7,560	0.42	21.0	22.7	159.8
5	15	50	10,080	0.40	20.0	22.1	181.9
4	15	50	12,600	0.40	20.0	22.1	204.0
3	15	50	15,120	0.40	20.0	22.1	226.1
2	15	50	17,640	0.40	20.0	22.1	248.2
1	15	50	20,160	0.40	20.0	22.1	270.3

N = nonreducible live load, R = reducible live load
* Roof live load reduced in accordance with IBC 1607.12.2
** Live load > 100 psf is not permitted to be reduced (IBC 1607.10.1.2)

- Roof

    The reduced roof live load, L_r, is determined by IBC Equation 16-26:

    $L_r = L_o R_1 R_2 = 20 R_1 R_2$

    The tributary area, A_t, of column B3 $= 28 \times 22.5 = 630$ square feet

    Since $A_t > 600$ square feet, R_1 is determined by IBC Equation 16-29: $R_1 = 0.6$.

    Since $F = 1/2 < 4$, $R_2 = 1$ (IBC Equation 16-30)

    Thus, $L_r = 20 \times 0.6 \times 1 = 12.0$ psf

    Axial load $= 12.0 \times 630 / 1,000 = 7.6$ kips

- Ninth floor

    Since the ninth floor is storage with a live load of 125 psf, which exceeds 100 psf, the live load is not permitted to be reduced (IBC 1607.10.1.2).

    Axial load $= 125 \times 630 / 1,000 = 78.8$ kips

- Typical floors

    Reducible nominal live load = 50 psf

58 Chapter 3

Since column B3 is an interior column, the live load element factor $K_{LL} = 4$ (IBC Table 1607.10.1). Equivalently, K_{LL} = influence area/tributary area = $2[(28 \times 25) + (28 \times 20)]/630 = 4$.

Reduced live load, L, is determined by IBC Equation 16-23:

$$L = L_o\left(0.25 + \frac{15}{\sqrt{K_{LL}A_T}}\right)$$

$\geq 0.50L_o$ for members supporting one floor

$\geq 0.40L_o$ for members supporting two or more floors

The reduction multiplier is equal to 0.40 where $K_{LL}A_T \geq 10,000$ square feet (see Figure 3.5).

Axial load = $(L + 15)A_T = 630(L + 15)$

Part 3: Determine reduced live load for column C1

A summary of the reduced live loads is given in Table 3.5. Detailed calculations for various floor levels follow.

Table 3.5 Summary of Reduced Live Loads for Column C1

Story	Live Load (psf) N	Live Load (psf) R	$K_{LL}A_T$ (sq. ft.)	Reduction Multiplier	Reduced Live Load (psf)	N + R (kips)	Cumulative N + R (kips)
10	---	20	---*	---	20.0	3.5	3.5
9	125	---	---**	---	---	21.9	25.4
8	15	50	700	0.82	41.0	9.8	35.2
7	15	50	1,400	0.65	32.5	8.3	43.5
6	15	50	2,100	0.58	29.0	7.7	51.2
5	15	50	2,800	0.53	26.5	7.3	58.5
4	15	50	3,500	0.50	25.0	7.0	65.5
3	15	50	4,200	0.48	24.0	6.8	72.3
2	15	50	4,900	0.46	23.0	6.7	79.0
1	15	50	5,600	0.45	22.5	6.6	85.6

N = nonreducible live load, R = reducible live load
* Roof live load reduced in accordance with IBC 1607.12.2
** Live load > 100 psf is not permitted to be reduced (IBC 1607.10.1.2)

- Roof

    The reduced roof live load, L_r, is determined by IBC Equation 16-26:

    $L_r = L_o R_1 R_2 = 20 R_1 R_2$

    The tributary area, A_t, of column C1 = $28 \times 25 / 4 = 175$ square feet

    Since $A_t < 200$ square feet, R_1 is determined by IBC Equation 16-27: $R_1 = 1$.

    Since $F = 1/2 < 4$, $R_2 = 1$ (IBC Equation 16-30)

Thus, $L_r = 20 \times 1 \times 1 = 20.0$ psf

Axial load = $20.0 \times 175 / 1,000 = 3.5$ kips

- Ninth floor

    Since the ninth floor is storage with a live load of 125 psf, which exceeds 100 psf, the live load is not permitted to be reduced (IBC 1607.10.1.2).

    Axial load = $125 \times 175 / 1,000 = 21.9$ kips

- Typical floors

    Reducible nominal live load = 50 psf

    Since column C1 is an exterior column without a cantilever slab, the live load element factor $K_{LL} = 4$ (IBC Table 1607.10.1). Equivalently, K_{LL} = influence area/tributary area = $(28 \times 25)/175 = 4$.

    Reduced live load, L, is determined by IBC Equation 16-23:

    $$L = L_o\left(0.25 + \frac{15}{\sqrt{K_{LL}A_T}}\right)$$

    $\geq 0.50 L_o$ for members supporting one floor

    $\geq 0.40 L_o$ for members supporting two or more floors

    Axial load = $(L + 15)A_T = 175(L + 15)$

Part 4: Determine reduced live load for column C4

A summary of the reduced live loads is given in Table 3.6. Detailed calculations for various floor levels follow.

Table 3.6 Summary of Reduced Live Loads for Column C4

Story	Live Load (psf) N	Live Load (psf) R	$K_{LL}A_T$ (sq. ft.)	Reduction Multiplier	Reduced Live Load (psf)	N + R (kips)	Cumulative N + R (kips)
10	---	20	---*	---	14.2	7.0	7.0
9	125	---	---**	---	---	61.3	68.3
8	15	50	1,960	0.59	29.5	21.8	90.1
7	15	50	3,920	0.49	24.5	19.4	109.5
6	15	50	5,880	0.45	22.5	18.4	127.9
5	15	50	7,840	0.42	21.0	17.6	145.5
4	15	50	9,800	0.40	20.0	17.2	162.7
3	15	50	11,760	0.40	20.0	17.2	179.9
2	15	50	13,720	0.40	20.0	17.2	197.1
1	15	50	15,680	0.40	20.0	17.2	214.3

N = nonreducible live load, R = reducible live load
* Roof live load reduced in accordance with IBC 1607.12.2
** Live load > 100 psf is not permitted to be reduced (IBC 1607.10.1.2)

- Roof

    The reduced roof live load, L_r, is determined by IBC Equation 16-26:

    $$L_r = L_o R_1 R_2 = 20 R_1 R_2$$

    The tributary area, A_t, of column C4 = $(28 \times 20)/4 + (28 \times 25)/2 = 490$ square feet

    Since 200 square feet $< A_t < 600$ square feet, R_1 is determined by IBC Equation 16-28:

    $$R_1 = 1.2 - 0.001 A_t = 1.2 - (0.001 \times 490) = 0.71$$

    Since $F = 1/2 < 4$, $R_2 = 1$ (IBC Equation 16-30)

    Thus, $L_r = 20.0 \times 0.71 \times 1 = 14.2$ psf

    Axial load = $14.2 \times 490 / 1{,}000 = 7.0$ kips

- Ninth floor

    Since the ninth floor is storage with a live load of 125 psf, which exceeds 100 psf, the live load is not permitted to be reduced (IBC 1607.10.1.2).

    Axial load = $125 \times 490 / 1{,}000 = 61.3$ kips

- Typical floors

    Reducible nominal live load = 50 psf

    Since column C4 is an exterior column without a cantilever slab, the live load element factor $K_{LL} = 4$ (IBC Table 1607.10.1). Equivalently, K_{LL} = influence area/tributary area = $28[20 + (2 \times 25)]/490 = 4$.

    Reduced live load, L, is determined by IBC Equation 16-23:

    $$L = L_o \left(0.25 + \frac{15}{\sqrt{K_{LL} A_T}} \right)$$

    $\geq 0.50 L_o$ for members supporting one floor

    $\geq 0.40 L_o$ for members supporting two or more floors

    The reduction multiplier is equal to 0.40 where $K_{LL} A_T \geq 10{,}000$ square feet (see Figure 3.5).

    Axial load = $(L + 15) A_T = 490(L + 15)$

Part 5: Determine reduced live load for column B6

A summary of the reduced live loads is given in Table 3.7. Detailed calculations for various floor levels follow.

Table 3.7 Summary of Reduced Live Loads for Column B6

Story	Live Load (psf) N	Live Load (psf) R	$K_{LL}A_T$ (sq. ft.)	Reduction Multiplier	Reduced Live Load (psf)	N + R (kips)	Cumulative N + R (kips)
10	---	20	---*	---	14.2	7.0	7.0
9	125	---	---**	---	---	61.3	68.3
8	15	50	1,470	0.64	32.0	23.0	91.3
7	15	50	2,940	0.53	26.5	20.3	111.6
6	15	50	4,410	0.48	24.0	19.1	130.7
5	15	50	5,880	0.45	22.5	18.4	149.1
4	15	50	7,350	0.42	21.0	17.6	166.7
3	15	50	8,820	0.41	20.5	17.4	184.1
2	15	50	10,290	0.40	20.0	17.2	201.3
1	15	50	11,760	0.40	20.0	17.2	218.5

N = nonreducible live load, R = reducible live load
* Roof live load reduced in accordance with IBC 1607.12.2
** Live load > 100 psf is not permitted to be reduced (IBC 1607.10.1.2)

- Roof

    The reduced roof live load, L_r, is determined by IBC Equation 16-26:

    $$L_r = L_o R_1 R_2 = 20 R_1 R_2$$

    The tributary area, A_t, of column B6 = $28 \times (25/2 + 5) = 490$ square feet

    Since 200 square feet $< A_t <$ 600 square feet, R_1 is determined by IBC Equation 16-28:

    $$R_1 = 1.2 - 0.001 A_t = 1.2 - (0.001 \times 490) = 0.71$$

    Since $F = 1/2 < 4$, $R_2 = 1$ (IBC Equation 16-30)

    Thus, $L_r = 20 \times 0.71 \times 1 = 14.2$ psf

    Axial load = $14.2 \times 490 / 1,000 = 7.0$ kips

- Ninth floor

    Since the ninth floor is storage with a live load of 125 psf, which exceeds 100 psf, the live load is not permitted to be reduced (IBC 1607.10.1.2).

    Axial load = $125 \times 490 / 1,000 = 61.3$ kips

- Typical floors

    Reducible nominal live load = 50 psf

    Column B6 is an exterior column with a cantilever slab; thus, the live load element factor $K_{LL} = 3$ (IBC Table 1607.10.1). Note that the actual influence area/tributary area = $2[(28 \times 25) + (5 \times 28)]/490 = 3.4$. IBC Table 1607.10.1 requires $K_{LL} = 3$, which is slightly conservative.

62   Chapter 3

Reduced live load, L, is determined by IBC Equation 16-23:

$$L = L_o\left(0.25 + \frac{15}{\sqrt{K_{LL}A_T}}\right)$$

$\geq 0.50L_o$ for members supporting one floor

$\geq 0.40L_o$ for members supporting two or more floors

The reduction multiplier is equal to 0.40 where $K_{LL}A_T \geq 10{,}000$ square feet (see Figure 3.5).

Axial load $= (L + 15)A_T = 490(L + 15)$

Part 6: Determine reduced live load for two-way slab AB23

- Roof

    The reduced roof live load, L_r, is determined by IBC Equation 16-26:

    $L_r = L_o R_1 R_2 = 20 R_1 R_2$

    The tributary area, A_t, of this slab $= 25 \times 28 = 700$ square feet

    Since $A_t > 600$ square feet, R_1 is determined by IBC Equation 16-29: $R_1 = 0.6$

    Since $F = 1/2 < 4$, $R_2 = 1$ (IBC Equation 16-30)

    Thus, $L_r = 20 \times 0.6 \times 1 = 12.0$ psf

- Ninth floor

    Since the ninth floor is storage with a live load of 125 psf, which exceeds 100 psf, the live load is not permitted to be reduced (IBC 1607.10.1.2).

    Live load = 125 psf

- Typical floors

    Reducible nominal live load = 50 psf

    According to IBC Table 1607.10.1, $K_{LL} = 1$ for a two-way slab.

    Reduced live load L is determined by IBC Equation 16-23:

    $$L = L_o\left(0.25 + \frac{15}{\sqrt{K_{LL}A_T}}\right) = L_o\left(0.25 + \frac{15}{\sqrt{700}}\right) = 0.82 L_o$$

    $= 41.0$ psf

    $> 0.50 L_o$ for members supporting one floor

    Total live load $= 41 + 15 = 56$ psf

3.6.2 Example 3.2 – Alternative Uniform Live Load Reduction, IBC 1607.10.2

Determine the reduced live loads for the elements in Example 3.1 using the alternative uniform live load reduction method of IBC 1607.10.2. Assume a nominal dead-to-live load ratio of 2.

SOLUTION

Part 1: Determine reduced live load for column A3

A summary of the reduced live loads is given in Table 3.8. Detailed calculations for various floor levels follow.

- Roof

    The reduced roof live load, L_r, is determined by IBC Equation 16-26:

    $$L_r = L_o R_1 R_2 = 20 R_1 R_2$$

    The tributary area, A_t, of column A3 = $(28/2) \times 22.5 = 315$ square feet

    Since 200 square feet $< A_t <$ 600 square feet, R_1 is determined by IBC Equation 16-28:

    $$R_1 = 1.2 - 0.001 A_t = 1.2 - (0.001 \times 315) = 0.89$$

    Since $F = 1/2 < 4$, $R_2 = 1$ (IBC Equation 16-30)

    Thus, $L_r = 20 \times 0.89 \times 1 = 17.8$ psf

    Axial load = $17.8 \times 315 / 1{,}000 = 5.6$ kips

Table 3.8 Summary of Reduced Live Loads for Column A3

Story	Live Load (psf) N	Live Load (psf) R	A (sq. ft.)	Reduction Factor, R (%)	Reduced Live Load (psf)	N + R (kips)	Cumulative N + R (kips)
10	---	20	---*	---	17.8	5.6	5.6
9	125	---	---**	---	---	39.4	45.0
8	15	50	315	13	43.5	18.4	63.4
7	15	50	630	38	31.0	14.5	77.9
6	15	50	945	60	20.0	11.0	88.9
5	15	50	1,260	60	20.0	11.0	99.9
4	15	50	1,575	60	20.0	11.0	110.9
3	15	50	1,890	60	20.0	11.0	121.9
2	15	50	2,205	60	20.0	11.0	132.9
1	15	50	2,520	60	20.0	11.0	143.9

N = nonreducible live load, R = reducible live load
* Roof live load reduced in accordance with IBC 1607.12.2
** Live load > 100 psf is not permitted to be reduced [IBC 1607.10.2(1)]

- Ninth floor

    Since the ninth floor is storage with a live load of 125 psf, which exceeds 100 psf, the live load is not permitted to be reduced [IBC 1607.10.2(1)].

    Axial load $125 \times 315 / 1{,}000 = 39.4$ kips

- Typical floors

    Reducible nominal live load = 50 psf

    Reduction factor, R, is given by IBC Equation 16-24:

    $R = 0.08(A - 150)$

    $\leq$ the smallest of $\begin{cases} 60 \text{ percent for vertical members (governs)} \\ 23.1(1 + D/L_o) = 23.1(1 + 2) = 69 \text{ percent} \end{cases}$

    Axial load = $[L_o(1 - 0.01R) + 15]A = 315[50(1 - 0.01R) + 15]$

Part 2: Determine reduced live load for column B3

A summary of the reduced live loads is given in Table 3.9. Detailed calculations for various floor levels follow.

Table 3.9 Summary of Reduced Live Loads for Column B3

Story	Live Load (psf) N	Live Load (psf) R	A (sq. ft.)	Reduction Factor, R (%)	Reduced Live Load (psf)	N + R (kips)	Cumulative N + R (kips)
10	---	20	---*	---	12.0	7.6	7.6
9	125	---	---**	---	---	78.8	86.4
8	15	50	630	38	31.0	29.0	115.4
7	15	50	1,260	60	20.0	22.1	137.5
6	15	50	1,890	60	20.0	22.1	159.6
5	15	50	2,520	60	20.0	22.1	181.7
4	15	50	3,150	60	20.0	22.1	203.8
3	15	50	3,780	60	20.0	22.1	225.9
2	15	50	4,410	60	20.0	22.1	248.0
1	15	50	5,040	60	20.0	22.1	270.1

N = nonreducible live load, R = reducible live load
* Roof live load reduced in accordance with IBC 1607.12.2
** Live load > 100 psf is not permitted to be reduced [IBC 1607.10.2(1)]

- Roof

    The reduced roof live load, L_r, is determined by IBC Equation 16-26:

    $L_r = L_o R_1 R_2 = 20 R_1 R_2$

    The tributary area, A_t, of column B3 = $28 \times 22.5 = 630$ square feet

    Since $A_t > 600$ square feet, R_1 is determined by IBC Equation 16-29: $R_1 = 0.6$.

    Since $F = 1/2 < 4$, $R_2 = 1$ (IBC Equation 16-30)

    Thus, $L_r = 20 \times 0.6 \times 1 = 12.0$ psf

    Axial load = $12.0 \times 630 / 1,000 = 7.6$ kips

- Ninth floor

    Since the ninth floor is storage with a live load of 125 psf, which exceeds 100 psf, the live load is not permitted to be reduced [IBC 1607.10.2(1)].

    Axial load = $125 \times 630 / 1,000 = 78.8$ kips

Dead, Live, Rain and Soil Lateral Loads   65

- Typical floors

    Reducible nominal live load = 50 psf

    Reduction factor, R, is given by IBC Equation 16-24:

    $R = 0.08(A - 150)$

    $\leq$ the smallest of $\begin{cases} 60 \text{ percent for vertical members (governs)} \\ 23.1(1 + D/L_o) = 23.1(1 + 2) = 69 \text{ percent} \end{cases}$

    Axial load = $[L_o(1 - 0.01R) + 15]A = 630[50(1 - 0.01R) + 15]$

Part 3: Determine reduced live load for column C1

A summary of the reduced live loads is given in Table 3.10. Detailed calculations for various floor levels follow.

Table 3.10 Summary of Reduced Live Loads for Column C1

Story	Live Load (psf) N	Live Load (psf) R	A (sq. ft.)	Reduction Factor, R (%)	Reduced Live Load (psf)	N + R (kips)	Cumulative N + R (kips)
10	---	20	---*	---	20.0	3.5	3.5
9	125	---	---**	---	---	21.9	25.4
8	15	50	175	2	49.0	11.2	36.6
7	15	50	350	16	42.0	10.0	46.6
6	15	50	525	30	35.0	8.8	55.4
5	15	50	700	44	28.0	7.5	62.9
4	15	50	875	58	21.0	6.3	69.2
3	15	50	1,050	60	20.0	6.1	75.3
2	15	50	1,225	60	20.0	6.1	81.4
1	15	50	1,400	60	20.0	6.1	87.5

N = nonreducible live load, R = reducible live load
* Roof live load reduced in accordance with IBC 1607.12.2
** Live load > 100 psf is not permitted to be reduced [IBC 1607.10.2(1)]

- Roof

    The reduced roof live load, L_r, is determined by IBC Equation 16-26:

    $L_r = L_o R_1 R_2 = 20 R_1 R_2$

    The tributary area, A_t, of column C1 = $28 \times 25 / 4 = 175$ square feet

    Since $A_t < 200$ square feet, R_1 is determined by IBC Equation 16-27: $R_1 = 1$

    Since $F = 1/2 < 4$, $R_2 = 1$ (IBC Equation 16-30)

    Thus, $L_r = 20 \times 1 \times 1 = 20.0$ psf

    Axial load = $20 \times 175 / 1,000 = 3.5$ kips

- Ninth floor

    Since the ninth floor is storage with a live load of 125 psf, which exceeds 100 psf, the live load is not permitted to be reduced [IBC 1607.10.2(1)].

    Axial load = $125 \times 175 / 1,000 = 21.9$ kips

- Typical floors

    Reducible nominal live load = 50 psf

    Reduction factor, R, is given by IBC Equation 16-24:

    $R = 0.08(A - 150)$

    $\leq$ the smallest of $\begin{cases} 60 \text{ percent for vertical members (governs)} \\ 23.1(1 + D/L_o) = 23.1(1 + 2) = 69 \text{ percent} \end{cases}$

    Axial load = $[L_o(1 - 0.01R) + 15]A = 175[50(1 - 0.01R) + 15]$

Part 4: Determine reduced live load for column C4

A summary of the reduced live loads is given in Table 3.11. Detailed calculations for various floor levels follow.

Table 3.11 Summary of Reduced Live Loads for Column C4

Story	Live Load (psf) N	Live Load (psf) R	A (sq. ft.)	Reduction Factor, R (%)	Reduced Live Load (psf)	N + R (kips)	Cumulative N + R (kips)
10	---	20	---*	---	14.2	7.0	7.0
9	125	---	---**	---	---	61.3	68.3
8	15	50	490	27	36.5	25.2	93.5
7	15	50	980	60	20.0	17.2	110.7
6	15	50	1,470	60	20.0	17.2	127.9
5	15	50	1,960	60	20.0	17.2	145.1
4	15	50	2,450	60	20.0	17.2	162.3
3	15	50	2,940	60	20.0	17.2	179.5
2	15	50	3,430	60	20.0	17.2	196.7
1	15	50	3,920	60	20.0	17.2	213.9

N = nonreducible live load, R = reducible live load
* Roof live load reduced in accordance with IBC 1607.12.2
** Live load > 100 psf is not permitted to be reduced [IBC 1607.10.2(1)]

- Roof

    The reduced roof live load, L_r, is determined by IBC Equation 16-26:

    $L_r = L_o R_1 R_2 = 20 R_1 R_2$

    The tributary area, A_t, of column C4 = $(28 \times 20) / 4 + (28 \times 25) / 2 = 490$ square feet

    Since 200 square feet $< A_t <$ 600 square feet, R_1 is determined by IBC Equation 16-28:

    $R_1 = 1.2 - 0.001 A_t = 1.2 - (0.001 \times 490) = 0.71$

    Since $F = 1/2 < 4$, $R_2 = 1$ (IBC Equation 16-30)

    Thus, $L_r = 20.0 \times 0.71 \times 1 = 14.2$ psf

    Axial load = $14.2 \times 490 / 1,000 = 7.0$ kips

- Ninth floor

    Since the ninth floor is storage with a live load of 125 psf, which exceeds 100 psf, the live load is not permitted to be reduced [IBC 1607.10.2(1)].

    Axial load = $125 \times 490 / 1,000 = 61.3$ kips

Dead, Live, Rain and Soil Lateral Loads 67

- Typical floors

    Reducible nominal live load = 50 psf

    Reduction factor, R, is given by IBC Equation 16-24:

    $R = 0.08(A - 150)$

    $\leq$ the smallest of $\begin{cases} 60 \text{ percent for vertical members (governs)} \\ 23.1(1 + D/L_o) = 23.1(1 + 2) = 69 \text{ percent} \end{cases}$

    Axial load = $[L_o(1 - 0.01R) + 15]A = 490[50(1 - 0.01R) + 15]$

Part 5: Determine reduced live load for column B6

A summary of the reduced live loads is given in Table 3.12. Detailed calculations for various floor levels follow.

Table 3.12 Summary of Reduced Live Loads for Column B6

Story	Live Load (psf) N	Live Load (psf) R	A (sq. ft.)	Reduction Factor, R (%)	Reduced Live Load (psf)	N + R (kips)	Cumulative N + R (kips)
10	---	20	---*	---	14.2	7.0	7.0
9	125	---	---**	---	---	61.3	68.3
8	15	50	490	27	36.5	25.2	93.5
7	15	50	980	60	20.0	17.2	110.7
6	15	50	1,470	60	20.0	17.2	127.9
5	15	50	1,960	60	20.0	17.2	145.1
4	15	50	2,450	60	20.0	17.2	162.3
3	15	50	2,940	60	20.0	17.2	179.5
2	15	50	3,430	60	20.0	17.2	196.7
1	15	50	3,920	60	20.0	17.2	213.9

N = nonreducible live load, R = reducible live load
* Roof live load reduced in accordance with IBC 1607.12.2
** Live load > 100 psf is not permitted to be reduced [IBC 1607.10.2(1)]

- Roof

    The reduced roof live load, L_r, is determined by IBC Equation 16-26:

    $L_r = L_o R_1 R_2 = 20 R_1 R_2$

    The tributary area, A_t, of column B6 = $28 \times (25/2 + 5) = 490$ square feet

    Since 200 square feet < A_t < 600 square feet, R_1 is determined by IBC Equation 16-28:

    $R_1 = 1.2 - 0.001 A_t = 1.2 - (0.001 \times 490) = 0.71$

    Since $F = 1/2 < 4$, $R_2 = 1$ (IBC Equation 16-30)

    Thus, $L_r = 20 \times 0.71 \times 1 = 14.2$ psf

    Axial load = $14.2 \times 490 / 1,000 = 7.0$ kips

- Ninth floor

    Since the ninth floor is storage with a live load of 125 psf, which exceeds 100 psf, the live load is not permitted to be reduced [IBC 1607.10.2(1)].

Axial load = 125 × 490 / 1,000 = 61.3 kips

- Typical floors

    Reducible nominal live load = 50 psf

    Reduction factor, R, is given by IBC Equation 16-24:

    $R = 0.08(A - 150)$

    $\leq$ the smallest of $\begin{cases} 60 \text{ percent for vertical members (governs)} \\ 23.1(1 + D/L_o) = 23.1(1 + 2) = 69 \text{ percent} \end{cases}$

    Axial load = $[L_o(1 - 0.01R) + 15]A = 490[50(1 - 0.01R) + 15]$

Part 6: Determine reduced live load for two-way slab AB23

- Roof

    The reduced roof live load, L_r, is determined by IBC Equation 16-26:

    $L_r = L_o R_1 R_2 = 20 R_1 R_2$

    The tributary area, A_t, of this slab = 25 × 28 = 700 square feet

    Since $A_t > 600$ square feet, R_1 is determined by IBC Equation 16-29: $R_1 = 0.6$

    Since $F = 1/2 < 4$, $R_2 = 1$ (IBC Equation 16-30)

    Thus, $L_r = 20 \times 0.6 \times 1 = 12.0$ psf

- Ninth floor

    Since the ninth floor is storage with a live load of 125 psf, which exceeds 100 psf, the live load is not permitted to be reduced [IBC 1607.10.2(1)].

    Live load = 125 psf

- Typical floors

    Reducible nominal live load = 50 psf

    Reduction factor, R, is given by IBC Equation 16-24:

    $R = 0.08(A - 150) = 0.08(700 - 150) = 44$ percent

    $\leq$ the smallest of $\begin{cases} 40 \text{ percent for horizontal members (governs)} \\ 23.1(1 + D/L_o) = 23.1(1 + 2) = 69 \text{ percent} \end{cases}$

    Reduced live load $L = L_o(1 - R) = 50(1 - 0.4) = 30$ psf

    Total live load = 30 + 15 = 45 psf

Note: the reduced live load on the shear walls in this example, as well as that in Example 3.1, can be determined using the same procedure as for columns. For example, the shear walls located at E5 can be collectively considered to be an edge column without a cantilever slab.

3.6.3 Example 3.3 – Live Load Reduction on a Girder

Determine the reduced live load on a typical interior girder of the warehouse shown in Figure 3.16. The roof is an ordinary flat roof.

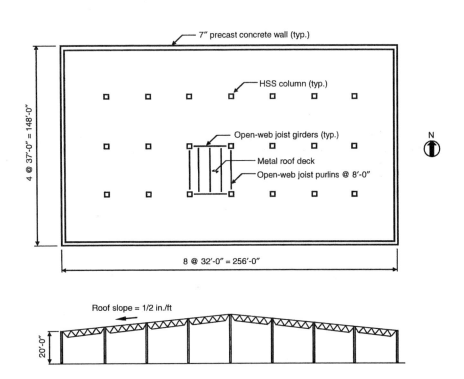

Figure 3.16 Plan and Elevation of Warehouse Building, Example 3.3

SOLUTION

The nominal roof live load is equal to 20 psf in accordance with IBC Table 1607.1, since the roof is an ordinary flat roof that is not occupiable.

The reduced roof live load, L_r, is determined by IBC Equation 16-26:

$L_r = L_o R_1 R_2 = 20 R_1 R_2$

The tributary area, A_t, of this girder $= 32 \times 37 = 1,184$ square feet

Since $A_t > 600$ square feet, R_1 is determined by IBC Equation 16-29: $R_1 = 0.6$

Since $F = 1/2 < 4$, $R_2 = 1$ (IBC Equation 16-30)

Thus, $L_r = 20 \times 0.6 \times 1 = 12.0$ psf

The required snow and wind loads on the roof of this warehouse are given in Chapters 4 and 5 of this publication, respectively.

3.6.4 Example 3.4 – Rain Load, IBC 1611

Determine the rain load, R, on a roof located in Madison, Wisconsin, similar to the one depicted in Figure 3.10 given the following design data:

- Tributary area of primary roof drain = 6,200 square feet
- Closed scupper size: 6 inches wide (b) by 4 inches high (h)

- Vertical distance from primary roof drain to inlet of scupper (static head distance, d_s) = 6 inches
- Rainfall rate i = 3.0 inches/hour (IBC Figure 1611.1)

SOLUTION

To determine the rain load, R, the hydraulic head, d_h, must be determined, based on the required flow rate.

Using Equation 3.7, determine the required flow rate, Q:

$Q = 0.0104Ai = 0.0104 \times 6{,}200 \times 3 = 193.4$ gallons/minute

Assume that the hydraulic head, d_h, is determined by Equation 3.9, which is applicable for closed scuppers where the free surface of the water is above the top of the scupper (that is, where $d_h > h$):

$$Q = 2.9b[d_h^{1.5} - (d_h - h)^{1.5}]$$

where b = width of the scupper, h = depth of the scupper, and $d_h - h$ = distance from the free surface of the water to the top of the scupper.

For a required flow rate of 193.4 gallons/minute, b = 6 inches and h = 4 inches, d_h = 5.5 inches.

Alternatively, by interpolating the values in ASCE/SEI Table C8-1 for a flow rate of 193.4 gallons/minute, d_h = 5.6 inches.

Note that the assumption that d_h = 5.5 inches > h = 4 inches is correct and no additional calculations are required.

The rain load, R, is determined by IBC Equation 16-36:

$R = 5.2(d_s + d_h) = 5.2(6 + 5.5) = 59.8$ psf

3.6.5 Example 3.5 – Rain Load, IBC 1611

Determine the rain load, R, on the roof depicted in Figure 3.17 given the following design data:

- Secondary drains: 6-inch diameter
- Vertical distance from primary roof drain to inlet of secondary drains (static head distance, d_s) = 3 inches (see Figure 3.11)
- Rainfall rate i = 4.5 inches/hour

Figure 3.17 Roof Plan, Example 3.5

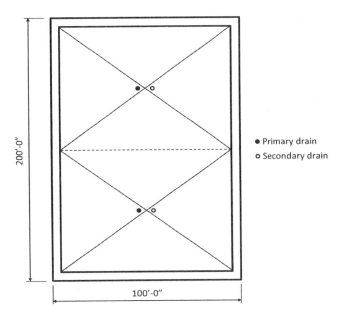

SOLUTION

To determine the rain load, R, the hydraulic head, d_h, must be determined, based on the required flow rate.

Using Equation 3.7, determine the required flow rate, Q, for each drain, which covers an area of 10,000 square feet:

$Q = 0.0104Ai = 0.0104 \times 10{,}000 \times 4.5 = 468.0$ gallons/minute

Interpolating the values in ASCE/SEI Table C8-1 for a 6-inch diameter drain with a flow rate of 468.0 gallons/minute yields a hydraulic head, $d_h = 3.3$ inches.

The rain load, R, is determined by IBC Equation 16-36:

$R = 5.2(d_s + d_h) = 5.2(3 + 3.3) = 32.8$ psf

3.7 References

3.1. International Conference of Building Officials. 1997. *Uniform Building Code*. Whittier, CA.

3.2. American Association of State Highway and Transportation Officials. 2010. *AASHTO LRFD Bridge Design Specifications*, 5th Ed. Washington, DC.

3.3. American Institute of Steel Construction. 2004. *Design Guide 7: Industrial Buildings— Roofs to Anchor Rods*, 2nd Ed. Chicago, IL.

3.4. International Code Council. 2011. 2012 *International Plumbing Code*. Washington, DC.

3.5. Factory Mutual Insurance Company. 2011. *Roof Loads for New Construction*. FM Global Property Loss Prevention Data Sheets 1-54. Johnston, RI.

3.8 Problems

3.1. Given the 10-story office building described in Example 3.1, determine reduced live loads for (a) column A3 and (b) column A6. Assume that the storage occupancy is on the eighth floor and all other floors are typical floors. Use the basic uniform live load reduction provisions of IBC 1607.10.1.

3.2. Given the information provided in Problem 3.1, determine the reduced live loads using the alternative uniform live load reduction provisions of IBC 1607.10.2. Assume a nominal dead-to-live load ratio of 2.

3.3. A steel transfer beam supports an interior column that supports a flat roof and four floors of office space. The tributary area of the column at each level is 600 square feet. The beam also supports a tributary floor area of 300 square feet of office space at the first elevated level. Using the basic uniform live load reduction provisions of IBC 1607.10.1, determine the reduced live loads on the beam.

3.4. Given the information provided in Problem 3.3, determine the reduced live loads on the transfer beam using the alternative uniform live load reduction provisions of IBC 1607.10.2. Assume a nominal dead-to-live load ratio of 0.75.

3.5. Given the information provided in Problem 3.3, determine the reduced live loads using the basic uniform live load reduction provisions of IBC 1607.10.1 assuming that all of the floors support patient rooms in a hospital.

3.6. Determine the reduced live load at each floor level of an edge column in an 8-story parking garage with a tributary area of 1,080 square feet at each level using the basic uniform live load reduction provisions of IBC 1607.10.1.

3.7. Determine the rain load, R, on a roof similar to the one depicted in Figure 3.10 given the following design data:

- Tributary area of primary roof drain = 3,000 square feet
- Rainfall rate $i = 3.75$ inches/hour
- 6-inch-wide (b) channel scupper
- Vertical distance from primary roof drain to inlet of scupper (static head distance, d_s) = 3 inches

3.8. Given the information provided in Problem 3.7, determine the rain load, R, assuming 4-inch diameter drains are used as the secondary drainage system.

CHAPTER 4

Snow and Ice Loads

4.1 Introduction

Structural members that are a part of roofs, balconies, canopies and similar structures that are exposed to the environment must be designed for the effects of snow loads in those geographic areas where snowfall can occur.

Loads on buildings and other structures due to snow are determined based on the anticipated ground snow load, the occupancy of the building, the exposure, the thermal resistance of the roof structure and the shape and slope of the roof. Partial loading, unbalanced snow loads due to roof configuration, drift loads on lower or adjacent roofs and on projections such as parapets and mechanical equipment, sliding snow loads and rain-on-snow loads must also be considered when designing for the effects from snow.

IBC 1608.1 requires that design snow loads be determined by the provisions of Chapter 7 of ASCE/SEI 7. In-depth information is given in the following sections on these provisions along with pertinent background information on the methodologies that are utilized.

In certain parts of the U.S., atmospheric ice loads must be considered in the design of all structures and structural elements that are exposed to the elements. Chapter 10 of ASCE/SEI 7-10 contains provisions on how to determine ice loads due to freezing rain on a variety of structural shapes, objects and configurations as a function of the design ice thickness. Provisions on how to determine atmospheric ice loads are presented at the end of this chapter.

4.2 Snow Loads

4.2.1 Ground Snow Loads

Ground snow load, p_g, is obtained from ASCE/SEI Figure 7-1 or IBC Figure 1608.2 for the conterminous U.S. and from ASCE/SEI Table 7-1 or IBC Table 1608.2 for locations in Alaska. The snow loads have been based on measurements taken at 204 National Weather Service stations and have a 2-percent annual probability of being exceeded (that is, a 50-year mean recurrence interval).

Table C7-1 in the commentary of ASCE/SEI 7 contains ground snow loads and the number of years measurements have been recorded at these stations. According to Note a in this table, it is not appropriate to use only the tabulated site-specific information to determine design snow loads. See ASCE/SEI C7.2 for more information on the methodology used to create the ground snow load maps.

In some areas of the U.S., the ground snow load is too variable to allow mapping. Such regions are noted on the maps as "CS," which indicates that a site-specific case study is required. Information on how to conduct a site-specific case study can be found in ASCE/SEI C7.2 and in Reference 4.1. It is always good practice to confirm ground snow loads with the authority having jurisdiction prior to design.

The maps also provide ground snow loads in mountainous areas based on elevation. Numbers in parentheses represent the upper elevation limits in feet for the ground snow load values that are given below the elevation. Where a building is located at an elevation greater than that shown on the maps, a site-specific case study must be conducted to establish the ground snow load.

Tabulated ground snow loads for Alaska are to be used for the specific locations noted in ASCE/SEI Table 7-1 or IBC Table 1608.2; these values do not necessarily represent the values that are relevant at nearby locations. The wide variability of snow loads in Alaska, which is evident from the tabulated values, precludes statewide mapping of ground snow loads similar to ASCE/SEI Figure 7-1 or IBC Figure 1608.2. Local records, experience and the authority having jurisdiction should also be utilized in such cases.

Additional information on snow loads for Rocky Mountain states can be obtained from the references given in ASCE/SEI C7.2. Since most of these references use mean recurrence intervals different than the 50-year interval used in ASCE/SEI 7, factors for converting annual probabilities of occurrence are given in Table C7-3.

4.2.2 Flat Roof Snow Loads

Overview

Once a ground snow load has been established, a *flat roof snow load*, p_f, is determined by Equation 7.3-1:

$$p_f = 0.7 C_e C_t I_s p_g \qquad (4.1)$$

This equation is applicable to the design of flat roofs, which is defined in 7.1 as a roof where the slope is less than or equal to 5 degrees (approximately 1 inch per foot).

Research has shown that the snow load on a roof is usually less than that on the ground in cases where drifting is not prevalent. The 0.7 factor in Equation 4.1 is a conservative ground-to-roof conversion factor that is used to account for this phenomenon.

The factors in this equation account for the thermal, aerodynamic, geometric and occupancy characteristics of a structure at its particular site, and are discussed in the following sections. Minimum snow loads, p_m, for low-slope roofs are also covered.

Exposure Factor

The exposure factor, C_e, accounts for the wind at the site and is related to the type of terrain and the exposure of the roof. Values of C_e are given in Table 7-2 as a function of the terrain category and the type of roof exposure. In general, unabated wind is more likely to blow snow off of a roof than wind that is impeded in some way; as such, snow loads are likely to be less.

Terrain categories B, C and D are the surface roughness categories defined in 26.7 for wind design (see Table 4.1 and Section 5.2.4 of this publication). A terrain category for a specific site should be chosen that represents the anticipated conditions during the life of the structure. Buildings located in terrain category D are more likely to have smaller snow loads on the roof than those located in terrain category B, since the roofs in the former are sheltered much less by the surrounding terrain than the latter. This is evident in Table 7-2 since the value of C_e decreases going from terrain category B to D for a given roof exposure.

Table 4.1 Terrain Categories

Terrain Category	Description
B	Urban and suburban areas, wooded areas or other terrain with numerous closely spaced obstructions having the size of single-family dwellings or larger
C	Open terrain with scattered obstructions having heights generally less than 30 feet; this category includes flat open country and grasslands
D	Flat, unobstructed areas and water surfaces; this category includes smooth mud flats, salt flats and unbroken ice

The terrain category that is identified as that above the tree line in windswept mountainous areas is to be used at appropriate locations other than high mountain valleys that receive little wind.

Roof exposures are defined as fully exposed, partially exposed and sheltered (see footnote a in Table 7-2). A fully exposed condition exists where a roof is exposed on all sides with no shelter provided by adjoining terrain, higher structures or trees. Roofs with large mechanical equipment, parapets that extend above the height of the balanced snow load or other similar obstructions are not considered to be fully exposed since such conditions can provide some shelter to the wind.

Obstructions are defined as providing shelter when they are located within a distance of $10h_o$ from the roof, where h_o is the height of the obstruction above the roof. The conifers depicted in Figure 4.1 provide shelter to the building roof if the distance, x, between the centerline of the building and the tree line is less than or equal to $10h_o$. In cases where deciduous trees that are leafless in winter surround the site, the fully exposed category is applicable.

Figure 4.1 Sheltered Roof Exposures for Snow Loads

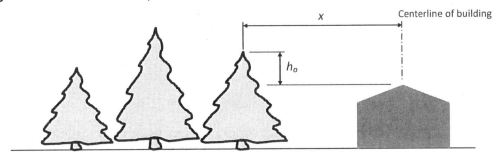

Obstruction provides shelter where $x \leq 10h_o$

Partially exposed roof exposures are to be used where fully exposed and sheltered conditions do not apply. This is generally the most common roof exposure.

Similar to terrain categories, a roof exposure condition must represent the conditions that are expected during the life of the building.

Although a single terrain category is specified at a particular site, buildings with multiple roofs may have different exposure conditions. For example, the upper roof of a building may have a fully exposed condition while the lower roof, which is sheltered by the higher portion of the building, may be partially exposed.

It is evident from Table 7-2 that for a given terrain category, the value of C_e and, thus, the corresponding snow load, increases going from a fully exposed roof exposure to a sheltered roof exposure.

Thermal Factor

The thermal factor, C_t, accounts for the amount of heat loss through the roof. In general, more snow will be present on cold roofs than warm roofs. Values of C_t are given in Table 7-3 as a function of the thermal condition. Like terrain categories and exposure conditions, the thermal condition that is chosen must represent the anticipated conditions during winters for the life of the building.

It can be seen in Table 7-3 that values of C_t are larger for colder roofs; thus, the snow load will be larger in such cases. It is possible for the flat roof snow load determined by Equation 4.1 for ordinary structures that are intentionally kept below freezing to be larger than the ground snow load without considering drift or sliding ($C_t = 1.3$). Observations of this phenomenon have been documented.

Importance Factor

The importance factor, I_s, adjusts the snow load based on the occupancy of the structure. Risk categories for buildings and other structures are defined in IBC Table 1604.5 and values of I_s are given in ASCE/SEI Table 1.5-2. It is evident from this table that the importance factor, and, thus, the corresponding snow load, is larger for more important structures.

An importance factor of 0.8 that is to be used for Risk Category I structures (low risk to human life) corresponds to a mean recurrence interval of about 25 years (approximately a 4 percent annual probability of exceedance) while an importance factor of 1.2 corresponds to about a 100-year mean recurrence interval (approximately a 1 percent annual probability of exceedance). Because of consistent nationwide statistical data for these mean recurrence intervals, only one snow load map is needed for design purposes; lower and higher risk uses and occupancies are accounted for by I_s.

Minimum Snow Loads for Low-slope Roofs

ASCE/SEI 7.3.4 contains provisions for minimum snow loads, p_m, for low-slope roofs, which are defined as follows (see Figure 4.2):

- Monoslope, hip and gable roofs with slopes less than 15 degrees

- Curved roofs where the vertical angle from the eaves to the crown is less than 10 degrees

Figure 4.2 Low-slope Roofs

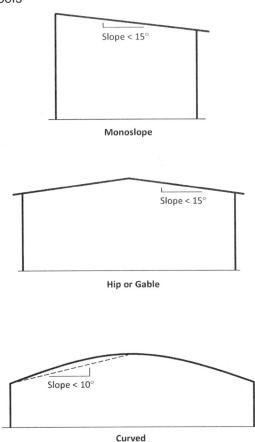

The purpose of the minimum snow loads is to account for important situations that can develop on roofs that are relatively flat. In regions where the ground snow load, p_g, is less than or equal to 20 psf, a single storm event can result in loading where the ground-to-roof conversion factor of 0.7 and the exposure and thermal factors (C_e and C_t, respectively) are not applicable. In other words, it is possible that a single storm in such regions can result in both ground and roof snow loads that approach the 50-year design snow load. This loading can occur especially on low-slope roofs where snow sliding is unlikely to happen and on relatively calm days with little to no wind following the storm where snow is unlikely to blow off of the roof. In such cases, $p_m = I_s p_g$.

In regions where the ground snow load exceeds 20 psf, it is also possible for a single storm to result in the ground and roof loads to be the same, but it is unlikely that these loads will be equal to the 50-year design snow load. Since it is improbable for a single storm to produce a load greater than 20 psf, then $p_m = 20 I_s$.

The roof slope limits for consideration of minimum loads are related to the roof slope limits for consideration of unbalanced loads. Unbalanced roof snow loads need to be considered for roofs based on the slopes in 7.6.1 (see Section 4.2.5 of this publication). In general, if a roof slope is steep enough so that unbalanced loads must be considered, it is unlikely that minimum snow loads will be applicable.

It is important to note that the minimum snow load is a uniform load case that is to be considered separately from any of the other applicable load cases. It need not be used in determining or in combination with drifting, sliding, unbalanced or partial snow loads.

Flowchart 4.1 in Section 4.4 of this publication can be used to determine the flat roof snow load, p_f.

4.2.3 Sloped Roof Snow Loads

Overview

Design snow loads for all structures are based on the *sloped roof snow load*, p_s, which is determined by Equation 7.4-1:

$$p_s = C_s p_f \qquad (4.2)$$

This snow load, which is also referred to as the *balanced snow load*, is assumed to act on the horizontal projection of a sloping roof surface. In general, snow loads decrease as the slope of a roof increases: snow is more likely to slide and be blown off of sloping roofs compared to those that are flat. In the case of flat roofs, $p_s = p_f$.

The slope factor, C_s, depends on the slope and temperature of the roof, the presence or absence of obstructions and the degree of slipperiness of the roof surface. Figure 7-2 contains graphs of C_s for warm and cold roof conditions and C7.4 contains equations for C_s.

The thermal factor, C_t, is used to determine whether a roof is warm or cold (see Section 4.2.2 of this publication). In particular, a roof is defined as warm where $C_t \leq 1.0$ and is defined as cold where $C_t > 1.0$. Warm roofs are more likely to shed snow than colder ones. Values of C_s in Figures 7-2b and 7-2c for cold roofs are greater than or equal to values of C_s in Figures 7-2a for warm roofs for a given roof angle and surface condition.

The ability of a sloped roof to shed snow also depends on the presence of obstructions on the roof and the degree of slipperiness of the roof surface. An obstruction can be considered as anything that impedes snow from sliding off of a roof. Large vent pipes, snow guards, parapet walls and large rooftop equipment are a few common examples of obstructions that could prevent snow from sliding off the roof. Ice dams and icicles along eaves can also possibly inhibit snow from sliding off of two types of warm roofs, which are described in 7.4.5 (see the discussion below).

An obstruction may also occur at the lower portions of sloping roofs that are near the ground; snow loads can concentrate near the lower portion of the roof since the snow may not be able to completely slide off of the roof due to the proximity of the ground.

Roof materials that are considered to be slippery and those that are not are given in 7.4. The dashed lines in Figure 7-2 should be used in determining C_s only when the roof surface is unobstructed and slippery.

Curved Roofs

According to 7.4.3, $C_s = 0$ for portions of curved roofs that have a slope exceeding 70 degrees; thus, $p_s = 0$ in such cases. Balanced snow loads for curved roofs are determined from the loading diagrams in Figure 7-3 with C_s determined from the appropriate curve in Figure 7-2.

Additional information on the determination of balanced and unbalanced snow loads for curved roofs is given in Section 4.2.5 of this publication.

Multiple Folded Plate, Sawtooth and Barrel Vault Roofs

Multiple folded plate, sawtooth and barrel vault roofs are to be designed using $C_s = 1$; thus, $p_s = p_f$ for these roof geometries (7.4.4). These types of roofs collect additional snow in their valleys by wind drifting and snow sliding, so no reduction in snow load based on roof slope is applied.

Information on the determination of balanced and unbalanced snow loads for sawtooth roofs is given in Section 4.2.5 of this publication.

Ice Dams and Icicles along Eaves

Certain types of warm roofs that drain water over their eaves can form relatively heavy loads on the cold, overhanging portions of the roof due to ice accumulation. Ice dams can form on an unventilated roof with a thermal resistance value (that is, R-value) less than 30 ft²·hr·°F/Btu or on a ventilated roof with an R-value less than 20 ft²·hr·°F/Btu.

To account for this phenomenon, 7.4.5 requires a uniformly distributed load of $2p_f$ plus the dead load to be applied on the overhanging portion; the flat roof snow load is determined based on the heated portion of the roof and no other loads are required to be applied.

This provision is intended for roof overhangs with a horizontal extent of less than or equal to 5 feet. In cases where the horizontal extent is greater than 5 feet, the uniformly distributed load of $2p_f$ need only be applied a distance of 5 feet from the eave of the heated structure; the remainder of the overhang is loaded with the unheated flat snow load.

Figure 4.3 illustrates the load cases for ice dams with overhanging roofs that are less than or equal to 5 feet and those that are greater than 5 feet (also see Figure C7.4).

Figure 4.3
Load Cases for Ice Dams

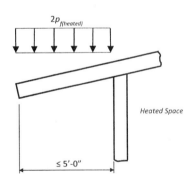

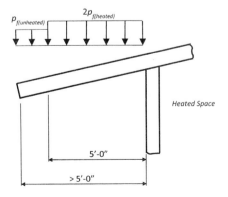

Flowcharts 4.2 and 4.3 in Section 4.4 of this publication can be used to determine the roof slope factor, C_s, and the sloped roof snow load, p_s, respectively.

4.2.4 Partial Loading

The partial loading provisions of 7.5 must be satisfied for continuous roof framing systems and all other roof systems where removal of snow load on one span (by wind or thermal effects) causes an increase in stress or deflection in an adjacent span. For example, an increase in bending moment and deflection will occur in the span of a cantilevered roof member that is adjacent to the cantilever span where half the snow load is removed.

For simplicity, only the three load cases given in Figure 7-4 need to be investigated; comprehensive alternate span (or checkerboard) loading analyses are not required:

- Full balanced snow load on either of the exterior spans and half of the balanced snow load on all other spans

- Half of the balanced snow load on either exterior span and full balanced snow load on all other spans

- All possible combinations of full balanced snow load on any two adjacent spans and half of the balanced snow load on all other spans. There will be $(n - 1)$ possible combinations for this case where n is equal to the number of spans in the continuous beam system.

Case 1 can occur when two different snow events that correspond to half of the balanced snow load are separated by an event that prevents the lower half of the roof snow from drifting. The intervening event can be sleet or freezing rain, for example. After the second snowfall occurs, a strong wind blows across the roof, removing snow from all of the spans and depositing it on all of the downwind spans. Case A in Figure 7-4 depicts the case for wind blowing from right to left.

Wind blowing from left to right must also be considered, and this is covered in Case 2 for a wind of shorter duration than that in Case 1. In particular, the wind event in Case 2 is long enough to remove the snow from the first span and deposit it on the adjoining spans.

The load pattern depicted for Case 3 in Figure 7-4 could correspond to a case where there is an obstruction on the roof (such as a parapet) that blocks the wind from blowing the snow behind the obstruction.

Partial loading provisions need not be considered for structural members that span perpendicular to the ridgeline of gable roofs with slopes greater than or equal to 2.38 degrees ($1/2$ on 12) since this loading case is addressed in the unbalanced snow load provisions of 7.6.1, which are covered in Section 4.2.5 of this publication.

4.2.5 Unbalanced Roof Snow Loads

Overview

Unbalanced snow load occurs on sloped roofs from wind and sunlight and in most cases can be considered a drift load. Wind tends to reduce the snow load on the windward portion of a roof and increase the snow load on the leeward portion. This is unlike partial loading where snow is removed on one portion of the roof and is not added to another portion. Provisions for unbalanced snow loads are given in 7.6.1 for hip and gable roofs, in 7.6.2 for curved roofs, in 7.6.3 for multiple folded plate, sawtooth and barrel vault roofs, and in 7.6.4 for dome roofs. All of these provisions are discussed in the following sections.

When determining unbalanced snow loads on a roof, it is important to consider wind from all directions.

Hip and Gable Roofs

Figure 4.4 depicts an unbalanced snow load on a hip and gable roof of a building due to wind. The shape of the roof drift can typically be characterized as a triangle located close to the ridgeline: the surcharge is essentially zero at the ridge and the top surface of the surcharge is basically horizontal. Loading conditions meant to represent this behavior are given in 7.6.1.

Figure 4.4 Unbalanced Snow Load on a Hip or Gable Roof

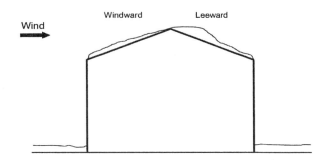

Unbalanced snow loads must be considered for roofs with slopes of $1/2$ on 12 (2.38 degrees) through 7 on 12 (30.2 degrees). Drifts typically do not form on roofs with slopes less than and greater than these limiting values. Figure 7-5 summarizes the balanced and unbalanced load conditions that must be addressed for hip and gable roofs (see Figure 4.5).

Figure 4.5 Balanced and Unbalanced Snow Loads for Hip and Gable Roofs

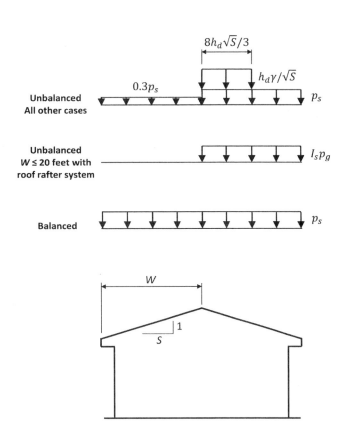

Two unbalanced load conditions are identified in the provisions. The first load condition is applicable to roofs with an eave-to-ridge distance, W, that is less than or equal to 20 feet where simply supported prismatic members span from the ridge to the eave (for example, a roof system consisting of wood or light gage roof members that are supported by a ridge beam). In such cases, the load on the windward portion of the roof is taken equal to zero and the load on the leeward portion is taken as a uniform load equal to $I_s p_g$. It can be shown that the moments and shear forces attributed to this loading condition are greater than those from the general loading condition, which is covered next. It is important to note that buildings with roof trusses should not be designed using this loading condition.

The second unbalanced load condition is applicable to all other hip and gable roofs where unbalanced loads must be investigated. The load on the windward portion of the roof is taken as a uniform load equal to $0.3p_s$. The load on the leeward portion is a combination of the balanced snow load across the entire width of the leeward roof plus a drift load equal to the drift height, h_d, multiplied by the density of snow, γ, and divided by the square root of the roof slope run for a rise of one S. The drift load acts over the length indicated in Figure 4.5. The drift height, h_d, in this case is determined by the equation given in Figure 7-9 where W is substituted for the upper roof length, ℓ_u:

$$h_d = 0.43(W)^{1/3}(p_g + 10)^{1/4} - 1.5 \qquad (4.3)$$

In cases where W is less than 20 feet, a value of W equal to 20 feet is used in Equation 4.3.

The density of snow, γ, is determined by Equation 7.7-1:

$$\gamma = 0.13p_g + 14 \leq 30 \text{ pcf} \qquad (4.4)$$

Additional information on snow drifts can be found in Section 4.2.6 of this publication.

Flowchart 4.4 in Section 4.4 of this publication can be used to determine the unbalanced roof snow loads for hip and gable roofs.

Curved Roofs

Overview

Provisions for balanced and unbalanced snow loads on curved roofs are given in 7.6.2. Any portion of a curved roof that has a slope that exceeds 70 degrees can be considered free of snow loads; in other words, it is assumed that snow will not be able to accumulate on portions of a roof that have such a steep slope. Note that the roof slope is measured from the horizontal to the tangent of the curved roof at that point. In cases where the roof slope exceeds 70 degrees, the point on the roof at a slope of 70 degrees is considered to be the eave (see discussion below).

Balanced and unbalanced load cases that are applicable to curved roofs are given in Figure 7-3 as a function of the slope of the roof at the eave. The three cases given in that figure are discussed below.

According to 7.6.2, unbalanced snow loads on curved roofs need not be considered where the slope of a straight line from the eaves or from a point on the roof where the tangent slope is equal to 70 degrees to the crown is less than 10 degrees or is greater than 60 degrees.

It is important to note that these provisions are not applicable to curved roofs that are concave upward. In such cases, as well as in cases for other roof geometries and complicated site conditions, wind tunnel model studies should be used to establish design snow loads (see C7.13).

Case 1 – Slope at Eaves Less than 30 Degrees

The balanced and unbalanced loads of Case 1 are depicted in Figure 4.6 for curved roofs that have a slope at the eave that is less than 30 degrees. In this case, the balanced load is trapezoidal near the eaves and uniform at the crown. At all locations, the magnitude of the balanced load is equal to $C_s p_f$ (see Equation 4.2).

Figure 4.6 Balanced and Unbalanced Loads for Curved Roofs—Slope at Eaves Less than 30 Degrees

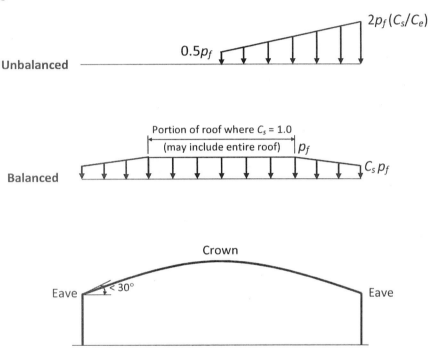

At the eaves, the roof slope factor, C_s, is determined from Figure 7-2 based on the thermal and surface conditions of the roof and on the slope of the roof at the eave. For shallow curved roofs, C_s may be equal to 1.0 across the entire roof. For curved roofs that are not shallow, C_s may be equal to 1.0 across a portion of the curved roof near the crown and may be less than 1.0 elsewhere. For example, for an unobstructed, cold roof ($C_t = 1.2$) with a slippery surface and a slope of 20 degrees at the eaves, C_s would be equal to 1.0 over the center portion of the roof where the slope is less than or equal to 15 degrees and would decrease linearly to 0.91 at the eaves (see Figure 7-2c).

In the unbalanced load case, the windward portion of the roof between the windward eave and the crown is assumed to be free of snow. A trapezoidal load distribution that varies from $0.5p_f$ at the crown to $2p_f(C_s/C_e)$ at the leeward eave is assumed to act on the leeward portion of the roof where C_s is determined from Figure 7-2 based on the slope at the eaves.

Case 2 – Slope at Eaves 30 Degrees to 70 Degrees

Case 2 is applicable to curved roofs with a slope at the eave from 30 through 70 degrees. The balanced load in this case is uniform near the crown (p_f) with middle and edge trapezoidal loads near the eaves (see Figure 4.7). The edge trapezoidal load extends from the eave to the point where the slope is equal to 30 degrees. The magnitude of the balanced load at the eave is equal to $C_{s|eave} p_f$ where $C_{s|eave}$ is determined from Figure 7-2 based on the slope at the eaves. The magnitude of the load linearly increases to a value of $C_{s|30} p_f$ at the point where the slope is 30 degrees; $C_{s|30}$ is determined from Figure 7-2 corresponding to a slope of 30 degrees. The middle trapezoidal load extends from the point where the slope is equal to 30 degrees to the point where $C_s = 1.0$.

Figure 4.7 Balanced and Unbalanced Loads for Curved Roofs—Slope at Eaves 30 Degrees through 70 Degrees

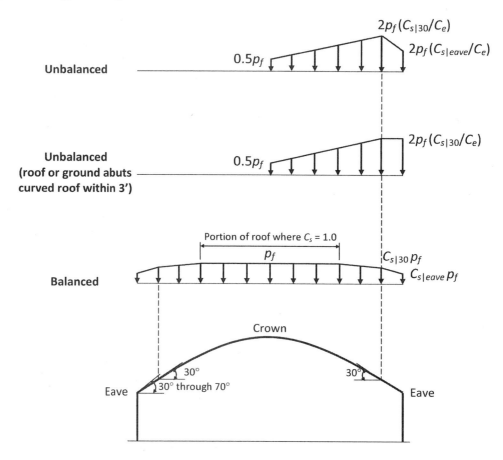

Like in the unbalanced load case of Case 1, the windward portion of the roof between the windward eave and the crown is assumed to be free of snow. There are two cases to consider for the leeward portion of the roof. In the first case, the ground or another roof abuts the curved roof within 3 feet of its eaves. At the crown, the magnitude of the unbalanced load is equal to $0.5p_f$, which is the same load as in Case 1. The load linearly increases away from the crown to a magnitude of $2p_f(C_{s|30}/C_e)$ at the point where the slope is equal to 30 degrees and remains a constant from that point to the eaves. In the second case, the ground or another roof does not abut the curved roof within 3 feet of its eaves. The loading is the same as that in the first case, except for the segment between the point where the slope is 30 degrees and the eaves: the load of $2p_f(C_{s|30}/C_e)$ decreases linearly to $2p_f(C_{s|eave}/C_e)$ at the eaves instead of remaining a constant.

Case 3 – Slope at Eaves Greater than 70 Degrees

The balanced and unbalanced loads are depicted in Figure 4.8 for cases where the slope at the eave of the curved roof exceeds 70 degrees. The balanced load is uniform at the crown (p_f) with a middle trapezoidal load and an end triangular load. Like in Case 2, the middle trapezoidal load extends from the point where the slope is equal to 30 degrees to the point where $C_s = 1.0$. The triangular load extends from the point where the roof slope is equal to 70 degrees to the point where it is 30 degrees. The load is equal to zero between the eave and the point where the slope is equal to 70 degrees, that is, it is zero in the segment where the roof slope is greater than or equal to 70 degrees.

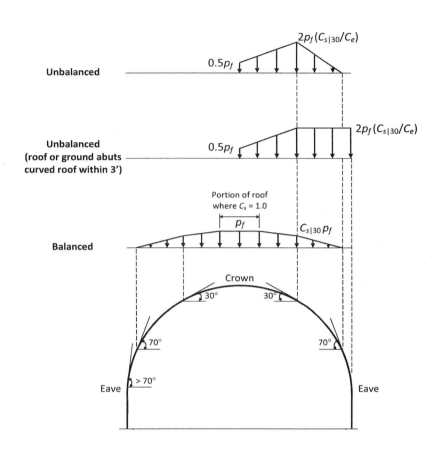

Figure 4.8
Balanced and Unbalanced Loads for Curved Roofs—Slope at Eaves Greater than 70 Degrees

The unbalanced load is very similar to Case 2: the windward portion of the roof between the windward eave and the crown is assumed to be free of snow, and there are two cases to consider for the leeward portion of the roof depending on whether or not the ground or another roof abuts the curved roof within 3 feet of its eaves (see Figure 4.8).

It is evident from the preceding discussion that the geometry of the roof must be known in order to locate the main points where a change in C_s occurs. For roofs that are arcs of constant radius, R, (that is, portions of a circle), the following equation can be used to evaluate the slope of the roof, θ, at any location, x:

$$\theta = \arctan\left(\frac{x}{\sqrt{R^2 - x^2}}\right) \qquad (4.5)$$

In this equation, x is the position on the horizontal projection of the roof that is measured from the crown (that is, $x = 0$ at the crown).

Flowchart 4.5 in Section 4.4 of this publication can be used to determine the unbalanced roof snow loads for curved roofs.

Multiple Folded Plate, Sawtooth and Barrel Vault Roofs

Provisions for unbalanced snow loads on folded plate, sawtooth and barrel vault roofs are given in 7.6.3. Unbalanced loads need to be considered when the slope exceeds $^3/_8$ inch/foot (1.79 degrees). These types of roof systems are especially vulnerable to heavy snow loads since they can accumulate snow in the valleys without having a means for wind to remove it.

It was discussed in Section 4.2.3 of this publication that $C_s = 1.0$ for these types of roofs; no reduction in snow load based on roof slope is applied. Therefore, the balanced snow load, p_s, is equal to p_f.

Similar to curved roofs, unbalanced loads are equal to $0.5p_f$ at the crown or ridge of the roof and are equal to $2p_f/C_e$ at the valleys. Illustrated in Figure 7-6 are the balanced and unbalanced loads for a sawtooth roof (see Figure 4.9).

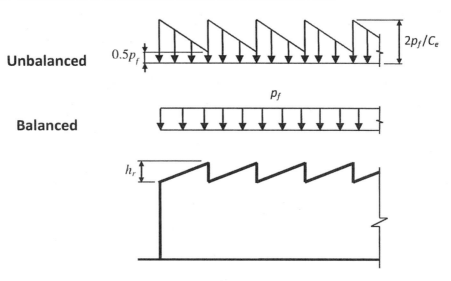

Figure 4.9 Balanced and Unbalanced Loads for a Sawtooth Roof

The snow load at a valley is limited by the space that is available for snow accumulation. The maximum permissible snow load at this location is equal to the load at the ridge ($0.5p_f$) plus the load corresponding to a snow depth equal to the distance, h_r, from the valley to the ridge (γh_r). If the unbalanced snow load ($2p_f/C_e$) at the valley is less than this maximum permissible load, the load at the valley is $2p_f/C_e$; otherwise, the load is set equal to the maximum permissible load.

Flowchart 4.6 in Section 4.4 of this publication can be used to determine the unbalanced roof snow loads for folded plate, sawtooth and barrel vault roofs.

Dome Roofs

According to 7.6.4, unbalanced snow loads for dome roofs are to be determined in the same manner as for curved roofs (see above discussion). Unbalanced loads are to be applied to the downwind 90-degree sector of the dome in plan (see Figure 4.10). The load decreases linearly to zero over 22.5-degree sectors on either side of this sector. No snow load is taken on the remaining 225-degree upwind sector.

Figure 4.10
Unbalanced Loads for a Dome Roof

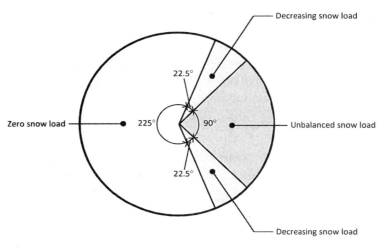

Plan of Dome Roof

The balanced and unbalanced load distributions depend on the slope of the roof at the eaves (see Cases 1 through 3 in Figures 4.6 through 4.8, respectively). In the unbalanced load case, the snow load at the eave or where the slope is equal to 70 degrees decreases linearly to zero over the 22.5-degree sector on each side of the 90-degree downwind sector of the roof.

Flowchart 4.5 in Section 4.4 of this publication can be used to determine the unbalanced roof snow loads for dome roofs.

4.2.6 Drifts on Lower Roofs

Overview

Section 7.7 contains provisions for snow drifts that can occur on lower roofs of a building due to

1. Wind depositing snow from higher portions of the same building or an adjacent building or terrain feature (such as a hill) to a lower roof; and

2. Wind depositing snow from the windward portion of a lower roof to the portion of the lower roof adjacent to a taller part of the building.

The first type of drift is called a leeward drift and the second is a windward drift. Both types of drifts are illustrated in Figure 7-8 (see Figure 4.11).

Figure 4.11
Windward and Leeward Snow Drifts

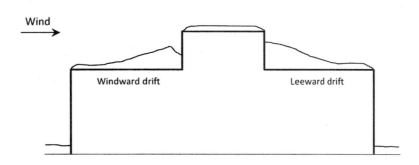

Also contained in this section are provisions for drifting on lower structures that are adjacent to higher ones. All of these requirements are discussed in the following sections.

Lower Roof of a Structure

As was discussed previously, wind will deposit snow from one area of a roof to another. In cases where there is a change in elevation between roofs, snow will have a tendency to accumulate at this location (see Figure 4.11). Depending on wind direction, either a windward or leeward drift will form. In the case of a windward drift, the wind blows across the length of the lower roof and deposits the snow on the lower roof adjacent to the wall. A leeward drift is formed when the wind blows the snow off of the upper roof on to the lower roof. Since wind can blow in any direction, both types of drifts must be investigated.

Leeward drifts are generally triangular in shape (see Figure 4.11). Windward drifts usually have more complex shapes than leeward ones depending on the height of the wall (or, roof step). For simplicity, a triangular shape is used to characterize windward drifts as well.

Figure 7-8 depicts the configuration that is to be taken for snow drifts on lower roofs (see Figure 4.12). It is evident from the figure that drift loads are superimposed on the balanced snow load on the roof, that is, the total load over the extent of the drift is equal to the balanced snow load on the lower roof plus the surcharge load due to drifting.

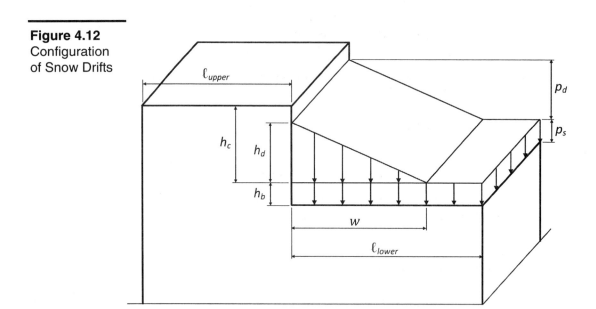

Figure 4.12
Configuration of Snow Drifts

In the case of leeward drifts, the height of the drift, h_d, at the roof step is determined from Figure 7-9 by substituting the length of the upper roof, ℓ_{upper}, for ℓ_u:

$$h_d|_{leeward} = 0.43(\ell_{upper})^{1/3}(p_g + 10)^{1/4} - 1.5 \qquad (4.6)$$

It is assumed that the snow over the entire area of the upper roof is available to form a drift on the leeward roof.

For windward drifts, the length of the lower roof, ℓ_{lower}, is used to determine the drift height, which is equal to three-quarters of h_d from Figure 7-9:

$$h_d|_{windward} = 0.75[0.43(\ell_{lower})^{1/3}(p_g + 10)^{1/4} - 1.5] \quad (4.7)$$

Similar to the leeward drift, the entire lower roof area is available to form the windward drift. The reduced height is based on case studies that show that windward roof steps are not as efficient at capturing snow as leeward roof steps. A review of these case studies revealed that the three-quarter reduction factor for windward drift heights is reasonable (Reference 4.2).

The larger of the windward and leeward drift heights is used to determine the drift load since wind can come from any direction. It is evident from Equations 4.6 and 4.7 that if the upper and lower roofs have the same length (or, fetch), the leeward drift height governs.

The maximum intensity of the drift surcharge load, p_d, occurs at the roof step and is equal to the governing drift height times the snow density, γ, which is determined by Equation 7.7-1. The magnitude of p_d decreasing linearly to zero over the drift width w, which is defined below.

The clear height, h_c, shown in Figure 4.12 is equal to the height of the roof step minus the height of the balanced snow, h_b. It is determined in this way assuming that the upper roof is blown clear of snow in the vicinity of the drift. While this assumption is generally valid for windward drifting, it is not necessarily accurate for leeward drifting since there may still be snow on the upper level roof when the drift has stopped growing. For simplicity, the same assumption is used for the clear height regardless of the type of drift that is formed.

The height of the balanced snow, h_b, is determined using the magnitude of the balanced snow load, p_s, and the density of the snow, γ:

$$h_b = \frac{p_s}{\gamma} = \frac{p_s}{0.13p_g + 14} \quad (4.8)$$

where γ need not be taken greater than 30 pcf (see Equation 7.7-1).

The width of the drift, w, depends on the clear height, h_c, as follows:

$$w = \begin{cases} 4h_d \text{ where } h_d \leq h_c \\ 4h_d^2/h_c \leq 8h_c \text{ where } h_d > h_c \end{cases} \quad (4.9)$$

If it is found that the governing drift height, h_d, determined from Figure 7-9 exceeds the clear height, h_c, then the drift height is set equal to h_c. The drift width in this case has been established by equating the cross-sectional area of the triangular drift that is limited to h_c (that is, $h_c \times w/2$) with the cross-sectional area of the triangular drift that is not limited to h_c (that is, $[h_d \times 4h_d/2]$). The upper limit of drift width $8h_c$ is based on studies that showed that additional snow accumulation is not expected within a rise-to-run ratio range of 1:6.5 to 1:10.

There may be cases where the drift width, w, exceeds the length of the lower roof, ℓ_{lower}. In these situations, the drift load is truncated at the far edge of the roof and the magnitude of the load at this location is not reduced to zero. Figure 4.13 shows the proper load distribution to use where $w > \ell_{lower}$.

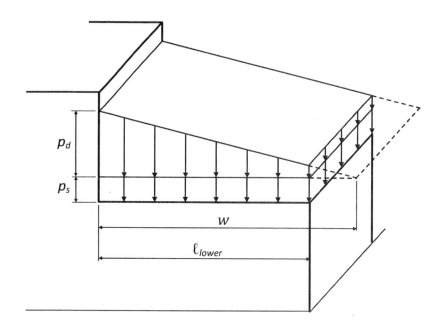

Figure 4.13 Load Configuration Where the Drift Width is Greater than the Length of the Lower Roof

According to 7.7.1, drift loads need not be considered where the ratio h_c/h_b is less than 0.2. In such cases, the roof step is relatively small so that the drift formation is negligible.

Flowchart 4.7 in Section 4.4 of this publication can be used to determine the drift load on the lower roof of a structure.

Adjacent Structures

Leeward drifts can form on the roof of a structure that is close enough to an adjacent one that has a higher roof. According to 7.7.2, leeward drifts form on lower roofs of adjacent buildings when the horizontal separation distance, s, between the two is less than 20 feet and is less than 6 times the vertical separation distance, h. The drift surcharge is determined using the provisions of 7.7.1 based on the smaller of the following two drift heights: (1) h_d based on the length of the adjacent higher structure and (2) $(6h - s)/6$. The drift width is equal to the smaller of six times the two aforementioned drift heights, that is, $6h_d$ or $(6h - s)$. Figure 4.14, which is based on the information in Figure C7-2, illustrates the leeward drift on the lower roof. Note that the total snow load on the lower roof is equal to the balanced snow load plus the drift load over its horizontal extent, w.

Figure 4.14 Leeward Snow Load on an Adjacent Roof

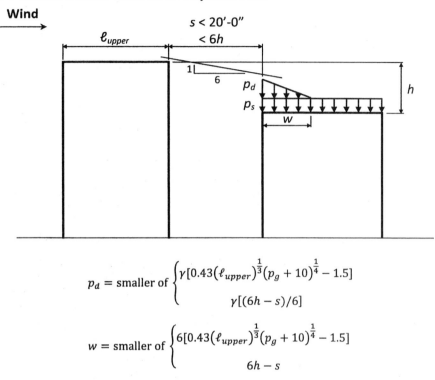

$$p_d = \text{smaller of} \begin{cases} \gamma[0.43(\ell_{upper})^{\frac{1}{3}}(p_g + 10)^{\frac{1}{4}} - 1.5] \\ \gamma[(6h - s)/6] \end{cases}$$

$$w = \text{smaller of} \begin{cases} 6[0.43(\ell_{upper})^{\frac{1}{3}}(p_g + 10)^{\frac{1}{4}} - 1.5] \\ 6h - s \end{cases}$$

The requirements of 7.7.1 are also used to determine the windward drift on the lower roof. In doing so, it is permitted to eliminate that portion of the drift that is located in the horizontal separation region, that is, in the zone where there is no roof. Figure 4.15, which is based on Figure C7-3, illustrates these requirements. The drift width $4h_d$ extends from the face of the building with the higher roof to the point on the lower roof where the drift load is equal to zero. Note that the drift height, h_d, is based on the length of the lower roof. The magnitude of the truncated drift load at the face of the building with the lower roof is equal to $1 - (s/4h_d)$ times that at the face of the building with the higher roof.

Figure 4.15 Windward Snow Load on an Adjacent Roof

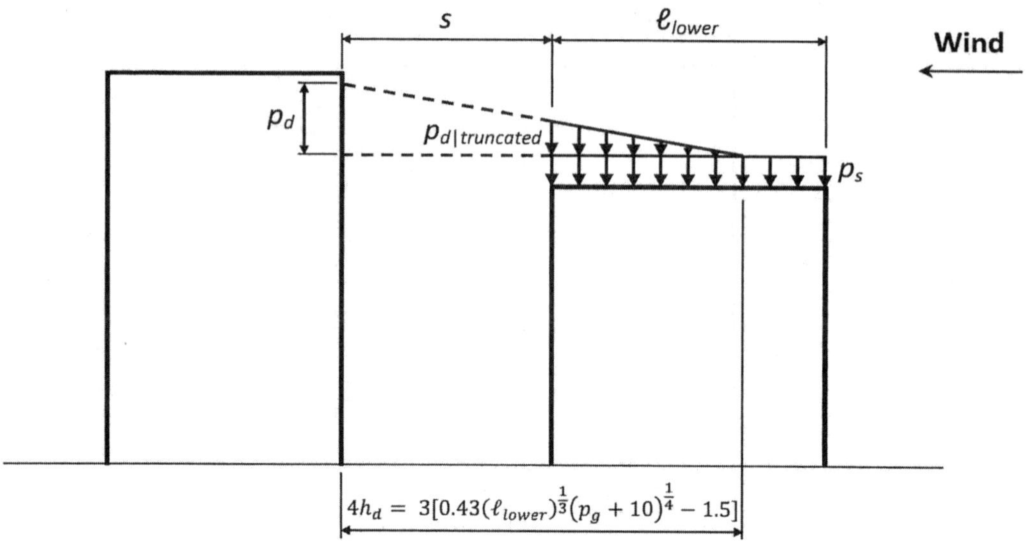

$$p_d = 0.75\gamma[0.43(\ell_{lower})^{\frac{1}{3}}(p_g + 10)^{\frac{1}{4}} - 1.5]$$

$$p_{d|truncated} = \left(1 - \frac{s}{4h_d}\right)p_d$$

4.2.7 Roof Projections and Parapets

Drift loads on sides of roof projections (including rooftop equipment) and at parapet walls are determined by the provisions of 7.8, which are based on the drift requirements of 7.7.1. The drift that is formed adjacent to a parapet wall is a windward drift: the only source of snow is from wind blowing the snow from the roof to the side of the parapet. The drifts that are formed adjacent to rooftop units and other projections are also assumed to be windward drifts even though a leeward drift can form due to snow that is blown from the roof upwind of the projection and due to snow that is blown off of the top of the projection. This leeward drift is usually insignificant, so for simplicity, it is not considered.

In the case of roof projections, the height of the drift is equal to three-quarters of h_d from Figure 7-9:

$$h_d = 0.75[0.43(\ell_u)^{1/3}(p_g + 10)^{1/4} - 1.5] \qquad (4.10)$$

The length, ℓ_u, in the direction of analysis is taken equal to the larger of the length of the roof upwind or length of the roof downwind of the projection in that direction (that is, the larger of the two lengths is conservatively used to determine the drift height for both sides of the projection). A drift load is not required on any side of a roof projection that is less than 15 feet long. In such cases, drifts will form, but their impact is generally negligible.

Equation 4.10 is also used to determine the drift height for parapets where ℓ_u is the length of the roof upwind of the parapet wall.

4.2.8 Sliding Snow

The load caused by snow sliding off of a sloped roof onto a lower roof is determined by the provisions of 7.9. Such loads are superimposed on the balanced snow load of the lower roof and need not be used in combination with drift, unbalanced, partial or rain-on-snow loads.

Sliding snow loads are applicable for upper roofs that are slippery with slopes greater than $1/4$ on 12 and for upper roofs that are not slippery with slopes greater than 2 on 12.

For structures that are adjacent to each other with no separation, the total sliding load per unit length of eave is equal to $0.4 p_f W$ where p_f and W are the flat roof snow load and the horizontal distance from the ridge to the eave of the upper roof, respectively (see Figure 4.16). The length over which this surcharge load acts on the lower roof is 15 feet, which is measured from the upper roof eave. If the length of the lower roof is less than 15 feet, then the sliding load is permitted to be reduced proportionally, that is, the sliding snow load on the lower roof is equal to $0.4 p_f W$ times the length of the lower roof divided by 15. This reduced load is uniformly distributed over the entire length of the lower roof.

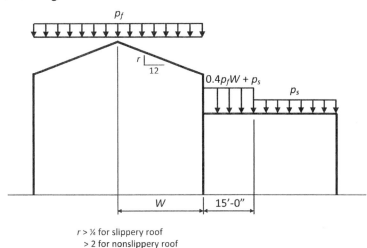

Figure 4.16 Sliding Snow Loads on a Lower Roof

$r > 1/4$ for slippery roof
> 2 for nonslippery roof

If the calculated total snow depth on the lower roof exceeds the distance from the upper roof eave to the top of the lower roof, sliding snow is blocked and a fraction of the sliding snow is forced to remain on the upper roof. In such cases, the total load on the lower roof near the upper roof eave is equal to the density of the snow, γ, multiplied by the distance from the upper roof eave to the top of the lower roof. This load is uniformly distributed over a distance of 15 feet or the width of the lower roof, whichever is less.

For adjacent structures that are horizontally separated by a distance, s, sliding loads on the lower roof must be considered where $s < 15$ feet and the vertical separation distance, h, between the eave of the upper roof and the eave of the lower roof is greater than s. The sliding snow load in such cases is equal to $0.4 p_f W (15 - s)/15$, which is uniformly distributed over a length equal to $(15 - s)$ on the lower roof (see Figure 4.17).

Figure 4.17 Sliding Snow Loads on an Adjacent Lower Roof

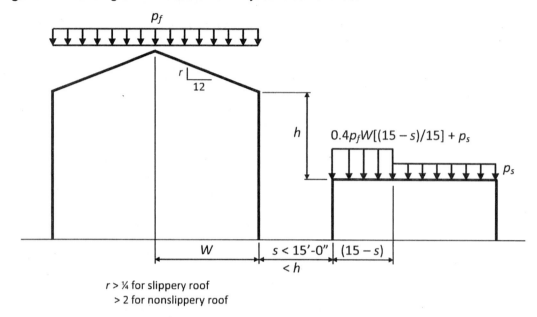

$r > ¼$ for slippery roof
> 2 for nonslippery roof

4.2.9 Rain-on-snow Surcharge Load

The snow load provisions in the discussions above consider load effects due to light rain on snow; effects due to heavy rain are not directly taken into account in the 50-year ground snow loads. At locations where the ground snow load is greater than 20 psf, it is assumed that because of the relatively deep snow pack, heavy rains have a less likely chance of permeating through the snow pack and draining away; as such, rain-on-snow load effects have been captured in the ground snow load measurements and an additional surcharge to account for this is not required.

A rain-on-snow surcharge load of 5 psf is to be added on all roofs that meet the conditions of 7.10: the building is located where p_g is greater than zero and less than 20 psf and the slope of the roof is less than $W/50$ where W is the horizontal distance from the ridge to the eave. In these situations, it is assumed that heavy rain permeates through the relatively shallow snow pack and drains away without being measured; a separate rain-on-snow surcharge is therefore needed. This surcharge load applies only to the balanced load case and need not be used in combination with drift, sliding, unbalanced, minimum or partial loads.

4.2.10 Ponding Instability

In cases where roofs do not have adequate slope or have insufficient and/or blocked drains to remove water due to rain or melting snow, water will tend to pond in low areas, which will cause the roof structure to deflect. These low areas will subsequently attract even more water, leading to additional deflection. Sufficient stiffness must be provided so that deflections will not continually increase until instability occurs, resulting in localized failure.

Provisions for ponding instability and progressive deflection of roofs are given in 7.11 and 8.4 (also see Chapter 3 of this publication for the determination of rain loads). Susceptible bays, that is, bays with a roof slope less than $1/4$ inch per foot or those where the primary drain system is blocked and the secondary drain system is functional, must be analyzed for the effects from the larger of the snow load or the rain load assuming the primary drainage system is blocked. The roof structure in these bays must be designed with adequate stiffness to preclude ponding instability.

Roof surfaces with a slope greater than or equal to $1/4$ inch per foot towards points of free drainage need not be considered susceptible to ponding instability.

4.2.11 Existing Roofs

Requirements for increased snow loads on existing roofs due to additions and alterations are covered in 7.12. New structures built adjacent to existing ones within a horizontal distance of 20 feet have the potential to increase the snow loads on the existing roof when the new roof is higher than the existing one. In particular, both drift and sliding snow loads must be considered on the existing roof due to the presence of the new roof. Additionally, the existing roof will most likely be partially exposed or sheltered by the new building or alteration; if it were fully exposed previously, an increase in snow load will be realized due to this change in exposure.

ASCE/SEI 7.12 requires that owners or agents for owners of an existing building with a lower roof be notified for the potential of increased snow loads when a new building or alteration with a higher roof is to be located within 20 feet of the existing building.

4.2.12 General Procedure to Determine Snow Loads

The following general procedure, which is based on that given in C7.0, can be used to determine design snow loads in accordance with Chapter 7 of ASCE/SEI 7:

1. Determine ground snow load, p_g, (7.2).

2. Determine flat roof snow load, p_f, by Equation 7.3-1 (7.3).

3. Determine sloped roof snow load, p_s, by Equation 7.4-1 (7.4).

4. Consider partial loading (7.5).

5. Consider unbalanced snow loads (7.6).

6. Consider snow drifts on lower roofs (7.7) and roof projections (7.8).

7. Consider sliding snow (7.9).

8. Consider rain-on-snow loads (7.10).

9. Consider ponding instability (7.11).

10. Consider existing roofs (7.12).

It is possible that snow loads in excess of the design values computed by Chapter 7 may occur on a building or structure. The snow-load-to-dead-load ratio of a roof structure is an important consideration when evaluating the implications of excess loads. Section C7.0 provides additional information on this topic.

Section C7.13 gives information on wind tunnel tests and other experimental and computational methods that have been employed to establish design snow loads for roof geometries and complicated sites not addressed in the provisions.

4.3 Ice Loads

4.3.1 Overview

An *ice-sensitive structure* is defined as one in which the effects due to atmospheric ice loading governs the design of part or all of the structure (IBC 202 and ASCE/SEI 10.2). Examples include the following: (1) lattice structures, (2) guyed masts, (3) overhead lines, (4) light suspension and cable-stayed bridges, (5) aerial cable systems (for example, ski lifts), (6) amusement rides, (7) open catwalks and platforms, (8) flagpoles and (9) signs.

Freezing rain is rain or drizzle that falls into a layer of subfreezing air in the earth's surface and freezes on contact with the ground or any other exposed surface to form glaze (clear, high-density) ice. Compared to *in-cloud icing* (which occurs when a supercooled cloud or fog droplets that are carried by the wind freeze on impact with objects) and snow, freezing rain is considered the cause of the most severe ice loads in most of the contiguous United States. Since values of ice thickness for in-cloud icing and snow are not currently available in a form that is suitable for inclusion in ASCE/SEI 7, only data for freezing rain are given in Chapter 10.

In areas where it has been determined by records or experience that in-cloud icing or snow produce larger loads than freezing rain, a site-specific study must be performed. Information on how to perform such as study is given in 10.1.1.

Ice that is formed on exposed surfaces increases the size of the surface, which increases the projected area that is exposed to wind. Chapter 10 contains requirements that address the proper wind loads that must be used on ice-covered structures.

The provisions of Chapter 10 do not apply to structures that are covered by national standards (for example, electric transmission systems and communication towers and masts). In such cases, the standards and guidelines listed in 10.1.3 are to be used where applicable.

Design for dynamic load effects resulting from galloping, ice shedding and aeolian vibrations, to name a few, are not covered in Chapter 10. Such effects must be considered in certain types of ice-sensitive structures.

4.3.2 Ice Loads due to Freezing Rain

Nominal Ice Thickness

Figures 10-2 through 10-6 provide an equivalent uniform radial thickness, t, of ice due to freezing rain at a height of 33 feet above the ground for the contiguous 48 states and Alaska based on a 50-year return period. The concurrent 3-second gust wind speeds are also given in the figures; these wind speeds correspond to the winds that occur during the freezing rain storm and those that occur between the time the freezing rain stops and the temperature rises to above freezing.

The data given in these figures are based on studies using an ice accretion model and historical data from 540 National Weather Service (NWS), military, Federal Aviation Administration (FAA) and Environment Canada weather stations (which are indicated in Figures C10-4 and 10-6) and from the U.S. Army Cold Regions Research and Engineering Laboratory (CRREL). The models utilize the measured weather and precipitation data to simulate the accretion of ice on horizontal cylinders that are located 33 feet above the ground level and oriented perpendicular to the direction of wind in freezing rain storms. It is assumed that the ice remains on the cylinder during the duration of the storm and remains there until after the temperature increases to at least 32°F.

Special icing regions are also identified on the maps (gray shaded areas) and occur in the western mountainous regions and in the Appalachian Mountains. In the west mountain areas, ice thicknesses may exceed the mapped value in foothills and passes, while in the Appalachian

Mountains, ice thicknesses may vary significantly over short distances because of local variations in elevation, topography and exposure. The thicknesses given in Figure 10-2 should be adjusted in mountainous regions to account for both freezing rain and in-cloud icing. Historical records and local experience, including local building officials, should be consulted when making such adjustments.

Height Factor

As noted above, the equivalent radial ice thicknesses due to freezing rain given in Figures 10-2 through 10-6 are based on a height above ground equal to 33 feet. The height factor, f_z, which is determined by Equation 10.4-4, is used to increase the radial thickness of ice for any height above ground, z:

$$f_z = \left(\frac{z}{33}\right)^{0.10} \quad \text{for } 0 \text{ feet} < z \leq 900 \text{ feet}$$

$$= 1.4 \quad \text{for } z > 900 \text{ feet} \tag{4.11}$$

This factor is similar to the velocity pressure exposure coefficient, K_z, that modifies wind velocity with respect to exposure and height above ground (see ASCE/SEI Chapter 27 and Chapter 5 of this publication).

Importance Factor

Importance factors, I_i, adjust the nominal ice thickness and concurrent wind pressure based on the occupancy of the structure. Risk categories for buildings and other structures are defined in IBC Table 1604.5 and values of I_i are given in ASCE/SEI Table 1.5-2. It is evident from this table that the importance factor, and thus, the ice thickness, is larger for more important structures.

An importance factor of 0.8 that is to be used for Risk Category I structures (low risk to human life) corresponds to a mean recurrence interval of about 25 years, while an importance factor of 1.25 for Risk Category III and IV structures corresponds to about a 100-year mean recurrence interval. Results from an extreme value analysis show that the concurrent wind speed does not change significantly with mean recurrence interval.

Topographic Factor

Because of wind speed-up effects, the ice thickness and concurrent wind speed are larger for buildings and structures that are situated on hills, ridges and escarpments compared to those located on level terrain. To account for these effects, the nominal ice thickness is modified by $(K_{zt})^{0.35}$ where K_{zt} is the topographic factor determined by Equation 26.8-1. Additional information on how to determine K_{zt} is given in Section 5.2.5 of this publication.

Design Ice Thickness for Freezing Rain

The design ice thickness, t_d, that is to be used in calculating ice weight is the nominal ice thickness, t, multiplied by the modification factors noted above (see Equation 10.4-5):

$$t_d = 2.0 t I_i f_z (K_{zt})^{0.35} \tag{4.12}$$

The factor of 2.0 in this equation adjusts the design ice thickness from a 50-year mean recurrence interval to a 500-year mean recurrence interval, which is consistent with the mean recurrence interval used for design wind loads (see Chapter 5 of this publication). As such, the load factors for load and resistance factor design for atmospheric ice loads are equal to 1.0 in the appropriate load combination equations (see ASCE/SEI 2.3.4 and Chapter 2 of this publication).

Chapter 4

Table C10-1 provides factors that can be used to adjust the 50-year ice thickness to other mean recurrence intervals.

Ice Weight

Ice load is determined using the ice weight, D_i, that is formed on all exposed surfaces of structural members, guys, components, appurtenances and cable systems. The cross-sectional area of ice, A_i, to be used in the determination of D_i in such cases is determined by Equation 10.4-1:

$$A_i = \pi t_d (D_c + t_d) \tag{4.13}$$

As noted previously, the equivalent radial ice thickness, t, due to freezing rain and the corresponding design ice thickness, t_d, have been established using a horizontal cylinder oriented perpendicular to the wind, and, thus, they are not directly applicable to structural shapes, prismatic members, or other similar shapes that are not round. However, the ice area from Equation 4.13 is the same for all shapes for which the circumscribed circles have equal diameters. The diameter of a cylinder circumscribing a shape or object, D_c, is given in Figure 10-1 for a variety of shapes. It is assumed in Equation 4.13 that the maximum dimension of the cross-section is perpendicular to the path of the raindrops.

The volume of ice on flat plates and large three-dimensional objects, such as domes and spheres, is determined by Equation 10.4-2:

$$V_i = \pi t_d A_s \tag{4.14}$$

For flat plates, A_s is the surface area of one side of the plate. For domes or spheres, A_s is the projected area of the dome or sphere, and is determined by Equation 10.4-3:

$$A_s = \pi r^2 \tag{4.15}$$

In this equation, r is the radius of the maximum cross-section of a dome or a radius of a sphere.

The ice volume given by Equation 4.14 is for a flat plate or projected surface that is oriented perpendicular to the path of the raindrops. ASCE/SEI 10.4.1 permits Equation 4.14 to be multiplied by 0.8 for vertical plates and 0.6 for horizontal plates.

Once A_i or V_i have been computed, the corresponding ice weight, D_i, is determined by multiplying A_i or V_i by the density of ice. According to 10.4.1, the density of ice shall not be taken less than 56 pcf.

4.3.3 Wind on Ice-covered Structures

Overview

Ice that has formed on structural members, components and appurtenances increases the projected area that is exposed to wind and changes the structure's wind drag coefficients. Ice accretions tend to round sharp edges thereby reducing drag coefficients for members like angles and rectangular bars.

Ice-sensitive structures must be designed for the wind loads determined by the provisions in ASCE/SEI Chapters 26 through 31 using increased projected area and the modifications set forth in 10.5.1 through 10.5.5, which are discussed next. The loads determined in this fashion are defined as the wind-on-ice loads, W_i (see Chapter 2 of this publication). It is assumed in the following discussions that the reader is familiar with the wind load provisions contained in Chapter 5 of this publication.

Wind on Ice-covered Chimneys, Tanks and Similar Structures

For chimneys, tanks and other similar types of structures, wind loads are determined by Equation 29.5-1:

$$F = q_z G C_f A_f \qquad (4.16)$$

The velocity pressure, q_z, is determined by Equation 29.3-1 using the topographic factor, K_{zt}, determined in accordance with 10.4.5 and the concurrent wind speed, V_c:

$$q_z = 0.00256 K_z K_{zt} K_d V_c^2 \qquad (4.17)$$

The velocity pressure exposure coefficient, K_z, is determined from Table 29.3-1 and the wind directionality factor, K_d, is defined in 26.6.

The gust factor, G, in Equation 4.16 is determined in accordance with 26.9.

For structures with square, hexagonal and octagonal cross-sections, the force coefficients, C_f, given in Figure 29.5-1 are to be used in Equation 4.16. For structures with round cross-sections, the force coefficients, C_f, in Figure 29.5-1 corresponding to round cross-sections with $D\sqrt{q_z} \leq 2.5$ are to be used for all ice thicknesses, wind speeds and structure diameters. As noted previously, the projected area, A_f, must take into consideration the increase due to ice thickness.

Wind on Ice-covered Solid Freestanding Walls and Solid Signs

Wind loads are determined for solid freestanding walls and solid signs using Equation 29.4-1:

$$F = q_h G C_f A_s \qquad (4.18)$$

where q_h is determined by Equation 4.17 at height h, which is defined in Figure 29.4-1, and G is determined in accordance with 26.9.

The force coefficients, C_f, given in Figure 29.4-1 are to be used in Equation 4.18 and the gross area, A_s, of the solid freestanding wall or solid sign must be based on the dimensions including ice.

Wind on Ice-covered Open Signs and Lattice Frameworks

Wind loads are determined for these types of structures using Equation 29.5-1 (see Equation 4.16).

For structures with flat-sided members, the force coefficients, C_f, given in Figure 29.5-2 are to be used in this equation. For rounded members and for the additional projected area due to ice on both flat and rounded members, the force coefficients, C_f, in this figure corresponding to round cross-sections with $D\sqrt{q_z} \leq 2.5$ are to be used for all ice thicknesses, wind speeds and member diameters.

In all cases, the solidity ratio ϵ (which is equal to the ratio of solid area to gross area) that is defined in Figure 29.5-2 is to be based on the projected area, including ice.

Wind on Ice-covered Trussed Towers

Wind loads are determined using Equation 29.5-1 (see Equation 4.16) using the force coefficients, C_f from Figure 29.5-3 and the gross area, A_f, based on the dimensions, including ice.

It is permitted to reduce the values of C_f for the projected area due to ice on both round and flat members by the factor for rounded members, which is given in Note 3 of Figure 29.5-3. In all cases, the solidity ratio ϵ is to be based on the projected area, including ice.

Wind on Ice-covered Guys and Cables

Wind loads are determined using Equation 29.5-1 (see Equation 4.16). Since there is very little published experimental data for force coefficients for ice-covered guys and cables, a single value of 1.2 is permitted to be used in all cases (10.5.5). The rationale behind using this value is given in C10.5.5.

4.3.4 Design Temperatures for Freezing Rain

Some ice-sensitive structures, especially those that utilize overhead cable systems, are also sensitive to changes in temperature. While maximum load effects usually occur at the lowest temperature when the structure is loaded with ice, it is also possible in some types of structures for maximum load effects to occur at or around the melting point of ice (32°F).

Figures 10-7 and 10-8 give temperatures concurrent with ice thickness due to freezing rain for the contiguous 48 states and Alaska, respectively. These temperatures were obtained from the freezing rain model that was described previously that was used for the determination of nominal ice thickness. The model tracked the temperature during each modeled icing event and the minimum temperature that occurred with maximum ice thickness was recorded for each event. The minimum temperatures for all of the freezing rain events that were used in the extreme value analysis of ice thickness were analyzed and eventually incorporated in Figures 10-7 and 10-8.

The design temperature for ice and wind-on-ice that is to be used is the temperature from Figures 10-7 and 10-8 or 32°F, whichever gives the maximum load effects. For Hawaii, the temperature shall be 32°F. This temperature is used for all mean recurrence intervals and is considered to be concurrent with the design ice load and the concurrent wind load.

4.3.5 Partial Loading

It has been found that variations in ice thickness due to freezing rain at a given elevation are usually small over distances of about 1,000 feet. Thus, partial loading from freezing rain does not usually produce maximum load effects. However, in certain types of structures, partial ice loads can produce maximum effects and this must be considered in design. Additional information on this topic can be found in C10.7.

4.3.6 General Procedure to Determine Atmospheric Ice Loads

The following general procedure, which is based on that given in 10.8, can be used to determine design ice loads in accordance with Chapter 10 of ASCE/SEI 7:

1. Determine nominal ice thickness, t, the concurrent wind speed, V_c, and the concurrent temperature from Figures 10-2 through 10-8 or from a site-specific study in accordance with 10.1.1 (10.4.2, 10.6).

2. Determine height factor, f_z, for each design segment of the structure (10.4.3).

3. Determine importance factor, I_i, (10.4.4).

4. Determine topographic factor, K_{zt}, (10.4.5).

5. Determine design ice thickness, t_d, by Equation 10.4-5 (10.4.6).

6. Determine the weight of ice, D_i, using the applicable equations in 10.4.1 (10.4.1).

7. Determine the velocity pressure, q_z, for wind speed, V_c, in accordance with 29.3.

8. Determine the applicable wind force coefficients, C_f, in accordance with 10.5.

9. Determine the gust effect factor, G, in accordance with 26.9.

10. Determine the design wind force, W_i, in accordance with the applicable provisions of Chapter 29 (10.5).

4.4 Flowcharts

A summary of the flowcharts provided in this chapter is given in Table 4.2.

Table 4.2 Summary of Flowcharts Provided in Chapter 4

Flowchart	Title
Flowchart 4.1	Flat Roof Snow Load, p_f
Flowchart 4.2	Roof Slope Factor, C_s
Flowchart 4.3	Sloped Roof Snow Load, p_s
Flowchart 4.4	Unbalanced Roof Snow Loads – Hip and Gable Roofs
Flowchart 4.5	Unbalanced Roof Snow Loads – Curved and Dome Roofs
Flowchart 4.6	Unbalanced Roof Snow Loads – Multiple Folded Plate, Sawtooth, and Barrel Vault Roofs
Flowchart 4.7	Drifts on Lower Roof of a Structure

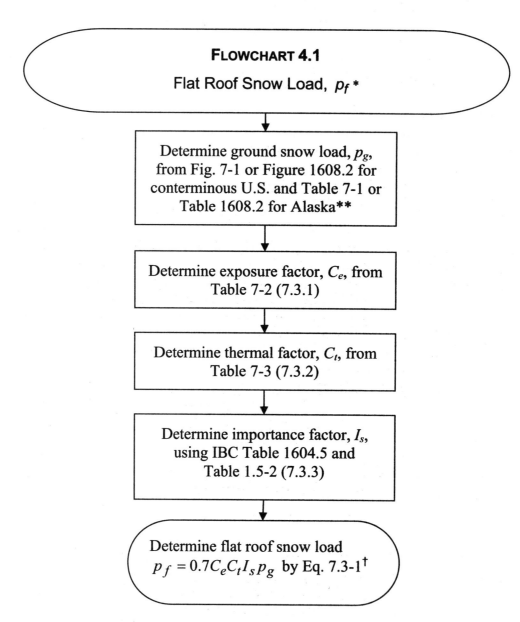

* A flat roof is defined as a roof with a slope that is less than or equal to 5 degrees.

** "CS" in the maps signifies areas where a site-specific study must be conducted to determine p_g. Numbers in parentheses represent the upper elevation limit in feet for the ground snow load values given below. Site-specific studies are required at elevations not covered in the maps.

† Minimum snow loads for low-slope roofs, p_m, are specified in 7.3.4. Low-slope roofs are defined in 7.3.4.

Snow and Ice Loads

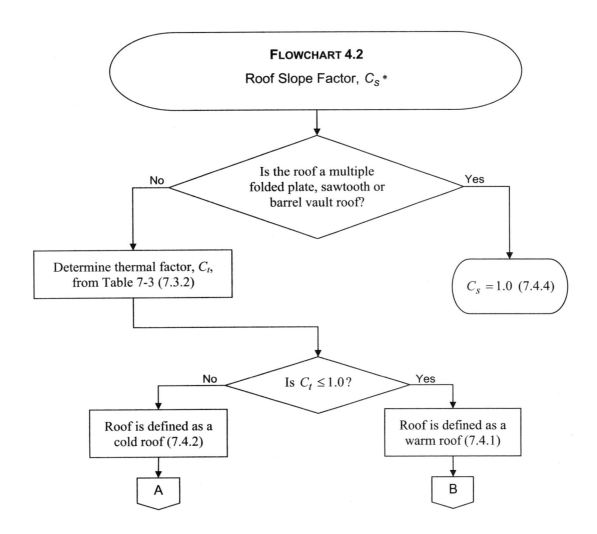

FLOWCHART 4.2
Roof Slope Factor, C_s *

* Portions of curved roofs having a slope exceeding 70 degrees shall be considered free of snow load, that is, $C_s = 0$ (7.4.3).

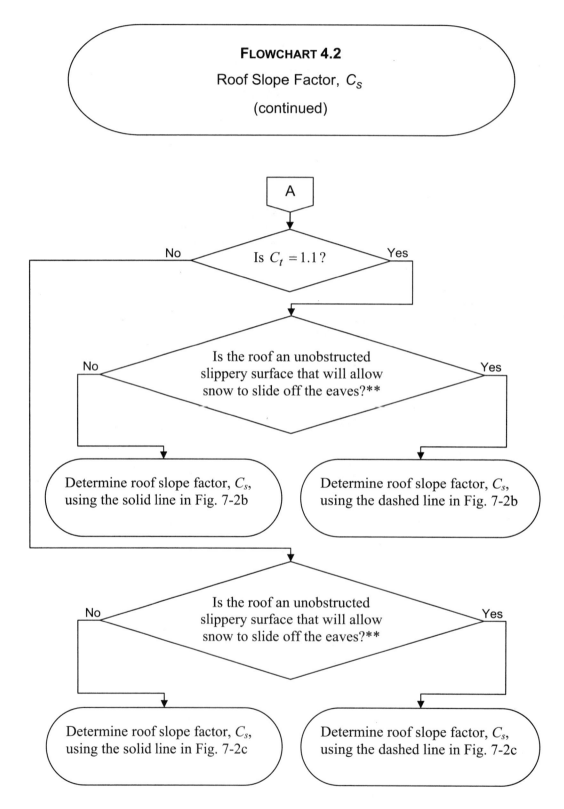

** See 7.4 for definitions of unobstructed and slippery surfaces.

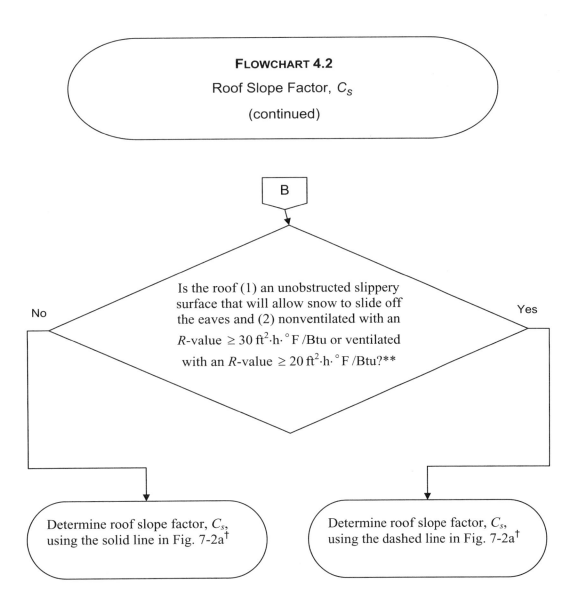

FLOWCHART 4.2

Roof Slope Factor, C_s

(continued)

** See 7.4 for definitions of unobstructed and slippery surfaces. An R-value for a roof is defined as its thermal resistance.

† See 7.4.5 for an additional uniformly distributed load that is to be applied on overhanging portions of warm roofs due to formation of ice dams and icicles along eaves.

106 Chapter 4

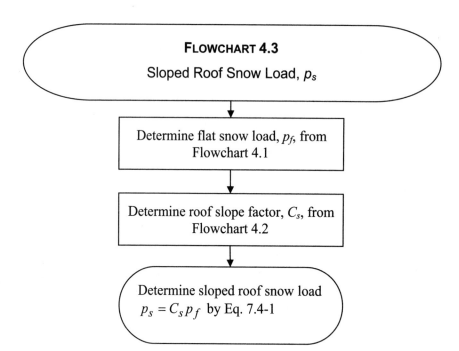

Snow and Ice Loads   107

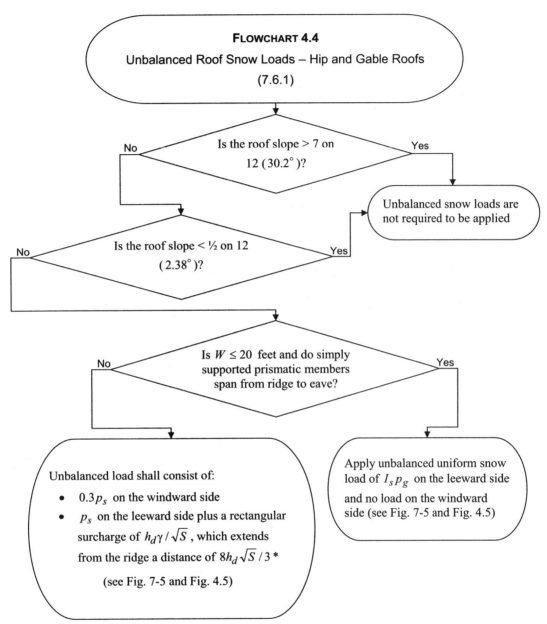

* h_d is the drift height from Fig. 7-9 with W substituted for ℓ_u, γ = snow density determined by Eq. 7.7-1, and S = roof slope run for a rise of one

108 Chapter 4

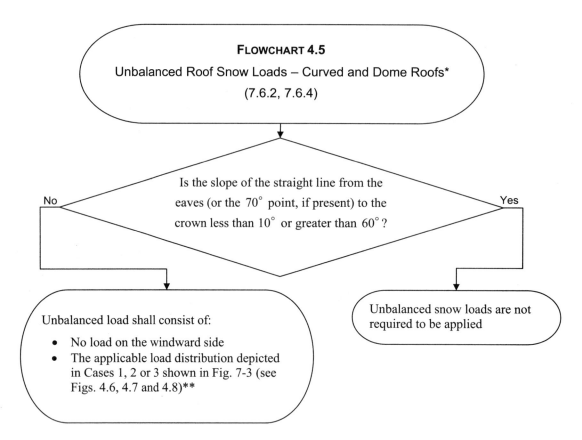

* Portions of curved roofs having a slope > 70 degrees shall be considered free of snow.
** See Fig. 4.10 for application of unbalanced snow loads on a dome roof.

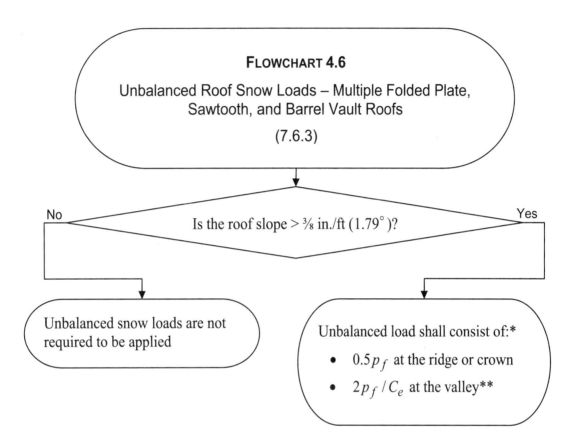

* Figure 7-6 and Fig. 4.9 illustrate balanced and unbalanced snow loads for a sawtooth roof.

** Snow surface above the valley shall not be at an elevation higher than the snow above the ridge. Snow depths shall be determined by dividing the snow load by the snow density given by Eq. 7.7-1.

110 Chapter 4

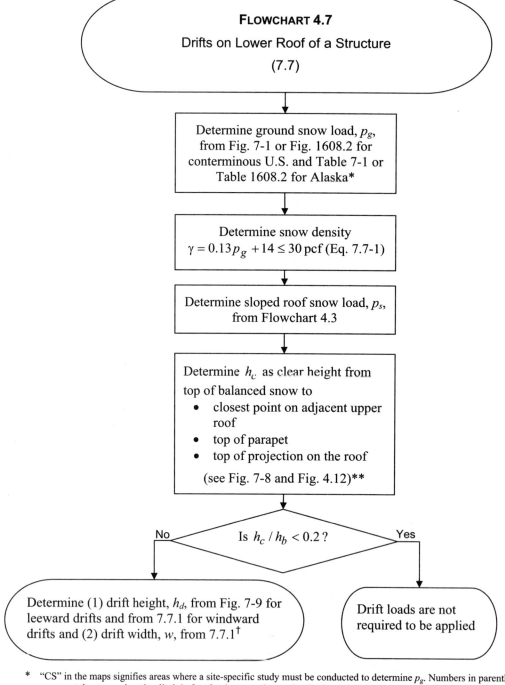

* "CS" in the maps signifies areas where a site-specific study must be conducted to determine p_g. Numbers in parentheses represent the upper elevation limit in feet for the ground snow load values given below. Site-specific studies are required at elevations not covered in the maps.

** Height of balanced snow $h_b = p_s/\gamma$ or p_f/γ (7.7.1)

† See 7.7.2 and Figs. 4.14 and 4.15 for drift loads caused by adjacent structures and terrain features. See 7.8 for drift loads on roof projections and parapet walls.

4.5 Examples

The following sections contain examples that illustrate the snow load design provisions of Chapter 7 in ASCE/SEI 7-10.

4.5.1 Example 4.1 – Snow Loads, Warehouse Building, Roof Slope of $^1/_2$ on 12

Determine the design snow loads for the one-story warehouse illustrated in Figure 4.18.

Figure 4.18 Plan and Elevation of Warehouse Building, Example 4.1

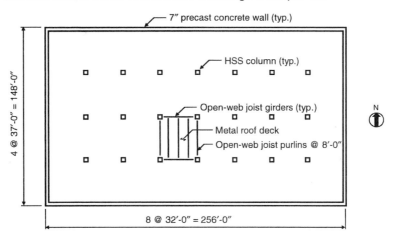

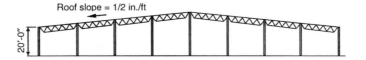

	DESIGN DATA
Location:	St. Louis, MO
Terrain category:	C (open terrain with scattered obstructions less than 30 feet in height)
Occupancy:	Warehouse use. Less than 300 people congregate in one area and the building is not used to store hazardous or toxic materials
Thermal condition:	Structure is kept just above freezing
Roof exposure condition:	Partially exposed
Roof surface:	Rubber membrane
Roof framing:	All members are simply supported

SOLUTION

1. Determine ground snow load, p_g.

   From Figure 7-1 or Figure 1608.2, the ground snow load is equal to 20 psf for St. Louis, MO.

2. Determine flat roof snow load, p_f, by Equation 7.3-1.

   Use Flowchart 4.1 to determine p_f.

   a. Determine exposure factor, C_e, from Table 7-2.

   From the design data, the terrain category is C and the roof exposure is partially exposed. Therefore, $C_e = 1.0$ from Table 7-2.

   b. Determine thermal factor, C_t, from Table 7-3.

   From the design data, the structure is kept just above freezing during the winter, so $C_t = 1.1$ from Table 7-3.

   c. Determine the importance factor, I_s, from Table 1.5-2.

   From IBC Table 1604.5, the Risk Category is II, based on the occupancy given in the design data. Thus, $I_s = 1.0$ from Table 1.5-2.

   Therefore,

   $p_f = 0.7 C_e C_t I_s p_g = 0.7 \times 1.0 \times 1.1 \times 1.0 \times 20 = 15.4$ psf

   Check if the minimum snow load requirements are applicable:

   A minimum roof snow load, p_m, in accordance with 7.3.4 applies to hip and gable roofs with slopes less than 15 degrees. Since the roof slope in this example is equal to 2.38 degrees, minimum roof snow loads must be considered.

   Since $p_g = 20$ psf, $p_m = I_s p_g = 20$ psf

3. Determine sloped roof snow load, p_s, by Equation 7.4-1.

   Use Flowchart 4.2 to determine roof slope factor, C_s.

   a. Determine thermal factor, C_t, from Table 7-3.

   From item 2 above, thermal factor $C_t = 1.1$.

   b. Determine if the roof is warm or cold.

   Since $C_t = 1.1$, the roof is defined as a cold roof in accordance with 7.4.2.

   c. Determine if the roof is unobstructed or not and if the roof is slippery or not.

   There are no obstructions on the roof that inhibit the snow from sliding off the eaves. Also, the roof surface is a rubber membrane. According to 7.4, rubber membranes are considered to be slippery surfaces.

Snow and Ice Loads   113

Since this roof is unobstructed and slippery, use the dashed line in Figure 7-2b to determine C_s:

For a roof slope of 2.38 degrees, $C_s = 1.0$.

Therefore, $p_s = C_s p_f = 1.0 \times 15.4 = 15.4$ psf. This is the balanced snow load for this roof.

4. Consider partial loading.

   Since all of the members are simply supported, partial loading is not considered (7.5).

5. Consider unbalanced snow loads.

   Flowchart 4.4 is used to determine if unbalanced loads on this gable roof need to be considered or not.

   Unbalanced snow loads must be considered for this roof, since the slope is equal to 2.38 degrees.

   Since $W = 128$ feet > 20 feet, the unbalanced load consists of the following (see Figure 7-5 and Figure 4.5):

   - Windward side: $0.3p_s = 0.3 \times 15.4 = 4.6$ psf
   - Leeward side: $p_s = 15.4$ psf along the entire leeward length plus a uniform pressure of $h_d \gamma / \sqrt{S} = (3.6 \times 16.6)/\sqrt{24} = 12.2$ psf, which extends from the ridge a distance of $8h_d\sqrt{S}/3 = (8 \times 3.6 \times \sqrt{24})/3 = 47.0$ feet where

   h_d = drift length from Figure 7-9 with $W = 128$ feet substituted for ℓ_u

   = $0.43(W)^{1/3}(p_g + 10)^{1/4} - 1.5 = 3.6$ feet

   γ = snow density (Eq. 7.7-1)

   = $0.13p_g + 14 = 16.6$ pcf < 30 pcf

   S = roof slope run for a rise of one = 24

6. Consider snow drifts on lower roofs and roof projections.

   Not applicable.

7. Consider sliding snow.

   Not applicable.

8. Consider rain-on-snow loads.

   In accordance with 7.10, a rain-on-snow surcharge of 5 psf is required for locations where the ground snow load, p_g, is 20 psf or less (but not zero) with roof slopes less than $W/50$.

In this example, p_g = 20 psf and $W/50$ = 128/50 = 2.56 degrees, which is greater than the roof slope of 2.38 degrees. Thus, an additional 5 psf must be added to the balanced load of 15.4 psf.

9. Consider ponding instability.

   Since the roof slope in this example is greater than $1/4$ inch/foot, progressive roof deflection and ponding instability from rain-on-snow or from snow meltwater need not be investigated (7.11 and 8.4).

10. Consider existing roofs.

    Not applicable.

In this example, the uniform load of 15.4 + 5 = 20.4 psf (balanced plus rain-on-snow) governs, since it is greater than the minimum roof snow load of 20 psf. The balanced and unbalanced snow loads are depicted in Figure 4.19.

Figure 4.19 Balanced and Unbalanced Snow Loads for Warehouse Building, Example 4.1

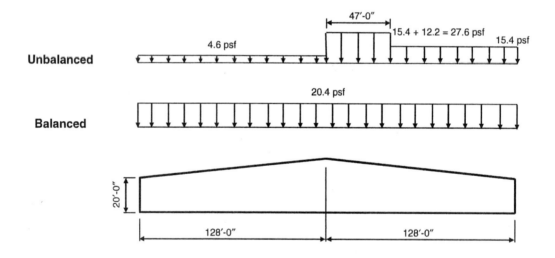

4.5.2 Example 4.2 – Snow Loads, Warehouse Building, Roof Slope of $1/4$ on 12

For the one-story warehouse depicted in Figure 4.18, determine the design snow loads for a roof slope of $1/4$ on 12. Use the same design data given in Example 4.1.

SOLUTION
---

1. Determine ground snow load, p_g.

   From Figure 7-1 or Figure 1608.2, the ground snow load is equal to 20 psf for St. Louis, MO.

2. Determine flat roof snow load, p_f, by Equation 7.3-1.

   It was determined in item 2 of Example 4.1 that p_f = 15.4 psf.

Check if the minimum snow load requirements are applicable:

A minimum roof snow load, p_m, in accordance with 7.3.4 applies to hip and gable roofs with slopes less than 15 degrees. Since the roof slope in this example is equal to 1.19 degrees, minimum roof snow loads must be considered.

Since $p_g = 20$ psf, $p_m = I_s p_g = 20$ psf

3. Determine sloped roof snow load, p_s, by Equation 7.4-1.

   It was determined in item 3 of Example 4.1 that $C_s = 1.0$ for a roof slope of 2.38 degrees. Using the dashed line in Figure 7-2b, $C_s = 1.0$ for a roof slope of 1.19 degrees as well.

   Therefore, $p_s = 1.0 \times 15.4 = 15.4$ psf.

4. Consider partial loading.

   Since all of the members are simply supported, partial loading is not considered (7.5).

5. Consider unbalanced snow loads.

   Flowchart 4.4 is used to determine if unbalanced loads on this gable roof need to be considered or not.

   Unbalanced snow loads need not be considered for this roof, since the slope is less than 2.38 degrees.

6. Consider snow drifts on lower roofs and roof projections.

   Not applicable.

7. Consider sliding snow.

   Not applicable.

8. Consider rain-on-snow loads.

   In accordance with 7.10, a rain-on-snow surcharge of 5 psf is required for locations where the ground snow load, p_g, is 20 psf or less (but not zero) with roof slopes less than $W/50$.

   In this example, $p_g = 20$ psf and $W/50 = 128/50 = 2.56$ degrees, which is greater than the roof slope of 1.19 degrees. Thus, an additional 5 psf must be added to the sloped roof snow load of 15.4 psf.

9. Consider ponding instability.

   Since the roof slope in this example is not less than $1/4$ inch/foot, progressive roof deflection and ponding instability from rain-on-snow or from snow meltwater need not be investigated (7.11 and 8.4).

10. Consider existing roofs.

    Not applicable.

In this example, the uniform load of 15.4 + 5 = 20.4 psf (balanced plus rain-on-snow) governs, since it is greater than the minimum roof snow load of 20 psf. The 20.4 psf snow load is uniformly distributed over the entire length of the roof, as depicted in Figure 4.19. This is the only load that needs to be considered in this example.

4.5.3 Example 4.3 – Snow Loads, Warehouse Building (Roof Slope of $1/2$ on 12) and Adjoining Office Building (Roof Slope of $1/2$ on 12)

A new one-story office building is to be constructed adjacent to the existing one-story warehouse in Example 4.1 (see Figure 4.20). Determine the design snow loads on the roof of the office building. Both structures have a roof slope of $1/2$ on 12. A summary of the design snow loads for the warehouse is given in Figure 4.19.

Figure 4.20 Elevation of Warehouse and Office Buildings, Example 4.3

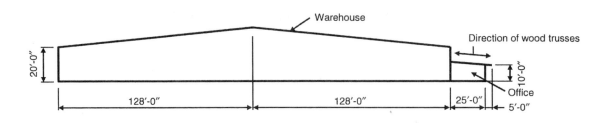

DESIGN DATA FOR OFFICE BUILDING

Location:	St. Louis, MO
Terrain category:	C (open terrain with scattered obstructions less than 30 feet in height)
Occupancy:	Business (less than 300 people congregate in one area)
Thermal condition:	Heated with unventilated roof (*R*-value less than 30 ft²·h·°F/Btu)
Roof exposure condition:	Partially exposed (due in part to the presence of the adjacent taller warehouse building)
Roof surface:	Asphalt shingles
Roof framing:	Wood trusses spaced 25 feet on center that overhang a masonry wall and wood purlins spaced 5 feet on center that frame between the trusses (see Figure 4.20)

SOLUTION

1. Determine ground snow load, p_g.

   From Figure 7-1 or Figure 1608.2, the ground snow load is equal to 20 psf for St. Louis, MO.

2. Determine flat roof snow load, p_f, by Equation 7.3-1.

   Use Flowchart 4.1 to determine p_f.

   a. Determine exposure factor, C_e, from Table 7-2.

      From the design data, the terrain category is C and the roof exposure is partially exposed. Therefore, $C_e = 1.0$ from Table 7-2.

   b. Determine thermal factor, C_t, from Table 7-3.

      From the design data, the structure is heated with an unventilated roof, so $C_t = 1.0$ from Table 7-3.

   c. Determine the importance factor, I_s, from Table 1.5-2.

      From IBC Table 1604.5, the Risk Category is II, based on the occupancy given in the design data. Thus, $I_s = 1.0$ from Table 1.5-2.

   Therefore, $p_f = 0.7 C_e C_t I_s p_g = 0.7 \times 1.0 \times 1.0 \times 1.0 \times 20 = 14.0$ psf

   Check if the minimum snow load requirements are applicable:

   A minimum roof snow load, p_m, in accordance with 7.3.4 applies to hip and gable roofs with slopes less than 15 degrees. Since the roof slope in this example is equal to 2.38 degrees, minimum roof snow loads must be considered.

   Since $p_g = 20$ psf, $p_m = I_s p_g = 20$ psf

3. Determine sloped roof snow load, p_s, by Equation 7.4-1.

   Use Flowchart 4.2 to determine roof slope factor C_s.

   a. Determine thermal factor C_t from Table 7-3.

      From item 2 above, thermal factor $C_t = 1.0$.

   b. Determine if the roof is warm or cold.

      Since $C_t = 1.0$, the roof is defined as a warm roof in accordance with 7.4.1.

   c. Determine if the roof is unobstructed or not and if the roof is slippery or not.

      In accordance with the design data, the roof surface is asphalt shingles. According to 7.4, asphalt shingles are not considered to be slippery.

      Also, since the roof is unventilated with an R-value less than 30 ft²·h·°F/Btu, it is possible for an ice dam to form at the eave, which can prevent the snow from sliding off of the roof (7.4.5). This is considered to be an obstruction.

   Thus, use the solid line in Figure 7-2a to determine C_s:

   For a roof slope of 2.38 degrees, $C_s = 1.0$.

   Therefore, $p_s = C_s p_f = 1.0 \times 14.0 = 14.0$ psf.

In accordance with 7.4.5, a uniformly distributed load of $2p_f = 2 \times 14.0 = 28.0$ psf must be applied on the 5-foot overhanging portion of the roof to account for ice dams. Only the dead load is to be present when this uniformly distributed load is applied.

4. Consider partial loading.

It is assumed that the roof purlins are connected to the wood trusses by metal hangers, which are essentially simple supports, so partial loads do not have to be considered for the roof purlins. Therefore, with a spacing of 5 feet, the uniform snow load on a purlin is equal to $20.0 \times 5.0 = 100$ plf (minimum snow load governs).

The roof trusses are continuous over the masonry wall; thus, partial loading must be considered (7.5). The balanced snow load to be used in partial loading cases is that determined by Equation 7.4-1, which is equal to 14.0 psf. With a spacing of 25 feet, the balanced and partial loads on a typical roof truss are

Balanced load = $14.0 \times 25.0 = 350$ plf

Partial load = one-half of balanced load = 175 plf

Shown in Figure 4.21 are the balanced and partial load cases that must be considered for the roof trusses in this example, including the ice dam load on the overhang, which was determined in item 3 above. Note that the minimum snow load of 20 psf is not applicable in the partial load cases and in the ice dam load case.

Figure 4.21
Balanced and Partial Load Cases for Roof Trusses, Example 4.3

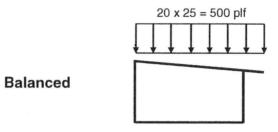
Balanced

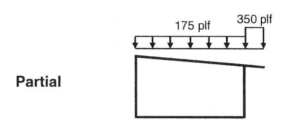
Partial

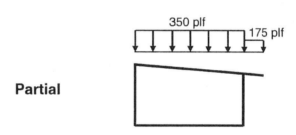
Partial

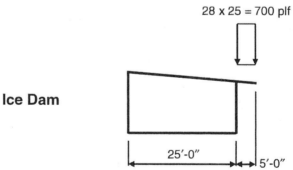
Ice Dam

5. Consider unbalanced snow loads.

   Not applicable.

6. Consider snow drifts on lower roofs and roof projections.

   Use Flowchart 4.7 to determine the leeward and windward drifts that form on the lower (office) roof.

120 Chapter 4

a. Determine ground snow load, p_g.

From item 1 above, the ground snow load is equal to 20 psf for St. Louis, MO.

b. Determine snow density, γ, by Equation 7.7-1.

$\gamma = 0.13p_g + 14 = (0.13 \times 20) + 14 = 16.6$ pcf < 30 pcf

c. Determine sloped roof snow load, p_s, from Flowchart 4.3.

From item 3 above, $p_s = 14.0$ psf.

d. Determine clear height, h_c.

In this example, the clear height, h_c, is from the top of the balanced snow to the top of the warehouse eave (see Figures 4.12 and 4.20). The height of the balanced snow is $h_b = p_s/\gamma = 14.0/16.6 = 0.8$ feet.

Thus, $h_c = (10 - 25 \tan 2.38°) - 0.8 = 8.2$ feet

e. Determine if drift loads are required or not.

Drift loads are not required where $h_c/h_b < 0.2$ (7.7.1). In this example, $h_c/h_b = 8.2/0.8 = 10.3 > 0.2$, so drift loads must be considered.

f. Determine drift load.

Both leeward and windward drift heights, h_d, must be determined by the provisions of 7.7.1. The larger of these two heights is used to determine the drift load.

- Leeward drift

  A leeward drift occurs when snow from the warehouse roof is deposited by wind to the office roof (wind from left to right in Figure 4.20).

  For leeward drifts, the drift height, h_d, is determined from Figure 7-9 using the length of the upper roof, $\ell_u = \ell_{upper}$. In this example, $\ell_u = 256$ feet and the ground snow load $p_g = 20$ psf. Using the equation in Figure 7-9:

  $h_d = 0.43(\ell_u)^{1/3}(p_g + 10)^{1/4} - 1.5 = 0.43(256)^{1/3}(20 + 10)^{1/4} - 1.5 = 4.9$ feet

- Windward drift

  A windward drift occurs when snow from the office roof is deposited adjacent to the wall of the warehouse building (wind from right to left in Figure 4.20).

  For windward drifts, the drift height, h_d, is 75 percent of that determined from Figure 7-9 using $\ell_u = \ell_{lower}$:

  $h_d = 0.75[0.43(\ell_u)^{1/3}(p_g + 10)^{1/4} - 1.5] = 0.75[0.43(30)^{1/3}(20 + 10)^{1/4} - 1.5] = 1.2$ feet

  Thus, the leeward drift controls and $h_d = 4.9$ feet.

  Since $h_d = 4.9$ feet $< h_c = 8.2$ feet, the drift width $w = 4h_d = 4 \times 4.9 = 19.6$ feet.

The maximum surcharge drift load $p_d = h_d \gamma = 4.9 \times 16.6 = 81.3$ psf.

The total load at the step is the balanced load on the office roof plus the drift surcharge = 14.0 + 81.3 = 95.3 psf, which is illustrated in Figure 4.22.

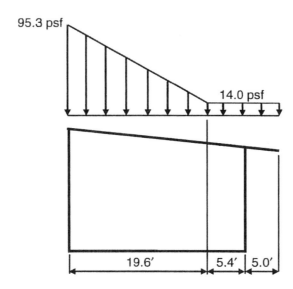

Figure 4.22
Balanced and Drift Loads on Office Roof, Example 4.3

The snow loads on the purlins and trusses are obtained by multiplying the loads depicted in Figure 4.22 by the respective tributary widths. As expected, the purlins closest to the warehouse have the largest loads.

If the office and warehouse were separated, the drift load on the office roof would be determined using the provisions of 7.7.2.

For example, if the buildings were separated by a horizontal distance of 3 feet, drifts would need to be considered since $s = 3$ feet is less than 20 feet and is less than $6h = 6 \times (10 - 25 \tan 2.38°) = 53.8$ feet.

The unmodified leeward and windward drift heights for the stepped roof without a separation were determined above as 4.9 feet and 1.2 feet, respectively.

For the separated structures, the leeward drift height is the smaller of the following (see Figure 4.14): $h_d = 4.9$ feet or $(6h - s)/6 = (53.8 - 3)/6 = 8.5$ feet. Therefore, the surcharge height is equal to 4.9 feet, and the surcharge load is $4.9 \times 16.6 = 81.3$ psf. The horizontal extent of the leeward drift surcharge is the smaller of $6h_d = 29.4$ feet and $6h - s = 50.8$ feet.

The windward drift height for the separated structures is equal to 1.2 feet, and p_d at the face of the warehouse is equal to $1.2 \times 16.6 = 19.9$ psf (see Figure 4.15). The truncated magnitude of p_d at the face of the office is $[1 - (3.0/4.8)] \times 19.9 = 7.5$ psf, which acts over a length equal to $4h_d - s = 1.8$ feet.

7. Consider sliding snow.

The provisions of 7.9 are used to determine if a load due to snow sliding off of the warehouse roof on to the office roof must be considered.

Load caused by snow sliding must be considered, since the warehouse roof is slippery with a slope greater than $1/4$ on 12. This load is in addition to the balanced load acting on the lower roof.

The total sliding load per unit length of eave is equal to $0.4 p_f W$ where p_f is the flat roof snow load of the warehouse and W is the horizontal distance from the eave to the ridge of the warehouse roof. From Example 4.1, $p_f = 15.4$ psf and $W = 128$ feet. Thus, the sliding load = $0.4 \times 15.4 \times 128 = 789$ plf. This load is to be uniformly distributed over a distance of 15 feet from the warehouse eave (see Figure 4.16). Thus, the sliding load is equal to $789/15 = 52.6$ psf.

The total load over the 15-foot width is equal to the balanced snow load on the office roof plus the sliding load = $14.0 + 52.6 = 66.6$ psf. The total depth of snow for the total load is equal to $66.6/16.6 = 4.0$ feet, which is less than the distance from the warehouse eave to the top of the office roof at the interface, which was determined previously as approximately 9 feet. Thus, sliding snow is not blocked and the full load can be developed over the 15-foot length.

Depicted in Figure 4.23 is the load case including sliding snow. The balanced snow load plus the sliding snow load is less than the balanced snow load plus the drift snow load (see Figure 4.22). However, the distributions are different and both must be considered in the design of the roof members.

Figure 4.23
Balanced and Sliding Snow Loads on Office Roof, Example 4.3

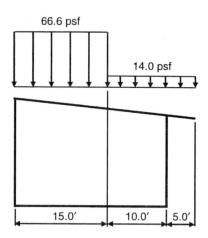

If the width of the lower roof were less than 15 feet, the sliding snow load would be reduced proportionally (7.9). For example, if the width of the office building in this example was 12 feet, the reduced sliding snow load = $(12/15) \times 789 = 631$ plf. This load would be applied uniformly over the 12-foot width.

8. Consider rain-on-snow loads.

In accordance with 7.10, a rain-on-snow surcharge of 5 psf is required for locations where the ground snow load, p_g, is 20 psf or less (but not zero) with roof slopes less than $W/50$.

In this example, $p_g = 20$ psf and $W/50 = 25/50 = 0.5$ degree, which is less than the roof slope of 2.38 degrees. Thus, rain-on-snow loads are not considered.

9. Consider ponding instability.

Since the roof slope in this example is greater than $1/4$ inch/foot, progressive roof deflection and ponding instability from rain-on-snow or from snow meltwater need not be investigated (7.11 and 8.4).

10. Consider existing roofs.

Not applicable.

When designing the roof purlins and trusses, the snow loads depicted in Figures 4.21, 4.22 and 4.23 must be combined with other applicable loads using the appropriate load combinations.

4.5.4 Example 4.4 – Snow Loads, Six-story Hotel with Parapet Walls

Determine the design snow loads for the six-story hotel depicted in Figure 4.24. Parapet walls are on all four sides of the building and the roof is nominally flat except for localized areas around roof drains that are sloped to facilitate drainage.

Figure 4.24 Plan and Elevation of Six-story Hotel with Parapet Walls, Example 4.4

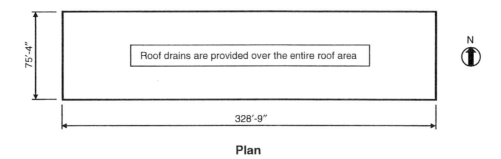

Plan

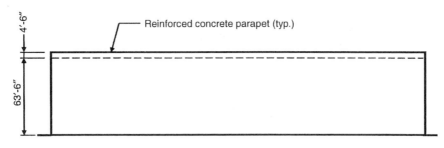

North/South Elevation

124 Chapter 4

DESIGN DATA	
Ground snow load, p_g:	40 psf
Terrain category:	B (urban area with numerous closely spaced obstructions having the size of single-family dwellings or larger)
Occupancy:	Residential (less than 300 people congregate in one area)
Thermal condition:	Cold, ventilated roof (R-value between the ventilated space and the heated space exceeds 25 ft²·h·°F/Btu)
Roof exposure condition:	Fully exposed
Roof surface:	Concrete slab with waterproofing

SOLUTION

1. Determine ground snow load, p_g.

   From the design data, the ground snow load, p_g, is equal to 40 psf.

2. Determine flat roof snow load, p_f, by Equation 7.3-1.

   Use Flowchart 4.1 to determine p_f.

   a. Determine exposure factor, C_e, from Table 7-2.

   From the design data, the terrain category is B and the roof exposure is fully exposed. Therefore, $C_e = 0.9$ from Table 7-2.

   b. Determine thermal factor, C_t, from Table 7-3.

   From the design data, the roof is cold and ventilated with an R-value between the ventilated space and the heated space that exceeds 25 ft²·h·°F/Btu, so $C_t = 1.1$ from Table 7-3.

   c. Determine the importance factor, I_s, from Table 1.5-2.

   From IBC Table 1604.5, the Risk Category is II, based on the occupancy given in the design data. Thus, $I_s = 1.0$ from Table 1.5-2.

   Therefore, $p_f = 0.7 C_e C_t I p_g = 0.7 \times 0.9 \times 1.1 \times 1.0 \times 40 = 27.7$ psf

3. Consider drift loading at parapet walls.

   According to 7.8, drift loads at parapet walls and other roof projections are determined using the provisions of 7.7.1.

   Windward drifts occur at parapet walls, and Flowchart 4.7 is used to determine the windward drift load.

a. Determine snow density, γ, by Equation 7.7-1.

$\gamma = 0.13 p_g + 14 = (0.13 \times 40) + 14 = 19.2$ pcf < 30 pcf

b. Determine clear height, h_c.

The clear height, h_c, is from the top of the balanced snow to the top of the parapet wall. For a flat roof, the height of the balanced snow is equal to $h_b = p_f/\gamma = 27.7/19.2 = 1.4$ feet.

Thus, $h_c = 4.5 - 1.4 = 3.1$ feet.

c. Determine if drift loads are required or not.

Drift loads are not required where $h_c/h_b < 0.2$ (7.7.1). In this example, $h_c/h_b = 3.1/1.4 = 2.2 > 0.2$, so drift loads must be considered.

d. Determine drift load.

Windward drift height, h_d, must be determined by the provisions of 7.7.1 using three-quarters of the drift height, h_d, from Figure 7-9 with ℓ_u equal to the length of the roof upwind of the parapet wall (7.8). Wind in both the north-south and east-west directions must be examined.

- Wind in north-south direction

    The equation in Figure 7-9 yields the following for the drift height, h_d, based on a ground snow load $p_g = 40$ psf and an upwind fetch $\ell_u = 75.33$ feet:

    $h_d = 0.75[0.43(\ell_u)^{1/3}(p_g + 10)^{1/4} - 1.5] = 0.75[0.43(75.33)^{1/3}(40 + 10)^{1/4} - 1.5] = 2.5$ feet

    Since $h_d = 2.5$ feet $< h_c = 3.1$ feet, the drift width $w = 4h_d = 4 \times 2.5 = 10.0$ feet.

    The maximum surcharge drift load $p_d = h_d \gamma = 2.5 \times 19.2 = 48.0$ psf.

    The total load at the face of the parapet wall is the balanced load plus the drift surcharge $= 27.7 + 48.0 = 75.7$ psf.

- Wind in east-west direction

    The equation in Figure 7-9 yields the following for the drift height, h_d, based on an upwind fetch $\ell_u = 328.75$ feet:

    $h_d = 0.75[0.43(\ell_u)^{1/3}(p_g + 10)^{1/4} - 1.5] = 0.75[0.43(328.75)^{1/3}(40 + 10)^{1/4} - 1.5] = 4.8$ feet

    Since $h_d = 4.8$ feet $> h_c = 3.1$ feet, the drift height is limited to 3.1 feet and the drift width $w = 4h_d^2/h_c = (4 \times 4.8^2)/3.1 = 29.7$ feet $> 8h_c = 24.8$ feet. Therefore, use $w = 24.8$ feet.

    The total load at the face of the parapet wall is $4.5 \times 19.2 = 86.4$ psf.

Balanced and drift loads at the parapet walls in both directions are shown in Figure 4.25.

Figure 4.25 Balanced and Drift Snow Loads at Parapet Walls, Example 4.4

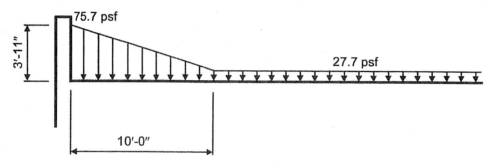

Parapet Walls at North and South Faces

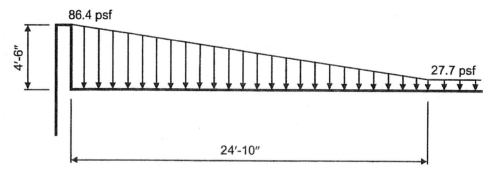

Parapet Walls at East and West Faces

Since the ground snow load, p_g, exceeds 20 psf, the minimum snow load is $20I_s = 20 \times 1.0 = 20$ psf (7.3), which is less than the flat roof snow load $p_f = 27.7$ psf. Also, a rain-on-snow surcharge load is not required in accordance with 7.10.

The only other load cases that need to be considered are the partial load cases of 7.5. For illustration purposes, assume that in the N-S direction the framing consists of a 3-span moment frame (cast-in-place concrete columns and beams) with 25 foot-2 inch exterior spans and a 25 foot-0 inch interior span. Balanced and partial loading diagrams for the concrete beams are illustrated in Figure 4.26. Partial loads are determined in accordance with 7.5 and Figure 7-4.

Figure 4.26 Balanced and Partial Loading Diagrams for the Concrete Beams Spanning in the N-S Direction, Example 4.4

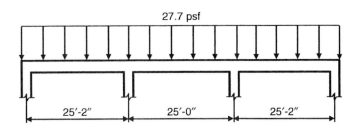

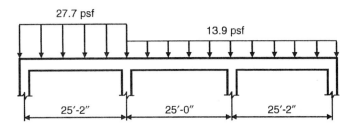

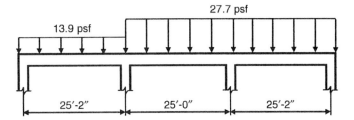

4.5.5 Example 4.5 – Snow Loads, Six-story Hotel with Rooftop Unit

For the six-story hotel in Example 4.4, determine the drift loads at the rooftop unit depicted in Figure 4.27. Use the same design data as in Example 4.4 and assume that the roof has no parapets.

Figure 4.27 Plan and Elevation of Six-story Hotel with Rooftop Unit, Example 4.5

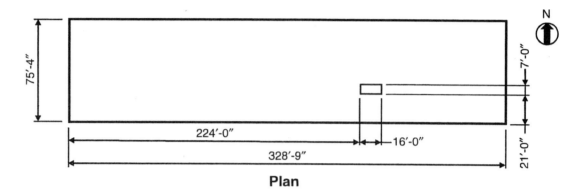

Plan

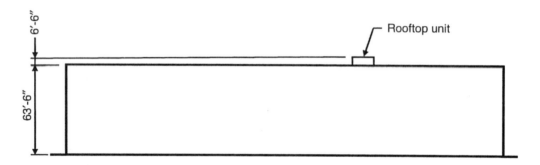

North/South Elevation

Solution

The following were determined in Example 4.4 and are used in this example:

   Sloped roof snow load $p_s = 27.7$ psf

   Snow density $\gamma = 19.2$ pcf

   Height of the balanced snow $h_b = p_s/\gamma = 27.7/19.2 = 1.4$ feet

The clear height to the top of the rooftop unit $h_c = 6.5 - 1.4 = 5.1$ feet

Drift loads are not required where $h_c/h_b < 0.2$ (7.7.1). In this example, $h_c/h_b = 5.1/1.4 = 3.6 > 0.2$, so drift loads must be considered.

Since the plan dimension of the rooftop unit in the N-S direction is less than 15 feet, a drift load is not required to be applied to those sides for wind in the E-W direction (7.8). Drift loads must be considered for the other sides of the rooftop unit, since those sides are greater than 15 feet.

For a N-S wind, the larger of the upwind fetches is 75.33 − 21.0 − 3.5 = 50.83 feet. For simplicity, this fetch is used for drift on both sides of the rooftop unit.

The equation in Figure 7-9 yields the following for the drift height, h_d, based on a ground snow load p_g = 40 psf and an upwind fetch ℓ_u = 50.83 feet:

$$h_d = 0.75[0.43(\ell_u)^{1/3}(p_g + 10)^{1/4} - 1.5] = 0.75[0.43(50.83)^{1/3}(40 + 10)^{1/4} - 1.5] = 2.1 \text{ feet}$$

Since h_d = 2.1 feet < h_c = 5.1 feet, the drift width $w = 4h_d = 4 \times 2.1 = 8.4$ feet.

The maximum surcharge drift load $p_d = h_d \gamma = 2.1 \times 19.2 = 40.3$ psf.

The total load at the face of the rooftop unit is the balanced load plus the drift surcharge = 27.7 + 40.3 = 68.0 psf.

Drift loads at the rooftop unit are illustrated in Figure 4.28.

Figure 4.28 Balanced and Drift Loads at Rooftop Unit, Example 4.5

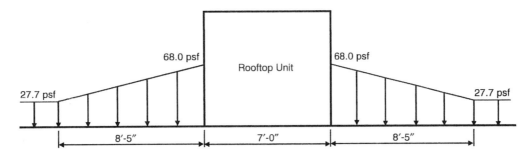

4.5.6 Example 4.6 – Snow Loads, Agricultural Building

Determine the design snow loads for the agricultural building depicted in Figure 4.29.

Figure 4.29 Agricultural Building, Example 4.6

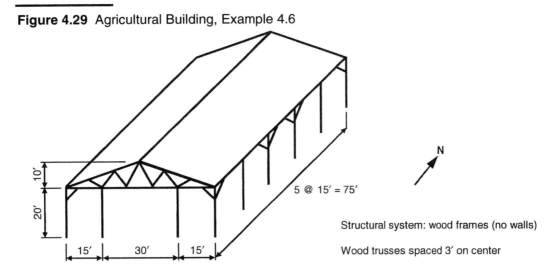

130   Chapter 4

DESIGN DATA	
Ground snow load, p_g:	30 psf
Terrain category:	C (open terrain with scattered obstructions having heights less than 30 feet)
Occupancy:	Utility and miscellaneous occupancy
Thermal condition:	Unheated structure
Roof exposure condition:	Sheltered
Roof surface:	Wood shingles

SOLUTION

1. Determine ground snow load, p_g.

    From the design data, the ground snow load, p_g, is equal to 30 psf.

2. Determine flat roof snow load, p_f, by Equation 7.3-1.

    Use Flowchart 4.1 to determine p_f.

    a. Determine exposure factor, C_e, from Table 7-2.

    From the design data, the terrain category is C and the roof exposure is sheltered. Therefore, $C_e = 1.1$ from Table 7-2.

    b. Determine thermal factor, C_t, from Table 7-3.

    From the design data, this open air structure is unheated, so $C_t = 1.2$ from Table 7-3.

    c. Determine the importance factor, I_s, from Table 1.5-2.

    From IBC Table 1604.5, the Risk Category is I for an agricultural facility. Thus, $I_s = 0.8$ from Table 1.5-2.

    Therefore, $p_f = 0.7 C_e C_t I p_g = 0.7 \times 1.1 \times 1.2 \times 0.8 \times 30 = 22.2$ psf

    Check if the minimum snow load requirements are applicable:

    A minimum roof snow load, p_m, in accordance with 7.3.4 applies to hip and gable roofs with slopes less than 15 degrees. Since the roof slope in this example is equal to 18.4 degrees, minimum roof snow loads do not apply.

3. Determine sloped roof snow load, p_s, by Equation 7.4-1.

    Use Flowchart 4.2 to determine roof slope factor, C_s.

    a. Determine thermal factor, C_t, from Table 7-3.

    From item 2 above, thermal factor $C_t = 1.2$.

    b. Determine if the roof is warm or cold.

    Since $C_t = 1.2$, the roof is defined as a cold roof in accordance with 7.4.2.

c. Determine if the roof is unobstructed or not and if the roof is slippery or not.

There are no obstructions on the roof that inhibit the snow from sliding off the eaves. Also, the roof surface has wood shingles. According to 7.4, wood shingles are not considered to be slippery.

Since this roof is unobstructed and not slippery, use the solid line in Figure 7-2c to determine C_s:

For a roof slope of 18.4 degrees, $C_s = 1.0$.

Therefore, $p_s = C_s p_f = 1.0 \times 22.2 = 22.2$ psf. This is the balanced snow load for this roof.

4. Consider partial loading.

Partial loads need not be applied to structural members that span perpendicular to the ridgeline in gable roofs with slopes greater than or equal to 2.38 degrees ($^1/_2$ on 12).

Since the roof slope is greater than 2.83 degrees, partial loading is not considered (partial loads on individual members of roof trusses, such as those illustrated in Figure 4.29, are generally not considered).

5. Consider unbalanced snow loads.

Flowchart 4.4 is used to determine if unbalanced loads on this gable roof need to be considered or not.

Unbalanced snow loads must be considered for this roof, since the slope is between 2.83 degrees and 30.2 degrees.

Since $W = 30$ feet > 20 feet, the unbalanced load consists of the following (see Figures 7-5 and 4.5):

- Windward side: $0.3 p_s = 0.3 \times 22.2 = 6.7$ psf
- Leeward side: $p_s = 22.2$ psf along the entire leeward length plus a uniform pressure of $h_d \gamma / \sqrt{S} = (1.9 \times 17.9)/\sqrt{3} = 19.6$ psf, which extends from the ridge a distance of $8 h_d \sqrt{S}/3 = (8 \times 1.9 \times \sqrt{3})/3 = 8.8$ feet where

    h_d = drift length from Figure 7-9 with $W = 30$ feet substituted for ℓ_u

    $\quad = 0.43(W)^{1/3}(p_g + 10)^{1/4} - 1.5 = 1.9$ feet

    γ = snow density (Equation 7.7-1)

    $\quad = 0.13 p_g + 14 = 17.9$ pcf < 30 pcf

    S = roof slope run for a rise of one = 3

132 Chapter 4

6. Consider snow drifts on lower roofs and roof projections.

    Not applicable.

7. Consider sliding snow.

    Not applicable.

8. Consider rain-on-snow loads.

    In accordance with 7.10, a rain-on-snow surcharge of 5 psf is required for locations where the ground snow load, p_g, is 20 psf or less (but not zero) with roof slopes less than $W/50$.

    In this example, $p_g = 30$ psf so an additional 5 psf need not be added to the balanced load of 22.2 psf.

9. Consider ponding instability.

    Since the roof slope in this example is greater than $1/4$ in./ft, progressive roof deflections and ponding instability from rain-on-snow or from snow meltwater need not be investigated (7.11 and 8.4).

10. Consider existing roofs.

    Not applicable.

The balanced and unbalanced snow loads are depicted in Figure 4.30.

Figure 4.30 Balanced and Unbalanced Snow Loads for Agricultural Building, Example 4.6

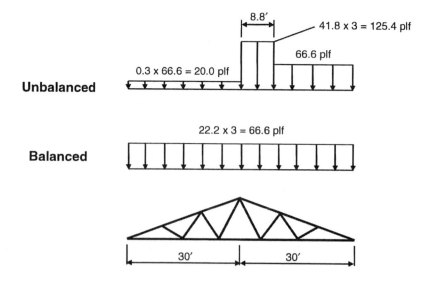

4.5.7 Example 4.7 – Snow Loads, University Facility with Sawtooth Roof

Determine the design snow loads for the university facility depicted in Figure 4.31.

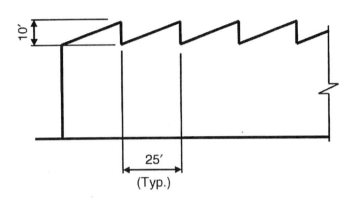

Figure 4.31 Elevation of University Facility, Example 4.7

DESIGN DATA

Ground snow load, p_g:	25 psf
Terrain category:	C (open terrain with scattered obstructions having heights less than 30 feet)
Occupancy:	Educational with an occupant load greater than 500
Thermal condition:	Cold, ventilated roof (R-value between the ventilated space and the heated space exceeds 25 ft²·h·°F/Btu)
Roof exposure condition:	Partially exposed
Roof surface:	Glass

SOLUTION

1. Determine ground snow load, p_g.

   From the design data, the ground snow load, p_g, is equal to 25 psf.

2. Determine flat roof snow load, p_f, by Equation 7.3-1.

   Use Flowchart 4.1 to determine p_f.

   a. Determine exposure factor, C_e, from Table 7-2.

   From the design data, the terrain category is C and the roof exposure is partially exposed. Therefore, $C_e = 1.0$ from Table 7-2.

   b. Determine thermal factor, C_t, from Table 7-3.

   From the design data, the roof is cold and ventilated roof with an R-value between the ventilated space and the heated space that exceeds 25 ft²·h·°F/Btu. Thus, $C_t = 1.1$ from Table 7-3.

c. Determine the importance factor, I_s, from Table 1.5-2.

   From IBC Table 1604.5, the Risk Category is III for this educational facility that has an occupant load greater than 500 people. Thus, $I_s = 1.1$ from Table 1.5-2.

   Therefore, $p_f = 0.7 C_e C_t I_s p_g = 0.7 \times 1.0 \times 1.1 \times 1.1 \times 25 = 21.2$ psf

3. Determine sloped roof snow load, p_s, from Equation 7.4-1.

   Use Flowchart 4.2 to determine roof slope factor, C_s. In accordance with 7.4.4, $C_s = 1.0$ for sawtooth roofs.

   Thus, $p_s = p_f = 21.2$ psf.

4. Consider unbalanced snow loads.

   Flowchart 4.6 is used to determine if unbalanced loads on this sawtooth roof need to be considered or not.

   Unbalanced snow loads must be considered, since the slope is greater than 1.79 degrees (7.6.3).

   In accordance with 7.6.3, the load at the ridge or crown is equal to $0.5 p_f = 0.5 \times 21.2 = 10.6$ psf. At the valley, the load is $2 p_f / C_e = 2 \times 21.2 / 1.0 = 42.4$ psf.

   The load at the valley is limited by the space that is available for snow accumulation. The unit weight of the snow is determined by Equation 7.7-1:

   $\gamma = 0.13 p_g + 14 = (0.13 \times 25) + 14 = 17.3$ pcf < 30 pcf

   The maximum permissible load is equal to the load at the ridge plus the load corresponding to 10 feet of snow: $10.6 + (10 \times 17.3) = 183.6$ psf. Since the unbalanced load of 42.4 psf at the valley is less than 183.6 psf, the load at the valley is not reduced.

   Balanced and unbalanced snow loads are illustrated in Figure 4.32.

Figure 4.32
Balanced and Unbalanced Snow Loads for University Facility, Example 4.7

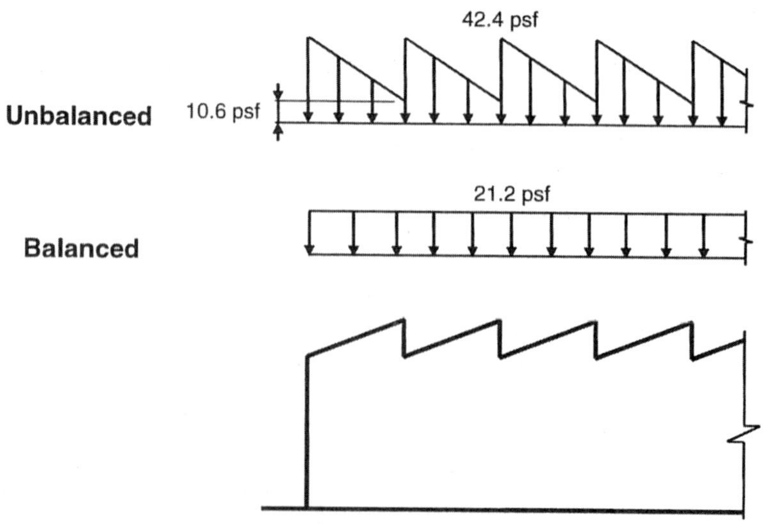

4.5.8 Example 4.8 – Snow Loads, Public Utility Facility with Curved Roof

Determine the design snow loads for the public utility facility depicted in Figure 4.33. The facility is required to remain operational during an emergency.

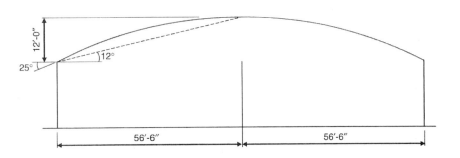

Figure 4.33 Elevation of Public Utility Facility, Example 4.8

DESIGN DATA	
Ground snow load, p_g:	60 psf
Terrain category:	D (flat unobstructed area near water)
Occupancy:	Essential facility
Thermal condition:	Unheated structure
Roof exposure condition:	Fully exposed
Roof surface:	Rubber membrane

SOLUTION

1. Determine ground snow load, p_g.

   From the design data, the ground snow load, p_g, is equal to 60 psf.

2. Determine flat roof snow load, p_f, by Equation 7.3-1.

   Use Flowchart 4.1 to determine p_f.

   a. Determine exposure factor, C_e, from Table 7-2.

   From the design data, the terrain category is D and the roof exposure is fully exposed. Therefore, $C_e = 0.8$ from Table 7-2.

   b. Determine thermal factor, C_t, from Table 7-3.

   From the design data, the structure is unheated. Thus, $C_t = 1.2$ from Table 7-3.

   c. Determine the importance factor, I_s, from Table 1.5-2.

   From IBC Table 1604.5, the Risk Category is IV for this essential facility. Thus, $I_s = 1.2$ from Table 1.5-2.

Therefore, $p_f = 0.7 C_e C_t I_s p_g = 0.7 \times 0.8 \times 1.2 \times 1.2 \times 60 = 48.4$ psf

Check if the minimum snow load requirements are applicable:

A minimum roof snow load, p_m, in accordance with 7.3 applies to curved roofs where the vertical angle from the eaves to the crown is less than 10 degrees. Since that slope in this example is equal to 12 degrees, minimum roof snow loads need not be considered.

3. Determine sloped roof snow load, p_s, from Eq. 7.4-1.

   Use Flowchart 4.2 to determine roof slope factor, C_s.

   a. Determine thermal factor, C_t, from Table 7-3.

      From item 2 above, thermal factor $C_t = 1.2$.

   b. Determine if the roof is warm or cold.

      Since $C_t = 1.2$, the roof is defined as a cold roof in accordance with 7.4.2.

   c. Determine if the roof is unobstructed or not and if the roof is slippery or not.

      There are no obstructions on the roof that inhibit the snow from sliding off the eaves. Also, the roof surface is a rubber membrane. According to 7.4, rubber membranes are considered to be slippery.

   Since this roof is unobstructed and slippery, use the dashed line in Figure 7-2c to determine C_s.

   For the tangent slope of 25 degrees at the eave, the roof slope factor C_s is determined by the equation in C7.4 for cold roofs with $C_t = 1.2$:

   $$C_s = 1.0 - \frac{(\text{slope} - 15°)}{55°} = 1.0 - \frac{(25° - 15°)}{55°} = 0.82$$

   Therefore, $p_s = C_s p_f = 0.82 \times 48.4 = 39.7$ psf, which is the balanced snow load at the eaves.

   Away from the eaves, the roof slope factor, C_s, is equal to 1.0 where the tangent roof slope is less than or equal to 15 degrees (see dashed line in Figure 7-2c). This occurs at distances of approximately 20.7 feet from the eaves at both ends of the roof. Therefore, in the center portion of the roof, $p_s = C_s p_f = 1.0 \times 48.4 = 48.4$ psf.

   The balanced snow load is depicted in Figure 4.34, which is based on Case 1 in Figure 7-3 for a slope at the eaves less than 30 degrees.

Figure 4.34
Balanced and Unbalanced Snow Loads for Public Utility Facility, Example 4.8

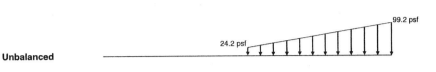
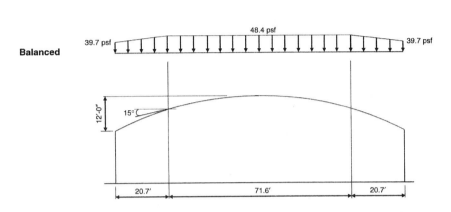

4. Consider unbalanced snow loads.

   Flowchart 4.5 is used to determine if unbalanced loads on this curved roof need to be considered or not.

   Since the slope of the straight line from the eaves to the crown is greater than 10 degrees and is less than 60 degrees, unbalanced snow loads must be considered (7.6.2).

   Unbalanced loads for this roof are given in Case 1 of Figure 7-3 (see Figure 4.6). No snow loads are applied on the windward side. On the leeward side, the snow load is equal to $0.5p_f = 0.5 \times 48.4 = 24.2$ psf at the crown and $2p_f C_s/C_e = 2 \times 48.4 \times 0.82/0.8 = 99.2$ psf at the eaves where C_s is based on the slope at the eaves. The unbalanced snow loads are shown in Figure 4.34.

4.5.9 Example 4.9 – Ice Loads, Support Structure of Sign on Commercial Building

Determine the weight of ice that can form on the structural members of a sign that is located on the top of a commercial building in Indianapolis, IN. The building is situated on a relatively flat site and the structural members of the sign consist of structural steel angles, which are L8×6×1/2. The uppermost structural members are located 70 feet above the ground level.

SOLUTION

The general procedure in 10.8 is used to determine the ice weight, D_i.

1. Determine nominal ice thickness, t, and the concurrent wind speed, V_c.

   From Figure 10-2, the nominal ice thickness, t, is equal to 1.0 inch and the concurrent wind speed, V_c, is equal to 40 mph for the structure located in Indianapolis, IN.

2. Determine the topographic factor, K_{zt}.

Since the building is located on a relatively flat site, $K_{zt} = 1.0$ in accordance with 10.4.5.

3. Determine the importance factor, I_i.

   Since the commercial building is classified under Risk Category II, $I_i = 1.0$ from Table 1.5-2.

4. Determine height factor, f_z.

   The height factor is determined by Equation 10.1-4:

   $$f_z = \left(\frac{z}{33}\right)^{0.10} = \left(\frac{70}{33}\right)^{0.10} = 1.08 \quad \text{for } 0 < z \le 900 \text{ feet}$$

5. Determine the design ice thickness, t_d.

   The design ice thickness is determined by Equation 10.4-5:

   $t_d = 2.0 t I_i f_z (K_{zt})^{0.35} = 2.0 \times 1.0 \times 1.0 \times 1.08 \times (1.0)^{0.35} = 2.2$ inches

6. Determine the weight of ice, D_i.

   The weight of ice for a structural shape is determined by multiplying the area of ice determined by Equation 10.4-1 and the density of ice, which is taken as 56 pcf (10.4.1).

   Referring to Figure 10-1, the diameter of the cylinder circumscribing the structural steel angle in this case is the hypotenuse of the right triangle formed by the legs of the angle:

   $$D_c = \sqrt{(8)^2 + (6)^2} = 10.0 \text{ inches}$$

   The area of ice is determined by Equation 10.4-1:

   $A_i = \pi t_d (D_c + t_d) = \pi \times 2.2 \times (10.0 + 2.2) = 84.3$ square inches

   Thus, the weight of ice is the following:

   $D_i = (84.3/144) \times 56 = 32.8$ plf

4.5.10 Example 4.10 – Ice Loads, Sign on Commercial Building

Determine the wind-on-ice load on the solid freestanding sign described in Example 4.9. Assume that the building is located at an Exposure B site and that the sign is 10 feet tall by 15 feet wide.

SOLUTION

The general procedure in 10.8 is used to determine the wind-on-ice load, W_i.

The information determined in steps 1 through 6 in Example 4.9 can be used in this example and is not repeated here.

1. Determine velocity pressure, q_h, for the concurrent wind speed, V_c, at the top of the sign.

   Equation 29.3-1 is used to determine q_h:

   $q_h = 0.00256 K_h K_{zt} K_d V_c^2$

The velocity pressure exposure coefficient, K_h, is determined from Table 29.3-1. At a height of 70 feet and Exposure B, $K_h = 0.89$.

For a solid freestanding sign, the wind directionality factor $K_d = 0.85$ (see Table 26.6-1).

Thus, $q_h = 0.00256 \times 0.89 \times 1.0 \times 0.85 \times 40^2 = 3.1$ psf

2. Determine wind force coefficient, C_f.

   The net force coefficient, C_f, for solid freestanding signs is determined in accordance with Figure 29.4-1. Load Cases A, B and C must be considered.

   - Case A: Resultant wind force acts normal to the face of the sign through the geometric center.

     Clearance ratio $s/h = 10/70 = 0.14$

     Aspect ratio $B/s = 15/10 = 1.50$

     Thus, $C_f = 1.80$ from Figure 29.4-1.

   - Case B: Resultant wind force acts normal to the face of the sign at a distance from the geometric center toward the windward edge equal to 0.2 times the average width of the sign.

     From Figure 29.4-1, C_f for Cases A and B are the same; thus, $C_f = 1.80$.

   - Case C: Resultant wind forces act normal to the face of the sign through the geometric centers of each region indicated in Figure 29.4-1.

     Since, $B/s < 2$, Case C need not be considered (see Note 3 in Figure 29.4-1).

3. Determine the gust factor, G, in accordance with 26.9.

   Assuming the structure is rigid, $G = 0.85$.

4. Determine the wind-on-ice load, W_i, in accordance with 29.4.1.

   The wind-on-ice load, W_i, is determined by Equation 29.4-1:

   $W_i = F = q_h G C_f A_s$

   The gross area of the solid freestanding sign, A_s, must include the design ice thickness, t_d, (10.5.2). It was determined in Example 4.9 that t_d is equal to 2.2 inches at a height of 70 feet. Thus,

   $$A_s = \left(10 + \frac{2 \times 2.2}{12}\right) \times \left(15 + \frac{2 \times 2.2}{12}\right) = 159.3 \text{ square feet}$$

   Therefore, for Load Cases A and B:

   $W_i = 3.1 \times 0.85 \times 1.80 \times 159.3 = 756$ lbs

   See Figure 29.4-1 for the location of this resultant force on the sign for Cases A and B.

4.6 References

4.1. O'Rourke, M. 2010. *Snow Loads: Guide to the Snow Load Provisions of ASCE 7-10*. American Society of Civil Engineers, Reston, VA.

4.2. O'Rourke, M. and DeAngelis, C. 2002. "Snow Drift at Windward Roof Steps." *Journal of Structural Engineering*, 128(10): 1330–1336.

4.7 Problems

4.1. Given the one-story warehouse described in Example 4.1, determine design snow loads for a roof slope of (a) 3 on 12 and (b) 6 on 12. Use the design data given in the example.

4.2. Given the one-story warehouse building and adjoining office building in Example 4.3, determine the design snow loads on the roof of the office building. Assume that the height of the office building at the eave is (a) 15 feet and (b) 18 feet instead of 10 feet. Use the design data given in the example.

4.3. Given the information provided in Problem 4.2, determine the design snow loads on the roof of the office building for parts (a) and (b). Assume that the warehouse and office building are separated horizontally a distance of 2 feet.

4.4. Determine the design snow loads for the buildings in Figure 4.35 given the design data in Table 4.3. The width of both roofs is 75 feet, and both roofs have a $1/4$ on 12 slope into the page. All roof members are simply supported.

Figure 4.35 Buildings in Problem 4.4

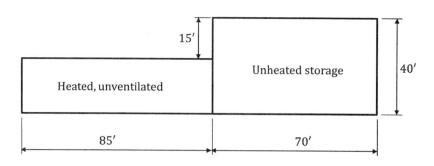

Table 4.3 Design Data for Problem 4.4

Ground snow load:	40 psf
Terrain category:	C
Occupancy:	Ordinary
Thermal condition:	As noted in Figure 4.35
Roof exposure condition:	Fully exposed
Roof surface:	Rubber membrane

4.5. Determine the design snow loads for the building in Figure 4.36 using the design data in Table 4.4. All roof members are simply supported.

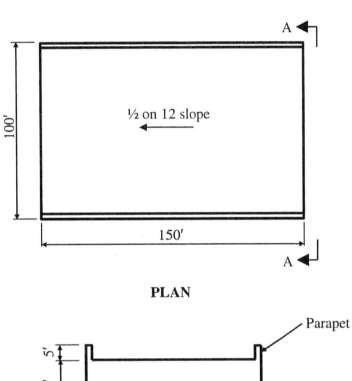

Figure 4.36
Building in Problem 4.5

PLAN

SECTION A-A

Table 4.4 Design Data for Problem 4.5

Ground snow load:	30 psf
Terrain category:	B
Occupancy:	Office
Thermal condition:	Heated
Roof exposure condition:	Partially exposed
Roof surface:	Bituminous

4.6. Determine the design snow loads for the building in Figure 4.37 using the design data in Table 4.5. All roof members are simply supported.

Figure 4.37 Building in Problem 4.6

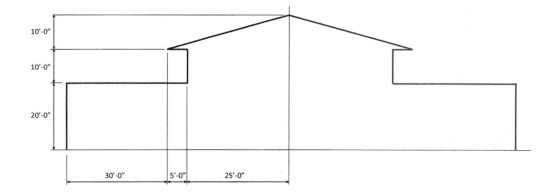

Table 4.5 Design Data for Problem 4.6

Location:	Fairbanks, Alaska
Terrain category:	C
Occupancy:	Residential
Thermal condition:	Ventilated roof with $R > 25$
Roof exposure condition:	Fully exposed
Roof surface:	Asphalt

4.7. Determine the design snow loads for the building in Figure 4.38 using the design data in Table 4.6. The roof profile is an arc of a circle that has a radius of 75 feet. All roof members are simply supported.

Figure 4.38
Building in Problem 4.7

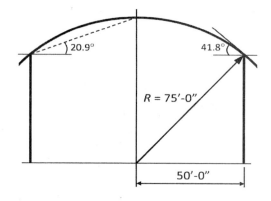

Table 4.6 Design Data for Problem 4.7

Ground snow load:	100 psf
Terrain category:	B
Occupancy:	Assembly
Thermal condition:	Intermittently heated
Roof exposure condition:	Fully exposed
Roof surface:	Slippery

4.8. Determine the design snow loads for the building in Problem 4.7. Assume that another roof abuts it within 3 feet of its eaves.

4.9. Determine the design snow loads for the building in Problem 4.7. Assume the following: $R = 150$ feet instead of 75 feet, and the distance to the eave = 143 feet instead of 50 feet.

4.10. Determine the design snow loads on the roof of the building and the canopy in Figure 4.39. Use the same design data as in Problem 4.6.

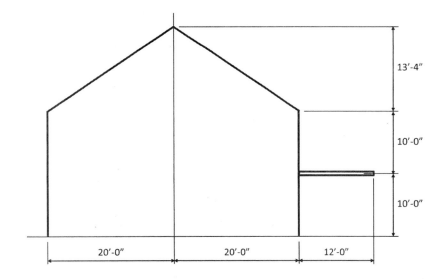

Figure 4.39
Building in Problem 4.10

4.11. Determine the weight of ice that can form on the support structure in Example 4.9. Assume the structural members are (a) HSS6.000 × 0.500 and (b) HSS6 × 4 × $1/2$. Use the same design data in Example 4.9.

4.12. Determine the wind-on-ice load of a round chimney that is 60 feet tall and has an outside diameter of 6 feet. Assume the chimney is located in Milwaukee, WI, on a relatively flat site in Exposure C.

CHAPTER 5
Wind Loads

5.1 Introduction

5.1.1 Nature of Wind Loads

In general, wind loading is the effect of the atmosphere passing by a stationary structure attached to the earth's surface. An in-depth discussion on the mechanics of atmospheric circulations can be found in *Wind Effects on Structures: Fundamentals and Applications to Design* (Reference 5.1).

Loads on buildings and other structures due to the effects from wind are determined by considering both atmospheric and aerodynamic effects. These effects form the basis of the methodologies given in ASCE/SEI 7 for the determination of wind loads. An elementary discussion of each effect is given below. The effects due to wind gust, which are considered a part of both atmospheric and aerodynamic effects, also play an important role in the calculation of wind loads and are covered as well.

Atmospheric Effects

The atmospheric factors that have a direct impact on the magnitude of wind loading on a building or other structure are obtained from meteorological and boundary layer effects. A brief discussion that is pertinent to wind loads on buildings and other structures follows.

Meteorology

Meteorology is the study of the atmosphere, and wind climatology is a branch within meteorology that focuses on the prediction of storm conditions. In particular, extreme wind speeds associated with different types of storms and the probability of occurrence of such extreme values are analyzed at specific geographical locations. This information is used in developing design wind speed maps in the IBC and ASCE/SEI 7. Wind velocity is used in calculating the maximum design wind loads that can be expected on a building or structure during its lifespan.

Wind speed measurements that are utilized in creating the design wind speed maps in the IBC and ASCE/SEI 7 are obtained from essentially two different sources. In nonhurricane regions, wind speed data is collected from standard 3-cup anemometers that are located 33 feet above the ground in open terrain (such as at airports) at 485 National Weather Service weather stations throughout the U.S. that have at least 5 years of available data. Design wind speeds in hurricane regions are obtained from statistical simulations based on historical data, since actual weather data on hurricanes at any particular location are generally limited.

The wind speed that was recorded at the aforementioned weather stations for many years was the *fastest-mile wind speed*, which is the maximum wind speed averaged over one mile of wind passing through an anemometer. From this definition, it follows that the corresponding averaging time is equal to 3,600 divided by the velocity of wind in miles per hour. As the National Weather Service phased out the older equipment, the fastest-mile wind speed was replaced with a *3-second gust speed*. The three-second averaging time is based on the response characteristics of the newer instruments that are utilized at the weather stations.

Thunderstorms, hurricanes, tornadoes and special regional effects are the climatological events that are of primary interest when designing buildings and structures in the U.S. Figure 5.1 illustrates the controlling climatological events that produce extreme wind speeds in regions of the conterminous U.S. It is evident from the figure that the prevailing wind speeds are generated by thunderstorms in most of the country.

Figure 5.1
Controlling Climatological Events for Extreme Wind Speeds

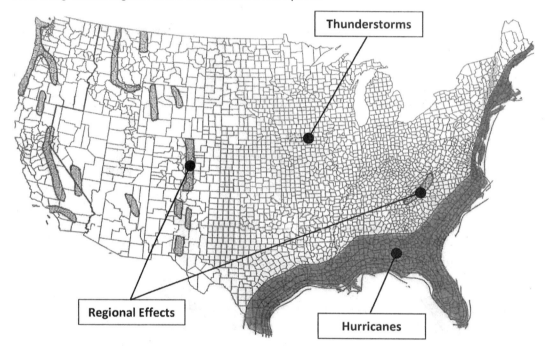

Wind speeds generated by tornadoes are not included in the wind speed maps in the IBC or ASCE/SEI 7. The primary reason for this has to do with the significantly lower probability of occurrence of tornadoes compared to that for basic wind speeds. Additional information on how to design for tornadoes is covered in Section 5.2.2 of this publication.

Special wind regions can have wind speeds significantly greater than those in surrounding areas. Such regional effects include wind blowing over mountain ranges or through gorges or valleys. These regions are identified as special wind regions on the basic wind speed maps.

Boundary Layer Fluid Dynamics

The layer of the earth's atmosphere that is located from the surface of the earth to approximately 3,300 feet above the surface is known as the boundary layer. The fluid dynamic effects that occur within this layer have an important impact on the magnitude of wind loads on buildings and other structures.

In general, the surface of the earth exerts a horizontal drag force on wind, which impedes its flow. More frictional resistance is experienced the closer the wind flow is to the surface; thus, wind velocity is smaller at or near the ground level compared to levels above the surface.

Similarly, at a given height above the surface, wind velocity is smaller over rougher surfaces compared to smoother ones because of friction. Figure 5.2 depicts the variation of wind speed with respect to height and surface roughness. Both phenomena are captured in the current wind load provisions using a modified version of the power-law methodology that was first introduced in the *Journal of Structural Division* (Reference 5.2). Note that the wind velocities become constant above certain heights for the different roughness categories; these heights are defined as gradient heights in ASCE/SEI 7.

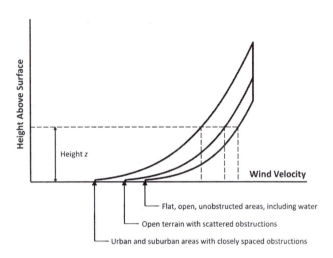

Figure 5.2
Variation of Wind Velocity with Respect to Height and Surface Roughness

Aerodynamic Effects

Overview

Wind flow is disturbed due to the presence of a building or structure in its path. The resulting responses due to this disturbance are governed by the laws of aerodynamics. In the case of a building or other structure that has an essentially block-like shape (which is referred to as a bluff body), bluff body aerodynamics is used to predict the effects caused by placing the bluff body in the flow of wind.

In general, when wind comes into contact with a building or other structure, the following pressures are created:

External pressures, which act on all exterior surfaces; these pressures are caused by the effects that are generated when the wind strikes the building or other structure.

Internal pressures, which act on all interior surfaces; these pressures are due to leakage of air through the exterior surface to the interior space.

It is assumed that the external and internal pressures act perpendicular to the exterior and interior surfaces of the building or other structure, respectively. A pressure is defined as positive when it acts towards the surface and negative when it acts away from the surface. Positive pressure is commonly referred to as just pressure and negative pressure is also identified as suction.

External Pressure

Figure 5.3 depicts idealized wind flow around a gable roof building. When the wind strikes the windward wall, it creates a positive pressure on that surface, which varies with respect to height (as discussed previously, wind velocity increases at distances above the surface and, as shown in Equation 5.1 below, wind pressure is directly proportional to the square of wind velocity). Located mid-width of wall at a distance approximately two-thirds the windward wall height above the surface is the *stagnation point*. The pressure, p_s, at this point can be theoretically determined using Bernoulli's equation:

$$p_s = \frac{1}{2}\rho V^2 \tag{5.1}$$

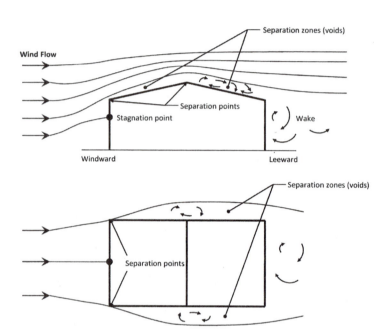

Figure 5.3 Wind Flow around a Gable Roof Building

In this equation, ρ is the atmospheric air density and V is the wind velocity at the elevation of the stagnation point.

A location where the wind flow is deflected by one surface of the building and separates and loses contact with the building is referred to as a *separation point* (see Figure 5.3). The void between the separated wind flow and the surface of the building or structure—commonly referred to as the *separation zone*—contains turbulent wind flow, which causes negative pressure on that surface.

The leeward wall and the side walls all have negative pressure (the negative pressure on the leeward wall is referred to as the wake). In the case of a sloped roof like the one depicted in Figure 5.3, the size of the void depends on the angle of the roof: for relatively shallow slopes, the void is relatively large and negative pressure acts over the windward portion of the roof. As the roof slope increases, the void area on the windward roof decreases. The pressure on the windward roof becomes positive when the roof angle matches or exceeds the angle of the separated wind flow; in other words, positive pressure is realized when the void of turbulent flow no longer exists on the windward area. Negative pressure occurs on the leeward area of the roof regardless of the roof slope. In the case of a flat roof, the entire roof has negative pressure (the void of turbulent flow acts over the entire area of the roof).

When the dimension of the building or structure is relatively long in the direction of wind flow, it is possible for the flow to reattach itself to the side walls or to the roof, or both. This has the effect of reducing the negative pressure on these surfaces as well as on the leeward wall. The reduction of pressure for such buildings is reflected in the wind design provisions in ASCE/SEI 7.

A generic representation of the external wind pressures on the walls and roof of the building depicted in Figure 5.3 is given in Figure 5.4.

Figure 5.4
Wind Pressures on a Gable Roof Building

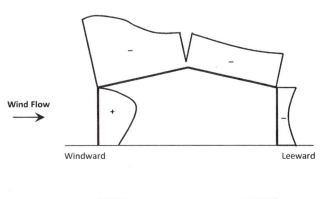

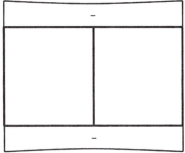

It is evident from the figure that the positive pressure on the windward wall varies with respect to height and that the negative pressures on all other surfaces are essentially constant. Also note that the negative pressure at the windward corner of the roof can be significantly larger than the pressures on any of the other surfaces.

The dimensionless pressure, C_p, which is commonly referred to as a *pressure coefficient*, is defined as the pressure, p, at any point on the building or structure divided by the stagnation pressure:

$$C_p = \frac{p}{\rho V^2/2} \tag{5.2}$$

In general, the external pressures due to wind vary on each surface of a building or structure based on the structural configuration and the direction of wind. For simplicity, pressure coefficients for the most part are taken as a constant over an entire surface area. One notable exception is for roofs that are relatively long in the direction of wind flow: in such cases, the pressure coefficients away from the windward edge of the roof are permitted to be smaller than that at the edge (see discussion on flow reattachment above).

Internal Pressure and Enclosure Classification

As noted previously, internal pressures act on all interior surfaces of a building or structure and are due to leakage of air (pressure) through the exterior surface to the interior space. The number and size of openings in the envelope of the structure play key roles when determining the effects caused by internal pressures. ASCE/SEI 7 defines an *opening* as an aperture or hole in the building envelope that allows air to flow through the building envelope and that is designed as open during design winds. Doors, operable windows, exhausts for ventilation systems and gaps (deliberate or otherwise) around cladding are just a few examples of openings.

A number of different cases need to be examined in order to properly understand the effects that internal pressure can have on a building or structure. In the case of a building that has a relatively large opening on the windward wall, the wind flow will try to inflate the building resulting in internal pressure that is positive (see Figure 5.5). Conversely, an opening on the leeward wall (as illustrated in Figure 5.5), side walls or roof will try to deflate the building and will result in negative internal pressure.

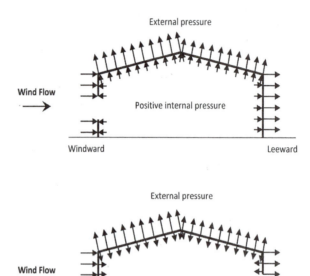

Figure 5.5 Effects of Openings on Internal Pressure Distribution

Typically, there will be more than just one opening in a building envelope. The size and location of these openings, which is commonly referred to as the *background porosity* of the building envelope, dictates whether the internal pressure is positive or negative. The *enclosure classifications* defined in ASCE/SEI 7 are based on background porosity.

Consider the case of a building with openings on all surfaces that are large enough to affect the internal pressure. Wind will be able to flow through the building without a buildup of any internal pressure. Such buildings are classified as *open*, and ASCE/SEI 7 defines such structures as having each wall at least 80 percent open.

A *partially enclosed* building has openings that are large enough to affect internal pressure and have background porosity low enough to allow internal pressure to build up. A partially enclosed condition typically exists when there is a relatively large opening on the windward wall compared to the openings on all of the other surfaces of the envelope. The internal pressures in such buildings can be as large as or greater than the external pressures.

An *enclosed* building is one that is neither open nor partially enclosed. A building or other structure can be classified as enclosed if it has relatively small infiltrations through the envelope and low background porosity; this combination results in a relatively small level of positive or negative internal pressure. Moreover, a building can also be enclosed if it has relatively large openings and sufficient background porosity to allow the pressure to escape as fast as it enters; this also results in a relatively small level of positive or negative internal pressure.

The effects of internal pressure are accounted for in ASCE/SEI 7 by internal pressure coefficients, the magnitudes of which are based on the enclosure classification of the building or other structure. It is evident from Figure 5.5 that internal pressure does not contribute to the overall horizontal wind pressures that act on a structure because such pressures cancel out. However, internal pressures must be considered in the design of individual components, such as walls and roof framing, and cladding; the type of internal pressure (positive or negative) that results in the critical load combination must be used.

It is important to point out that both the IBC and ASCE/SEI 7 have specific requirements for the protection of openings in *wind-borne debris regions* (see Section 5.2.7 of this publication for the definition of wind-borne debris regions). In short, openings in such regions must have glazing that is impact resistant or protected with an impact-resistant covering that meets the requirements of impact-resistant standards.

Gust Effects

Wind velocity typically changes dramatically with time; numerous peaks and valleys normally occur over relatively short time spans. In general, the average wind speed obtained from a wind event is larger when a shorter averaging time is used. The peaks in wind velocity are called *gusts*, and these effects must be considered in design.

Gust-effect factors are used in ASCE/SEI 7 to account for this phenomenon. In short, a gust-effect factor relates the peak to mean response in terms of an equivalent static design load or load effect. The method that is used to determine such factors is based on the following assumption: the fundamental mode of vibration of a building or structure has an approximately linear mode shape. Additional information on the pioneering work in gust-effect factors can be found in *Tall Building Structures: Analysis and Design* (Reference 5.3).

When the fastest-mile wind was utilized, the averaging time was generally between 30 and 60 seconds (see the discussion on fastest-mile wind speed above). As such, the corresponding gust-effect factors were greater than 1.0 in order to sufficiently capture the effects due to gust. In the later provisions that are based on a 3-second gust wind speed, the gust-effect factors are typically less than 1.0 (except for certain types of flexible buildings) because the 3-second averaging time corresponds to a peak gust whose effects are greater than those from the gust level that has been deemed reasonable in design.

It is important to note that the gust-effect factors in ASCE/SEI 7 for both rigid and flexible buildings (that is, buildings with a fundamental frequency greater than or equal to 1 Hz and less than 1 Hz, respectively) account for the effects in the along-wind direction only. In the case of flexible buildings or structures, along-wind effects due to dynamic amplification are also accounted for in the gust-effect factor.

The mass distribution, flexibility and damping of a building or structure can have a significant impact on its response to large gusts. It is possible for the fundamental frequency of lighter, more flexible buildings and structures with relatively small amounts of inherent damping to be in the same range as the average frequencies of large gusts. When this occurs, large resonant motions can occur, which must be considered in design.

Certain types of buildings and structures—especially those that are relatively tall and slender—are susceptible to one or more of the following: (1) across-wind load effects, (2) vortex shedding, (3) instability due to galloping or flutter or (4) dynamic torsional effects. The gust-effect factors in ASCE/SEI 7 do not account for the loading effects caused by these phenomena; wind tunnel tests should be performed in such cases to properly capture these effects.

5.1.2 Overview of Code Requirements

According to IBC 1609.1.1, wind loads on buildings and structures are to be determined by the provisions of Chapters 26 to 30 of ASCE/SEI 7-10 or by the alternate all-heights method of IBC 1609.6. Five exceptions are given in IBC 1609.1.1 that permit wind loads to be determined on certain types of structures using industry standards other than ASCE/SEI 7, and one exception is given that permits the use of wind tunnel tests that conform to the provisions of Chapter 31 of ASCE/SEI 7.

Wind is assumed to come from any horizontal direction and its effects are applied in the form of pressures that act normal to the surfaces of a building or other structure (IBC 1609.1.1). As discussed in Section 5.1.1 of this publication, positive wind pressure acts toward the surface and is commonly referred to as just pressure. Negative wind pressure, which is also called suction, acts away from the surface.

Positive pressure acts on the windward wall of a building and negative pressure acts on the leeward wall, the side walls, and the leeward portion of the roof (see Figure 5.6). Either positive pressure or negative pressure acts on the windward portion of the roof, depending on the slope of the roof (flatter roofs will be subjected to negative pressure while more sloped roofs will be subjected to positive pressure). Note that the wind pressure on the windward face varies with respect to height and that the pressures on all other surfaces are assumed to be constant.

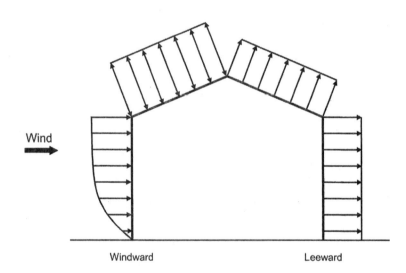

Figure 5.6
Application of Wind Pressures on a Building with a Gable or Hip Roof in Accordance with the IBC and ASCE/SEI 7

In general, pressures must be considered on the *main wind-force resisting system* (MWFRS) and *components and cladding* (C&C) of a building or other structure. The MWFRS consists of structural elements that have been assigned to resist the effects from the wind loads for the overall structure. Shear walls, moment frames and braced frames are a few examples of different types of MWFRSs.

The elements of the building envelope that do not qualify as part of the MWFRS are the C&C. The C&C receive the wind loads either directly or indirectly (for example, from the cladding) and transfer the loads to the MWFRS. Examples of C&C are individual exterior walls, roof decking, roof members (joists or purlins), curtain walls, windows or doors. A discussion on how wind load propagates in a building is given in Chapter 8 of this publication. Certain members must be designed for more than one type of loading; for example, a roof beam that is part of the MWFRS must be designed for the load effects associated with the MWFRS and those associated with the C&C.

Table 5.1 contains the procedures that are available in ASCE/SEI 7-10 to determine wind loads on MWFRSs and C&C. In the *directional procedure*, wind loads for specific wind directions are determined using external pressure coefficients that are based on wind tunnel tests of prototypical building models for the corresponding direction of wind. The provisions of this procedure cover a wide range of buildings and other structures.

Table 5.1 Summary of Wind Load Procedures in ASCE/SEI 7

System	ASCE/SEI 7 Chapter	Description
MWFRS	27	Directional procedure for buildings of all heights
	28	Envelope procedure for low-rise buildings[1]
	29	Directional procedure for building appurtenances and other structures
	31	Wind tunnel procedure for any building or other structure
C&C	30	• Envelope procedure in Part 1 or 2, or • Directional procedure in Parts 3, 4 and 5 • Building appurtenances in Part 6
	31	Wind tunnel procedure for any building or other structure

1. Low-rise buildings are defined in ASCE/SEI 26.2 as enclosed or partially enclosed buildings that have (1) a mean roof height less than or equal to 60 feet and (2) a mean roof height that is less than or equal to the least horizontal dimension of the building.

In the *envelope procedure*, pseudo-external pressure coefficients derived from wind tunnel tests on prototypical building models that are rotated 360 degrees in a wind tunnel are used to determine wind loads; the pressure coefficients envelope the maximum values that are obtained from all possible wind directions.

As noted previously, the alternate all-heights method of IBC 1609.6, which is a simplification of the directional procedure in ASCE/SEI 7, is another method that is available to determine wind loads on buildings and other structures that meet the conditions of that section. Contrary to its name, this method is applicable to a certain class of buildings that have a height that is less than or equal to 75 feet. More information is given in Section 5.6 of this publication.

With the exception of the wind tunnel procedure, all of these procedures and methods in Table 5.1 and the alternate all-heights method are static methods for estimating wind pressures: the magnitude of wind pressure on a building or structure depends on its size, openness, occupancy, location and the height above ground level. Also accounted for are wind gust and local extreme pressures at various locations on a building or structure. Static methods generally yield very accurate results for low-rise buildings.

Figure 26.1-1 provides an outline of the process that is required for determining wind loads. In-depth information on all of these procedures and methods, including the general requirements in ASCE/SEI Chapter 26 and the alternate all-heights method of IBC 1609.6, is given in the following sections.

5.2 General Requirements

5.2.1 Overview

Chapter 26 of ASCE/SEI 7 contains the following general requirements for determining wind loads on MWFRS and C&C:

- Basic wind speed (Figures 26.5-1A, 26.5-1B and 26.5-1C)

- Wind directionality (26.6)

- Exposure (26.7)
- Topographic effects (26.8)
- Gust effects (26.9)
- Enclosure classification (26.10)
- Internal pressure coefficients (26.11)

These requirements are used in conjunction with the methods and procedures contained in Chapters 27 through 31.

The following sections discuss these requirements and provide additional background information on fundamental concepts.

5.2.2 Wind Hazard Map

Regardless of the wind load procedure that is employed to determine wind pressures, the basic wind speed, V, must be determined at the location of the building or other structure (note that in the IBC, the basic wind speed is designated V_{ult} and is defined as the ultimate design wind speed).

Figures 1609A, 1609B and 1609C in the IBC and Figures 26.5-1A, 26.5-1B and 26.5-1C in ASCE/SEI 7 are identical and provide basic wind speeds based on 3-second gusts at 33 feet above ground for Exposure C for Risk Categories II, III and IV, and I, respectively.

These maps are a departure from the single map with importance factors that were used in prior editions of these documents. In particular, these new maps provide ultimate design wind speeds for different categories of building occupancies (more precisely, risk categories) defined in IBC Table 1604.5; as a result of this, the importance factor, I, is no longer needed.

Since the wind speeds are ultimate design wind speeds, the wind load factors in the design load combinations are equal to 1.0 (see Chapter 2 of this publication for information on load combinations). Table 5.2 provides a summary of the information associated with these maps.

Table 5.2 Summary of Basic Wind Speed Maps in the 2012 IBC and ASCE/SEI 7-10

Figure No.	Risk Category*	Return Period (years)
1609A 26.5-1A	II	700
1609B 26.5-1B	III, IV	1,700
1609C 26.5-1C	I	300

*See Table 1604.5 for definitions of risk categories.

The shaded areas on the wind speed maps are designated as special wind regions. As was discussed in Section 5.1.1 of this publication, these are areas where unusual wind conditions exist. The local authority having jurisdiction over the project should be consulted to obtain the local design wind speed (ASCE/SEI 26.5.2). Information on how to estimate basic wind speeds from regional climatic data can be found in ASCE/SEI 26.5.3.

More information on the decision to move to multiple design wind speed maps and the selection of the corresponding return periods can be found in ASCE/SEI C26.5.1. Figure C26.5-1 in ASCE/SEI C26.5.1 shows the variation of the maximum wind speed over time for the nonhurricane wind speeds contained in these maps.

An aid has been developed to assist in determining the basic wind speeds in the IBC and ASCE/SEI 7: www.atcouncil.org/windspeed (Reference 5.4). Wind speeds corresponding to the provisions of ASCE/SEI 7-10, ASCE/SEI 7-05 and ASCE/SEI 7-93 can be obtained for a particular location based on its latitude and longitude. Wind velocities for mean recurrence intervals of 10, 25, 50 and 100 years are also provided; these are useful in various applications, including serviceability (see discussion below).

The ultimate design wind speeds on the wind hazard maps do not include effects of tornadoes (ASCE/SEI 26.5.4). However, ASCE/SEI C26.5.4 contains references and a tornadic gust wind speed map of the U.S. that corresponds to a return period of 100,000 years. The information presented in this section can be used as a guide in developing and designing buildings and other structures for the effects of tornadoes. *Taking Shelter From the Storm: Building a Safe Room For Your Home or Small Business* (Reference 5.5) also contains design guidance for tornadoes.

When the provisions of the standards referenced in Exceptions 1 through 5 of IBC 1609.1.1 are used to determine wind loads, the ultimate design wind speeds, V_{ult}, given in Figures 1609A, 1609B and 1609C must be converted to nominal design wind speeds, V_{asd}, using Equation 16-33 in IBC 1609.3.1:

$$V_{asd} = V_{ult}\sqrt{0.6} \tag{5.3}$$

Values of V_{asd} are tabulated in IBC Table 1609.3.1 for various V_{ult}.

Some referenced standards and some product evaluation reports have been developed and/or evaluated based on fastest-mile wind speed, which was the wind speed utilized in the 1993 and earlier editions of ASCE/SEI 7 and in the legacy codes. To facilitate coordination between the various wind speeds, ASCE/SEI Table C26.5-6 provides wind speeds corresponding to the 1993, 2005 and 2010 editions of ASCE/SEI 7.

When checking serviceability due to wind effects, using a 700-year or 1,700-year return period is generally considered to be excessively conservative. The load combination given in Equation CC-3 of the Commentary section of Appendix C of ASCE/SEI 7 can be used to check short-term serviceability effects based on the serviceability wind speeds in Figures CC-1, CC-2, CC-3 and CC-4:

$$D + 0.5L + W_a \tag{5.4}$$

In this equation, W_a is the wind load effect based on the wind speeds in Figures CC-1 through CC-4, which give peak gust wind speeds at 33 feet above ground in Exposure C for return periods of 10, 25, 50 and 100 years, respectively.

5.2.3 Wind Directionality

The wind directionality factor, K_d, that is given in ASCE/SEI 26.6 accounts for the statistical nature of wind flow and the probability of the maximum effects occurring at any particular time for any given wind direction. In particular, this factor accounts for the following:

1. The reduced probability that the maximum wind from a certain direction aligns with the controlling load case.

2. The reduced probability that the maximum pressure coefficients occur for any given wind direction.

Table 26.6-1 contains values of K_d as function of structure type. This factor is equal to 0.85 for the MWFRS and C&C of buildings.

One of the requirements in ASCE/SEI 26.6 is that K_d must only be included in determining the effects due to wind loads when the strength design and allowable stress design load combinations of 2.3 and 2.4, respectively, are used in the design of the structural members. The reason behind this has to do with how the directionality factor is accounted for in the design equations. In the editions of ASCE/SEI 7 prior to 2005 and in the legacy codes, the directionality factor was included in the load factor for wind, which was equal to 1.3 for strength design. Once this factor was extracted from the wind load factor in the design provisions in 2005, the load factor was increased to 1.6 (which was rounded up from $1.3/0.85 = 1.53$). The move to strength-level wind effects in the 2010 edition with a wind load factor of 1.0 automatically considers the explicit inclusion of K_d in the design equations for both strength level and allowable stress design.

5.2.4 Exposure

Wind Direction and Sectors

According to IBC 1609.4 and ASCE/SEI 26.7, an exposure category must be determined upwind of a building or other structure for each wind direction that is considered in design. Wind must be assumed to come from any horizontal direction when determining wind loads (IBC 1609.1.1 and ASCE/SEI 26.5.1). One rational way of satisfying this requirement is to assume that there are eight wind directions: four that are perpendicular to the main axes of the building or other structure and four that are at 45-degree angles to the main axes. Figure C26.7-5, which is reproduced here as Figure 5.7, shows the sectors that are to be used to determine the exposure for a selected wind direction (IBC 1609.4.1 and ASCE/SEI 26.7.1).

Figure 5.7
Sectors for Determining Exposure

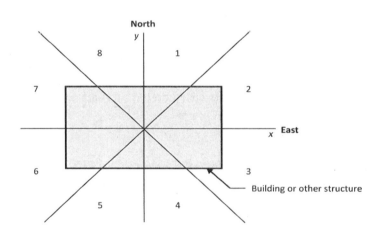

Upwind exposure in a particular direction is determined based on the terrain categories that are in the 45-degree sectors on each side of the wind direction axis. For wind blowing north to south, the surface roughness is determined for sectors 1 and 8 and the corresponding exposure that results in the higher wind load effects is used; these wind loads are applied to the building in that direction. The exposures for wind coming from the east, south and west are determined in a similar fashion. For wind coming from the northeast (or, similarly from any of the other diagonal directions), sectors 1 and 2 are used to determine the critical exposure; full individual wind loading in the x and y directions are determined based on that exposure, and a percentage of these loads are then applied simultaneously to the building (see Section 5.3.2 of this publication on load cases that must be considered in wind design).

Surface Roughness Categories

Surface roughness categories are defined in IBC 1609.4.2 and ASCE/SEI 26.7.2 and are summarized in Table 5.3. These definitions are descriptive and have been purposely expressed this way so that they can be applied easily—while still being sufficiently precise—in most practical applications.

Table 5.3 Surface Roughness Categories

Surface Roughness Category	Description
B	Urban and suburban areas, wooded areas or other terrain with numerous closely spaced obstructions having the size of single-family dwellings or larger
C	Open terrain with scattered obstructions having heights generally less than 30 feet; this category includes flat open country and grasslands
D	Flat, unobstructed areas and water surfaces; this category includes smooth mud flats, salt flats and unbroken ice

Exposure Categories

Exposure categories are based on the surface roughness categories defined above and essentially account for the boundary layer concept of surface roughness discussed in Section 5.1.1 of this publication.

Table 5.4 contains definitions of the three exposure categories given in IBC 1609.4.3 and ASCE/SEI 26.7.3. Definitions of Exposure B and Exposure D are illustrated in Figures C26.7-1 and C26.7-2, respectively.

Table 5.4 Exposure Categories

Exposure Category	Definition
B	• Mean roof height $h \leq 30$ feet Surface Roughness Category B prevails in the upwind direction for a distance > 1,500 feet • Mean roof height $h > 30$ feet Surface Roughness Category B prevails in the upwind direction for a distance > 2,600 feet or 20 times the height of the building, whichever is greater
C	Applies for all cases where Exposures B and D do not apply
D	• Surface Roughness Category D prevails in the upwind direction for a distance > 5,000 feet or 20 times the height of the building, whichever is greater • Surface roughness immediately upwind of the site is B or C, and the site is within a distance of 600 feet or 20 times the building height, whichever is greater, from an Exposure D condition as defined above

Aerial photographs that illustrate each exposure type are provided at the end of Chapter C26 in ASCE/SEI 7. Note that sites located along the shoreline of hurricane-prone regions are to be assigned to Exposure D (see C26.7 for a summary of the research findings that led to this requirement).

According to ASCE/SEI 26.7.3, the exposure category that results in the largest wind loads must be used for sites that are located in transition zones between exposure categories. Consider the building in Figure 5.8, which is located between Exposure D and Exposure B or C. Exposure D must extend the larger of 600 feet or 20 times the building height from the point where Exposure D ends (see definition in Table 5.4). In other words, Exposure D transitions to Exposure B or C over the larger of those two lengths. Thus, Exposure D is required to be used in this transition zone in accordance with 26.7.3; however, the exception in 26.7.3 permits an

158   Chapter 5

intermediate exposure category to be used provided that it is determined by a rational analysis method defined in recognized literature. An example of such an analysis is given in C26.7.

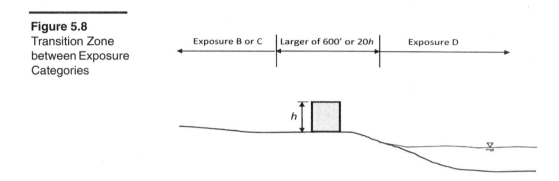

Figure 5.8
Transition Zone between Exposure Categories

Two other topics of interest are covered in C26.7. The first pertains to a mathematical procedure that can be used to make a more detailed assessment of surface roughness and exposure category. Ground surface roughness is defined in terms of a roughness length parameter, z_o. Surface Roughness Categories B, C and D correspond to a range of values of z_o, which are tabulated in Table C26.7-1 (this table also includes values for the now defunct Exposure A as a reference for wind tunnel studies). Additional information on z_o can be found in Table C26.7-2, and Equation C26.7-1 can be used to estimate z_o for a particular terrain.

The second topic has to do with the effects of "open patches" in Exposure B. It is possible for patches of Exposure C or D to occur within Exposure B. Large parking lots or lakes, freeways and tree clearings are a few examples of such patches. If open patches are large enough, they can have a significant impact on the determination of the exposure category. According to the commentary, open patches with sides smaller than 164 feet do not need to be considered in the determination of exposure category. The effects of open patches of Surface Roughness C or D on the use of Exposure Category B are shown in Figures C26.7-3 and C26.7-4.

Exposure Requirements

ASCE/SEI 26.7.4 contains exposure requirements that must be satisfied for all of the wind load procedures that are available in ASCE/SEI 7. A summary of these requirements is given in Table 5.5.

It is evident from the table that for C&C and for low-rise buildings, wind loads are determined using the upwind exposure for the single surface roughness in one of the eight sectors that gives the largest wind forces (see Figure 5.7). In all other cases, an exposure category is determined in each wind direction based on the upwind surface roughness, and wind loads in those directions are determined using the corresponding exposure categories.

Table 5.5 Exposure Requirements

Wind Load Procedure	Chapter	Requirements
Directional	27	• MWFRS of enclosed and partially enclosed buildings Use an exposure category determined in accordance with 26.7.3 in each wind direction • Open buildings with monoslope, pitched or troughed free roofs Use the exposure category determined in accordance with 26.7.3 from the eight sectors that results in the highest wind loads
Envelope	28	• MWFRS of all low-rise buildings designed using this procedure Use the exposure category determined in accordance with 26.7.3 from the eight sectors that results in the highest wind loads
Directional	29	• Building appurtenances and other structures Use an exposure category determined in accordance with 26.7.3 in each wind direction
C&C	30	• C&C Use the exposure category determined in accordance with 26.7.3 from the eight sectors that results in the highest wind loads

It is common practice to use only the critical exposure category obtained from all wind directions when determining the wind loads on the MWFRS of a building or other structure; in general, this reduces the calculations that need to be made and it typically yields results that are not unduly conservative.

5.2.5 Topographic Effects

Buildings or other structures that are sited on the upper half of an isolated hill, ridge or escarpment can experience significantly higher wind velocities than those sited on relatively level ground. The *topographic factor*, K_{zt}, in 26.8 accounts for this increase in wind speed, which is commonly referred to as wind speed-up.

A two-dimensional ridge or escarpment or a three-dimensional axisymmetrical hill is described by the parameters H and L_h:

- H is the height of the topographic feature or the difference in elevation between the crest and the windward terrain.

- L_h is the distance upwind of the crest to where the ground elevation is equal to one-half of the height of the topographic feature (see Figure 26.8-1).

Not every hill, ridge or escarpment requires an increase in wind velocity. Wind speed-up must be considered only when all of the five conditions in 26.8.1 are satisfied:

1. The hill, ridge or escarpment is isolated and is unobstructed upwind by similar topographic features of comparable height for a distance equal to the lesser of $100H$ or 2 miles.

2. The hill, ridge or escarpment protrudes above the height of all upwind terrain features within a 2-mile radius in any quadrant by a factor of two or more.

3. The structure is located (a) in the upper one-half of a hill or ridge or (b) near the crest of an escarpment (see Figure 26.8-1).

4. The ratio $H/L_n \geq 0.2$.

5. For Exposure B, $H \geq 60$ feet, and for Exposures C and D, $H \geq 15$ feet.

When all of these conditions are met, K_{zt} is determined by Equation 26.8-1:

$$K_{zt} = (1 + K_1 K_2 K_3)^2 \tag{5.5}$$

Values of K_1, K_2 and K_3 are obtained from Figure 26.8-1. These multipliers are based on the assumption that wind approaches the topographic feature along the direction of maximum slope, which causes the greatest increase in velocity at the crest. The multiplier K_1 accounts for the shape of the topographic feature and the maximum speed-up effect. The reduction in wind speed-up with respect to horizontal distance is accounted for in the multiplier K_2: values of K_2 decrease as the distance from the topographic feature increases (that is, as x increases). Finally, the multiplier K_3 accounts for the reduction of wind speed-up with respect to height above the local terrain. It is evident from Figure 26.8-1 that K_3 decreases as the height above the windward terrain, z, increases. When one or more of these conditions are not met, $K_{zt} = 1.0$ (26.8.2).

It is important to note that the provisions in 26.8 are not meant to address the general case of wind flowing over hilly or complex terrain. In such cases, a wind tunnel study should be performed. Also, these provisions do not include vertical wind speed-up, which is known to exist in such cases; vertical effects determined by rational methods should be included where appropriate.

5.2.6 Gust Effects

Gust-effect Factor

As was discussed in Section 5.1.2 of this publication, the effects of wind gusts must be included in the design of any building or other structure. The *gust-effect factor* defined in 26.9 accounts for both atmospheric and aerodynamic effects in the along-wind direction.

The gust-effect factor depends on the natural frequency, n_1, of the structure. In particular, the method in which the gust-effect is determined is contingent on whether the structure is rigid or flexible. By definition, a *rigid building or other structure* is one where $n_1 \geq 1$ Hz, and a *flexible building or other structure* is one where $n_1 < 1$ Hz (26.2). Note that low-rise buildings that satisfy the definition in 26.2 (that is, buildings with a mean roof height $h \leq 60$ feet and $h \leq$ least horizontal dimension of building) are permitted to be considered rigid (26.9.2).

For rigid structures, the gust-effect factor, G, is permitted to be taken as 0.85 (26.9.1) or may be calculated using the provisions of 26.9.4. The gust-effect factor for flexible structures, G_f, accounts for along-wind loading effects due to dynamic amplification; provisions to determine G_f are given in 26.9.5. The fundamental dynamic equations for maximum along-wind displacement, root-mean-square along-wind acceleration and maximum along-wind acceleration, which form the basis of the methodology for determining the gust-effect factor, can be found in C26.9. Also provided in this commentary section is a discussion on structural damping; recommendations on damping ratios, which are needed to determine G_f, are also included.

ASCE/SEI 26.9.6 permits the use of any rational analysis that is recognized in the literature to determine gust-effect factors and natural frequencies.

Flowchart 5.1 in Section 5.7.2 of this publication contains step-by-step procedures on how to determine the gust-effect factor for both rigid and flexible structures.

Approximate Natural Frequency

Many tools are available to determine the fundamental frequency, n_1, of a structure. Most computer programs that are used to analyze structures can provide an estimate of n_1 based on member sizes and material properties that are used in the model. In the preliminary design stages, this information may not be known. Thus, 26.9.3 provides equations to determine an approximate natural frequency, n_a, for buildings that meet the height and slenderness conditions of 26.9.2.1:

1. Building height must be less than or equal to 300 feet.

2. Building height must be less than four times its effective length, L_{eff}, which is determined by Equation 26.9-1:

$$L_{eff} = \frac{\sum_{i=1}^{n} h_i L_i}{\sum_{i=1}^{n} h_i} \tag{5.6}$$

In this equation, h_i is the height above grade of level i of the building, and L_i is the length of the building at level i parallel to the direction of wind. The summations are taken over the height of the building that has n levels. This equation is essentially a height-weighted average of the along-wind length of the building that accounts for slenderness in buildings with setbacks. The two limitations of 26.9.2.1 are needed because the equations for n_a that are provided in this section are based on regular buildings.

Table 5.6 contains the equations given in 26.9.3 for the determination of n_a for buildings and other structures that satisfy the conditions of 26.9.2.1.

Table 5.6 Approximate Natural Frequency, n_a

MWFRS	Eq. No.	Equation
Structural steel moment-resisting frame	26.9-2	$n_a = 22.2 / h^{0.8}$
Concrete moment-resisting frame	26.9-3	$n_a = 43.5 / h^{0.9}$
Steel and concrete buildings with lateral-force-resisting systems other than moment-resisting frames	26.9-4	$n_a = 75 / h$
Concrete or masonry shear walls	26.9-5	$n_a = 385 (C_w)^{0.5} / h$ where $C_w = \dfrac{100}{A_B} \sum_{i=1}^{n} \left(\dfrac{h}{h_i}\right)^2 \dfrac{A_i}{\left[1 + 0.83\left(\dfrac{h_i}{D_i}\right)^2\right]}$

In these equations, h is the mean roof height of the structure. According to 26.2, *mean roof height* is the average of the height at the roof eave and the height to the highest point on the roof surface; for roof angles less than or equal to 10 degrees, the mean roof height may be taken as the height at the roof eave. The variables in Equation 26.9-5 for concrete or masonry shear walls are defined in 26.9.3 and depend on the number of shear walls in the direction of analysis (n), the base area of the structure (A_B), the horizontal cross-sectional area of shear walls (A_i), the length of shear walls (D_i) and the height of shear walls (h_i). Equations 26.9-2 through 26.9-4 are plotted in Figure C26.9-1.

These lower-bound expressions for n_a are more appropriate than the empirical equations for natural frequency given in the seismic provisions of ASCE/SEI 7 because the latter provide higher estimates of natural frequency, which are conservative for determination of seismic loads but are generally unconservative for determination of wind loads. Equations for natural frequency based on other studies that are appropriate for wind design are provided in C26.9.

5.2.7 Enclosure Classification

The different types of enclosure classifications and their relationship to internal pressures were discussed under aerodynamic effects in Section 5.1.1 of this publication. The following discussion covers definitions for each type of classification and the requirements for protecting glazed openings in wind-borne debris regions.

Any building or other structure must be classified as enclosed, partially enclosed or open based on the definitions in 26.2. A summary of these definitions is given in Table 5.7.

Table 5.7 Enclosure Classifications

Classification	Definition
Open building	For each wall in the building, $A_o \geq 0.8 A_g$
Partially enclosed building	A building that complies with all of the following conditions: • $A_o > 1.1 A_{oi}$ • $A_o >$ lesser of 4 square feet or $0.01 A_g$ • $A_{oi} / A_{gi} \leq 0.2$
Enclosed building	A building that does not comply with the requirements for open or partially enclosed buildings

The quantities in Table 5.7 are as follows (see Figure 5.9):

A_o = total area of openings in a wall that receives positive external pressure

A_g = gross area of wall in which A_o is identified

A_{oi} = sum of the areas of openings in the building envelope (walls and roof) not including A_o

A_{gi} = sum of the gross surface areas of the building envelope (walls and roof) not including A_g

Figure 5.9
Definition of Wall Openings for Determination of Enclosure Classification

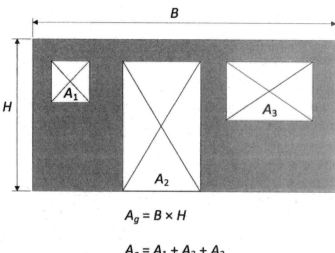

$A_g = B \times H$

$A_o = A_1 + A_2 + A_3$

Requirements for buildings that comply with more than one classification are given in 26.10.4. For a building that meets both open and partially enclosed definitions, the building is to be classified as open. Also, as noted in Table 5.7, a building must be classified as enclosed when the definitions of open or partially enclosed are not met.

Judgment must be used when determining the enclosure classification where large doors are incorporated in the building envelope (for example, overhead doors or hangar doors). If the doors are designed for the required design wind pressures, it is appropriate for the building to be classified as enclosed because it is unlikely that the doors will fail during a design windstorm. It is not always obvious what the exposure classification should be for buildings where the doors are not designed for the required wind pressure; in such cases, it may be appropriate to classify the building as partially enclosed after considering the size and number of openings.

It is common practice in regions that are not prone to hurricanes to assume that windows and other glazed units are not openings provided that they have been designed for the appropriate C&C wind pressures at a particular site.

The situation is significantly different in regions where hurricanes can occur. *Hurricane-prone regions* are located along the Atlantic Ocean and Gulf of Mexico where the basic wind speed for Risk Category II buildings is greater than 115 mph. Hawaii, Puerto Rico, Guam, the Virgin Islands and American Samoa are also classified as hurricane-prone regions. *Wind-borne debris regions* are in hurricane-prone regions and are located as follows (IBC 202 and ASCE/SEI 26.10.3.1):

1. Within 1 mile of the coastal mean high water line where the basic wind speed (ultimate design wind speed—IBC) is greater than or equal to 130 mph; or

2. In areas where the basic wind speed (ultimate design wind speed—IBC) is greater than or equal to 140 mph.

Actual locations of wind-borne debris regions are to be based on the wind speeds that are in the IBC and ASCE/SEI 7 figures, which are summarized in Table 5.8 (IBC 202 and ASCE/SEI 26.10.3.1).

Table 5.8 Wind-borne Debris Region Wind Speed Figures

Classification	Figure No.
• Risk Category II buildings • Risk Category III buildings, except health care facilities	Figure 1609A Figure 26.5-1A
• Risk Category III health care facilities • Risk Category IV buildings	Figure 1609B Figure 26.5-1B

Special requirements are given in IBC 1609.1.2 and ASCE/SEI 26.10.3.2 for the protection of glazed openings in wind-borne debris regions. These requirements are based primarily on observations during actual hurricanes: debris picked up by the wind can cause significant window breakage, which can lead to a substantial increase in internal pressure. With a few exceptions, the provisions require that all glazing must be protected with an impact-protective system, such as shutters, or the glazing itself must be impact-resistive for buildings in wind-borne debris regions.

Impact-protective systems and impact-resistant glazing must meet the requirements of an approved impact-resistant standard or ASTM E 1996, *Standard Specification for Performance of Exterior Windows, Curtain Walls, Doors, and Impact Protective Systems Impacted by Windborne Debris in Hurricanes* and ASTM E 1886 *Standard Test Method for Performance of Exterior Windows, Curtain Walls, Doors, and Impact Protective Systems Impacted by Missile(s) and Exposed to Cyclic Pressure Differentials*. Glazed openings that are located in the lower 30 feet of the building must meet the requirements of the large missile tests in ASTM E 1996, and those located more than 30 feet above grade must meet the requirements of the small missile tests.

Since ASTM E 1996 is based on the wind speed contours in ASCE/SEI 7-05, the wind zone designations in Section 6.2.2 of that document must be modified as outlined in IBC 1609.1.2.2 and ASCE/SEI C26.10 in order to provide conformance with the latest wind provisions.

Exceptions to the requirements in wind-borne debris regions are given in IBC 1609.1.2 and ASCE/SEI 26.10.3.1. Both the IBC and ASCE/SEI 7 allow glazing located over 60 feet above grade and over 30 feet above aggregate surface roofs (including roofs with gravel or stone ballast) that are located within 1,500 feet of the building to be unprotected. The IBC also contains an exception for Risk Category I buildings and provides specific requirements for opening protection that can be used in one- and two-story residential buildings. Requirements for louvers that are protecting intake and exhaust ventilation ducts are also given in the IBC (IBC 1609.1.2.1).

5.2.8 Internal Pressure Coefficients

Internal pressure coefficients (GC_{pi}) are given in Table 26.11-1 and are based on the enclosure classifications defined in 26.10. These coefficients have been obtained from wind tunnel tests and full-scale data, and they are assumed to be valid for a building of any height even though the wind tunnel tests were conducted primarily for low-rise buildings. Gust and aerodynamic effects are combined into one factor (GC_{pi}); in accordance with 26.9.7, the gust-effect factor shall not be determined separately in the analysis.

It is evident from Table 26.11-1 that the internal pressure coefficient for partially enclosed buildings is approximately three times that for enclosed buildings. Thus, it is imperative that the correct enclosure classification be established in order to capture the proper influence of internal pressure on the design of the appropriate members in the building.

For partially enclosed buildings that contain a single, relatively large volume without any partitions, the reduction factor, R_i, calculated by Equation 26.11-1 may be used to reduce the applicable internal pressure coefficient. This reduction factor is based on research that has shown that the response time of internal pressure increases as the volume of a building without partitions increases; as such, the gust factor associated with the internal pressure is reduced, resulting in lower internal pressure.

5.3 Main Windforce-resisting Systems (MWFRSs)

5.3.1 Overview

Chapters 27, 28, 29 and 31 in ASCE/SEI 7-10 contain design requirements for determining wind pressures and loads on MWFRSs of buildings and other structures (see Table 5.1). The provisions in Chapters 27 through 29 are discussed in the following sections. Chapter 31, which contains the requirements for wind tunnel procedures, is covered in Section 5.5 of this publication.

5.3.2 Directional Procedure for Buildings (Chapter 27)

Scope

The Directional Procedure of Chapter 27 applies to the determination of wind loads on the MWFRS of enclosed, partially enclosed and open buildings of all heights that meet the conditions and limitations given in 27.1.2 and 27.1.3, respectively. This procedure is the former "buildings of all heights" provision in Method 2 of ASCE/SEI 7-05 for MWFRSs.

A summary of the wind load procedures and their applicability for MWFRSs in accordance with Chapter 27 are given in Table 5.9.

Part 1 in this chapter is applicable to enclosed, partially enclosed and open buildings of all heights; a wide range of buildings is covered by the provisions in this part. In general, wind pressures are determined as a function of wind direction using equations that are appropriate for each surface of the building. A simplified method for a special class of buildings up to 160 feet in height is provided in Part 2, which is based on the provisions in Part 1.

In order to apply these provisions, buildings must be regularly-shaped (that is, must have no unusual geometrical irregularities in spatial form) and must exhibit essentially along-wind response characteristics. Buildings of unusual shape that do not meet these conditions must be designed by either recognized literature that documents such wind load effects or by the wind tunnel procedure in Chapter 31 (27.1.2).

Reduction in wind pressures due to apparent shielding by surrounding buildings, other structures or terrain is not permitted (27.1.4). Removal of such features around a building at a later date could result in wind pressures that are much higher than originally accounted for; as such, wind pressures must be calculated assuming that all shielding effects are not present.

Table 5.9 Summary of Wind Load Procedures in Chapter 27 of ASCE/SEI 7-10 for MWFRSs

ASCE/SEI Chapter	Part	Applicability		Conditions
		Building Type	Height Limit	
27	1	Enclosed	None	• Regular-shaped building • Building does not have response characteristics making it subject to across-wind loading, vortex shedding, instability due to galloping or flutter • Building is not located at a site where channeling effects or buffeting in the wake of upwind obstructions warrant special consideration
		Partially enclosed		
		Open		

	2	Enclosed, simple diaphragm	$h \leq 160$ feet	• Same conditions as in Part 1
				• Building must meet the conditions for either a Class 1 or Class 2 building:
				Class 1 Building:
				  1. $h \leq 60$ feet
				  2. $0.2 \leq L/B \leq 5.0$
				  3. K_{zt} is calculated in accordance with 26.8
				Class 2 Building:
				  1. 60 feet $< h \leq 160$ feet
				  2. $0.5 \leq L/B \leq 2.0$
				  3. $n_1 \geq 75/h$
				  4. K_{zt} is calculated in accordance with 26.8
				• Building having either a rigid or flexible diaphragm

Minimum design wind pressures and loads are given in 27.1.5, which are applicable to buildings designed using Part 1 or Part 2. In the case of enclosed or partially enclosed buildings, wind pressures of 16 psf and 8 psf must be applied simultaneously to the vertical plane normal to the assumed wind direction over the wall and roof area of the building, respectively. Application of these minimum wind pressures are illustrated in Figure 5.10 for wind along the two primary axes of the building. For open buildings, the minimum wind force is equal to 16 psf multiplied by the area of the open building either normal to the wind direction or projected on a plane normal to the wind direction. It is important to note that minimum design pressures or loads are load cases that must be considered separate from any other load cases that are specified in Part 1 or 2.

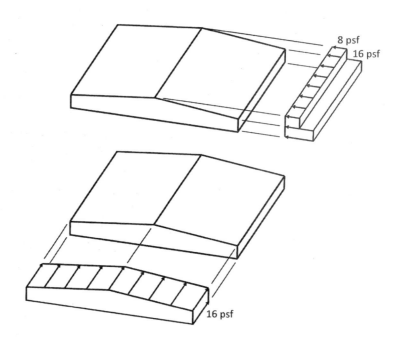

Figure 5.10
Application of Minimum Design Wind Pressures in Accordance with 27.1.5

Part 1 - Enclosed, Partially Enclosed and Open Buildings of All Heights
Overview

Part 1 of Chapter 27 is applicable to buildings with any general plan shape, height or roof geometry that matches the figures provided in this chapter. This procedure entails the determination of velocity pressures (which are determined as a function of exposure, height, topographic effects, wind directionality, wind velocity and building occupancy), gust-effect factors, external pressure coefficients and internal pressure coefficients for each surface of a rigid, flexible or open building. Table 27.2-1 contains the overall steps that can be used to determine wind pressures on such buildings.

Velocity Pressure

Velocity pressure is related to the atmospheric effects due to wind, which are described in Section 5.1.1 of this publication. The velocity pressure, q_z, at height z above the ground surface is determined by Equation 27.3-1; this is essentially Bernoulli's equation, and it converts the basic wind speed, V, to a velocity pressure:

$$q_z = 0.00256 K_z K_{zt} K_d V^2 \tag{5.7}$$

The variables in Equation 5.7 are discussed below. Note that at the mean roof height of the building, the velocity pressure is denoted q_h, and the velocity pressure coefficient is denoted K_h, that is, the subscript changes from z to h.

Air density. The constant 0.00256 in Equation 5.7 is related to the mass density of air for the standard atmosphere (59°F and sea level pressure of 29.92 inches of mercury), and it is obtained as follows (constant = one-half times the density of air times the velocity squared where the velocity is in miles per hour and the pressure is in pounds per square foot):

$$\text{Constant} = 0.5 \left[\frac{0.0765 \frac{\text{lb}}{\text{ft}^3}}{32.2 \frac{\text{ft}}{\text{sec}^2}} \right] \times \left[\left(1 \frac{\text{mi}}{\text{hr}}\right) \times 5,280 \frac{\text{ft}}{\text{mi}} \times \frac{1 \text{ hr}}{3,600 \text{ sec}} \right]^2 = 0.00256 \tag{5.8}$$

The numerical constant of 0.00256 should be used except where sufficient weather data is available to justify a different value. In general, the density of air varies with respect to altitude, latitude, temperature, weather and season. Minimum, average and maximum ambient air densities at various altitudes can be found in ASCE/SEI Table C27.3-2.

Velocity pressure exposure coefficient, K_z. This coefficient modifies wind velocity (or pressure) with respect to exposure and height above ground. Values of K_z for Exposures B, C and D at various heights above ground level are given in Table 27.3-1. In lieu of linear interpolation and for heights greater than 500 feet above the surface, K_z may be calculated at any height, z, using the equations at the bottom of that table:

$$K_z = \begin{cases} 2.01 \left(\frac{15}{z_g}\right)^{\frac{2}{\alpha}} & \text{for } z < 15 \text{ ft} \\ 2.01 \left(\frac{z}{z_g}\right)^{\frac{2}{\alpha}} & \text{for } 15 \text{ ft} \leq z \leq z_g \end{cases} \tag{5.9}$$

The constant α is the 3-second gust speed power law exponent, which defines the approximately parabolic shape of the wind speed profile for each exposure (see Figure 5.2). The nominal height of the atmospheric boundary layer, which is also referred to as the gradient height, is denoted as z_g. As was discussed in Section 5.1.1 of this publication, wind speeds become constant above the gradient heights. Thus, regardless of the exposure, $K_z = 2.01$ when $z = z_g$. The pressure for each exposure is identical when it is evaluated at the top of the boundary layer, and, with everything else being constant, it is equal to 2.01 times the stagnation pressure of $0.00256V^2$. Values of α and z_g are given in Table 26.9-1 as a function of exposure.

The above discussion on the determination of K_z is valid for the case of a single roughness category (that is, uniform terrain). Procedures on how to determine K_z for a single roughness change or multiple roughness changes are given in C27.3.1.

Topographic factor, K_{zt}. This factor modifies the velocity pressure exposure coefficients for buildings located on the upper half of an isolated hill or escarpment. See Section 5.2.5 of this publication for information on how to determine K_{zt}.

Wind directionality factor, K_d. This factor accounts for the statistical nature of wind flow and the probability of the maximum effects occurring at any particular time for any given wind direction. See Table 26.6-1 and Section 5.2.3 of this publication for more information on how to determine K_d.

Basic wind speed, V. As discussed in Section 5.2.2 of this publication, V is the 3-second gust speed at 33 feet above the ground in Exposure C (see IBC 1609.3 or ASCE/SEI 26.5.1).

Flowchart 5.2 in Section 5.7.3 of this publication can be used to determine q_z and q_h.

Design Wind Pressures

Enclosed and partially enclosed rigid buildings. Design wind pressures, p, are calculated by Equation 27.4-1 for the MWFRS of enclosed and partially enclosed rigid buildings of all heights:

$$p = qGC_p - q_i(GC_{pi}) \tag{5.10}$$

This equation is used to calculate the wind pressures on each surface of the building: windward wall, leeward wall, side walls and roof. The pressures are applied simultaneously on the walls and roof, as depicted in Figure 27.4-1, 27.4-2 and 27.4-3 (see also Figure 5.6). The first part of the equation is the external pressure contribution and the second part is the internal pressure contribution. External pressure varies with height above ground on the windward wall and is a constant on all of the other surfaces based on the mean roof height. The quantities in this equation are discussed below.

The gust-effect factor, G, for rigid buildings may be taken equal to 0.85 or may be calculated by Equation 26.9-6 (see Section 5.2.6 of this publication for more information on G).

External pressure coefficients, C_p, capture the aerodynamic effects that are discussed in Section 5.1.1 of this publication, and have been determined experimentally through wind tunnel tests on buildings of various shapes and sizes. These coefficients reflect the actual wind loading on each surface of a building as a function of wind direction.

Figure 27.4-1 contains C_p values for windward walls, leeward walls, side walls, and roofs for buildings with gable and hip roofs, monoslope roofs and mansard roofs. Wall pressure coefficients are constant on windward and side walls, and they vary with the plan dimensions of the building (that is, vary with the aspect ratio of the building L/B) on the leeward wall. The table in the upper part of Figure 27.4-1 also designates which velocity pressure to use—q_z or q_h—on a particular wall surface. Roof pressure coefficients vary with the ratio of the mean roof height to the plan dimension of the building (h/L) and with the roof angle (θ) for a given wind direction (normal to ridge or parallel to ridge). All of these pressure coefficients are intended to be used with q_h, and the parallel-to-ridge wind direction is applicable for flat roofs. It is evident from the figure that negative roof pressures increase as the ratio h/L increases. Also, as θ increases, negative pressure decreases until a roof angle is reached where the pressure becomes positive; this is consistent with the aerodynamic effect of the separation zone that was described in Section 5.1.1 of this publication (see Figure 5.3). Where two values of C_p are listed in the figure, the windward roof is subjected to either positive or negative pressure and the structure must be designed for both. Other important information on the use of this figure is given in the notes below the tabulated pressure coefficients.

The external pressure coefficients in Figure 27.4-2 for dome roofs are adapted from the 1995 edition of the Eurocode (Reference 5.6) and are based on data obtained from a modeled atmospheric boundary layer flow that does not fully comply with the wind tunnel testing requirements given in Chapter 31. Two load cases must be considered. In Case A, pressure coefficients are determined between various locations on the dome by linear interpolation along arcs of the dome parallel to the direction of wind; this defines maximum uplift on the dome in many cases. In Case B, the pressure coefficient is assumed to be a constant value at a specific point on the dome for angles less than or equal to 25 degrees, and it is determined by linear interpolation from 25 degrees to other points on the dome; this properly defines positive pressures for some cases, which results in maximum base shear. Wind tunnel tests are recommended for domes that are larger than 200 feet in diameter and in cases where resonant response can be an issue (C27.4.1).

The pressure and force coefficients in Figure 27.4-3 for arched roofs are the same as those that were first introduced in 1972 (Reference 5.7). These coefficients were obtained from wind tunnel tests conducted under uniform flow and low turbulence. References 5.8 and 5.9 can be consulted for pressure coefficients that are not specified in this figure.

The velocity pressure for internal pressure determination, q_i, is used in capturing the effects caused by internal pressure. On all of the surfaces of enclosed buildings and for negative internal pressure evaluation in partially enclosed buildings, q_i is to be taken as the velocity pressure evaluated at the mean roof height, q_h. For positive internal pressure evaluation, 27.4.1 permits q_i to be set equal to q_z in partially enclosed buildings where q_z is the velocity pressure evaluated at the location of the highest opening in the building that could affect positive internal pressure. Note that it is conservative to set q_i equal to q_h in all cases where positive internal pressure is evaluated. In the case of low-rise buildings, the distance between the uppermost opening and the mean roof height is usually relatively small, and this approximation yields reasonable results. However, this approximation can be overly conservative in certain cases, especially for taller buildings where the distance between the uppermost opening and the mean roof height is relatively large. For buildings located in wind-borne debris regions with glazing that does not meet the protection requirements of 26.10.3.2, q_i is to be determined assuming that the glazing will be breached.

The velocity pressure, q_i, is multiplied by the internal pressure coefficient (GC_{pi}), which is discussed in Section 5.2.8 of this publication. Both positive and negative values of (GC_{pi}) must be considered in order to establish the critical load effects.

It is evident from Equation 5.10 and Figure 5.5 that the effects from internal pressure cancel out when evaluating the total horizontal wind pressure on the MWFRS of a building. Thus, the total horizontal pressure at any height, z, above ground in the direction of wind is equal to the external pressure, p_z, on the windward face at height z plus the external pressure, p_h, on the leeward face.

Enclosed and partially enclosed flexible buildings. Design wind pressures, p, are calculated by Equation 27.4-2 for the MWFRS of enclosed and partially enclosed flexible buildings of all heights:

$$p = qG_fC_p - q_i(GC_{pi}) \tag{5.11}$$

External pressures on the surfaces of flexible buildings are determined basically in the same way as for rigid buildings; the only difference is that the gust-effect factor for flexible buildings, G_f, is used instead of G. All of the other quantities in Equation 5.11 are determined in accordance with 27.4.1 as shown previously for rigid buildings. See 26.9.5 and Section 5.2.6 of this publication for more information on how to determine G_f.

Open buildings with monoslope, pitched or troughed roofs. Design wind pressures, p, are calculated by Equation 27.4-3 for the MWFRS of open buildings with monoslope, pitched or troughed roofs:

$$p = q_hGC_N \tag{5.12}$$

In this equation, q_h is the velocity pressure at the mean roof height determined by Equation 5.7, and G is the gust-effect factor determined in accordance with 26.9. Net pressure coefficients, C_N, are given in Figures 27.4-4 through 27.4-7, which are based on the results from wind tunnel studies. Two load cases are identified in the figures, Load Case A and Load Case B. Both load cases must be considered in order to obtain the maximum load effects for a particular roof slope and blockage configuration.

The magnitude of roof pressure in open buildings is highly dependent on the blockage configuration beneath the roof. Goods or materials stored under the roof can restrict air flow, which can introduce significant upward pressures on the bottom surface of the roof. The net pressure coefficients for clear wind flow are to be used in cases where blockage is less than or equal to 50 percent. Obstructed wind flow is applicable where the blockage is greater than 50 percent. In cases where the usage below the roof is not evident, both unobstructed and obstructed load cases should be investigated.

Roof overhangs. In the case of roof overhangs, the positive external pressure on the bottom surface of a windward roof overhang is determined using the external pressure coefficient for the windward wall ($C_p = 0.8$). This pressure is combined with the top surface pressures determined in accordance with Figure 27.4-1 (see Figure 5.11).

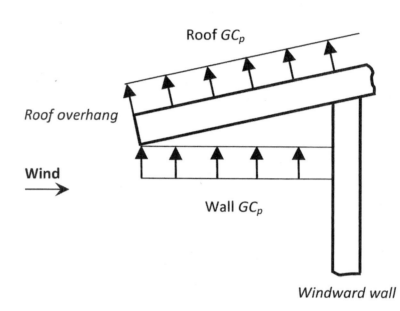

Figure 5.11
Application of Wind Pressures on a Roof Overhang, Part 1 of Chapter 27

Parapets. Design wind pressures, p_p, for the effects of parapets on the MWFRS of rigid or flexible buildings with flat, gable or hip roofs are calculated by Equation 27.4-4:

$$p_p = q_p (GC_{pn}) \tag{5.13}$$

In this equation, q_p is the velocity pressure evaluated at the top of the parapet using Equation 5.7 and the combined net pressure coefficient is (GC_{pn}), which is equal to +1.5 for a windward parapet and −1.0 for a leeward parapet. It is important to note that p_p is the combined net pressure due to the combination of the net pressures from the front and back surfaces of the parapet. The following discussion explains the methodology that was used to develop the combined net pressure coefficients.

Consider the parapets illustrated in Figure 5.12. In general, a windward parapet will experience positive wall pressure on the surface that is on the exterior side of the building (front surface) and negative roof pressure on the surface that is on the side of the roof (back surface). The behavior on the back surface is based on the assumption that the zone of negative pressure caused by the wind flow separation at the eave of the roof moves up to the top of the parapet; this results in the back side of the parapet having the same negative pressure as that of the roof.

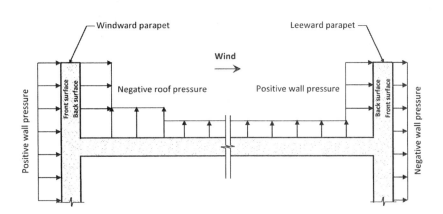

Figure 5.12
Application of Wind Pressures on Parapets, Part 1 of Chapter 27

A leeward parapet will experience a positive wall pressure on the roof side (back surface) and a negative wall pressure on the exterior side (front surface). It is assumed that the windward and leeward parapets are separated a sufficient distance so that shielding by the windward parapet does not decrease the positive wall pressure on the leeward parapet.

For simplicity, the pressures on the front and back of the parapet have been combined into one pressure, which is captured by the combined net pressure coefficients (GC_{pn}) for windward and leeward parapets. Since the wind can occur in any direction, a parapet must be designed for both sets of pressures. Note that the internal pressures inside the parapet cancel out in the determination of the combined pressure coefficient.

The pressures determined on the parapets are combined with the external pressures on the building to obtain the total wind pressures on the MWFRS.

Design wind load cases. Buildings subjected to the wind pressures determined by Chapter 27 must be designed for the load cases depicted in Figure 27.4-8, which are reproduced here in Figure 5.13 (in this figure, the subscripts x and y refer to the principal axes of the building, and w and l refer to the windward and leeward faces, respectively). In Load Case 1, design wind pressures are applied along the principal axes of a building separately.

Figure 5.13
Design Wind Load Cases in Part 1 of Chapter 27

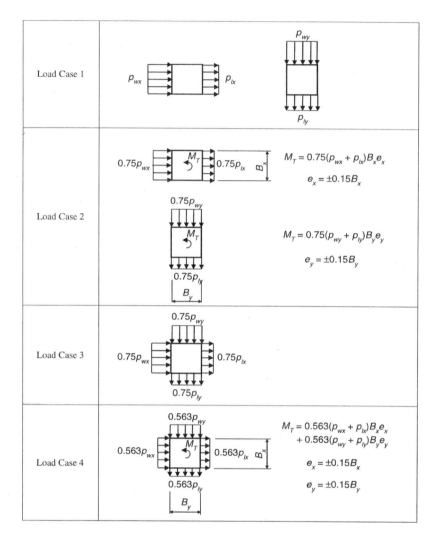

Load Cases 2 accounts for the effects of nonuniform pressure on different faces of the building due to wind flow; these pressure distributions have been documented in wind tunnel tests. Nonuniform pressures introduce torsion on the building, and this is accounted for in design by subjecting the building to 75 percent of the design wind pressures applied along the principal axis of the building plus a torsional moment, M_T, that is determined using an eccentricity equal to 15 percent of the appropriate plan dimension of the building. Torsional effects are determined in each principal direction separately.

A critical load case can occur when the design wind load acts diagonally to a building. This is accounted for in Load Case 3, where 75 percent of the maximum design wind pressures are applied along the principal axes of a building simultaneously.

Load Case 4 considers the effects due to diagonal wind loads and torsion. Seventy-five percent of the wind pressures in Load Case 2 are applied along the principal axes of a building simultaneously, and a torsional moment is applied, which is determined using 15 percent of the plan dimensions of the building.

The exception in 27.4.6 permits buildings that meet the requirements of D1.1 of Appendix D to be designed for Load Cases 1 and 3 only. In particular, the following buildings do not need to be designed for the effects from torsion:

- One-story buildings with a mean roof height of less than or equal to 30 feet;

- Buildings two stories or less framed with light-frame construction (that is, structural systems made up of repetitive wood or cold-formed steel framing members or subassemblies); and

- Buildings two stories or less with flexible diaphragms.

It has been demonstrated that buildings that meet one or more of these criteria are basically insensitive to torsional load effects.

In the case of flexible buildings, dynamic effects can increase the effects from torsion. Equation 27.4-5 accounts for these effects. The eccentricity, e, determined by this equation is to be used in the appropriate load cases in Figure 5.13 in lieu of the eccentricities e_x and e_y that are given in that figure for rigid structures. An eccentricity must be considered for each principal axis of the building, and the sign of the eccentricity must be plus or minus, whichever causes the more severe load case.

Flowchart 5.3 in Section 5.7.3 of this publication can be used to determine design wind pressures on the MWFRS of buildings in accordance with Part 1 of Chapter 27.

Part 2 - Enclosed, Simple Diaphragm Buildings with $h \leq 160$ feet
Overview

Part 2 of Chapter 27 is applicable to enclosed, simple diaphragm buildings with mean roof heights less than or equal to 160 feet that meet the additional conditions in 27.5.2 for either a Class 1 or Class 2 building. Based on the procedures in Part 1, it is meant to be a simplified method for determining wind pressures on such buildings.

According to 26.2, a *simple diaphragm building* is one in which both windward and leeward wind loads are transmitted by roof and vertically-spanning wall assemblies through continuous floor and roof diaphragms to the MWFRS. In other words, the wind loads are delivered to the elements of the MWFRS via roof and floor diaphragms. As such, internal pressures cancel out in the determination of the total wind load in the direction of analysis. Thus, in order for this approach to be valid, no structural expansion joints are permitted in the system: expansion joints interrupt the continuity of the diaphragm resulting in internal pressures that do not cancel out. Also, no girts or other horizontal members that transfer significant wind loads directly to the vertical members of the MWFRS should be present.

The conditions that define Class 1 and Class 2 buildings are summarized in Table 5.9 and are illustrated in Figure 27.5-1. Since Class 1 buildings are limited to a mean roof height that is less than or equal to 60 feet, it is assumed that these buildings are rigid and a gust-effect factor of 0.85 has been used to calculate the tabulated wind pressures in this part (see 26.9.1 and 26.9.2). For Class 2 buildings, the requirement that the natural frequency must be greater than or equal to $75/h$ (where h is the mean roof height in feet) is needed to ensure that the gust-effect factor, which has been calculated and built into the design procedure, is consistent with the tabulated wind pressures. This frequency ($75/h$) is meant to represent a reasonable lower bound to values of frequencies found in practice.

The design procedures in this part apply to buildings with either rigid or flexible diaphragms. For consideration of wind loading, diaphragms constructed of untopped metal deck, concrete-filled metal deck and concrete slabs can be idealized as rigid (27.5.4). Diaphragms constructed of wood panels can be considered flexible.

Table 27.5-1 contains the overall steps that can be used to determine wind pressures on Class 1 and Class 2 buildings.

Design Wind Pressures

Wall and roof surfaces. Net design wind pressures for the walls and roof surfaces of Class 1 and 2 buildings can be determined directly from Table 27.6-1 and 27.6-2, respectively. These pressures have been calculated using the procedures in Part 1, including the external pressure coefficients in Figure 27.4-1. Note that for Class 1 buildings with aspect ratio L/B values less than 0.5, the tabulated wind pressures for $L/B = 0.5$ are to be used. Similarly, for Class 1 buildings with L/B values greater then 2.0, the tabulated wind pressures for $L/B = 2.0$ are to be used.

Net wall pressures are tabulated for Exposures B, C and D as a function of wind velocity, V, mean roof height, h, and building aspect ratio, L/B. The top pressure in the table is defined as p_h, and the bottom pressure is defined as p_o. Interpolation between these values is permitted (see Note 5 in Table 27.6-1).

Along-wind net wind pressures are distributed over the height of the building as shown in Table 27.6-1 and Figure 27.6-1, which is reproduced here in Figure 5.14. The net pressures are applied normal to the projected area of the building walls in the direction of wind. This linear wall pressure distribution produces the same overturning moment as the one that is obtained using the pressures determined from Part 1. As noted previously, internal pressures cancel out when considering the net effect on the MWFRS of a simple diaphragm building. What is important to keep in mind is that the effects from internal pressure must be included in cases where the net wind pressure on any individual wall element is required.

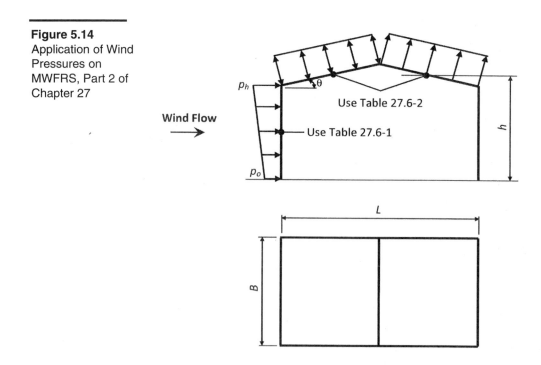

Figure 5.14
Application of Wind Pressures on MWFRS, Part 2 of Chapter 27

The pressure on the side walls is determined using Note 2 in Table 27.6-1. In particular, the side wall pressures are calculated as a percentage of the along-wind pressure, p_h, at the top of the building based on the aspect ratio, L/B. These pressures are applied to the projected area of the building walls normal to the wind direction. The percentages of p_h given in this note are obtained using the external pressure coefficients, C_p, in Part 1. In particular, C_p is equal to 0.8 on the windward wall for all aspect ratios, L/B; –0.5 on the leeward wall for values of $L/B = 0.5$ to 1.0; –0.3 on the leeward wall for $L/B = 2.0$; and –0.7 for side walls. Since the side wall pressures are constant over the height of the building, the side wall pressure can be calculated as a fraction of p_h as follows: $0.7/(0.8 + 0.5) = 0.54$ for $L/B = 0.5$ to 1.0. Similarly, the fraction is $0.7/(0.8 + 0.3) = 0.64$ for $L/B = 2.0$. These match the percentages given in Note 2.

A method of determining the distribution of the tabulated net wall pressures between the windward and leeward wall surfaces of a building is provided in Note 4 of Table 27.6-1. Having such a distribution can be useful when designing floor and roof diaphragm elements, such as collectors as well as shear walls that are a part of the MWFRS. Like in the case of side walls, the percentages of p_h given in this note are obtained using the external pressure coefficients, C_p, in Part 1. As noted above, C_p is equal to 0.8 on the windward wall for all aspect ratios, L/B; –0.5 on the leeward wall for values of $L/B = 0.5$ to 1.0; and –0.3 on the leeward wall for $L/B = 2.0$. Since the leeward wall pressure is constant over the entire height of the building, the leeward wall pressure can be calculated as a fraction of p_h as follows: $0.5/(0.8 + 0.5) = 0.38$ for $L/B = 0.5$ to 1.0. Similarly, the fraction is equal to $0.3/(0.8 + 0.3) = 0.27$ for $L/B = 2.0$. These match the percentages given in Note 4.

Tabulated roof pressures are given in Table 27.6-2 for Exposure C as a function of V, h and roof slope. Exposure adjustment factors are provided in the table for Exposures B and D; the tabulated roof pressures are to be multiplied by the appropriate adjustment factors. In the determination of the tabulated roof pressures, the external pressure coefficients, C_p, given in Part 1 have been multiplied by 0.85, which is deemed a reasonable gust-effect factor for common roof framing systems. These modified external pressure coefficients were then combined with the internal pressure coefficient of ±0.18 for enclosed buildings.

Roof pressures are applied perpendicularly to the roof surfaces as shown in Figure 27.6-1. The different zones over which these pressures are to be distributed are identified in Table 27.6-2 for flat, gable, hip, monoslope and mansard roofs.

Roof pressure is given for two load cases and both must be investigated where applicable. Load Case 2 is required when investigating maximum overturning effects on the building due to the wind pressures.

According to 27.6.1, pressures on the walls and the roof must be applied simultaneously to the building as shown in Figure 27.6-1. Also, the MWFRS must be designed for the load cases defined in Figure 27.4-8 (see Figure 5.13). Load Cases 2 and 4 (torsional load cases) need not be considered in the following cases:

- Class 1 buildings that meet the conditions of Cases A through F in Appendix D; and

- Class 2 buildings that meet the conditions of Cases A through E in Appendix D.

Torsional load cases need not be considered where the lines of resistance provided by the MWFRS are established in accordance with these cases, which are illustrated in Figures D1.5-1 and D1.5-2. Additional information can be found in the Commentary section of Appendix D.

Parapets. According to 27.6.2, the additional pressure on the MWFRS due to roof parapets is equal to 2.25 times the wall pressure tabulated in Table 27.6-1 using an aspect ratio of $L/B = 1.0$ and a height, h_p, equal to the distance from the ground to the top of the parapet. This net horizontal pressure accounts for both the windward and leeward parapet loading on both the windward and leeward building surfaces, and it is applied to the projected area of the parapet surface simultaneously with the net wall and roof pressures (see Figure 27.6-2 and Figure 5.15).

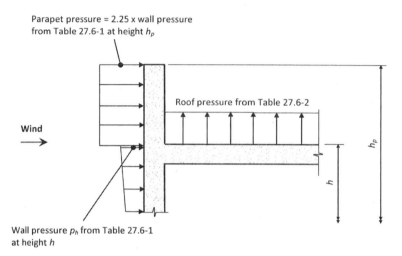

Figure 5.15
Application of Wind Pressure on Parapets, Part 2 of Chapter 27

It was shown in Part 1 above that the net pressure coefficient is +1.5 for the windward parapet and is −1.0 for the leeward parapet. A net combined pressure coefficient of +2.5 can be applied to the windward parapet to account for the cumulative effect on the MWFRS in a simple diaphragm building. The net pressure coefficient that is used in calculating the horizontal wall pressure, p_h, at the top of the building in Table 27.6-1 is equal to 1.3 times the gust-effect factor. Assuming a lower-bound gust-effect equal to 0.85, the ratio of the parapet pressure to the wall pressure is equal to $2.5/(1.3 \times 0.85) = 2.25$, which is the factor given in 27.6.2. This is assumed to be a reasonable factor to apply to the tabulated wall pressures at the top of the building to account for the additional wind pressure on the MWFRS from the parapets.

Roof overhangs. ASCE/SEI 27.6.3 contains provisions to account for the effects of vertical wind pressures on roof overhangs. According to this section, a positive wind pressure equal to 75 percent of the roof edge pressure from Table 27.6-2 for Zone 1 or Zone 3, whichever is applicable, must be applied to the underside of the windward overhang (see Figure 27.6-3, which is reproduced here in Figure 5.16).

Figure 5.16
Application of Wind Pressures on a Roof Overhang, Part 2 of Chapter 27

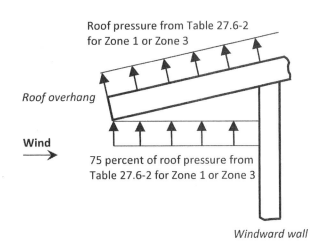

As discussed previously, 27.4.4 (Part 1) contains provisions for windward roof overhangs. A positive pressure coefficient of 0.8 is specified for the positive external pressure on the underside of the windward overhang. The pressure coefficient used to determine the roof pressures in Zone 3 of Table 27.6-2 is determined by multiplying the gust-effect factor (0.85) by the roof pressure coefficient in Figure 27.4-1 (−1.3, which is multiplied by the reduction factor of 0.8 that accounts for an effective wind area of 1,000 square feet or more) and then adding the internal pressure coefficient for enclosed buildings: $C_p = \{0.85 \times [0.8 \times (-1.3)]\} - (-0.18) = -1.06$. The ratio of pressure coefficients from Parts 1 and 2 is $-0.8/-1.06 = 0.755$, which is rounded to 0.75 in the provisions.

Flowchart 5.4 in Section 5.7.3 of this publication can be used to determine design wind pressures on the MWFRS of buildings in accordance with Part 2 of Chapter 27.

5.3.3 Envelope Procedure for Buildings (Chapter 28)

Scope

The Envelope Procedure of Chapter 28 applies to the determination of wind loads on the MWFRS of enclosed or partially enclosed low-rise buildings that meet the conditions and limitations given in 28.1.2 and 28.1.3, respectively. This procedure is the former "low-rise buildings" provision in Method 2 of ASCE/SEI 7-05 for MWFRSs. The simplified method in this chapter is based on the provisions of Method 1 of ASCE/SEI 7-05 for simple diaphragm buildings up to 60 feet in height.

A *low-rise building* is defined in 26.2 as an enclosed or partially enclosed building with a mean roof height less than or equal to 60 feet and a mean roof height that does not exceed the least horizontal dimension of the building.

A summary of the wind load procedures and their applicability for MWFRSs in accordance with Chapter 28 is given in Table 5.10.

Table 5.10 Summary of Wind Load Procedures in Chapter 28 of ASCE/SEI 7-10 for MWFRSs

ASCE/SEI Chapter	Part	Applicability		Conditions
		Building Type	Height Limit	
28	1	Enclosed, low-rise	$h \leq 60$ feet	• Regular-shaped building • Building does not have response characteristics making it subject to across-wind loading, vortex shedding, instability due to galloping or flutter • Building is not located at a site where channeling effects or buffeting in the wake of upwind obstructions warrant special consideration
		Partially enclosed, low-rise		
	2	Enclosed, simple diaphragm, low-rise	$h \leq$ least horizontal dimension	• Same conditions as in Part 1 • $n_1 \geq 1$ Hz • Building has an approximately symmetrical cross-section in each direction with either a flat roof or a gable or hip roof with $\theta \leq 45$ degrees • Building is exempted from torsional load cases as indicated in Note 5 of Figure 28.4-1, or the torsional load cases defined in Note 5 do not control the design of any of the MWFRSs of the building

Part 1 in this chapter is applicable to enclosed and partially enclosed low-rise buildings that have a flat, gable or hip roof. Like the method provided in Part 1 of Chapter 27, wind pressures are determined as a function of wind direction using equations that are appropriate for each surface of the building.

In order to apply these provisions, buildings must be regularly-shaped (that is, must have no unusual geometrical irregularities in spatial form) and must exhibit essentially along-wind response characteristics. Buildings of unusual shape that do not meet these conditions must be designed by either recognized literature that documents such wind load effects or by the wind tunnel procedure in Chapter 31 (28.1.2).

Reduction in wind pressures due to apparent shielding by surrounding buildings, other structures or terrain is not permitted (28.1.4). Removal of such features around a building at a later date could result in wind pressures that are much higher than originally accounted for; as such, wind pressures must be calculated assuming that all shielding effects are not present.

Part 1 - Enclosed and Partially Enclosed Low-rise Buildings
Overview

As noted above, Part 1 of this chapter is applicable to enclosed and partially enclosed low-rise buildings that have a flat, gable or hip roof. This procedure entails the determination of the velocity pressure at the mean roof height (which is determined as a function of exposure, height, topographic effects, wind directionality, wind velocity and building occupancy), combined gust-effect factors and external pressure coefficients and internal pressure coefficients. Table 28.2-1 contains the overall steps that can be used to determine wind pressures on such buildings.

Velocity Pressure

Velocity pressure, q_z, is determined by Equation 28.3-1, which is the same equation given in 27.3.2 under Part 1 of Chapter 27 (see Equation 27.3-1 and Equation 5.7 of this publication). The only difference between the two methods is that the velocity pressure coefficient, K_z, must be determined by Table 28.3-1 in Part 1 of Chapter 28 instead of by Table 27.3-1. Values of K_z in Table 28.3-1 for Exposures B, C and D are given up to a mean roof height of 60 feet. See the discussion under Part 1 of Section 5.3.2 of this publication for additional information on how to determine the velocity pressure.

Design Wind Pressures

MWFRS. Design wind pressures, p, are determined by Equation 28.4-1 for the MWFRS of low-rise buildings that satisfy the conditions and limitations of Part 1 of this chapter:

$$p = q_h[(GC_{pf}) - (GC_{pi})] \tag{5.14}$$

In this equation, q_h is the velocity pressure determined by Equation 28.3-1 evaluated at the mean roof height, h, (GC_{pf}) are external pressure coefficients and (GC_{pi}) are internal pressure coefficients, which are determined by Table 26.11-1. Wind pressures are determined on each of the building surfaces identified in Figure 28.4-1.

The external pressure coefficients (GC_{pf}) combine both a gust-effect factor and external pressure coefficients for low-rise buildings and are not allowed to be separated (28.4.1.1). Unlike the external pressure coefficients given in Figure 27.4-1 that reflect the actual pressure on each surface of a building as a function of wind direction, the coefficients in Figure 28.4-1 are essentially "pseudo" pressure conditions that, when applied to a building, envelop the desired structural actions independent of wind direction.

The "pseudo" values of (GC_{pf}) were determined from the output of wind tunnel tests, which measured bending moments, total horizontal forces, and total uplift as a building was rotated 360 degrees in the wind tunnel (see Reference 5.10 and Figure C28.4-1). Thus, values of (GC_{pf}) produce maximum measured structural actions and are not the actual surface pressures.

In order to capture all appropriate structural actions, a building must be designed for all wind directions by considering in turn each corner of the building as the windward (or, reference) corner; these conditions are illustrated in Figure 28.4-1. At each corner, two load cases must be considered (Load Case A and Load Case B), one for each range of wind direction. In general, a total of 16 separate load cases must be evaluated since both positive and negative internal pressure must be considered. For symmetrical buildings, some of these load cases will be repetitive and can be eliminated. Figure 5.17 illustrates Load Case A and Load Case B for the same windward corner of a low-rise building.

Figure 5.17 Basic Load Cases for Low-rise Buildings, Part 1 of Chapter 28

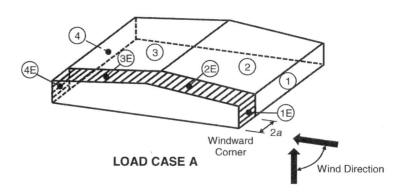

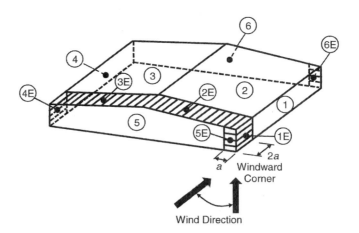

It is important to note that all of the original and subsequent research that was performed to develop and refine this methodology was done on low-rise buildings with gable roofs. A suggested method utilizing Figure 28.4-1 for low-rise buildings with hip roofs is given in Figure C28.4-2.

The torsional load cases given in Figure 28.4-1 must be considered in the design of all low-rise buildings except for the following (see Note 5 in this figure):

- One-story buildings with a mean roof height of less than or equal to 30 feet;

- Buildings two stories or less that are framed with light-frame construction (that is, structural systems made up of repetitive wood or cold-formed steel framing members or subassemblies); and

- Buildings two stories or less with flexible diaphragms.

It is evident from Equation 28.4-1 that the effects from internal pressure cancel out when evaluating the total horizontal wind pressure on the MWFRS of a low-rise building. It can be critical, however, in one-story buildings with moment-resisting frames and in the top story of buildings with moment-resisting frames.

Parapets. The design wind pressure for the effect of parapets on the MWFRS of low-rise buildings with flat, gable or hips roofs is determined by Equation 28.4-2, which is identical to Equation 27.4-4 in Part 1 of Chapter 27. See the discussion under Part 1 of Section 5.3.2 of this publication for information on how to determine the wind pressure on parapets.

Roof overhangs. Positive external pressures on the bottom surface of windward roof overhangs are to be determined using the pressure coefficient $C_p = 0.7$ in combination with the top surface pressures determined by Figure 28.4-1. Application of this pressure is similar to that shown in Figure 5.11.

Minimum design wind loads. ASCE/SEI 28.4.4 prescribes the minimum wind pressures in the design of a MWFRS for enclosed or partially enclosed low-rise buildings. The pressures of 16 psf on the projected area of the walls and 8 psf on the projected area of the roof are considered a separate load case from any of the other load cases specified in this part (see Figure 5.10).

Flowchart 5.5 in Section 5.7.4 of this publication can be used to determine design wind pressures on the MWFRS of buildings in accordance with Part 1 of Chapter 28.

Part 2 - Enclosed, Simple Diaphragm, Low-rise Buildings
Overview

Part 2 is applicable to enclosed, simple diaphragm, low-rise buildings with flat, gable or hip roofs that meet the conditions in 28.6.2. Although there are several conditions that need to be satisfied (see Table 5.10), a large number of typical low-rise buildings meet these criteria.

This method is based on Part 1 of Chapter 28 for simple diaphragm buildings. As was discussed in Part 2 of Section 5.3.2 of this publication, simple diaphragm buildings are structures where wind loads are delivered to the elements of the MWFRS via roof and floor diaphragms. As such, internal pressures cancel out in the determination of the total wind load in the direction of analysis. In order for this approach to be valid, no structural expansion joints are permitted in the system (expansion joints interrupt the continuity of the diaphragm resulting in internal pressures that do not cancel out) and no girts or other horizontal members should be present that transfer significant wind loads directly to the vertical members of the MWFRS.

According to condition number 8 in 28.6.2, only buildings that are exempted from torsional load cases as indicted in Note 5 of Figure 28.4-1 (see the discussion in Part 1 above) or those where the torsional load cases defined in this note do not control the design of any of the MWFRSs of the building are permitted to be designed by this simplified procedure. The torsional loading in Figure 28.4-1 was considered to be too complicated to include in a simplified method of determining wind pressures.

Design Wind Pressures

MWFRS. Simplified design wind pressures, p_{s30}, for walls and roofs located at various zones on a building are tabulated in Figure 28.6-1 as a function of the basic wind speed, V, and roof angle for buildings with a mean roof height of 30 feet that are located on primarily flat ground in Exposure B. Modifications are made to these tabulated pressures based on actual building height and exposure using the adjustment factor, λ, that is given in the figure. Such pressures must also be modified by the topographic factor, K_{zt}, where applicable.

Horizontal pressure, p_s, is determined by Equation 28.6-1:

$$p_s = \lambda K_{zt} p_{s30} \tag{5.15}$$

Horizontal wall pressures on Zones A and C are the net sum of the windward and leeward pressures on vertical projections of the wall (see Figure 28.6-1, which is reproduced here as Figure 5.18). Horizontal roof pressures on Zones B and D are the net sum of the windward and leeward pressures on the vertical projection of the roof, and the vertical roof pressures on Zones E, F, G and H are the net sum of the external and internal pressures (using an internal pressure coefficient of ±0.18 for enclosed buildings) on the horizontal projection of the roof. The pressure coefficients that were used to generate these pressures are from Figure 28.4-1.

Figure 5.18
Application of Wind Pressures for Low-rise Buildings, Part 2 of Chapter 28

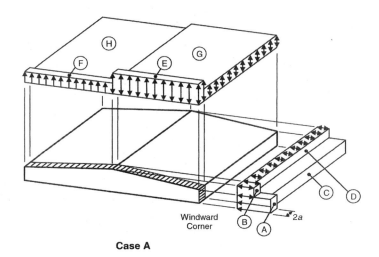

Case A

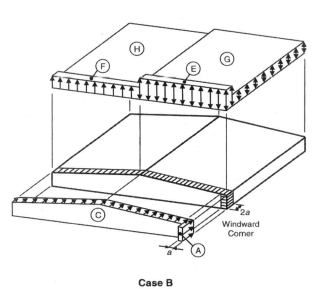

Case B

Note 7 in Figure 28.6-1 requires that the total horizontal load must not be decreased due to negative pressures on Zones B and D. When these pressures are found to be negative, use $p_s = 0$ in these zones. Other important information is contained in the notes of that figure.

Design wind pressures on windward and leeward walls can be obtained by multiplying p_s for Zones A and C by +0.85 and −0.70, respectively. For side walls, p_s for Zone C multiplied by −0.65 should be used.

Due to the enveloped nature of the loads obtained by this method for roof members, such members that are part of the MWFRS that span at least from the eave to the ridge or that support members spanning at least from the eave to the ridge need not be designed for the higher end zone loads depicted in Figure 5.18. The interior zone loads are applicable in such cases.

Minimum design wind loads. ASCE/SEI 28.6.4 prescribes the minimum wind pressures in the design of a MWFRS for buildings designed by this method. The minimum load cases is defined by assuming that the pressures, p_s, for Zones A and C are equal to +16 psf, for Zones B and D are equal to +8 psf and for Zones E, F, G and H are equal to zero.

Flowchart 5.6 in Section 5.7.4 of this publication can be used to determine design wind pressures on the MWFRS of buildings in accordance with Part 2 of Chapter 28.

5.3.4 Directional Procedure for Building Appurtenances and Other Structures (Chapter 29)

Scope

Chapter 29 is applicable to the determination of wind loads on the MWFRSs of building appurtenances (including rooftop structure and rooftop equipment) and other structures (including solid freestanding walls, freestanding solid signs, chimneys, tanks, open signs, lattice framework and trussed towers). The conditions of 29.1.2 and the limitations of 29.1.3 must be satisfied in order for these provisions to be applied.

Wind pressures are calculated using specific equations that are part of the Directional Procedure in Chapter 27 (see Section 5.3.2 of this publication). The methods presented in Chapter 29 were part of Method 2 in ASCE/SEI 7-05.

Solid Freestanding Walls and Solid Signs

Solid freestanding walls and solid freestanding signs. The design wind force, F, for solid freestanding walls and solid freestanding signs is determined by Equation 29.4-1:

$$F = q_h G C_f A_s \qquad (5.16)$$

In this equation, q_h is the velocity pressure determined by Equation 29.3-1 at height h defined in Figure 29.4-1; this is the same equation given in 27.3.2 under Part 1 of Chapter 27 (see Equation 27.3-1 and Equation 5.7 of this publication). The only difference between the two methods is that the velocity pressure coefficient, K_h, must be determined by Table 29.3-1 instead of by Table 27.3-1. See the discussion under Part 1 of Section 5.3.2 of this publication for additional information on how to determine the velocity pressure.

The gust-effect factor, G, is determined from 26.9 (see Section 5.2.6 of this publication for information on how to determine G). The term A_s is the gross area of the solid freestanding wall or sign.

Net force coefficients, C_f, are given in Figure 29.4-1 as a function of the geometrical properties of the wall or sign. These coefficients are based on the results of boundary layer wind tunnel studies (see C29.3.2 for more information on these studies as well as for an equation that can be used to determine C_f).

Three load cases must be investigated: Cases A, B and C. Case A and Case B consider the resultant wind load acting normal through the geometric center of the wall or sign and normal through a point at an eccentricity from the geometric center, respectively. Case C accounts for the higher pressures that occur near the windward edge of a freestanding wall or sign when it is subjected to an oblique wind direction (that is, at a direction that is not normal to the face of the wall or sign). Wind tunnel tests and full-scale test data have shown that the net pressures at the windward edges are significantly reduced where return corners are present. Reduction factors that can be used to account for this reduced pressure are given under Case C in the figure as a function of the return length.

Signs and walls are considered to be solid in cases where any openings comprise less than 30 percent of the gross area. Force coefficients are permitted to be reduced by the factor in Note 2 of Figure 29.4-1 where such openings occur.

Solid attached signs. The provisions of Chapter 30 (C&C) must be used to determine wind pressures on solid signs that are attached to the wall of a building and that meet the following conditions:

- Plane of the sign is parallel to and in contact with the plane of the wall; and

- The sign does not extend beyond the side or top edges of the wall.

This procedure is also applicable to signs that are not in direct contact with the wall provided that the gap between the sign and the wall is less than or equal to 3 feet and the edge of the sign is at least 3 feet in from the edges of the wall (that is, side and top edges and bottom edges of elevated walls). In any case, the internal pressure coefficient, (GC_{pi}), is to be set equal to zero when calculating the pressures in accordance with Chapter 30. The provisions of Chapter 30 are covered in Section 5.4 of this publication. In essence, the attached sign should experience approximately the same external pressure as the wall to which it is attached.

Other Structures

The design wind force, F, for other structures—including chimneys, tanks, rooftop equipment for buildings with $h > 60$ feet, open signs, lattice frameworks and trussed towers—is determined by Equation 29.5-1:

$$F = q_z G C_f A_f \tag{5.17}$$

In this equation, q_z is the velocity pressure determined by Equation 29.3-1 at height z of the centroid of the area, A_f, and G is the gust-effect factor determined in accordance with 26.9. The area, A_f, is either the projected area normal to the wind or the actual surface area depending on how the force coefficient, C_f, is specified.

Force coefficients, C_f, are given in the following figures:

- Figure 29.5-1: Chimneys, tanks, rooftop equipment and similar structures

- Figure 29.5-2: Open signs and lattice frameworks

- Figure 29.5-3: Trussed towers

The coefficients in Figures 29.5-1 and 29.5-2 are the same as those originally reported in Reference 5.7.

Values of C_f in Figure 29.5-1 are given for square, hexagonal and round cross-sections as a function of the height-to-cross-sectional dimension of the section. It is important to note that this figure is valid for rooftop equipment on buildings that have a mean roof height greater than 60 feet; the requirements of 29.5.1 must be used in cases where $h \leq 60$ feet (see below).

The force coefficients in Figure 29.5-2 are applicable to open signs, that is, signs with openings comprising more than 30 percent of the gross area. Signs that do not meet this criterion are classified as solid and the force coefficients in Figure 29.4-1 must be used (see above).

Force coefficients for trussed towers with square and triangular cross-sections are given in Figure 29.5-3. These simplified coefficients are consistent with those in Reference 5.11.

Rooftop Structures and Equipment for Buildings with $h \leq 60$ feet

Both lateral and vertical wind loads must be determined for structures and equipment located on the rooftop of buildings with mean roof heights less than or equal to 60 feet. The lateral force, F_h, is determined by Equation 29.5-2:

$$F_h = q_h(GC_r)A_f \tag{5.18}$$

In this equation, q_h is the velocity pressure determined by Equation 29.3-1 evaluated at the mean roof height of the building, and A_f is the vertical projected area of the rooftop structure or equipment on a plane normal to the direction of wind (see Figure 5.19 where $A_f = h_1B_1$).

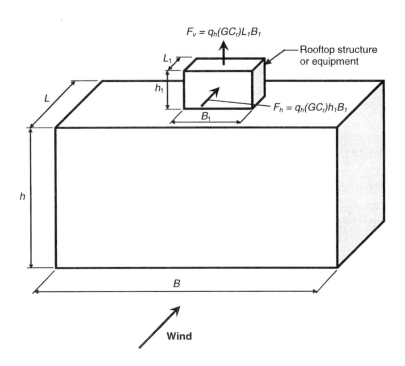

Figure 5.19
Lateral and Vertical Wind Loads on Rooftop Structures and Equipment

The combined gust-effect factor and pressure coefficient, (GC_r), accounts for higher wind pressures due to higher correlation of pressures across the structure surface, higher turbulence on the roof of the building, and accelerated wind speed on the roof. In the case of relatively small rooftop equipment where A_f is less than 10 percent of the windward building area, Bh, $(GC_r) = 1.9$. As the area, A_f, approaches Bh (that is, as the projected area of the rooftop building or equipment approaches that of the windward area of the building), (GC_r) is permitted to be reduced linearly to 1.0.

Equation 29.5-3 is used to calculate the uplift force, F_v, on rooftop structures and equipment:

$$F_v = q_h(GC_r)A_r \tag{5.19}$$

This force is distributed over the horizontal projected area of the rooftop structure or equipment A_r, which is equal to L_1B_1 in Figure 5.19.

The combined gust-effect factor and pressure coefficient, (GC_r), is equal to 1.5 where A_r is less than 10 percent of the building area, BL. A linear reduction to 1.0 is permitted as A_r increases to BL.

Parapets

ASCE/SEI 29.6 refers to 27.4.5 and 28.4.2 for wind pressures on parapets for buildings of all heights and for low-rise buildings, respectively. Refer to Sections 5.3.2 and 5.3.3 of this publication for more information on these methods.

Roof Overhangs

ASCE/SEI 29.7 refers to 27.4.4 and 28.4.3 for wind pressures on roof overhangs for buildings of all heights and for low-rise buildings, respectively. Refer to Sections 5.3.2 and 5.3.3 of this publication for more information on these methods.

Minimum Design Wind Loading

The minimum design wind load for other structures must be greater than or equal to 16 psf multiplied by the projected area normal to the wind A_f (29.8). This load case is to be applied as a separate load case in addition to the other load cases specified in Chapter 29.

Flowchart 5.7 in Section 5.7.5 of this publication can be used to determine design wind pressures on the MWFRS of other structures and building appurtenances in accordance with Chapter 29.

5.4 Components and Cladding

5.4.1 Overview

Chapter 30 contains design wind load provisions for C&C. These requirements may be used in the design of such elements provided the conditions and limitations of 30.1.2 and 30.1.3 are satisfied. The following subsections discuss the six parts that are contained in Chapter 30.

A summary of the wind load procedures and their applicability for C&C in accordance with Chapter 30 are given in Table 5.11.

Table 5.11 Summary of Wind Load Procedures in ASCE/SEI 7-10 for C&C

ASCE/SEI Chapter	Part	Applicability		Conditions
		Building/ Element Type	Height Limit	
30	1	Enclosed, low-rise	$h \leq 60$ feet and $h \leq$ least horizontal dimension	• Regular-shaped building • Building does not have response characteristics making it subject to across-wind loading, vortex shedding, instability due to galloping or flutter • Building is not located at a site where channeling effects or buffeting in the wake of upwind obstructions warrant special consideration • See additional conditions on selected figure(s) referenced in this part
		Partially enclosed, low-rise		
		Enclosed with $h \leq 60$ feet		
		Partially enclosed with $h \leq 60$ feet	$h \leq 60$ feet	
	2	Enclosed, low-rise	$h \leq 60$ feet and $h \leq$ least horizontal dimension	• Same first three conditions as in Part 1 • Building has either a flat roof, a gable roof with θ ≤ 45 degrees or a hip roof with θ ≤ 27 degrees
		Enclosed with $h \leq 60$ feet	$h \leq 60$ feet	
	3	Enclosed	$h > 60$ feet	• Same first three conditions as in Part 1 • See additional conditions on selected figure(s) referenced in this part
		Partially enclosed		
	4	Enclosed	$h \leq 160$ feet	• Same first three conditions as in Part 1
	5	Open	None	• Same first three conditions as in Part 1 • See additional conditions on selected figure(s) referenced in this part
	6	Parapets	---	• Same first three conditions as in Part 1 • All building types except enclosed buildings with $h \leq 160$ feet for which the provisions of Part 4 are used
		Roof overhangs	---	• Same first three conditions as in Part 1 • All building types except enclosed buildings with $h \leq 160$ feet for which the provisions of Part 4 are used
		Roof structures and equipment	$h \leq 60$ feet	• Same first three conditions as in Part 1

Similar to MWFRSs, reduction in wind pressures due to apparent shielding of surrounding buildings, other structures or terrain is not permitted in the design of C&C (30.1.4). Removal of such features around a building at a later date could result in wind pressures that are much higher than originally accounted for; as such, wind pressures must be calculated assuming that all shielding effects are not present.

Design wind pressures determined by Chapter 30 are permitted to be used in the design of air permeable roof or wall cladding. Examples of this type of cladding include siding, pressure-equalized rain screen walls, shingles, tiles, concrete roof pavers and aggregate roof surfacing. In general, this type of cladding allows partial air pressure equalization between the exterior and interior surfaces. Additional information can be found in C30.1.5.

5.4.2 General Requirements

Minimum Design Wind Pressures

According to 30.2.2, the design wind pressure for C&C shall not be less than a net pressure of 16 psf acting in either direction (positive or negative) normal to the surface. Like in the case of MWFRSs, this is a load case that needs to be considered in addition to the other required load cases in this chapter.

Tributary Areas Greater than 700 Square Feet

C&C elements that support a tributary area greater than 700 square feet are permitted to be designed for wind pressures using the provisions for MWFRSs. The 700-square-foot tributary area is deemed sufficiently large enough so that the localized wind effects are not pronounced as is the case of C&C; as such, the wind pressures on these elements are essentially equal to those determined by the method for MWFRSs.

External Pressure Coefficients

Numerous figures are provided in this chapter that give values for the combined gust-effect factor and pressure coefficient (GC_p) for C&C. The gust-effect factor and pressure coefficients are not permitted to be separated (30.2.4). Additional information on these coefficients is given in the following sections.

5.4.3 Velocity Pressure

The velocity pressure, q_z, evaluated at height z is determined by Equation 30.3-1, which is the same equation as that given for MWFRSs in 27.3.2. For C&C, the velocity pressure exposure coefficient, K_z, is determined by Table 30.3-1. Additional information on how to determine q_z is given in Section 5.3.2 of this publication.

5.4.4 Envelope Procedures

Scope

Parts 1 and 2 in 30.4 and 30.5, respectively, contain methods to determine wind loads on C&C of low-rise buildings that meet the conditions of 30.4.1 and 30.5.1, respectively. Both of these parts are based on the Envelope Procedure in Chapter 28 (see Section 5.3.3 of this publication). Different methods were utilized in the development of these procedures than those that were developed for MWFRSs. Additional information is provided in the following sections.

Part 1 – Low-rise Buildings

Overview

Part 1 is applicable to enclosed and partially enclosed low-rise buildings and buildings with a mean roof height less than or equal to 60 feet.

This procedure entails the determination of the velocity pressure at the mean roof height of the building (which is determined as a function of exposure, topographic effects, wind directionality, wind velocity and building occupancy), combined gust-effect factors and external pressure coefficients and internal pressure coefficients. Design wind pressures are obtained for various designated zones on the walls and roof of buildings.

Table 30.4-1 contains the overall steps that can be used to determine wind pressures on C&C of such buildings.

Design Wind Pressures

Design wind pressures, p, are determined by Equation 30.4-1:

$$p = q_h[(GC_p) - (GC_{pi})] \tag{5.20}$$

In this equation, q_h is the velocity pressure evaluated at the mean roof height of the building as determined by Equation 30.3-1, and (GC_{pi}) are the internal pressure coefficients given in Table 26.11-1 (see Section 5.2.8 of this publication for information on internal pressure coefficients).

Combined gust-effect factor and pressure coefficients (GC_p) are given in the following figures for walls and roofs:

- Walls: Figure 30.4-1
- Roofs

   — Flat, gable and hip roofs: Figures 30.4-2A through 30.4-2C
   — Stepped roofs: Figure 30.4-3
   — Multispan gable roofs: Figure 30.4-4
   — Monoslope roofs: Figures 30.4-5A and 30.4-5B
   — Sawtooth roofs: Figure 30.4-6
   — Domed roofs: Figure 30.4-7
   — Arched roofs: Figure 27.4-3, footnote 4

Except for domed roofs, external pressure coefficients are provided in these figures for various surfaces of the walls and roofs as a function of the *effective wind area*, which is defined in 26.2. This area does not necessarily correspond to the area of the building surface that contributes to the load under consideration.

Two cases can arise when determining the effective wind area of C&C. In the first case, the effective wind area is equal to the total area that is tributary to the element. The second case occurs where components are spaced relatively close together (such as wall studs or roof trusses). Due to the close spacing, the load is distributed and shared amongst adjoining components. To account for this load distribution, an effective width equal to one-third the span length is used in determining the effective wind area, which in this case is equal to the span length multiplied by the effective width. This area is usually greater than the tributary area of the component, which is generally long and narrow. The larger of the effective areas from the two of these cases is used in determining the pressure coefficients.

Consider the basic representation of a glazing system depicted in Figure 5.20. Glazing panels that have a height, h_2, span between mullions that are spaced a distance, s, on center. The mullions span the full story height, h_1, between floor slabs. The following are the effective wind areas for these two components:

- Glazing panel: Larger of $(s \times h_2)$ or $(s \times s/3)$
- Mullion: Larger of $(s \times h_1)$ or $(h_1 \times h_1/3)$

Figure 5.20
Effective Wind Areas for C&C

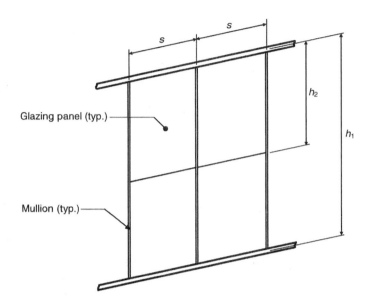

Effective wind area of glazing panel = larger of $(s \times h_2)$ or $(s \times s/3)$

Effective wind area of mullion = larger of $(s \times h_1)$ or $(h_1 \times h_1/3)$

For both components, the first area is the area tributary to the component and the second is that associated with the effective width. The effective wind area, which is defined as the larger of these two areas, depends on the relative magnitudes of the spans and heights for both components.

Flowchart 5.8 in Section 5.7.6 of this publication can be used to determine design wind pressures on C&C of buildings in accordance with Part 1 of Chapter 30.

Part 2 – Low-rise Buildings, Simplified

Overview

The simplified method in Part 2 is applicable to enclosed low-rise buildings and buildings with a mean roof height less than or equal to 60 feet with flat, gable or hip roofs. Wind pressures on C&C located at various building surfaces are determined from the tabulated values in Figure 30.5-1 for a building with a mean roof height of 30 feet located at a site classified as Exposure B. These pressures are adjusted for different building heights and exposures using the adjustment factor in the figure.

Table 30.5-1 contains the overall steps that can be used to determine wind pressures on C&C of such buildings.

Design Wind Pressures

Simplified design wind pressures, p_{net30}, for C&C located at various zones on a building are tabulated in Figure 30.5-1 as a function of the basic wind speed, V, and roof angle for buildings with a mean roof height of 30 feet that are located on primarily flat ground in Exposure B. Modifications are made to these tabulated pressures based on actual building height and exposure using the adjustment factor, λ, that is given in the figure. Such pressures must also be modified by the topographic factor, K_{zt}, evaluated at $0.33h$, where applicable.

Net design wind pressure, p_{net}, is determined by Equation 30.5-1:

$$p_{net} = \lambda K_{zt} p_{net30} \tag{5.21}$$

The methodology of determining the wind pressures on C&C is essentially the same as that for MWFRS (see Section 5.3.3 of this publication).

The tabulated wind pressures in the shaded areas of Figure 30.5-1 are less than the minimum wind pressure of 16 psf prescribed in 30.2.2. The note at the bottom of this figure points out that the final net pressure determined by Equation 5.21 must be greater than or equal to 16 psf.

Flowchart 5.9 in Section 5.7.6 of this publication can be used to determine design wind pressures on C&C of buildings in accordance with Part 2 of Chapter 30.

5.4.5 Directional Procedures

Scope

Parts 3, 4, 5 and 6 in 30.6, 30.7, 30.8 and 30.9, respectively, contain methods to determine wind loads on C&C of buildings, parapets, roof overhangs and rooftop structures and equipment that meet the conditions of 30.6.1, 30.7, 30.8.1, 30.9, 30.10 and 30.11, respectively. These parts are based on the Directional Procedure in Chapter 27 (see Section 5.3.2 of this publication).

Part 3 – Buildings with h > 60 feet

Overview

Part 3 is applicable to enclosed or partially enclosed buildings with a mean roof height greater than 60 feet with various types of roof shapes. Design wind pressures are calculated using the appropriate external pressure coefficients for walls and roofs given in Figures 30.6-1, 27.4-3, and 30.4-7.

This procedure entails the determination of the velocity pressure (which is determined as a function of exposure, height, topographic effects, wind directionality, wind velocity and building occupancy), combined gust-effect factors and external pressure coefficients and internal pressure coefficients. Design wind pressures are obtained for designated zones on the surfaces of buildings.

Table 30.6-1 contains the overall steps that can be used to determine wind pressures on C&C of buildings designed by this method.

Design Wind Pressures

Design wind pressures, p, on C&C are determined by Equation 30.6-1:

$$p = q(GC_p) - q_i(GC_{pi}) \tag{5.22}$$

All of the terms in this equation, except for the external pressure coefficients, (GC_p), are the same as those for MWFRSs and have been defined in Section 5.3.2 of this publication. External pressure coefficients can be found in Figure 30.6-1 for walls and flat roofs, in footnote 4 of Figure 27.4-3 for arched roofs, in Figure 30.4-7 for domed roofs and in Note 6 of Figure 30.6-1 for other roof angles and geometries.

The exception in 30.6.2 permits design wind pressures on the C&C of buildings with mean roof heights greater than 60 feet and less than 90 feet to be determined using the external pressures coefficients, (GC_p), from Figures 30.4-1 through 30.4-6 in Part 1 provided the height-to-width ratio of the building is less than or equal to one. The external pressure coefficients from the Envelope Procedure have been deemed adequate to predict design wind pressures for buildings of these proportions.

192 Chapter 5

Flowchart 5.10 in Section 5.7.6 of this publication can be used to determine design wind pressures on C&C of buildings in accordance with Part 3 of Chapter 30.

Part 4 – Buildings with $h \leq 160$ feet
Overview

Part 4 provides a simplified method of determining wind loads on C&C of enclosed buildings with a mean roof height less than or equal to 160 feet. Wind pressures on C&C located on various surfaces can be read directly from Table 30.7-2 for a building site classified as Exposure C and an effective wind area of 10 square feet. These pressures are modified by an effective area reduction factor, exposure adjustment factor and the topographic factor where applicable (see Table 30.7-2 and Equation 30.7-1).

Table 30.7-1 contains the overall steps that can be used to determine wind pressures on C&C of buildings designed by this method.

Design Wind Pressures

Wall and roof surfaces. Design wind pressures on designated zones of wall and roof surfaces are determined from Table 30.7-2 as a function of the basic wind speed, V, mean roof height and roof angle for buildings that are located on primarily flat ground in Exposure C. These tabulated pressures are valid for an effective wind area of 10 square feet and have been determined using the applicable external pressure coefficients from Part 3 (namely, Figure 30.6-1 for flat roofs, Figure 30.4-2A, 2B and 2C for gable and hip roofs and Figure 30.4-5A and 5B for monoslope roofs) and an internal pressure coefficient of ±0.18 for enclosed buildings. Modifications are made to these tabulated pressures based on the actual exposure and effective wind area.

Design wind pressures, p, are determined by Equation 30.7-1:

$$p = p_{table} (\text{EAF})(\text{RF})K_{zt} \qquad (5.23)$$

In this equation, (EAF) is the exposure adjustment factor given in Table 30.7-2, which modifies the tabulated pressures in cases where the exposure at the site is different than Exposure C.

The effective area reduction factor (RF) is also given in Table 30.7-2 and modifies the tabulated pressures for effective wind areas greater than 10 square feet. Reduction factors, which are based on the graphs of the external pressure coefficients in the figures in Part 3, are provided for designated zones on walls and roofs for five different roof shapes and for roof overhangs.

Information on how to determine the topographic factor, K_{zt}, can be found in Section 5.2.5 of this publication.

Flowchart 5.11 in Section 5.7.6 of this publication can be used to determine design wind pressures on C&C of wall and roof surfaces in accordance with Part 4 of Chapter 30.

Parapets. Equation 30.7-1 in conjunction with the applicable edge and corner pressures, p_{table}, in Table 30.7-2 are used to determine parapet C&C wind pressures. Pressures are applied to the parapet in accordance with Figure 30.7-1, which is reproduced here as Figure 5.21. Two load cases must be considered, namely, Load Case A, which is applicable to windward parapets, and Load Case B, which is applicable to leeward parapets:

- Load Case A: Positive wall pressure from Zone 4 or 5 is applied to the windward (front) surface of the parapet, and negative roof pressure from Zone 2 or 3 is applied to the leeward (back) surface of the parapet.

- Load Case B: Positive wall pressure from Zone 4 or 5 is applied to the windward (back) surface of the parapet, and negative wall pressure from Zone 4 or 5 is applied to the leeward (front) surface of the parapet.

Figure 5.21 Application of Parapet Wind Pressures for C&C, Part 4 of Chapter 30

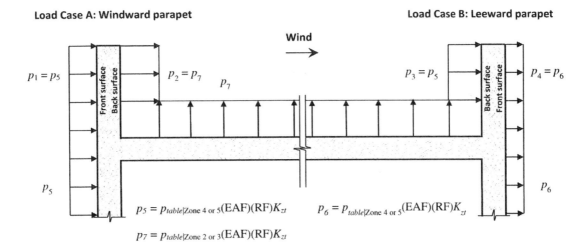

The height that is to be used in determining p_{table} from Table 30.7-2 is the height to the top of the parapet.

The pressures obtained by this method are slightly conservative compared to those determined by Part 3.

Roof overhangs. Equation 30.7-1 in conjunction with the applicable pressures, p_{table}, in Table 30.7-2 are used to determine roof overhang C&C wind pressures. Pressures are applied to the overhang in accordance with Figure 30.7-2, which is reproduced here as Figure 5.22.

Figure 5.22 Application of Roof Overhang Wind Pressures for C&C, Part 4 of Chapter 30

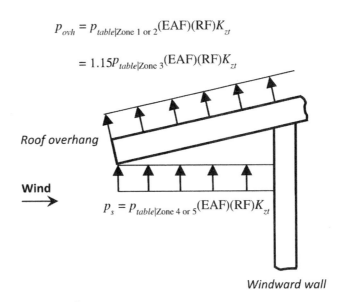

In Zones 1 and 2, the pressures on the top surface of the roof overhang are set equal to the tabulated roof pressures in those zones, and in Zone 3 the pressures are set equal to 1.15 times the tabulated roof pressures. On the underside of the overhang, the pressure is set equal to the adjacent wall pressure. In all cases, these pressures are slightly conservative and are based on the external pressure coefficients in Figures 30.4-2A to 30.4-2C of Part 3.

Part 5 – Open Buildings

Overview

Part 5 is applicable to open buildings with pitched, monoslope or troughed roofs. Net design wind pressures on C&C are determined by Equation 30.8-1 using the appropriate net pressure coefficients in Figures 30.8-1, 30.8-2 and 30.8-3.

Table 30.8-1 contains the overall steps that can be used to determine wind pressures on C&C of buildings designed by this method.

Design Wind Pressures

Net design wind pressures on the C&C of open buildings are determined by Equation 30.8-1:

$$p = q_h G C_N \tag{5.24}$$

In this equation, q_h is velocity pressure evaluated at the mean roof height of the building and G is the gust-effect factor determined in accordance with 26.9 (see Sections 5.3.2 and 5.2.6 of this publication for information on these quantities, respectively).

The net pressure coefficients, C_N, are given in Figure 30.8-1 for monosloped roofs, Figure 30.8-2 for pitched roofs and Figure 30.8-3 for troughed roofs based on the effective wind area. These coefficients include contributions from both the top and bottom surfaces of the roof, which implies that the element receives pressure from both surfaces.

Flowchart 5.12 in Section 5.7.6 of this publication can be used to determine design wind pressures on C&C of wall and roof surfaces in accordance with Part 5 of Chapter 30.

Part 6 – Building Appurtenances and Rooftop Structures and Equipment

Overview

Part 6 contains methods to determine wind pressures on C&C of parapets (30.9), roof overhangs (30.10) and rooftop structures and equipment for buildings with $h \leq 60$ feet (30.11).

This procedure entails the determination of velocity pressure (which is determined as a function of exposure, height, topographic effects, wind directionality, wind velocity and building occupancy), combined gust-effect factors and external pressure coefficients and internal pressure coefficients. Design wind pressures are obtained for designated zones on the surfaces of buildings.

Design Wind Pressures

Parapets. Design wind pressures, p, for C&C elements of parapets for all building types and heights, except for enclosed buildings with a mean roof height less than or equal to 160 feet (see Part 4) are calculated by Equation 30.9-1:

$$p = q_p[(GC_p) - (GC_{pi})] \tag{5.25}$$

In this equation, q_p is the velocity pressure evaluated at the top of the parapet and (GC_{pi}) are the internal pressure coefficients from Table 26.11-1 based on the porosity of the envelope of the parapet. The external pressure coefficients, (GC_p), are determined from the same figures as those in Parts 1 and 3 for walls and various roof configurations (see 30.9 for a comprehensive list of applicable figures).

Similar to the requirements of Part 4, Load Case A and Load Case B must be considered for the windward and leeward parapets, respectively. Figure 30.9-1, which illustrates these load cases, is essentially the same as Figure 30.7-1 in Part 4 with the exception that the pressures in Part 6 must be determined from the applicable figures noted in 30.9.

Table 30.9-1 contains the overall steps that can be used to determine wind pressures on C&C of parapets designed by this method.

Flowchart 5.13 in Section 5.7.6 of this publication can be used to determine design wind pressures on C&C of parapets in accordance with Part 6 of Chapter 30.

Roof overhangs. Design wind pressures, p, for C&C elements of roof overhangs are determined by Equation 30.10-1:

$$p = q_h[(GC_p) - (GC_{pi})] \tag{5.26}$$

In this equation, q_h is the velocity pressure measured at the mean roof height and (GC_{pi}) are the internal pressure coefficients from Table 26.11-1.

The external pressure coefficients, (GC_p), for overhangs are given in Figures 30.4-2A to 30.4-2C (flat roofs, gable roofs, and hip roofs, respectively), and they include contributions from top and bottom surfaces of overhang.

Table 30.10-1 contains the overall steps that can be used to determine wind pressures on C&C of roof overhangs designed by this method.

Flowchart 5.14 in Section 5.7.6 of this publication can be used to determine design wind pressures on C&C of roof overhangs in accordance with Part 6 of Chapter 30.

Rooftop structures and equipment for buildings with $h \leq 60$ feet. The provisions of 29.5.1 that are applicable in the design of MWFRSs are to be used in determining the design wind pressures on the C&C of rooftop

structures and equipment for buildings with a mean roof height less than or equal to 60 feet.

The design wind pressures on the walls and roof are determined by dividing the horizontal and vertical wind loads obtained from Equations 5.18 and 5.19, respectively, by the corresponding projected areas of the structure or equipment. The wall pressures can act inward or outward while the roof pressure acts outward.

5.5 Wind Tunnel Procedure

5.5.1 Overview

The Wind Tunnel Procedure in Chapter 31 can be utilized to determine wind loads on MWFRSs and C&C of any building or other structure in lieu of any of the procedures in Chapters 27 through 30, and it must be used where the conditions of these procedures are not satisfied (in particular, where a structure contains any of the characteristics defined in 27.1.3, 28.1.3, 29.1.3 or 30.1.3).

Wind tunnel tests should be seriously considered where buildings or other structures are not regularly-shaped, are flexible and/or slender, have the potential to be buffeted by upwind buildings or other structures or have the potential to be subjected to accelerated wind flow from channeling by buildings or topographic features. In the case of tall, slender buildings, only a wind tunnel test can properly capture any possible effects due to vortex shedding, galloping or flutter. For buildings in the heart of a city, a wind tunnel test is mandatory since Exposures B through D cannot properly capture the conditions in such cases. Every project has its own unique characteristics, and engineering judgment also plays a role in the decision-making process. When determining whether a wind tunnel test is required or not, it is always very important to keep in mind the limitations in Chapters 27 through 30, especially the general one related to along-wind response.

Of all of the methods that are contained in ASCE/SEI 7, the wind tunnel procedure is generally considered to produce the most accurate results. For certain types of buildings, the results from a wind tunnel test will be significantly smaller than those from any of the other methods. On the other hand, wind tunnel tests will yield results that are greater than those obtained from the other methods under certain conditions; as such, it is important to understand when such tests are required in order to adequately design the building or other structure for the effects of wind.

Information on the three basic types of wind tunnel test models that are commonly used is given in C31. Wind tunnel tests can also provide valuable information on snow loads, the effects of wind on pedestrians, and concentrations of air-pollutant emissions, to name a few. References that provide detailed information and guidance for the determination of wind loads and other types of design data by wind tunnel tests are provided in C31.

5.5.2 Test Conditions

Basic requirements for test conditions of wind tunnel tests or any other tests that employ a fluid other than air are given in 31.2. These seven conditions must be followed when any such test is conducted. Additional information on the basic procedures of conducting a wind tunnel test can be found in the references in C31.

5.5.3 Dynamic Response

The test conditions of 31.2 are to be used when determining the dynamic response of a building or other structure. Mass distribution, stiffness and damping must be properly accounted for in the model and in the subsequent analysis.

5.5.4 Load Effects

ASCE/SEI 31.4.2 prescribes limitations on the wind speed used in the tests, and 31.4.3 gives lower limits on the magnitude of the principal loads that are to be applied to a building or structure for both the MWFRS and C&C. Two conditions are given that permit the limiting values to be reduced. C31 provides a comprehensive discussion on the reasons behind these limitations.

5.6 Alternate All-heights Method

5.6.1 Overview

The alternate all-heights method in IBC 1609.6, which is based on the Directional Procedure of ASCE/SEI 7, can be used to determine wind pressures on buildings and other structures that meet the five conditions listed in IBC 1609.6.1:

1. The building or other structure is less than or equal to 75 feet in height with a height-to-least-width ratio of 4 or less, or the building or other structure has a fundamental frequency greater than or equal to 1 Hz.

2. The building or other structure is not sensitive to dynamic effects.

3. The building or other structure is not located on a site for which channeling effects or buffeting in the wake of upwind obstructions warrant special consideration.

4. The building meets the requirements of a simple diaphragm building according to the provisions of ASCE/SEI 26.2 where wind loads are transmitted to the MWFRS through the diaphragms.

5. For open buildings, multispan gable roofs, stepped roofs, sawtooth roofs, domed roofs, roofs with slopes greater than 45 degrees, solid freestanding walls and solid signs and rooftop equipment, the applicable provisions of ASCE/SEI 7 must be used.

It is evident from the above conditions that this method is applicable to regularly-shaped, low-rise buildings (contrary to what the title of the method suggests) that are rigid. The following sections outline the methods of how to determine design wind pressures on MWFRSs and C&C.

5.6.2 Design Wind Pressures

Net design wind pressures, p_{net}, on MWFRSs and C&C are determined by IBC Equation 16-35:

$$p_{net} = 0.00256 V^2 K_z C_{net} K_{zt} \qquad (5.27)$$

In this equation, 0.00256 is the air density constant and V is the basic wind velocity in miles per hour (see Sections 5.3.2 and 5.2.2 of this publication, respectively).

The velocity exposure coefficient, K_z, and the topographic factor, K_{zt}, are determined in accordance with ASCE/SEI 27.3.1 and 26.8, respectively. For the windward face of the building, these terms are determined as a function the height, z, above ground level, while for the leeward wall, side walls and roof, they are determined based on the mean roof height, h.

The net pressure coefficient, C_{net}, is equal to $K_d[GC_p - (GC_{pi})]$, which are the same terms in the design pressure equation in Part 1 of Chapter 27:

- The directionality factor, K_d, is equal to 0.85 for buildings (see ASCE/SEI 26.6 and Section 5.2.3 of this publication)

- The gust effect factor, G, is equal to 0.85 for rigid buildings (see ASCE/SEI 26.9 and Section 5.2.6 of this publication)

- The pressure coefficients, C_p, are given in ASCE/SEI Figure 27.4-1 (see Section 5.3.2 of this publication)

- The internal pressure coefficients, (GC_{pi}), are given in ASCE/SEI Table 26.11-1 (see Section 5.2.8 of this publication).

Tabulated values of C_{net} are given in IBC Table 1609.6.2 for the walls and roofs of enclosed and partially enclosed buildings. For example, on the windward wall of an enclosed building with positive internal pressure, $C_{net} = 0.85[(0.85 \times 0.8) - (0.18)] = 0.43$, which matches the value in IBC Table 1609.6.2. In cases where more than one value of C_{net} is listed in the table, the more severe wind load condition shall be used for design.

The net pressures determined by Equation 5.27 are applied simultaneously on, and in a direction normal to, all wall and roof surfaces. IBC 1609.6.4.1 requires that the torsional load cases identified in ASCE/SEI Figure 27.4-8 must be considered in design.

IBC 1609.6.3 requires a minimum design pressure of 16 psf projected on a plane normal to the assumed wind direction for the MWFRS. For C&C, a minimum design wind pressure of 16 psf is also required, which acts in either direction normal to the surface. The minimum load case is a separate load case, which is in addition to any of the other load cases required by this method.

Flowchart 5.15 in Section 5.7.7 of this publication can be used to determine design wind pressures on MWFRSs and C&C of buildings in accordance with the alternate all-heights method.

5.7 Flowcharts

5.7.1 Overview

Fifteen flowcharts that can be used in determining design wind pressures and loads are provided in this section. These flowcharts provide step-by-step procedures on how to determine design wind pressures and forces on MWFRSs and C&C of buildings and other structures based on the provisions of Chapters 26 through 30 of ASCE/SEI 7 and the all-heights method in the 2012 IBC.

A summary of the flowcharts provided in this chapter is given in Table 5.12. Included is a description of the content of each flowchart.

Table 5.12 Summary of Flowcharts Provided in Chapter 5

Flowchart	Title	Description
ASCE/SEI 26.9—Gust-effect Factors		
Flowchart 5.1	Gust-effect Factors, G and G_f	Provides step-by-step procedures for determining gust-effect factors that are defined in Chapter 26.
ASCE/SEI Chapter 27—MWFRS, Directional Procedure		
Flowchart 5.2	Velocity Pressures, q_z and q_h	Provides step-by-step procedures for determining velocity pressures.
Flowchart 5.3	Part 1—Buildings, MWFRS	Provides step-by-step procedures on how to determine design wind pressures on MWFRSs of enclosed, partially enclosed and open buildings of all heights.
Flowchart 5.4	Part 2—Buildings, MWFRS	Provides step-by-step procedures on how to determine design wind pressures on MWFRSs of enclosed, simple diaphragm buildings with $h \leq 160$ feet.
ASCE/SEI Chapter 28—MWFRS, Envelope Procedure		
Flowchart 5.5	Part 1—Buildings, MWFRS	Provides step-by-step procedures on how to determine design wind pressures on MWFRSs of enclosed and partially enclosed low-rise buildings.
Flowchart 5.6	Part 2—Buildings, MWFRS	Provides step-by-step procedures on how to determine design wind pressures on MWFRSs of enclosed, simple diaphragm, low-rise buildings.
ASCE/SEI Chapter 29—Other Structures and Building Appurtenances		
Flowchart 5.7	Structures Other than Buildings	Provides step-by-step procedures on how to determine design wind loads on solid freestanding walls, freestanding solid signs, chimneys, tanks, open signs, lattice framework, trussed towers, rooftop structures and rooftop equipment.
ASCE/SEI Chapter 30—C&C		
Flowchart 5.8	Part 1—Buildings, C&C	Provides step-by-step procedures on how to determine design wind pressures on C&C of enclosed and partially enclosed (1) low-rise buildings and (2) buildings with $h \leq 60$ feet.
Flowchart 5.9	Part 2—Buildings, C&C	Provides step-by-step procedures on how to determine design wind pressures on C&C of enclosed (1) low-rise buildings and (2) buildings with $h \leq 60$ feet.
Flowchart 5.10	Part 3—Buildings, C&C	Provides step-by-step procedures on how to determine design wind pressures on C&C of enclosed and partially enclosed buildings with $h > 60$ feet.
Flowchart 5.11	Part 4—Buildings, C&C	Provides step-by-step procedures on how to determine design wind pressures on C&C of enclosed buildings with $h \leq 160$ feet.
Flowchart 5.12	Part 5—Buildings, C&C	Provides step-by-step procedures on how to determine design wind pressures on C&C of open buildings.
Flowchart 5.13	Part 6—Parapets, C&C	Provides step-by-step procedures on how to determine design wind pressures on C&C of parapets.
Flowchart 5.14	Part 6—Roof Overhangs, C&C	Provides step-by-step procedures on how to determine design wind pressures on C&C for roof overhangs of enclosed and partially enclosed buildings.
IBC 1609.6 Alternate All-heights Method		
Flowchart 5.15	Buildings, MWFRS and C&C	Provides step-by-step procedures on how to determine design wind pressures on MWFRSs and C&C.
MWFRS = main windforce-resisting system C&C = components and cladding		

5.7.2 Gust-effect Factors

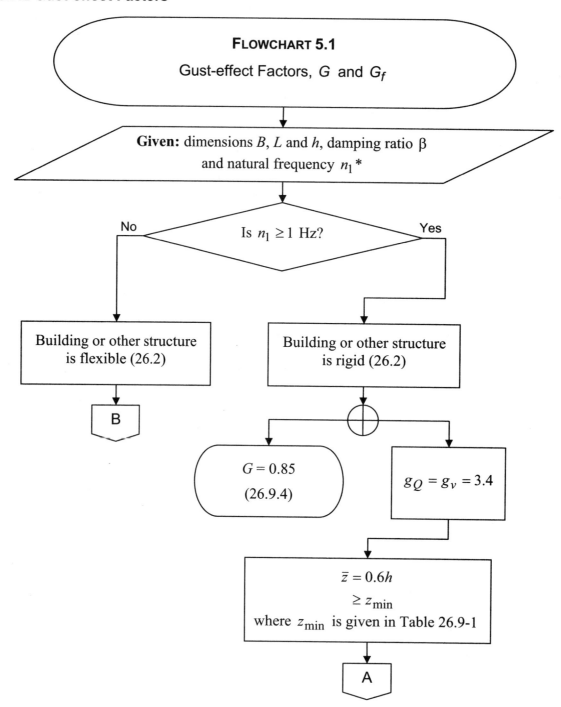

* Notes:
1. Information on structural damping can be found in C26.9.
2. n_1 can be determined from a rational analysis or estimated from approximate equations given in 26.9.2.

FLOWCHART 5.1

Gust-effect Factors, G and G_f

(continued)

A

$$I_{\bar{z}} = c\left(\frac{33}{\bar{z}}\right)^{1/6} \quad \text{Eq. 26.9-7}$$

where c is given in Table 26.9-1

$$L_{\bar{z}} = \ell\left(\frac{\bar{z}}{33}\right)^{\bar{\epsilon}} \quad \text{Eq. 26.9-9}$$

where ℓ and $\bar{\epsilon}$ are given in Table 26.9-1

$$Q = \sqrt{\frac{1}{1+0.63\left(\frac{B+h}{L_{\bar{z}}}\right)^{0.63}}} \quad \text{Eq. 26.9-8}$$

$$G = 0.925\left(\frac{1+1.7g_Q I_{\bar{z}} Q}{1+1.7g_v I_{\bar{z}}}\right) \quad \text{Eq. 26.9-6}$$

FLOWCHART 5.1

Gust-effect Factors, G and G_f

(continued)

B

$$g_Q = g_v = 3.4$$

$$g_R = \sqrt{2\ln(3{,}600n_1)} + \frac{0.577}{\sqrt{2\ln(3{,}600n_1)}} \quad \text{Eq. 26.9-11}$$

$$\bar{z} = 0.6h$$
$$\geq z_{min}$$

where z_{min} is given in Table 26.9-1

$$I_{\bar{z}} = c\left(\frac{33}{\bar{z}}\right)^{1/6} \quad \text{Eq. 26.9-7}$$

where c is given in Table 26.9-1

C

FLOWCHART 5.1

Gust-effect Factors, G and G_f

(continued)

C

$$L_{\bar{z}} = \ell \left(\frac{\bar{z}}{33}\right)^{\bar{\epsilon}} \quad \text{Eq. 26.9-9}$$

where ℓ and $\bar{\epsilon}$ are given in Table 26.9-1

$$Q = \sqrt{\frac{1}{1 + 0.63\left(\frac{B+h}{L_{\bar{z}}}\right)^{0.63}}} \quad \text{Eq. 26.9-8}$$

Determine the risk category of the building or structure using IBC Table 1604.5

Determine the basic wind speed, V, for the applicable risk category from Fig. 1609A, B or C, or Fig. 26.5-1A, B or C*

D

* See 26.5.2 and 26.5.3 for basic wind speed in special wind regions and estimation of basic wind speeds from regional climatic data. Tornadoes have not been considered in developing basic wind speed distributions shown in the figures.

FLOWCHART 5.1
Gust-effect Factors, G and G_f
(continued)

D

$$\overline{V}_{\overline{z}} = \overline{b}\left(\frac{\overline{z}}{33}\right)^{\overline{\alpha}}\left(\frac{88}{60}\right)V \qquad \text{Eq. 26.9-16}$$

where $\overline{b}$ and $\overline{\alpha}$ are given in Table 26.9-1

$$N_1 = \frac{n_1 L_{\overline{z}}}{\overline{V}_{\overline{z}}} \qquad \text{Eq. 26.9-14}$$

$$R_n = \frac{7.47 N_1}{(1+10.3 N_1)^{5/3}} \qquad \text{Eq. 26.9-13}$$

$$R_h = \frac{1}{\eta} - \frac{1}{2\eta^2}\left(1 - e^{-2\eta}\right) \text{ for } \eta > 0$$

$$R_h = 1 \text{ for } \eta = 0$$

where $\eta = 4.6 n_1 h / \overline{V}_{\overline{z}}$

E

FLOWCHART 5.1

Gust-effect Factors, G and G_f

(continued)

E

$$R_B = \frac{1}{\eta} - \frac{1}{2\eta^2}\left(1 - e^{-2\eta}\right) \text{ for } \eta > 0$$

$$R_B = 1 \text{ for } \eta = 0$$

where $\eta = 4.6 n_1 B / \overline{V}_{\bar{z}}$

$$R_L = \frac{1}{\eta} - \frac{1}{2\eta^2}\left(1 - e^{-2\eta}\right) \text{ for } \eta > 0$$

$$R_L = 1 \text{ for } \eta = 0$$

where $\eta = 15.4 n_1 L / \overline{V}_{\bar{z}}$

$$R = \sqrt{\frac{1}{\beta} R_n R_h R_B (0.53 + 0.47 R_L)} \quad \text{Eq. 26.9-12}$$

$$G_f = 0.925 \left(\frac{1 + 1.7 I_{\bar{z}} \sqrt{g_Q^2 Q^2 + g_R^2 R^2}}{1 + 1.7 g_v I_{\bar{z}}} \right) \quad \text{Eq. 26.9-10}$$

5.7.3 Chapter 27 – MWFRS, Directional Procedure

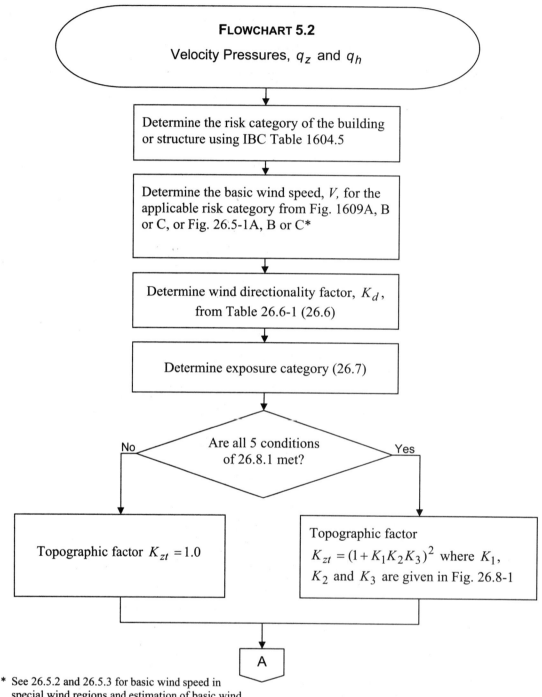

* See 26.5.2 and 26.5.3 for basic wind speed in special wind regions and estimation of basic wind speeds from regional climatic data. Tornadoes have not been considered in developing basic wind speed distributions shown in the figures.

FLOWCHART 5.2

Velocity Pressures, q_z and q_h
(continued)

A

Determine velocity pressure exposure coefficients, K_z and K_h, from Table 27.3-1 (27.3.1)

Determine velocity pressure at height z and h by Eq. 27.3-1:

$$q_z = 0.00256 K_z K_{zt} K_d V^2$$

$$q_h = 0.00256 K_h K_{zt} K_d V^2 **$$

** Notes:
1. q_z = velocity pressure evaluated at height z
2. q_h = velocity pressure evaluated at mean roof height h
3. The numerical constant of 0.00256 should be used except where sufficient weather data are available to justify a different value (see C27.3.2)

208 Chapter 5

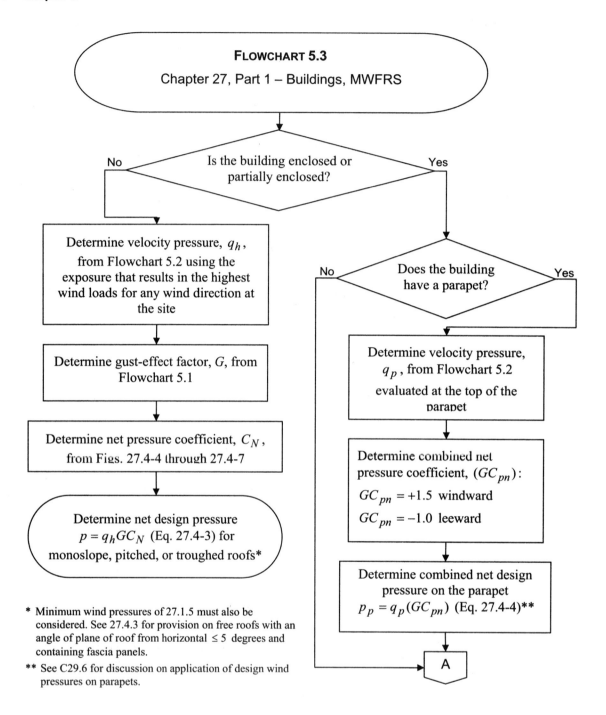

FLOWCHART 5.3
Chapter 27, Part 1 – Buildings, MWFRS
(continued)

Determine whether the building is rigid or flexible in accordance with 26.9.2

Determine the gust-effect factor (G or G_f) from Flowchart 5.1

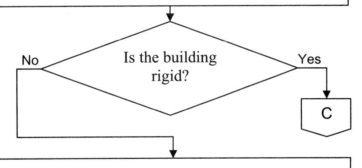

Determine velocity pressure: q_z for windward walls along the height of the building, and q_h for leeward walls, side walls and roof using Flowchart 5.2

Determine pressure coefficients, C_p, for the walls and roof from Fig. 27.4-1, 27.4-2 or 27.4-3

Determine q_i for the walls and roof using Flowchart 5.2[‡]

[‡] $q_i = q_h$ or $q_i = q_z$ depending on enclosure classification (see 27.4.1). q_i may conservatively be evaluated at height h ($q_i = q_h$) where applicable.

FLOWCHART 5.3

Chapter 27, Part 1 – Buildings, MWFRS

(continued)

Determine internal pressure coefficients, (GC_{pi}), from Table 26.11-1 based on enclosure classification

Determine design wind pressures by Eq. 27.4-2:
- Windward walls: $p_z = q_z G_f C_p - q_i (GC_{pi})$
- Leeward walls, side walls, and roofs: $p_h = q_h G_f C_p - q_i (GC_{pi})$ [+]

[+] Notes:
1. See 27.4.6 and Fig. 27.4-8 for the load cases that must be considered.
2. Minimum wind pressures of 27.1.5 must also be considered.

FLOWCHART 5.3
Chapter 27, Part 1 – Buildings, MWFRS
(continued)

Determine velocity pressure: q_z for windward walls along the height of the building, and q_h for leeward walls, side walls and roof using Flowchart 5.2

Determine pressure coefficients, C_p, for the walls and roof from Fig. 27.4-1, 27.4-2 or 27.4-3

Determine q_i for the walls and roof using Flowchart 5.2‡‡

Determine internal pressure coefficients, (GC_{pi}), from Table 26.11-1 based on enclosure classification

Determine design wind pressures by Eq. 27.4-1:
- Windward walls: $p_z = q_z GC_p - q_i(GC_{pi})$
- Leeward walls, side walls, and roofs: $p_h = q_h GC_p - q_i(GC_{pi})$ ++

‡‡ $q_i = q_h$ or $q_i = q_z$ depending on enclosure classification (see 27.4.1). q_i may conservatively be evaluated at height h ($q_i = q_h$) where applicable.

++ Notes:
1. See 27.4.6 and Fig. 27.4-8 for the load cases that must be considered.
2. Minimum wind pressures of 27.1.5 must also be considered.
3. See 27.4.4 for wind pressure on roof overhangs.

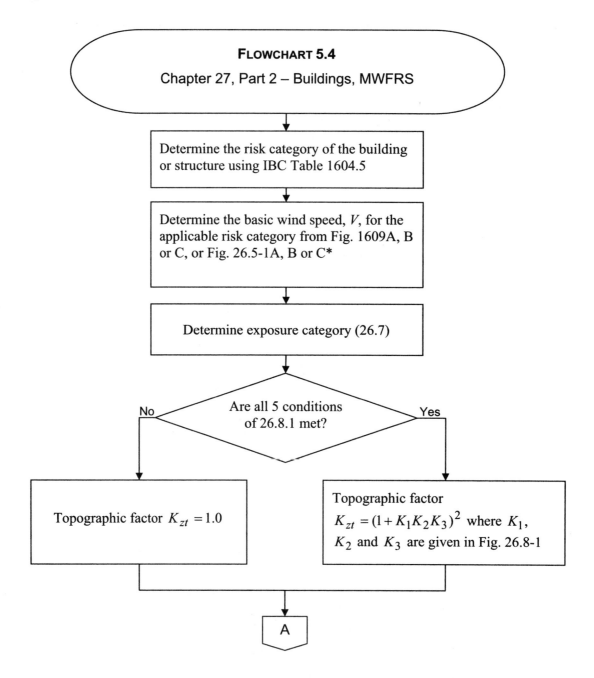

* See 26.5.2 and 26.5.3 for basic wind speed in special wind regions and estimation of basic wind speeds from regional climatic data. Tornadoes have not been considered in developing basic wind speed distributions shown in the figures.

FLOWCHART 5.4
Chapter 27, Part 2 – Buildings, MWFRS
(continued)

Determine the net pressures on the walls at the top p_h and at the base p_o of the building using Table 27.6-1**

↓

Determine the net pressures on the roof for the applicable zones using Table 27.6-2‡

↓

Multiply the net pressures on the walls and roof by K_{zt}

↓

Apply the final net pressures on the walls and roof simultaneously (Fig. 27.6-1).+,++

** Notes:
  1. See Table 27.6-1 for distribution of net pressure over the height of the building.
  2. See the notes in Table 27.6-1 for pressures on side walls and for other important information.

‡ The net pressures from the table must be multiplied by the exposure adjustment factors in this table for Exposures B and D.

+ Minimum wind pressures of 27.1.5 must also be considered.

++ See 27.6.2 and Fig. 27.6-2 for additional load on MWFRS from parapets, and 27.6.3 and Fig. 27.6-3 for wind on roof overhangs.

5.7.4 Chapter 28 – MWFRS, Envelope Procedure

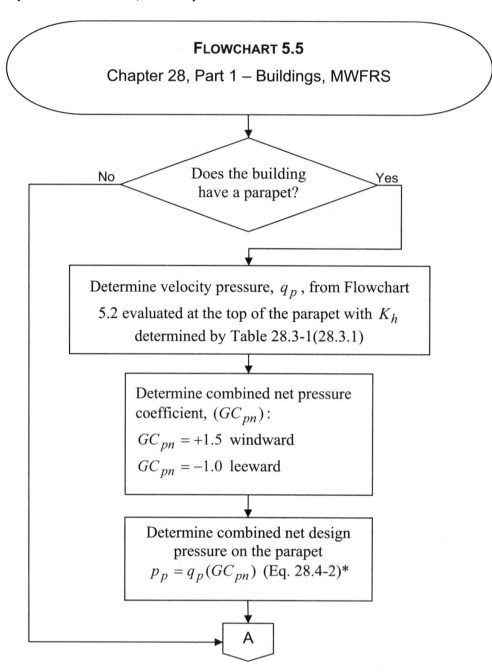

* See C29.6 for discussion on application of design wind pressures on parapets.

FLOWCHART 5.5

Chapter 28, Part 1 – Buildings, MWFRS
(continued)

A

Determine velocity pressure, q_h, using Flowchart 5.2 with K_h determined by Table 28.3-1 (28.3.1)

Determine external pressure coefficients, (GC_{pf}), from Fig. 28.4-1

Determine internal pressure coefficients, (GC_{pi}), from Table 26.11-1 based on enclosure classification

Determine design wind pressures by Eq. 28.4-1:
$p = q_h[(GC_{pf}) - (GC_{pi})]$**

** Notes:
1. See Fig. 28.4-1 for the basic and torsional load cases that must be considered.
2. Minimum wind pressures of 28.4.4 must also be considered.
3. See 28.4.3 for wind pressure on roof overhangs.

216   Chapter 5

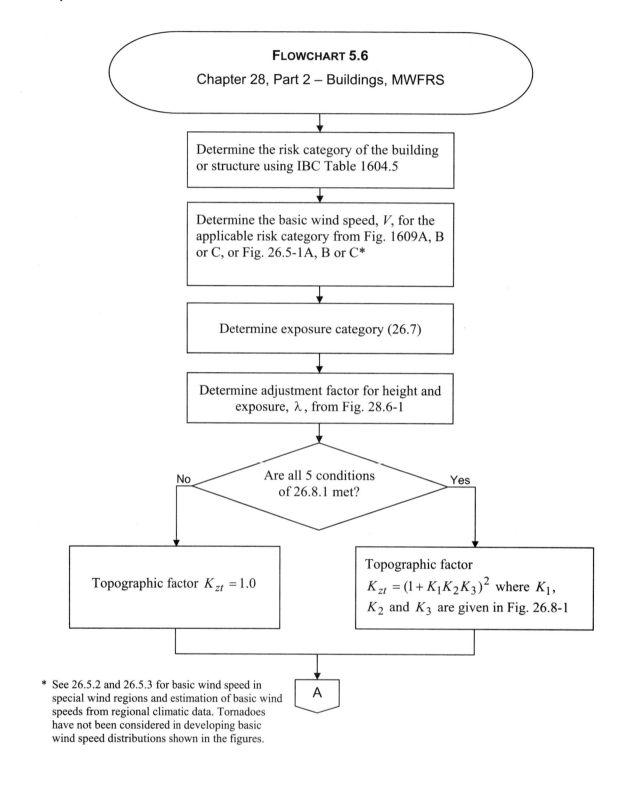

FLOWCHART 5.6

Chapter 28, Part 2 – Buildings, MWFRS

(continued)

Determine design wind pressures, p_{s30}, from Fig. 28.6-1 for Zones A through H on the building

Determine net design wind pressures $p_s = \lambda K_{zt} p_{s30}$ (Eq. 28.6-1) for Zones A through H**

** Notes:
1. For horizontal pressure zones, p_s is the sum of the windward and leeward net (sum of internal and external) pressures on vertical projection of Zones A, B, C and D. For vertical pressure zones, p_s is the net (sum of internal and external) pressure on horizontal projection of Zones E, F, G and H.

2. The load patterns shown in Fig. 28.6-1 shall be applied to each corner of the building in turn as the reference corner. See other notes in Fig. 28.6-1.

3. Load effects of the design wind pressures determined by Eq. 28.6-1 shall not be less than those from the minimum load case of 28.6.4. It is assumed that the pressures, p_s, for Zones A and C are equal to +16 psf, Zones B and D equal to +8 psf and the pressures for Zones E, F, G and H are equal to 0 psf.

5.7.5 Chapter 29 – Other Structures and Building Appurtenances

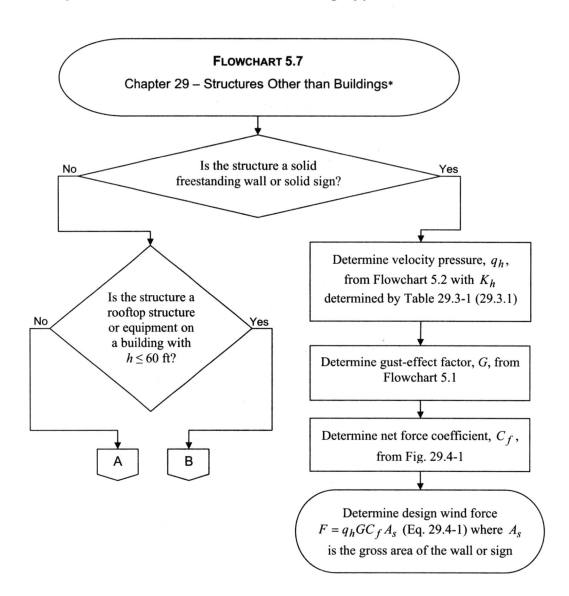

* See 29.6 for parapets and 29.7 for roof overhangs.

FLOWCHART 5.7
Chapter 29 – Structures Other than Buildings
(continued)

Determine velocity pressure, q_z, evaluated at height z of the centroid of area A_f from Flowchart 5.2 with K_z determined by Table 29.3-1 (29.3.1)

Determine gust-effect factor, G, from Flowchart 5.2

Determine force coefficient, C_f, from Figs. 29.5-1 through 29.5-3

Determine design wind force $F = q_z G C_f A_f$ (Eq. 29.5-1) where A_f is the projected area normal to the wind except where C_f is specified for the actual surface area**

** The design wind force for other structures shall be not less than 16 psf multiplied by A_f (29.8).

FLOWCHART 5.7

Chapter 29 – Structures Other than Buildings

(continued)

↓ B

Determine velocity pressure, q_h, from Flowchart 5.2 with K_h determined by Table 29.3-1 (29.3.1)

↓

Determine force coefficient, (GC_r), for lateral force as follows:
- $(GC_r) = 1.9$ for rooftop structures and equipment with $A_f < 0.1Bh$
- (GC_r) may be reduced linearly from 1.9 to 1.0 as the value of A_f is increased from $0.1Bh$ to Bh

↓

Determine lateral force F_h by Eq. 29.5-2:
$F_h = q_h(GC_r)A_f$

↓

Determine force coefficient, (GC_r), for vertical uplift force as follows:
- $(GC_r) = 1.5$ for rooftop structures and equipment with $A_r < 0.1BL$
- (GC_r) may be reduced linearly from 1.5 to 1.0 as the value of A_r is increased from $0.1BL$ to BL

↓

Determine vertical force, F_v, by Eq. 29.5-3:
$F_v = q_h(GC_r)A_r$

5.7.6 Chapter 30 – C&C

FLOWCHART 5.8

Chapter 30, Part 1 – Buildings, C&C*

↓

Determine velocity pressure, q_h, from Flowchart 5.2 with K_h determined by Table 30.3-1 (30.3.1)

↓

Determine external pressure coefficient, (GC_p), for zones on the walls and roof from Figs. 30.4-1 through 30.4-7 and 27.4-3 based on the effective wind area**

↓

Determine internal pressure coefficients, (GC_{pi}), from Table 26.11-1 based on enclosure classification

↓

Determine design wind pressures by Eq. 30.4-1:
$p = q_h[(GC_p) - (GC_{pi})]$ ‡

---

* C&C elements with tributary areas greater than 700 square feet may be designed using the provisions for MWFRSs (30.2.3).

** See 26.2 for definition of effective wind area.

‡ Minimum wind pressures of 30.2.2 must also be considered.

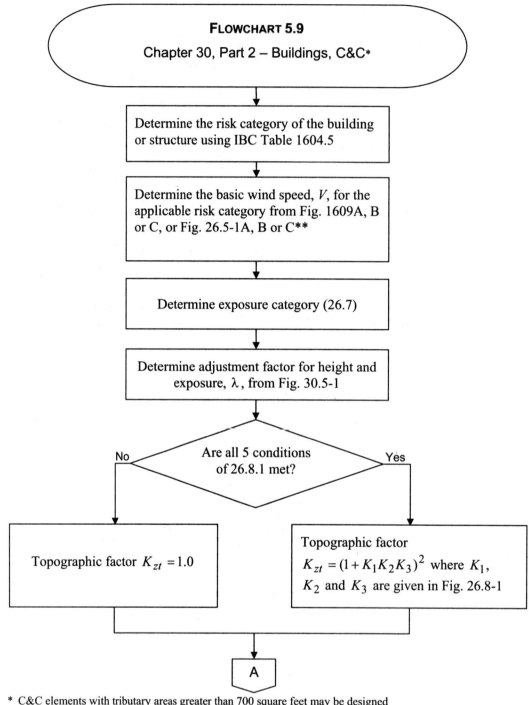

* C&C elements with tributary areas greater than 700 square feet may be designed using the provisions for MWFRSs (30.2.3).

** See 26.5.2 and 26.5.3 for basic wind speed in special wind regions and estimation of basic wind speeds from regional climatic data. Tornadoes have not been considered in developing basic wind speed distributions shown in the figures.

FLOWCHART 5.9

Chapter 30, Part 2 – Buildings, C&C

(continued)

A

Determine design wind pressures, p_{net30}, from Fig. 30.5-1 for Zones 1 through 5 on the building based on the effective wind area[+]

Determine net design wind pressures
$p_{net} = \lambda K_{zt} p_{net30}$ (Eq. 30.5-1) for Zones 1 through 5[‡]

[+] See 26.2 for definition of effective wind area.
[‡] Minimum wind pressures of 30.2.2 must also be considered.

FLOWCHART 5.10

Chapter 30, Part 3 – Buildings, C&C*

↓

Determine velocity pressure: q_z for windward walls along the height of the building, and q_h for leeward walls, side walls and roof using Flowchart 5.2 with K_z determined by Table 30.3-1 (30.3.1)

↓

Determine external pressure coefficients, (GC_p), for zones on the walls and roof from Figs. 30.6-1, 27.4-3, 30.4-7 or Note 6 of 30.6-1 based on the effective wind area**

↓

Determine q_i for the walls and roof using Flowchart 5.2 with K_z determined by Table 30.3-1 (30.3.1)‡

↓

Determine internal pressure coefficients, (GC_{pi}), from Table 26.11-1 based on enclosure classification

↓

Determine design wind pressures by Eq. 30.6-1:
$p = q(GC_p) - q_i(GC_{pi})$ +

* C&C elements with tributary areas greater than 700 square feet may be designed using the provisions for MWFRSs (30.2.3).

** See 26.2 for definition of effective wind area. Also see the exception in 30.6.2 for buildings with 60 feet < h < 90 feet.

‡ $q_i = q_h$ or $q_i = q_z$ depending on enclosure classification (see 30.6.2). q_i may conservatively be evaluated at height h ($q_i = q_h$) where applicable.

+ Minimum wind pressures of 30.2.2 must also be considered.

Wind Loads 225

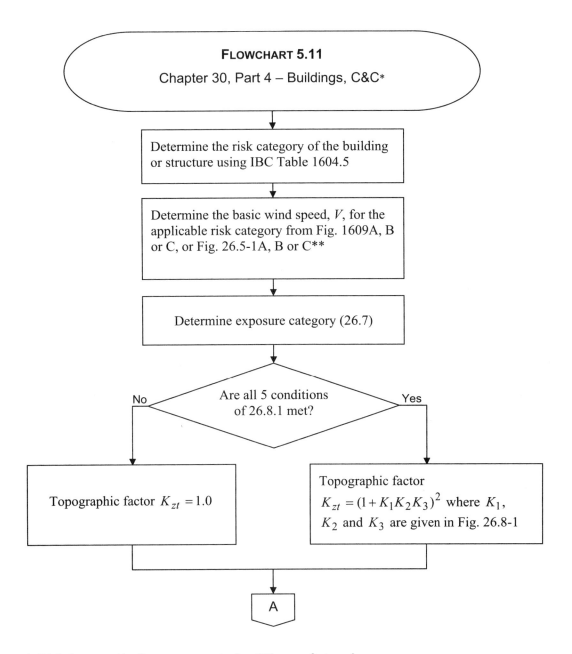

* C&C elements with tributary areas greater than 700 square feet may be designed using the provisions for MWFRSs (30.2.3). See 30.7.1.2 for parapets and 30.7.1.3 for roof overhangs.

** See 26.5.2 and 26.5.3 for basic wind speed in special wind regions and estimation of basic wind speeds from regional climatic data. Tornadoes have not been considered in developing basic wind speed distributions shown in the figures.

FLOWCHART 5.11
Chapter 30, Part 4 – Buildings, C&C
(continued)

Determine design wind pressures, p_{table}, from Table 30.7-2 for Zones 1 through 5 on the building based on effective wind area of 10 sq ft[+]

Determine exposure adjustment factor, EAF, from Table 30.7-2 if the exposure is different than Exposure C

Determine effective area reduction factor, RF, from Table 30.7-2 if the effective wind area is greater than 10 sq ft

Determine net design wind pressures
$p = p_{table}(EAF)(RF)K_{zt}$ (Eq. 30.7-1) for Zones 1 through 5[‡]

[+] See 26.2 for definition of effective wind area.
[‡] Minimum wind pressures of 30.2.2 must also be considered.

Wind Loads 227

FLOWCHART 5.12

Chapter 30, Part 5 – Buildings, C&C*

↓

Determine velocity pressure, q_h, from Flowchart 5.2 with K_h determined by Table 30.3-1 (30.3.1)

↓

Determine the gust-effect factor, G, from Flowchart 5.1

↓

Determine net pressure coefficient, C_N, for Zones 1 through 3 from Figs. 30.8-1 through 30.8-3 based on the effective wind area**

↓

Determine design wind pressures by Eq. 30.8-1:
$p = q_h G C_N$ ‡

---

* C&C elements with tributary areas greater than 700 square feet may be designed using the provisions for MWFRSs (30.2.3).

** See 26.2 for definition of effective wind area.

‡ Minimum wind pressures of 30.2.2 must also be considered.

FLOWCHART 5.13

Chapter 30, Part 6 – Parapets, C&C*

↓

Determine velocity pressure, q_p, at top of parapet from Flowchart 5.2 with K_h determined by Table 30.3-1 (30.3.1)

↓

Determine external pressure coefficient, (GC_p), for wall and roof surfaces adjacent to parapet using Figs. 30.4-1 through 30.4-7, 30.6-1 and 27.4-3 based on effective wind area**

↓

Determine internal pressure coefficients, (GC_{pi}), from Table 26.11-1 based on enclosure classification

↓

Determine design wind pressures by Eq. 30.9-1:
$p = q_p[(GC_p) - (GC_{pi})]$ ‡

---

* C&C elements with tributary areas greater than 700 square feet may be designed using the provisions for MWFRSs (30.2.3).

** See 26.2 for definition of effective wind area.

‡ Two load cases must be considered (see Fig. 30.9-1). Also, minimum wind pressures of 30.2.2 must also be considered.

FLOWCHART 5.14

Chapter 30, Part 6 – Roof Overhangs, C&C*

↓

Determine velocity pressure, q_h, from Flowchart 5.2 with K_h determined by Table 30.3-1 (30.3.1)

↓

Determine external pressure coefficient, (GC_p), for overhangs in Figs. 30.4-2A to 30.4-2C based on effective wind area**

↓

Determine internal pressure coefficients, (GC_{pi}), from Table 26.11-1 based on enclosure classification

↓

Determine design wind pressures by Eq. 30.10-1:
$p = q_h[(GC_p) - (GC_{pi})]$ ‡

---

* C&C elements with tributary areas greater than 700 square feet may be designed using the provisions for MWFRSs (30.2.3).

** See 26.2 for definition of effective wind area.

‡ Minimum wind pressures of 30.2.2 must also be considered.

5.7.7 IBC 1609.6 – Alternate All-heights Method

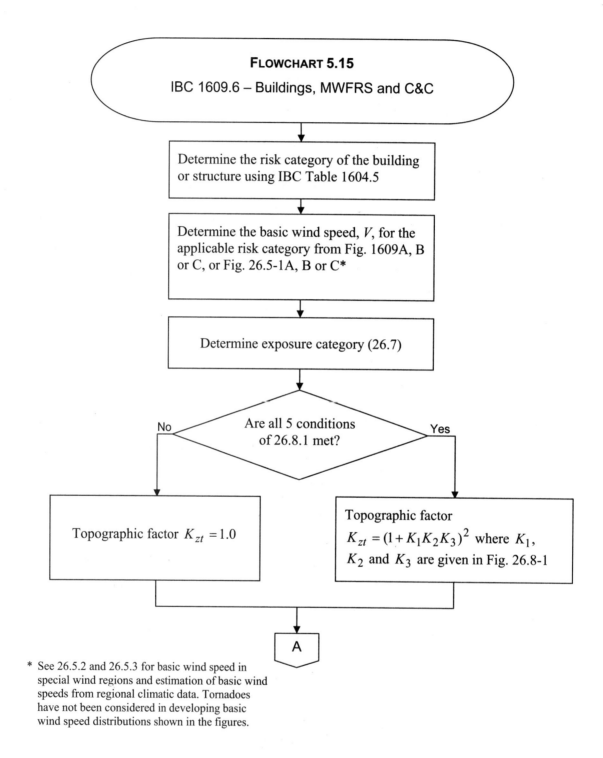

* See 26.5.2 and 26.5.3 for basic wind speed in special wind regions and estimation of basic wind speeds from regional climatic data. Tornadoes have not been considered in developing basic wind speed distributions shown in the figures.

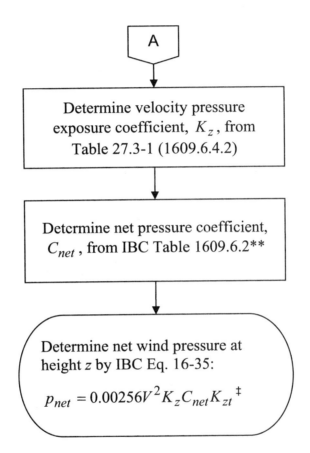

** Where C_{net} has more than one value in IBC Table 1609.6.2, the more severe wind load condition shall be used for design (1609.6.4.3(2)).

‡ Wind pressures are to be applied to the building envelope wall and roof surfaces in accordance with IBC 1609.6.4.4. Minimum wind pressures of 1609.6.3 must also be considered.

5.8 Examples

The following sections contain examples that illustrate the wind design provisions of IBC 1609 and Chapters 26 to 30 of ASCE/SEI 7.

5.8.1 Example 5.1—Warehouse Building Using Chapter 27, Part 1 (MWFRS)

For the one-story warehouse illustrated in Figure 5.23, determine design wind pressures on the MWFRS in both directions using Part 1 of Chapter 27 (Directional Procedure for buildings of all heights).

Figure 5.23 Plan and Elevation of Warehouse Building, Example 5.1

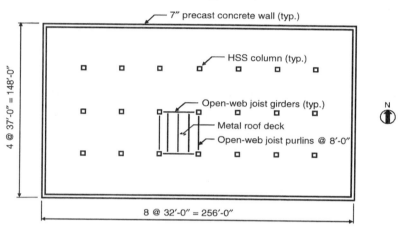

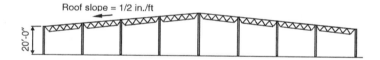

DESIGN DATA

Location:	St. Louis, MO
Surface Roughness:	C (open terrain with scattered obstructions less than 30 feet in height)
Topography:	Not situated on a hill, ridge or escarpment
Occupancy:	Less than 300 people congregate in one area and the building is not used to store hazardous or toxic materials

SOLUTION

- **Step 1:** Check if the building meets all of the conditions and limitations of 27.1.2 and 27.1.3 so that this procedure can be used to determine the wind pressures on the MWFRS.

The building is regularly-shaped, that is, it does not have any unusual geometric irregularities in spatial form. Also, the building does not have response characteristics that make it subject to across-wind loading or other similar effects, and it is not sited at a location where channeling effects or buffeting in the wake of upwind obstructions need to be considered.

Thus, Part 1 of Chapter 27 may be used to determine the design wind pressures on the MWFRS.

- **Step 2:** Determine the enclosure classification of the building.

The enclosure classification of the building depends on the number and size of openings in the precast walls.

Assume that there are two Type A precast panels on each of the east and west walls and two Type B precast panels on each of the north and south walls (see Figure 5.24). The openings in these walls are door openings. All other precast panels do not have door openings, and assume that there are no openings in the roof.

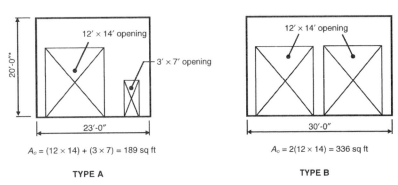

Figure 5.24
Door Openings in Precast Wall Panels, Example 5.1

* Elevation of top of walls on north and south faces vary with roof slope.

By definition, an open building is a building having each wall at least 80 percent open. It is evident that this building is not open since the area of openings on each wall is less than 80 percent of the gross area of the wall (see Figure 5.9). Therefore, it is enclosed or partially enclosed.

A building is defined as enclosed when it does not comply with the requirements for an open or partially enclosed building (26.2).

Calculate the area of each wall and the area of the openings in each wall:

- East and West walls

    $A_g = 20 \times 148 = 2,960$ square feet

    $A_o = 2[(12 \times 14) + (3 \times 7)] = 378$ square feet

- North and South walls

    $A_g = (20 \times 256) + [0.5 \times 256 \times (128 \times \tan 2.39°)] = 5,803$ square feet

    $A_o = 2 \times [(12 \times 14) + (12 \times 14)] = 672$ square feet

- Roof

  $A_g = 2 \times [148 \times (128/\cos 2.39°)] = 37{,}921$ square feet

  $A_o = 0$

- Totals

  $A_g = (2 \times 2{,}960) + (2 \times 5{,}803) + 37{,}921 = 55{,}447$ square feet

  $A_o = 2 \times (378 + 672) = 2{,}100$ square feet

Check the conditions for a partially enclosed building:

- East and West walls

  $A_o/A_{oi} = 378/(2{,}100 - 378) = 0.22 < 1.1$ (condition is not met)

  $A_o\ \ = 378$ square feet

  > 4 square feet (governs) or $0.01A_g = 0.01 \times 2{,}960 = 29.6$ square feet
  condition is met)

  $A_{oi}/A_{gi} = (2{,}100 - 378)/(55{,}447 - 2{,}960) = 0.03 < 0.20$ (condition is met)

- North and South walls

  $A_o/A_{oi} = 672/(2{,}100 - 672) = 0.47 < 1.1$ (condition is not met)

  $A_o\ \ = 672$ square feet

  > 4 square feet (governs) or $0.01A_g = 0.01 \times 5{,}803 = 58.0$ square feet
  (condition is met)

  $A_{oi}/A_{gi} = (2{,}100 - 672)/(55{,}447 - 5{,}803) = 0.03 < 0.20$ (condition is met)

Since all of the conditions for each set of walls are not met and since the building is not open, the building is defined as enclosed.

Note: Since this roof of the building in this example is basically flat, essentially the same results would be obtained by disregarding the slope of the roof.

- **Step 3:** Use Flowchart 5.3 to determine the design wind pressures, p, on the MWFRS.

  1. Determine whether the building is rigid or flexible in accordance with 26.9.2.

     A flexible building is defined in 26.2 as one in which the fundamental natural frequency of the building, n_1, is less than 1 Hz. Although it is evident by inspection that the building is not flexible, the natural frequency will be determined and compared to 1 Hz.

     In lieu of obtaining the natural frequency of the building from a dynamic analysis, check if the approximate lower–bound natural frequencies given in 26.9.3 can be used for this building (26.9.2.1):

     - Building height = 20 feet < 300 feet

- Building height = 20 feet < $4L_{eff}$ = 4 × 148 = 592 feet in N-S direction
  $\qquad\qquad\qquad\qquad\qquad$ = 4 × 256 = 1,024 feet in E-W direction

Note: For a one-story building, the effective building length, L_{eff}, is equal to the actual building length in the direction of analysis (see Equation 26.9-1).

Since both of these limitations are satisfied, Equation 26.9-5 can be used to determine an approximate value of n_1 in the N-S direction for concrete shear wall systems:

$$n_a = 385(C_w)^{0.5} / h$$

where $C_w = \dfrac{100}{A_B} \sum_{i=1}^{n} \left(\dfrac{h}{h_i}\right)^2 \dfrac{A_i}{\left[1 + 0.83\left(\dfrac{h_i}{D_i}\right)^2\right]}$

A_B = base area of the building = 148 × 256 = 37,888 square feet

h, h_i = mean roof height and height of shear wall, respectively = 20 feet

A_i = area of shear wall = (7 / 12) × 148 = 86.3 square feet assuming that there are no openings in the precast walls

D_i = length of shear wall = 148 feet

$$C_w = \dfrac{2 \times 100}{37,888} \dfrac{86.3}{\left[1 + 0.83\left(\dfrac{20}{148}\right)^2\right]} = 0.45$$

$n_a = 385(0.45)^{0.5} / 20 = 12.9$ Hz >> 1 Hz

Similar calculations in the E-W direction yield n_a = 17.1 Hz >> 1 Hz. Thus, the building is rigid.

2. Determine the gust-effect factor, G, using Flowchart 5.1.

   According to 26.9.4, gust-effect factor, G, for rigid buildings may be taken as 0.85 or can be calculated by Equation 26.9-6. For simplicity, use $G = 0.85$.

3. Determine velocity pressure, q_z, for windward walls along the height of the building, and q_h for leeward walls, side walls and roof using Flowchart 5.2.

   a. Determine the risk category of the building using IBC Table 1604.5.

      Due to the nature of its occupancy, this warehouse building falls under Risk Category II.

   b. Determine basic wind speed, V, for the applicable risk category from IBC Figure 1609A, B or C, or ASCE/SEI Figure 26.5-1A, B or C.

      For Risk Category II, use Figure 1609A or Figure 26.5-1A. From either of these figures, V = 115 mph for St. Louis, MO.

c. Determine wind directionality factor, K_d, from Table 26.6-1.

For the MWFRS of a building structure, $K_d = 0.85$.

d. Determine exposure category.

In the design data, the surface roughness is given as C. It is assumed that Exposures B and D are not applicable, so Exposure C applies (see 26.7.3).

e. Determine topographic factor, K_{zt}.

As noted in the design data, the building is not situated on a hill, ridge or escarpment. Thus, topographic factor $K_{zt} = 1.0$ (26.8.2).

f. Determine velocity pressure exposure coefficients K_z and K_h from Table 27.3-1.

Values of K_z and K_h for Exposure C are summarized in Table 5.13.

Table 5.13 Velocity Pressure Exposure Coefficient, K_z

Height above ground level, z (feet)	K_z
20	0.90
15	0.85

g. Determine velocity pressure q_z and q_h by Equation 27.3-1.

$$q_z = 0.00256 K_z K_{zt} K_d V^2 = 0.00256 \times K_z \times 1.0 \times 0.85 \times 115^2 = 28.78 K_z \text{ psf}$$

A summary of the velocity pressures is given in Table 5.14.

Table 5.14 Velocity Pressure, q_z

Height above ground level, z (feet)	K_z	q_z (psf)
20	0.90	25.9
15	0.85	24.5

4. Determine pressure coefficients, C_p, for the walls and roof from Figure 27.4-1.

For wind in the E-W direction:

Windward wall: $C_p = 0.8$ for use with q_z

Leeward wall ($L/B = 256/148 = 1.7$): $C_p = -0.35$ (from linear interpolation) for use with q_h

Side wall: $C_p = -0.7$ for use with q_h

Roof (normal to ridge with $\theta < 10$ degrees and $h/L = 20/256 = 0.08 < 0.5$):

$C_p = -0.9, -0.18$ from windward edge to $h = 20$ feet for use with q_h

$C_p = -0.5, -0.18$ from 20 feet to $2h = 40$ feet for use with q_h

$C_p = -0.3, -0.18$ from 40 feet to 256 feet for use with q_h

Note: The smaller uplift pressures on the roof due to $C_p = -0.18$ may govern the design when combined with roof live load or snow loads. This pressure is not shown in this example, but in general must be considered.

For wind in the N-S direction:

> Windward wall: $C_p = 0.8$ for use with q_z
>
> Leeward wall ($L/B = 148/256 = 0.6$): $C_p = -0.5$ for use with q_h
>
> Side wall: $C_p = -0.7$ for use with q_h
>
> Roof (parallel to ridge with $h/L = 20/148 = 0.14 < 0.5$):
>
> > $C_p = -0.9, -0.18$ from windward edge to $h = 20$ feet for use with q_h
> >
> > $C_p = -0.5, -0.18$ from 20 feet to $2h = 40$ feet for use with q_h
> >
> > $C_p = -0.3, -0.18$ from 40 feet to 148 feet for use with q_h

5. Determine q_i for the walls and roof using Flowchart 5.2.

   In accordance with 27.4.1, $q_i = q_h = 25.9$ psf for windward walls, side walls, leeward walls and roofs of enclosed buildings.

6. Determine internal pressure coefficients, (GC_{pi}), from Table 26.11-1.

   For an enclosed building, $GC_{pi} = +0.18, -0.18$.

7. Determine design wind pressures p_z and p_h by Equation 27.4-1.

   Windward walls:

   $p_z = q_z G C_p - q_h (GC_{pi})$

   $= (q_z \times 0.85 \times 0.8) - 25.9(\pm 0.18)$

   $= (0.68 q_z \mp 4.7)$ psf (external ± internal pressure)

   Leeward wall, side walls and roof:

   $p_h = q_h G C_p - q_h (GC_{pi})$

   $= (25.9 \times 0.85 \times C_p) - 25.9(\pm 0.18)$

   $= (22.0 C_p \mp 4.7)$ psf (external ± internal pressure)

   A summary of the maximum design wind pressures in the E-W and N-S directions is given in Tables 5.15 and 5.16, respectively.

Table 5.15 Design Wind Pressures, p, in E-W Direction

Location	Height above ground level, z (feet)	q (psf)	External pressure qGC_p (psf)	Internal pressure $q_h(GC_{pi})$ (psf)	Net pressure, p (psf) $+(GC_{pi})$	Net pressure, p (psf) $-(GC_{pi})$
Windward wall	20	25.9	17.6	±4.7	12.9	22.3
Windward wall	15	24.5	16.7	±4.7	12.0	21.4
Leeward wall	All	25.9	−7.7	±4.7	−12.4	−3.0
Side walls	All	25.9	−15.4	±4.7	−20.1	−10.7
Roof	20	25.9	−19.8*	±4.7	−24.5	−15.1
Roof	20	25.9	−11.0†	±4.7	−15.7	−6.3
Roof	20	25.9	−6.6‡	±4.7	−11.3	−1.9

* from windward edge to 20 feet
† from 20 to 40 feet
‡ from 40 to 256 feet

Illustrated in Figures 5.25 and 5.26 are the net design wind pressures in the E-W and N-S directions, respectively, for positive and negative internal pressure.

The MWFRS of buildings whose wind loads have been determined by Chapter 27 must be designed for the wind load cases defined in Figure 27.4-8 (see 27.4.6 and Figure 5.13).

In Case 1, the full design wind pressures act on the projected area perpendicular to each principal axis of the structure. These pressures are assumed to act separately along each principal axis. The wind pressures on the windward and leeward walls depicted in Figures 5.25 and 5.26 fall under Case 1.

Table 5.16 Design Wind Pressures, p, in N-S Direction

Location	Height above ground level, z (feet)	q (psf)	External pressure qGC_p (psf)	Internal pressure $q_h(GC_{pi})$ (psf)	Net pressure, p (psf) $+(GC_{pi})$	Net pressure, p (psf) $-(GC_{pi})$
Windward wall	20	25.9	17.6	±4.7	12.9	22.3
Windward wall	15	24.5	16.7	±4.7	12.0	21.4
Leeward wall	All	25.9	−11.0	±4.7	−15.7	−6.3
Side walls	All	25.9	−15.4	±4.7	−20.1	−10.7
Roof	20	25.9	−19.8*	±4.7	−24.5	−15.1
Roof	20	25.9	−11.0†	±4.7	−15.7	−6.3
Roof	20	25.9	−6.6‡	±4.7	−11.3	−1.9

* from windward edge to 20 feet
† from 20 to 40 feet
‡ from 40 to 148 feet

Figure 5.25 Design Wind Pressures on MWFRS in E-W Direction, Example 5.1

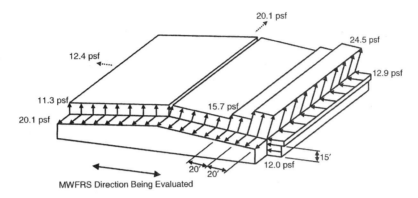

(a) Positive Internal Pressure

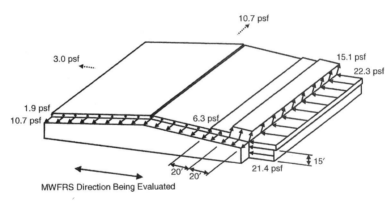

(b) Negative Internal Pressure

Figure 5.26 Design Wind Pressures on MWFRS in N-S Direction, Example 5.1

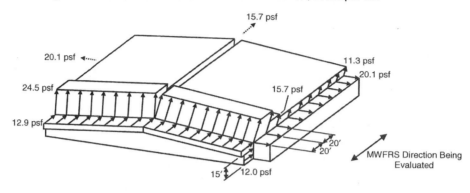

(a) Positive Internal Pressure

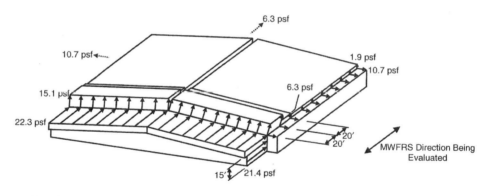

(b) Negative Internal Pressure

According to the exception in 27.4.6, buildings that meet the requirements of D1.1 need only be designed for Load Case 1 and Load Case 3. Since this is a one-story building with $h \leq 30$ feet, the requirements of D1.1 are satisfied, and this building can be designed for Load Cases 1 and 3 only.

In Case 3, 75 percent of the wind pressures on the windward and leeward walls of Case 1, which are shown in Figures 5.25 and 5.26, act simultaneously on the building. This load case, which needs to be considered in addition to the load cases in Figures 5.25 and 5.26, accounts for the effects due to wind along the diagonal of the building.

Finally, the minimum design wind loading prescribed in 27.1.5 must be considered as a load case in addition to the load cases described above (see Figure C27.4-1 and Figure 5.10).

5.8.2 Example 5.2 – Warehouse Building Using Chapter 27, Part 2 (MWFRS)

For the one-story warehouse illustrated in Figure 5.23, determine design wind pressures on the MWFRS in both directions using Part 2 of Chapter 27 (Directional Procedure for enclosed, simple diaphragm buildings with $h \leq 160$ feet).

SOLUTION

- **Step 1:** Check if the building meets all of the conditions and limitations of 27.1.2 and 27.1.3 and the additional conditions of 27.5.2 and 27.5.4 so that this procedure can be used to determine the wind pressures on the MWFRS.

  The building is regularly-shaped, that is, it does not have any unusual geometric irregularities in spatial form. Also, the building does not have response characteristics that make it subject to across-wind loading or other similar effects, and it is not sited at a location where channeling effects or buffeting in the wake of upwind obstructions need to be considered.

  Check the conditions for a Class 1 Building (27.5.2):

  1. The building is enclosed (see Example 5.1) and the building is a simple diaphragm building as defined in 26.2, since the windward and leeward wind loads are transmitted through the metal deck roof (diaphragm) to the precast walls (MWFRS), and there are no structural separations in the MWFRS.

  2. Mean roof height, $h = 20$ feet < 60 feet

  3. In the N-S direction, $L/B = 148/256 = 0.6$, which is greater than 0.2 and less than 5.0

     In the E-W direction, $L/B = 256/148 = 1.7$, which is greater than 0.2 and less than 5.0

  4. Topographic effect factor, $K_{zt} = 1.0$ (see Example 5.1)

  Therefore, the building is a Class 1 Building.

  Check the condition for diaphragm flexibility (27.5.4):

  Part 2 of Chapter 27 is to be applied to buildings with either rigid or flexible diaphragms. This condition is satisfied in this example since according to 27.5.4, diaphragms constructed of untopped metal deck with a span-to-depth ratio of 2 or less can be idealized as rigid for consideration of wind loading.

  Thus, Part 2 of Chapter 27 may be used to determine the design wind pressures on the MWFRS.

- **Step 2:** Use Flowchart 5.4 to determine the design wind pressures, p, on the MWFRS.

  1. Determine the risk category of the building using IBC Table 1604.5.

     Due to the nature of its occupancy, this warehouse building falls under Risk Category II.

  2. Determine basic wind speed, V, for the applicable risk category from IBC Figure 1609A, B or C, or ASCE/SEI Figure 26.5-1A, B or C.

For Risk Category II, use Figure 1609A or Figure 26.5-1A. From either of these figures, $V = 115$ mph for St. Louis, MO.

3. Determine exposure category.

   In the design data of Example 5.1, the surface roughness is given as C. It is assumed that Exposures B and D are not applicable, so Exposure C applies (see 26.7.3).

4. Determine topographic factor, K_{zt}.

   As noted in the design data of Example 5.1, the building is not situated on a hill, ridge or escarpment. Thus, topographic factor $K_{zt} = 1.0$ (26.8.2).

5. Determine the net pressures on the walls at the top (p_h) and at the base (p_o) using Table 27.6-1.

   a. E-W wind

   - Along-wind net wall pressures

     The along-wind net wall pressures are obtained by reading the values from Table 27.6-1 for Exposure C, a wind velocity of 115 mph, a mean roof height of 20 feet and $L/B = 256/148 = 1.7$:

     At the top of the wall: $p_h = 25.9$ psf (by linear interpolation)

     At the bottom of the wall: $p_o = 25.5$ psf (by linear interpolation)

     These pressures are applied to the projected area of the building walls in the direction of the wind (see Figure 27.6-1 and Figure 5.14).

     Note 4 in Table 27.6-1 gives information on how to distribute the along-wind net wall pressures between windward and leeward wall faces based on the ratio L/B.

   - Side wall external pressures

     According to Note 2 in Table 27.6-1, the uniform side wall external pressures, which are applied to the projected area of the building walls in the direction normal to the direction of the wind, are equal to $0.61p_h = 15.8$ psf (by linear interpolation).

     It is important to note that these pressures do not include the effects from internal pressure.

   - Roof pressures

     Roof pressures are obtained by reading the values from Table 27.6-2 for Exposure C, a wind velocity of 115 mph, a flat roof and a mean roof height of 20 feet. Note that "NA" is given in Table 27.6-2 for Zones 1 and 2 for a roof with a slope less than 2:12. Thus, these zones are not applicable on this flat gable roof when wind is blowing perpendicular to the ridge; however Zones 3 through 5 are applicable.

     Zone 3: $p_3 = -27.5$ psf applied normal to the roof area from the windward edge to $0.5h = 10$ feet from the windward edge (see Figure 27.6-1 and Table 27.6-2)

Zone 4: $p_4 = -24.5$ psf applied normal to the roof area from $0.5h = 10$ feet from the windward edge to $h = 20$ feet from the windward edge (see Figure 27.6-1 and Table 27.6-2)

Zone 5: $p_5 = -20.1$ psf applied normal to the remaining roof area (see Figure 27.6-1 and Table 27.6-2)

b. N-S wind

- Along-wind net wall pressures

   The along-wind net wall pressures are obtained by reading the values from Table 27.6-1 for Exposure C, a wind velocity of 115 mph, a mean roof height of 20 feet and $L/B = 148/256 = 0.6$:

   At the top of the wall: $p_h = 28.6$ psf for L/B between 0.5 and 1.0

   At the bottom of the wall: $p_o = 28.3$ psf for L/B between 0.5 and 1.0

   These pressures are applied to the projected area of the building walls in the direction of the wind (see Figure 27.6-1 and Figure 5.14).

- Side wall external pressures

   According to Note 2 in Table 27.6-1, the uniform side wall external pressures, which are applied to the projected area of the building walls in the direction normal to the direction of the wind, are equal to $0.54p_h = 15.4$ psf since $0.2 < L/B = 0.6 < 1.0$.

   It is important to note that these pressures do not include the effects from internal pressure.

- Roof pressures

   Since the wind pressures on the roof are independent of the ratio L/B, the wind pressures for wind in the N-S direction are the same as those determined previously from Table 27.6-2 for the E-W direction.

Illustrated in Figures 5.27 and 5.28 are the design wind pressures in the E-W and N-S directions, respectively. In accordance with 27.6.1, the wind pressures on the walls are applied simultaneously with those on the roof.

Figure 5.27
Design Wind Pressures on MWFRS in E-W Direction, Example 5.2

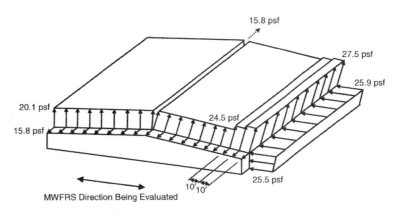

Note: dashed arrows represent uniformly distributed load over side surface.

Figure 5.28
Design Wind Pressures on MWFRS in N-S Direction, Example 5.2

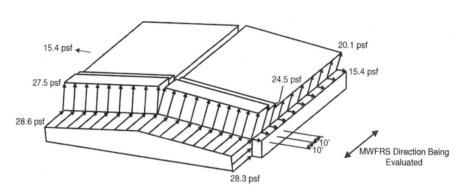

The MWFRS in each direction must be designed for the wind load cases defined in Figure 27.4-8 (see Figure 5.13). Since this warehouse is a one-story building that has a mean roof height less than or equal to 30 feet, it meets the requirements of D1.1 in Appendix D; thus, the torsional load cases in Figure 27.4-8 (Cases 2 and 4) need not be considered.

The minimum design wind loading prescribed in 27.1.5 must be considered as a load case in addition to those load cases described above (see Figure C27.4-1 and Figure 5.10).

5.8.3 Example 5.3 – Warehouse Building Using Chapter 28, Part 1 (MWFRS)

For the one-story warehouse illustrated in Figure 5.23, determine design wind pressures on the MWFRS in both directions using Part 1 of Chapter 28 (Envelope Procedure for enclosed and partially enclosed low-rise buildings).

SOLUTION

- **Step 1:** Check if the building meets all of the conditions and limitations of 28.1.2 and 28.1.3 and the additional condition that the building is a low-rise building so that this procedure can be used to determine the wind pressures on the MWFRS.

  The building is regularly-shaped, that is, it does not have any unusual geometric irregularities in spatial form. Also, the building does not have response characteristics that make it subject to across-wind loading or other similar effects, and it is not sited at a location where channeling effects or buffeting in the wake of upwind obstructions need to be considered.

  Two conditions must be checked to determine if a building is a low-rise building (26.2):

  1. Mean roof height = 20 feet < 60 feet

  2. Mean roof height = 20 feet < least horizontal dimension = 148 feet

     Therefore, this building is a low-rise building.

Thus, Part 1 of Chapter 28 may be used to determine the design wind pressures on the MWFRS.

- **Step 2:** Use Flowchart 5.5 to determine the net design wind pressures, p, on the MWFRS.

  1. Determine velocity pressure, q_h, using Flowchart 5.2.

     a. Determine the risk category of the building using IBC Table 1604.5.

        Due to the nature of its occupancy, this warehouse building falls under Risk Category II.

     b. Determine basic wind speed, V, for the applicable risk category from IBC Figure 1609A, B or C, or ASCE/SEI Figure 26.5-1A, B or C.

        For Risk Category II, use Figure 1609A or Figure 26.5-1A. From either of these figures, $V = 115$ mph for St. Louis, MO.

     c. Determine wind directionality factor, K_d, from Table 26.6-1.

        For the MWFRS of a building structure, $K_d = 0.85$.

     d. Determine exposure category.

        In the design data of Example 5.1, the surface roughness is given as C. It is assumed that Exposures B and D are not applicable, so Exposure C applies (see 26.7.3).

     e. Determine topographic factor, K_{zt}.

        As noted in the design data of Example 5.1, the building is not situated on a hill, ridge or escarpment. Thus, topographic factor $K_{zt} = 1.0$ (26.8.2).

     f. Determine velocity pressure exposure coefficients, K_h, from Table 28.3-1.

        For a mean roof height of 20 feet and Exposure C, $K_h = 0.9$.

     g. Determine velocity pressure, q_h, by Equation 28.3-1.

        $q_h = 0.00256 K_z K_{zt} K_d V^2 = 0.00256 \times 0.9 \times 1.0 \times 0.85 \times 115^2 = 25.9$ psf

  2. Determine external pressure coefficients, (GC_{pf}), from Figure 28.4-1.

     External pressure coefficients, (GC_{pf}), can be read directly from Figure 28.4-1 using a roof angle between 0 and 5 degrees in the E-W direction (Load Case A) and a roof angle between 0 and 90 degrees for wind in the N-S direction (Load Case B). Pressure coefficients for both load cases are summarized in Table 5.17.

Table 5.17 External Pressure Coefficients, (GC_{pf}), for MWFRS

Surface	(GC_{pf})	
	Load Case A	Load Case B
1	0.40	−0.45
2	−0.69	−0.69
3	−0.37	−0.37
4	−0.29	−0.45
5	---	0.40
6	---	−0.29
1E	0.61	−0.48
2E	−1.07	−1.07
3E	−0.53	−0.53
4E	−0.43	−0.48
5E	---	0.61
6E	---	−0.43

3. Determine internal pressure coefficients, (GC_{pi}), from Table 26.11-1.

For an enclosed building, $(GC_{pi}) = +0.18, -0.18$.

4. Determine design wind pressures, p, by Equation 28.4-1.

$$p = q_h[(GC_{pf}) - (GC_{pi})] = 25.9[(GC_{pf}) - (\pm 0.18)]$$

Calculations of design wind pressures are illustrated for surface 1 for wind in the E-W and N-S directions:

E-W direction (Load Case A):

  For positive internal pressure: $p = 25.9(0.40 - 0.18) = 5.7$ psf
  For negative internal pressure: $p = 25.9[0.40 - (-0.18)] = 15.0$ psf

N-S direction (Load Case B):

  For positive internal pressure: $p = 25.9(-0.45 - 0.18) = -16.3$ psf
  For negative internal pressure: $p = 25.9[-0.45 - (-0.18)] = -7.0$ psf

A summary of the design wind pressures is given in Table 5.18. Pressures are applicable to wind in both the E-W and N-S directions and are provided for both positive and negative internal pressures. These pressures act normal to the surface.

Table 5.18 Design Wind Pressures, p, on MWFRS

Surface	Design Pressure, p (psf)					
	Load Case A			Load Case B		
	(GC_{pf})	(GC_{pi})		(GC_{pf})	(GC_{pi})	
		+0.18	−0.18		+0.18	−0.18
1	0.40	5.7	15.0	−0.45	−16.3	−7.0
2	−0.69	−22.5	−13.2	−0.69	−22.5	−13.2
3	−0.37	−14.2	−4.9	−0.37	−14.2	−4.9
4	−0.29	−12.2	−2.8	−0.45	−16.3	−7.0
5	---	---	---	0.40	5.7	15.0
6	---	---	---	−0.29	−12.2	−2.8
1E	0.61	11.1	20.5	−0.48	−17.1	−7.8
2E	−1.07	−32.4	−23.1	−1.07	−32.4	−23.1
3E	−0.53	−18.4	−9.1	−0.53	−18.4	−9.1
4E	−0.43	−15.8	−6.5	−0.48	−17.1	−7.8
5E	---	---	---	0.61	11.1	20.5
6E	---	---	---	−0.43	−15.8	6.5

Determine distance, a, from Note 9 in Figure 28.4-1:

$$a = 0.1 \text{ (least horizontal dimension)} = 0.1 \times 148 = 14.8 \text{ feet}$$

$$= 0.4h = 0.4 \times 20 = 8 \text{ feet (governs)}$$

Minimum $a = 0.04$ (least horizontal dimension) $= 0.04 \times 148 = 5.9$ feet

$$= 3 \text{ feet}$$

According to Note 8 in Figure 28.4-1, when the roof pressure coefficients, (GC_{pf}), are negative in Zones 2 or 2E, they shall be applied in Zone 2/2E for a distance from the edge of the roof equal to 50 percent of the horizontal dimension of the building that is parallel to the direction of the MWFRS being designed or 2.5 times the eave height, h_e, at the windward wall, whichever is less. The remainder of Zone 2/2E extending to the ridge line must use the pressure coefficients, (GC_{pf}), for Zone 3/3E.

For this building:

Transverse direction: $0.5 \times 256 = 128$ feet

Longitudinal direction: $0.5 \times 148 = 74$ feet

$2.5h_e = 2.5 \times 20 = 50$ feet (governs in both directions)

Therefore, in the transverse direction, Zone 2/2E applies over a distance of 50 feet from the edge of the windward roof, and Zone 3/3E applies over a distance of $128 − 50 = 78$ feet in what is normally considered to be Zone 2/2E. In the longitudinal direction, Zone 3/3E is applied over a distance of $74 − 50 = 24$ feet.

The design pressures are to be applied on the building in accordance with Load Cases A and B illustrated in Figure 28.4-1. As shown in the figure, each corner of the building is considered a windward corner.

According to Note 4 in Figure 28.4-1, combinations of external and internal pressures are to be evaluated to obtain the most severe loading. Thus, when both positive and negative pressures are considered, a total of 16 separate loading patterns must be evaluated for this building (in general, the number of load cases can be reduced for symmetrical buildings).

Illustrated in Figures 5.29 and 5.30 are the design wind pressures for one load pattern in Load Case A and one load pattern in Load Base B, respectively, including positive and negative internal pressure.

Figure 5.29 Design Wind Pressures on MWFRS—Load Case A (E-W Wind), Example 5.3

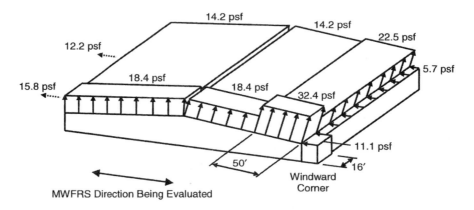

(a) Positive Internal Pressure

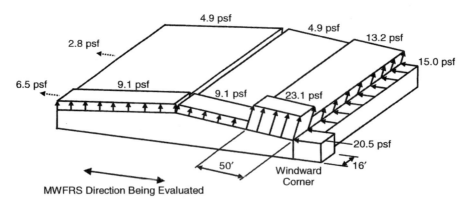

(b) Negative Internal Pressure

Figure 5.30 Design Wind Pressures on MWFRS—Load Case B (N-S Wind), Example 5.3

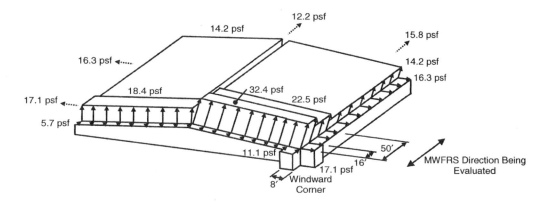

(a) Positive Internal Pressure

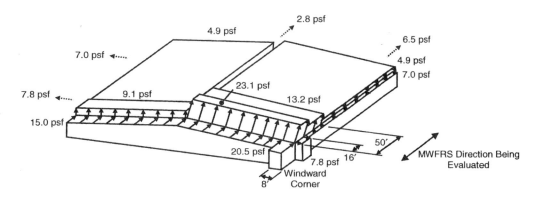

(b) Negative Internal Pressure

Torsional load cases, which are also given in Figure 28.4-1, must be considered in addition to Load Cases A and B noted above, unless the exception in Note 5 of the figure is satisfied. Since this building is one story with a mean roof height, h, less than 30 feet, the first item under this exception is satisfied and torsional load cases need not be considered in this example.

The minimum design loading of 28.4.4 must also be investigated (see Figure C27.4-1 and Figure 5.10).

5.8.4 Example 5.4 – Warehouse Building Using Chapter 28, Part 2 (MWFRS)

For the one-story warehouse illustrated in Figure 5.23, determine design wind pressures on the MWFRS in both directions using Part 2 of Chapter 28 (Envelope Procedure for enclosed, simple diaphragm low-rise buildings).

SOLUTION

- **Step 1:** Check if the building meets all of the conditions and limitations of 28.1.2 and 28.1.3 and the additional conditions in 28.6.2 so that this procedure can be used to determine the wind pressures on the MWFRS.

  The building is regularly-shaped, that is, it does not have any unusual geometric irregularities in spatial form. Also, the building does not have response characteristics that make it subject to across-wind loading or other similar effects, and it is not sited at a location where channeling effects or buffeting in the wake of upwind obstructions need to be considered.

  The following conditions are also relevant (see 28.6.2):

  - The building is a simple diaphragm building as defined in 26.2, since windward and leeward wind loads are transmitted through the metal deck roof (diaphragm) to the precast walls (MWFRS), and there are no separations in the MWFRS.

  - The building is a low-rise building as defined in 26.2, since the mean roof height of 20 feet is less than 60 feet and the mean roof height is less than the least horizontal dimension of 148 feet.

  - The building is enclosed.

  - It was shown in Example 5.1 that the building is not a flexible building as defined in 26.2.

  - The building has a symmetrical cross-section in each direction, and the slope of the roof is less than 45 degrees.

  - It was shown in Example 5.3 that the building is exempted from the torsional load cases indicated in Note 5 of Figure 28.4-1.

Thus, Part 2 of Chapter 28 may be used to determine the design wind pressures on the MWFRS.

- **Step 2:** Use Flowchart 5.6 to determine the net design wind pressures, p, on the MWFRS.

  1. Determine the risk category of the building using IBC Table 1604.5.

     Due to the nature of its occupancy, this warehouse building falls under Risk Category II.

  2. Determine basic wind speed, V, for the applicable risk category from IBC Figure 1609A, B or C, or ASCE/SEI Figure 26.5-1A, B or C.

     For Risk Category II, use Figure 1609A or Figure 26.5-1A. From either of these figures, $V = 115$ mph for St. Louis, MO.

  3. Determine exposure category.

In the design data of Example 5.1, the surface roughness is given as C. It is assumed that Exposures B and D are not applicable, so Exposure C applies (see 26.7.3).

4. Determine topographic factor, K_{zt}.

   As noted in the design data of Example 5.1, the building is not situated on a hill, ridge or escarpment. Thus, topographic factor $K_{zt} = 1.0$ (26.8.2).

5. Determine design wind pressures, p_{s30}, from Figure 28.6-1 for Zones A through H on the building.

   Wind pressures, p_{s30}, can be read directly from Figure 28.6-1 for $V = 115$ mph and a roof angle between 0 and 5 degrees. Since the roof is essentially flat, only Load Case 1 is considered (see Note 4 in Figure 28.6-1). These pressures, which are based on Exposure B, $h = 30$ feet and $K_{zt} = 1.0$, are given in Table 5.19.

Table 5.19 Wind Pressures, p_{s30}, on MWFRS

Horizontal pressures (psf)				Vertical pressures (psf)			
A	B	C	D	E	F	G	H
21.0	−10.9	13.9	−6.5	−25.2	−14.3	−17.5	−11.1

6. Determine net design wind pressures $p_s = \lambda K_{zt} p_{s30}$ by Equation 28.6-1 for Zones A through H.

   From Figure 28.6-1, the adjustment factor for building height and exposure, λ, is equal to 1.29 for a mean roof height of 20 feet and Exposure C. Thus,

   $p_s = 1.29 \times 1.0 \times p_{s30} = 1.29 p_{s30}$

   The horizontal pressures in Table 5.20 represent the combination of the windward and leeward net (sum of internal and external) pressures. Similarly, the vertical pressures represent the net (sum of internal and external) pressures.

Table 5.20 Wind Pressures, p_s, on MWFRS

Horizontal pressures (psf)				Vertical pressures (psf)			
A	B	C	D	E	F	G	H
27.1	−14.1	17.9	−8.4	−32.5	−18.5	−22.6	−14.3

The net design pressures, p_s, in Table 5.20 are to be applied to the surfaces of the building in accordance with Cases A and B in Figure 28.6-1.

According to Note 7 in Figure 28.6-1, the total horizontal load must not be less than that determined by assuming $p_s = 0$ in Zones B and D. Since the net pressures in Zones B and D in this example act in the direction opposite to those in A and C, they decrease the horizontal load. Thus, the pressures in Zones B and D are set equal to 0 when analyzing the structure for wind in the E-W direction (Case A).

According to Note 2 in Figure 28.6-1, the load patterns for Case A and Case B are to be applied to each corner of the building, that is, each corner of the building must be considered a reference (windward) corner. Eight different load cases need to be examined (four in Case A and four in Case B). One load pattern in the E-W direction (Case A) and one in the N-S direction (Case B) are illustrated in Figure 5.31.

Figure 5.31 Design Wind Pressures on MWFRS, Example 5.4

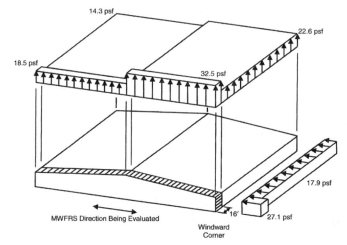

Case A (E-W Wind)

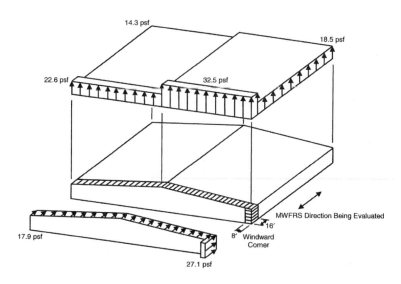

Case B (N-S Wind)

Determine distance, a, from Note 9 in Figure 28.4-1:

$$a = 0.1 \text{ (least horizontal dimension)} = 0.1 \times 148 = 14.8 \text{ feet}$$

$$= 0.4h = 0.4 \times 20 = 8 \text{ feet (governs)}$$

Minimum $a = 0.04$ (least horizontal dimension) $= 0.04 \times 148 = 5.9$ feet

$$= 3 \text{ feet}$$

The minimum design wind load case of 28.6.4 must also be considered: the load effects from the design wind pressures calculated above must not be less than the load effects assuming that $p_s = +16$ psf in Zones A and C, $p_s = +8$ psf in Zones B and D and $p_s = 0$ psf in Zones E through H (see Figure C27.4-1 and Figure 5.10 for application of these pressures).

5.8.5 Example 5.5 – Warehouse Building Using Alternate All-heights Method (MWFRS)

For the one-story warehouse illustrated in Figure 5.23, determine design wind pressures on the MWFRS in both directions using the alternate all-heights method of IBC 1609.6.

SOLUTION

- **Step 1:** Check if the provisions of IBC 1609.6 can be used to determine the design wind pressures on this building.

    The provisions of IBC 1609.6 may be used to determine design wind pressures on this regularly-shaped building provided the conditions of IBC 1609.6.1 are satisfied:

    1. The height of the building is 20 feet, which is less than 75 feet, and the height-to-least-width ratio = 20/148 = 0.14 < 4. Also, it was shown in Example 5.1 that the fundamental frequency $n_1 > 1$ Hz in both directions.

    2. As was discussed in Example 5.1, this building is not sensitive to dynamic effects.

    3. This building is not located on a site where channeling effects or buffeting in the wake of upwind obstructions need to be considered.

    4. It was shown in Example 5.2 that the building meets the requirements of a simple diaphragm building as defined in 26.2.

    5. The fifth condition is not applicable in this example.

The provisions of the alternate all-heights method of IBC 1609.6 can be used to determine the design wind pressures on the MWFRS.

- **Step 2:** Use Flowchart 5.15 to determine the net design wind pressures, p_{net}, on the MWFRS.

    1. Determine the risk category of the building using IBC Table 1604.5.

        Due to the nature of its occupancy, this warehouse building falls under Risk Category II.

    2. Determine basic wind speed, V, for the applicable risk category from IBC Figure 1609A, B or C, or ASCE/SEI Figure 26.5-1A, B or C.

        For Risk Category II, use Figure 1609A or Figure 26.5-1A. From either of these figures, $V = 115$ mph for St. Louis, MO.

3. Determine exposure category.

   In the design data of Example 5.1, the surface roughness is given as C. It is assumed that Exposures B and D are not applicable, so Exposure C applies (see 26.7.3).

4. Determine topographic factor, K_{zt}.

   As noted in the design data of Example 5.1, the building is not situated on a hill, ridge or escarpment. Thus, topographic factor $K_{zt} = 1.0$ (26.8.2).

5. Determine velocity pressure exposure coefficients, K_z, from Table 27.3-1.

   Values of K_z are summarized in Table 5.21.

Table 5.21 Velocity Pressure Exposure Coefficient, K_z

Height above ground level, z (feet)	K_z
20	0.90
15	0.85

6. Determine net pressure coefficients, C_{net}, for the walls and roof from IBC Table 1609.6.2 for an enclosed building.

   - For wind in the E-W direction:

     Windward wall: $C_{net} = 0.43$ for positive internal pressure
     $C_{net} = 0.73$ for negative internal pressure

     Leeward wall: $C_{net} = -0.51$ for positive internal pressure
     $C_{net} = -0.21$ for negative internal pressure

     Side walls: $C_{net} = -0.66$ for positive internal pressure
     $C_{net} = -0.35$ for negative internal pressure

     Leeward roof (wind perpendicular to ridge):
     $C_{net} = -0.66$ for positive internal pressure
     $C_{net} = -0.35$ for negative internal pressure

     Windward roof (wind perpendicular to ridge with roof slope < 2:12):
     $C_{net} = -1.09, -0.28$ for positive internal pressure
     $C_{net} = -0.79, 0.02$ for negative internal pressure

   - For wind in the N-S direction:

     Windward wall: $C_{net} = 0.43$ for positive internal pressure
     $C_{net} = 0.73$ for negative internal pressure

     Leeward wall: $C_{net} = -0.51$ for positive internal pressure
     $C_{net} = -0.21$ for negative internal pressure

     Side walls: $C_{net} = -0.66$ for positive internal pressure
     $C_{net} = -0.35$ for negative internal pressure

     Roof (wind parallel to ridge):
     $C_{net} = -1.09$ for positive internal pressure
     $C_{net} = -0.79$ for negative internal pressure

7. Determine net design wind pressures, p_{net}, by IBC Equation 16-35.

$$p_{net} = 0.00256 V^2 K_z C_{net} K_{zt} = 33.9 K_z C_{net}$$

According to IBC 1609.6.4.2, windward wall pressures are based on height, z, and leeward walls, side walls and roof pressures are based on mean roof height, h. A summary of the net design wind pressures in the E-W and N-S directions is given in Tables 5.22 and 5.23, respectively. Illustrated in Figures 5.32 and 5.33 are the net design wind pressures in the E-W and N-S directions, respectively, for positive and negative internal pressure. Note that for wind in the E-W direction, only the Condition 1 wind pressures on the roof are illustrated in Figure 5.32. Although the Condition 2 pressures are not shown in the figure, they must be considered in the overall design.

Table 5.22 Net Design Wind Pressures, p_{net}, in E-W Direction

Location		Height above ground level, z (feet)	K_z	C_{net}		Net design pressure, p_{net} (psf)	
				+ Internal pressure	− Internal pressure	+ Internal pressure	− Internal pressure
Windward wall		20	0.90	0.43	0.73	13.1	22.3
		15	0.85	0.43	0.73	12.4	21.0
Leeward wall		All	0.90	−0.51	−0.21	−15.6	−6.4
Side walls		All	0.90	−0.66	−0.35	−20.1	−10.7
Roof	Windward	20	0.90	−1.09	−0.79	−33.3	−24.1
		20	0.90	−0.28	0.02	−8.5	0.6
	Leeward	20	0.90	−0.66	−0.35	−20.1	−10.7

Table 5.23 Net Design Wind Pressures, p_{net}, in N-S Direction

Location	Height above ground level, z (feet)	K_z	C_{net}		Net design pressure, p_{net} (psf)	
			+ Internal pressure	− Internal pressure	+ Internal pressure	− Internal pressure
Windward wall	20	0.90	0.43	0.73	13.1	22.3
	15	0.85	0.43	0.73	12.4	21.0
Leeward wall	All	0.90	−0.51	−0.21	−15.6	−6.4
Side walls	All	0.90	−0.66	−0.35	−20.1	−10.7
Roof	20	0.90	−1.09	−0.79	−33.3	−24.1

Figure 5.32 Design Wind Pressures on MWFRS in E-W Direction, Example 5.5

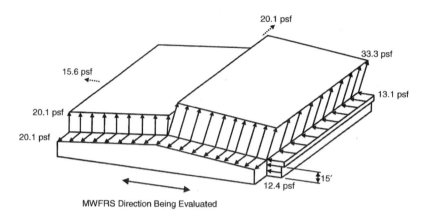

(a) Positive Internal Pressure

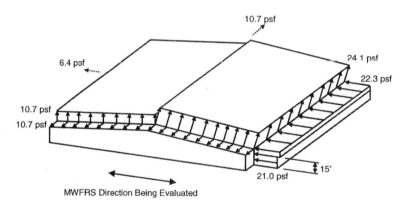

(b) Negative Internal Pressure

Figure 5.33 Design Wind Pressures on MWFRS in N-S Direction, Example 5.5

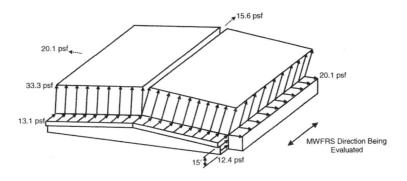

(a) Positive Internal Pressure

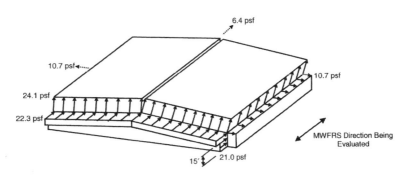

(b) Negative Internal Pressure

The MWFRS of buildings whose wind loads have been determined by IBC 1609.6 must be designed for the wind load cases identified in ASCE/SEI Figure 27.4-8 (IBC 1609.6.4.1). In Case 1, the full design wind pressures act on the projected area perpendicular to each principal axis of the structure. These pressures are assumed to act separately along each principal axis. The wind pressures on the windward and leeward walls depicted in Figures 5.32 and 5.33 fall under Case 1.

IBC 1609.6.4.1 requires consideration of torsional effects as indicated in Figure 27.4-8. According to the exception in 27.4.6, buildings that meet the requirements of D1.1 of Appendix D need only be designed for Load Case 1 and Load Case 3. The building in this example meets one of these requirements (one-story building less than or equal to 30 feet in height) so only Load Cases 1 and 3 need to be investigated.

In Case 3, 75 percent of the wind pressures on the windward and leeward walls of Case 1, which are shown in Figures 5.32 and 5.33, act simultaneously on the building (see ASCE/SEI Figure 27.4-8 and Figure 5.13). This load case, which needs to be considered in addition to the load cases illustrated in Figures 5.32 and 5.33, accounts for the effects due to wind along the diagonal of the building.

Finally, the minimum design wind loading prescribed in IBC 1609.6.3 must be considered as a load case in addition to those load cases described above. The minimum 16 psf wind pressure acts on the area of the building projected on a plane normal to the direction of wind.

5.8.6 Example 5.6 – Residential Building Using Chapter 27, Part 1 (MWFRS)

For the three-story residential building illustrated in Figure 5.34, determine design wind pressures on the MWFRS in both directions using Chapter 27, Part 1 (Directional Procedure for buildings of all heights).

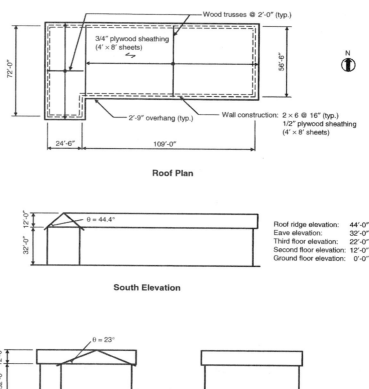

Figure 5.34 Roof Plan and Elevations of Three-story Residential Building, Example 5.6

DESIGN DATA	
Location:	Sacramento, CA
Surface Roughness:	B (suburban area with numerous closely spaced obstructions having the size of single-family dwellings and larger)
Topography:	Not situated on a hill, ridge or escarpment
Occupancy:	Residential building where less than 300 people congregate in one area

SOLUTION

- **Step 1:** Check if the building meets all of the conditions and limitations of 27.1.2 and 27.1.3 so that this procedure can be used to determine the wind pressures on the MWFRS.

  The building is regularly-shaped, that is, it does not have any unusual geometric irregularities in spatial form. Also, the building does not have response characteristics that make it subject to across-wind loading or other similar effects, and it is not sited at a location where channeling effects or buffeting in the wake of upwind obstructions need to be considered.

  Thus, Part 1 of Chapter 27 may be used to determine the design wind pressures on the MWFRS.

  Note: Even though the building is less than 60 feet in height, it is not recommended to use the provisions of Chapter 28, since L-, T-, and U-shaped buildings are considered to be outside of the scope of that method.

- **Step 2:** Determine the enclosure classification of the building.

  Assume that the building is enclosed.

- **Step 3:** Use Flowchart 5.3 to determine the design wind pressures, p, on the MWFRS.

  1. Determine whether the building is rigid or flexible in accordance with 26.9.2.

     Assume that the fundamental frequency of the building, n_1, has been determined to be greater than 1 Hz. Thus, the building is rigid.

  2. Determine the gust-effect factor, G, using Flowchart 5.1.

     According to 26.9.4, gust-effect factor, G, for rigid buildings may be taken as 0.85 or can be calculated by Equation 26.9-6. For simplicity, use $G = 0.85$.

  3. Determine velocity pressure: q_z for windward walls along the height of the building, and q_h for leeward walls, side walls and roof using Flowchart 5.2.

     a. Determine the risk category of the building using IBC Table 1604.5.

        Due to the nature of its occupancy, this residential building falls under Risk Category II.

     b. Determine basic wind speed, V, for the applicable risk category from IBC Figure 1609A, B or C, or ASCE/SEI Figure 26.5-1A, B or C.

        For Risk Category II, use Figure 1609A or Figure 26.5-1A. From either of these figures, $V = 110$ mph for Sacramento, CA.

     c. Determine wind directionality factor, K_d, from Table 26.6-1.

        For the MWFRS of a building structure, $K_d = 0.85$.

     d. Determine exposure category.

In the design data, the surface roughness is given as B. Assume that Exposures B is applicable in all directions.

e. Determine topographic factor, K_{zt}.

As noted in the design data, the building is not situated on a hill, ridge or escarpment. Thus, topographic factor $K_{zt} = 1.0$ (26.8.2).

f. Determine velocity pressure exposure coefficients K_z and K_h from Table 27.3-1.

Values of K_z and K_h for Exposure B are summarized in Table 5.24.

Mean roof height = $\dfrac{44 + 32}{2}$ = 38 feet

Table 5.24 Velocity Pressure Exposure Coefficient, K_z

Height above ground level, z (feet)	K_z
38	0.75
30	0.70
25	0.66
20	0.62
15	0.57

g. Determine velocity pressure, q_z and q_h, by Equation 27.3-1.

$$q_z = 0.00256 K_z K_{zt} K_d V^2 = 0.00256 \times K_z \times 1.0 \times 0.85 \times 110^2 = 26.33 K_z \text{ psf}$$

A summary of the velocity pressures is given in Table 5.25.

Table 5.25 Velocity Pressure, q_z

Height above ground level, z (feet)	K_z	q_z (psf)
38	0.75	19.8
30	0.70	18.4
25	0.66	17.4
20	0.62	16.3
15	0.57	15.0

4. Determine pressure coefficients, C_p, for the walls and roof from Figure 27.4-1.

Since the building is not symmetric, all four wind directions normal to the walls must be considered.

Figure 5.35 provides identification marks for each surface of the building.

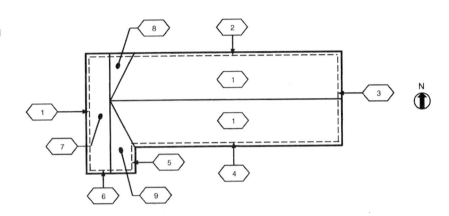

Figure 5.35
Identification Marks for Building Surfaces, Example 5.6

Tables 5.26 through 5.29 contain the external pressure coefficients for wind in all four directions.

Table 5.26 External Pressure Coefficients, C_p, for Wind from West to East

Surface(s)	Type		C_p	Use with
1	Windward wall		0.80	q_z
1	Overhang		0.80*	q_z
2, 4, 6	Side wall		−0.70	q_h
3, 5	Leeward wall		−0.32**	q_h
7	Windward roof		0.40†	q_h
8, 9	Leeward roof		−0.60†	q_h
10, 11	Roof parallel to ridge‡	Edge to 38′	−0.90, −0.18	q_h
		38′ to 76′	−0.50, −0.18	
		76′ to 133.5′	−0.30, −0.18	

* See 27.4.4
** Obtained by linear interpolation using $L/B = 133.5/72 = 1.9$
† Normal to ridge with θ = 44.4 degrees and $h/L = 38/133.5 = 0.3$
‡ The smaller uplift pressures on the roof due to $C_p = -0.18$ may govern the design when combined with roof live load or snow loads. This pressure is not shown in this example, but in general it must be considered.

Table 5.27 External Pressure Coefficients, C_p, for Wind from East to West

Surface(s)	Type		C_p	Use with
1	Leeward wall		−0.32*	q_h
2, 4, 6	Side wall		−0.70	q_h
3, 5	Windward wall		0.80	q_z
3, 5	Overhang		0.80**	q_h
7	Leeward roof		−0.60†	q_h
8, 9	Windward roof		0.40†	q_h
10, 11	Roof parallel to ridge‡	Edge to 38′	−0.90, −0.18	q_h
		38′ to 76′	−0.50, −0.18	
		76′ to 133.5′	−0.30, −0.18	

* Obtained by linear interpolation using $L/B = 133.5/72 = 1.9$
** See 27.4.4
† Normal to ridge with $\theta = 44.4$ degrees and $h/L = 38/133.5 = 0.3$
‡ The smaller uplift pressures on the roof due to $C_p = -0.18$ may govern the design when combined with roof live load or snow loads. This pressure is not shown in this example, but in general it must be considered.

Table 5.28 External Pressure Coefficients, C_p, for Wind from North to South

Surface(s)	Type		C_p	Use with
1, 3, 5	Side wall		−0.70	q_h
2	Windward wall		0.80	q_z
2	Overhang		0.80*	q_h
4, 6	Leeward wall		−0.50**	q_h
7, 8	Roof parallel to ridge†	Edge to 38′	−0.90, −0.18	q_h
		38′ to 72′	−0.50, −0.18	
9, 11	Leeward roof		−0.60‡	q_h
10	Windward roof		0.11, −0.35‡	q_h

* See 27.4.4
** $L/B = 72/133.5 = 0.54$
† For surface 8, $C_p = -0.90$ for the entire length. The smaller uplift pressures on the roof due to $C_p = -0.18$ may govern the design when combined with roof live load or snow loads. This pressure is not shown in this example, but in general it must be considered.
‡ Normal to ridge with $\theta = 23$ degrees and $h/L = 38/72 = 0.53$. On windward roof, values are obtained by linear interpolation.

Table 5.29 External Pressure Coefficients, C_p, for Wind from South to North

Surface(s)	Type		C_p	Use with
1, 3, 5	Side wall		−0.70	q_h
2	Leeward wall		−0.50*	q_h
4, 6	Windward wall		0.80	q_z
4, 6	Overhang		0.80**	q_h
7, 9	Roof parallel to ridge†	Edge to 38′	−0.90, −0.18	q_h
7, 9	Roof parallel to ridge†	38′ to 72′	−0.50, −0.18	q_h
8, 10	Leeward roof		−0.60‡	q_h
11	Windward roof		0.11, −0.35‡	q_h

* $L/B = 72/133.5 = 0.54$
** See 27.4.4
† For surface 9, $C_p = -0.90$ for the entire length. The smaller uplift pressures on the roof due to $C_p = -0.18$ may govern the design when combined with roof live load or snow loads. This pressure is not shown in this example, but in general it must be considered.
‡ Normal to ridge with $\theta = 23$ degrees and $h/L = 38/72 = 0.53$. Values are obtained by linear interpolation.

5. Determine q_i for the walls and roof using Flowchart 5.2.

   In accordance with 27.4.1, $q_i = q_h = 19.8$ psf for windward walls, side walls, leeward walls and roofs of enclosed buildings.

6. Determine internal pressure coefficients, (GC_{pi}), from Table 26.11-1.

   For an enclosed building, $(GC_{pi}) = +0.18, -0.18$.

7. Determine design wind pressures p_z and p_h by Equation 27.4-1.

   Windward walls:

   $p_z = q_z GC_p - q_h(GC_{pi})$

   $ = (q_z \times 0.85 \times 0.8) - 19.8(\pm 0.18)$

   $ = (0.68 q_z \mp 3.6)$ psf (external ± internal pressure)

   Leeward wall, side walls and roof:

   $p_h = q_h GC_p - q_h(GC_{pi})$

   $ = (19.8 \times 0.85 \times C_p) - 19.8(\pm 0.18)$

   $ = (16.8 C_p \mp 3.6)$ psf (external ± internal pressure)

   Overhangs:

   $p_h = q_h GC_p$

   $ = 19.8 \times 0.85 \times 0.8 = 13.5$ psf

   Maximum design wind pressures in all four wind directions are given in Tables 5.30 through 5.33.

The MWFRS of buildings, whose wind loads have been determined by Chapter 27, must be designed for the wind load cases defined in Figure 27.4-8 (27.4.6). Since the building in this example is not symmetrical, all four wind directions must be considered when combining wind loads according to Figure 27.4-8.

In Case 1, the full design wind pressures act on the projected area perpendicular to each principal axis of the structure. These pressures are assumed to act separately along each principal axis. The wind pressures on the windward and leeward walls given in Tables 5.30 through 5.33 fall under Case 1.

Table 5.30 Design Wind Pressures, p, for Wind from West to East

Surface(s)	Height above ground level, z (feet)	q (psf)	External pressure, qGC_p (psf)	Internal pressure, $q_h(GC_{pi})$ (psf)	Net pressure, p (psf) $+(GC_{pi})$	Net pressure, p (psf) $-(GC_{pi})$
1	38	19.8	13.5	±3.6	9.9	17.1
1	30	18.4	12.5	±3.6	8.9	16.1
1	25	17.4	11.8	±3.6	8.2	15.4
1	20	16.3	11.1	±3.6	7.5	14.7
1	15	15.0	10.2	±3.6	6.6	13.8
2, 4, 6	All	19.8	−11.8	±3.6	−15.4	−8.2
3, 5	All	19.8	−5.4	±3.6	−9.0	−1.8
7	38	19.8	6.7	±3.6	3.1	10.3
8, 9	38	19.8	−10.1	±3.6	−13.7	−6.5
10, 11	38	19.8	−15.1*	±3.6	−18.7	−11.5
10, 11	38	19.8	−8.4†	±3.6	−12.0	−4.8
10, 11	38	19.8	−5.0‡	±3.6	−8.6	−1.4

* from windward edge to 38 feet
† from 38 to 76 feet
‡ from 76 to 133.5 feet

Table 5.31 Design Wind Pressures, p, for Wind from East to West

Surface(s)	Height above ground level, z (feet)	q (psf)	External pressure, qGC_p (psf)	Internal pressure, $q_h(GC_{pi})$ (psf)	Net pressure, p (psf) $+(GC_{pi})$	Net pressure, p (psf) $-(GC_{pi})$
1	All	19.8	−5.4	±3.6	−9.0	−1.8
2, 4, 6	All	19.8	−11.8	±3.6	−15.4	−8.2
3, 5	38	19.8	13.5	±3.6	9.9	17.1
3, 5	30	18.4	12.5	±3.6	8.9	16.1
3, 5	25	17.4	11.8	±3.6	8.2	15.4
3, 5	20	16.3	11.1	±3.6	7.5	14.7
3, 5	15	15.0	10.2	±3.6	6.6	13.8
7	38	19.8	−10.1	±3.6	−13.7	−6.5
8, 9	38	19.8	6.7	±3.6	3.1	10.3
10, 11	38	19.8	−15.1*	±3.6	−18.7	−11.5
10, 11	38	19.8	−8.4†	±3.6	−12.0	−4.8
10, 11	38	19.8	−5.0‡	±3.6	−8.6	−1.4

* from windward edge to 38 feet
† from 38 to 76 feet
‡ from 76 to 133.5 feet

Table 5.32 Design Wind Pressures, p, for Wind from North to South

Surface(s)	Height above ground level, z (feet)	q (psf)	External pressure, qGC_p (psf)	Internal pressure, $q_h(GC_{pi})$ (psf)	Net pressure, p (psf) $+(GC_{pi})$	Net pressure, p (psf) $-(GC_{pi})$
1, 3, 5	All	19.8	−11.8	±3.6	−15.4	−8.2
2	38	19.8	13.5	±3.6	9.9	17.1
2	30	18.4	12.5	±3.6	8.9	16.1
2	25	17.4	11.8	±3.6	8.2	15.4
2	20	16.3	11.1	±3.6	7.5	14.7
2	15	15.0	10.2	±3.6	6.6	13.8
4, 6	All	19.8	−8.4	±3.6	−12.0	−4.8
7, 8	38	19.8	−15.1*	±3.6	−18.7	−11.5
7, 8	38	19.8	−8.4†	±3.6	−12.0	−4.8
9, 11	38	19.8	−10.1	±3.6	−13.7	−6.5
10	38	19.8	1.9	±3.6	−1.7	5.5
10	38	19.8	−5.9	±3.6	−9.5	−2.3

* from windward edge to 38 feet
† from 38 to 72 feet

Table 5.33 Design Wind Pressures, p, for Wind from South to North

Surface(s)	Height above ground level, z (feet)	q (psf)	External pressure, qGC_p (psf)	Internal pressure, $q_h(GC_{pi})$ (psf)	Net pressure, p (psf) $+(GC_{pi})$	Net pressure, p (psf) $-(GC_{pi})$
1, 3, 5	All	19.8	−11.8	±3.6	−15.4	−8.2
2	All	19.8	−8.4	±3.6	−12.0	−4.8
4, 6	38	19.8	13.5	±3.6	9.9	17.1
4, 6	30	18.4	12.5	±3.6	8.9	16.1
4, 6	25	17.4	11.8	±3.6	8.2	15.4
4, 6	20	16.3	11.1	±3.6	7.5	14.7
4, 6	15	15.0	10.2	±3.6	6.6	13.8
7, 9	38	19.8	−15.1*	±3.6	−18.7	−11.5
7, 9	38	19.8	−8.4†	±3.6	−12.0	−4.8
8, 10	38	19.8	−10.1	±3.6	−13.7	−6.5
11	38	19.8	1.9	±3.6	−1.7	5.5
11	38	19.8	−5.9	±3.6	−9.5	−2.3

* from windward edge to 38 feet
† from 38 to 72 feet

In Case 2, 75 percent of the design wind pressures on the windward and leeward walls are applied on the projected area perpendicular to each principal axis of the building along with a torsional moment. The wind pressures and torsional moment are applied separately for each principal axis.

The wind pressures and torsional moment at the mean roof height for Case 2 are as follows. (Please see Figure 5.13. Note: since the internal pressures cancel out in the horizontal direction, it makes no difference whether the net pressures based on positive or negative internal pressure are used in the load cases for the design of the MWFRS. In this example, the net pressures based on negative internal pressure are used.)

For west-to-east wind: $0.75p_{wx} = 0.75 \times 17.1 = 12.8$ psf (surface 1)

$$0.75p_{lx} = 0.75 \times 1.8 = 1.4 \text{ psf} \quad \text{(surfaces 3 and 5)}$$

$$M_T = 0.75(p_{wx} + p_{lx})B_x e_x$$

$$= 0.75(17.1 + 1.8) \times 72 \times (\pm 0.15 \times 72)$$

$$= \pm 11{,}023 \text{ ft-lb/ft}$$

For north-to-south wind: $0.75p_{wy} = 0.75 \times 17.1 = 12.8$ psf (surface 2)

$$0.75p_{ly} = 0.75 \times 4.8 = 3.6 \text{ psf} \quad \text{(surfaces 4 and 6)}$$

$$M_T = 0.75(p_{wx} + p_{ly})B_y e_y$$

$$= 0.75(17.1 + 4.8) \times 133.5 \times (\pm 0.15 \times 133.5)$$

$$= \pm 43{,}910 \text{ ft-lb/ft}$$

Similar calculations can be made for east-to-west wind and for south-to-north wind. All four of these load combinations must be considered for Case 2, since the building is not symmetrical. Similar calculations can also be made at other elevations below the roof.

In Case 3, 75 percent of the wind pressures of Case 1 are applied to the building simultaneously. This accounts for wind along the diagonal of the building. Like in Case 2, four load combinations must be considered for Case 3.

In Case 4, 75 percent of the wind pressures and torsional moments defined in Case 2 act simultaneously on the building. As with all of the other cases, four load combinations must be considered for this load case as well.

Finally, the minimum design wind loading prescribed in 27.1.5 must be considered as a load case in addition to those load cases described above (see Figure C27.4-1 and Figure 5.10).

5.8.7 Example 5.7 – Six-story Hotel Using Chapter 27, Part 1 (MWFRS)

For the six-story hotel illustrated in Figure 5.36, determine design wind pressures on the MWFRS in both directions using Chapter 27, Part 1 (Directional Procedure for buildings of all heights). Note that door and window openings are not shown in the figure.

Figure 5.36 Plan and Elevation of Six-story Hotel, Example 5.7

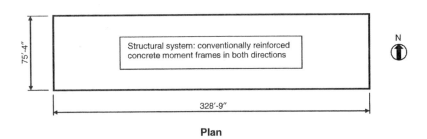

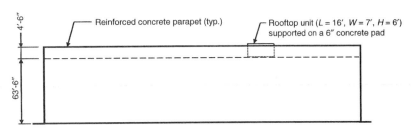

Design Data

Location:	Miami, FL
Surface Roughness:	D (adjacent to water surface)
Topography:	Not situated on a hill, ridge or escarpment
Occupancy:	Residential building where less than 300 people congregate in one area

Solution

- **Step 1:** Check if the building meets all of the conditions and limitations of 27.1.2 and 27.1.3 so that this procedure can be used to determine the wind pressures on the MWFRS.

  The building is regularly-shaped, that is, it does not have any unusual geometric irregularities in spatial form. Also, the building does not have response characteristics that make it subject to across-wind loading or other similar effects, and it is not sited at a location where channeling effects or buffeting in the wake of upwind obstructions need to be considered.

Thus, Part 1 of Chapter 27 may be used to determine the design wind pressures on the MWFRS.

Note: The mean roof height is greater than 60 feet, so the provisions of Chapter 28 cannot be used.

- **Step 2:** Determine the enclosure classification of the building.

  It is assumed in this example that the building is enclosed (see Step 3).

- **Step 3:** Use Flowchart 5.3 to determine the design wind pressures, p, on the MWFRS.

  1. Determine the design wind pressure effects of the parapet on the MWFRS.

     a. Determine the velocity pressure, q_p, at the top of the parapet from Flowchart 5.2.

        i. Determine the risk category of the building using IBC Table 1604.5.

           Due to the nature of its occupancy, this residential building falls under Risk Category II.

        ii. Determine basic wind speed, V, for the applicable risk category from IBC Figure 1609A, B or C, or ASCE/SEI Figure 26.5-1A, B or C.

           For Risk Category II, use Figure 1609A or Figure 26.5-1A. From either of these figures, $V = 170$ mph for Miami, FL.

           This building is located in a wind-borne debris region since the basic wind speed is greater than 140 mph (see 26.10.3.1). According to 26.10.3.2, glazing in buildings located in wind-borne debris regions must be protected with an impact-protective system or must be impact-resistant glazing.

           It is assumed that impact-resistant glazing is used over the entire height of the building; thus, the building is classified as enclosed.

        iii. Determine wind directionality factor, K_d, from Table 26.6-1.

           For the MWFRS of a building structure, $K_d = 0.85$.

        iv. Determine exposure category.

           In the design data, the surface roughness is given as D. Assume that Exposure D is applicable in all directions.

        v. Determine topographic factor, K_{zt}.

           As noted in the design data, the building is not situated on a hill, ridge or escarpment. Thus, topographic factor $K_{zt} = 1.0$ (26.8.2).

        vi. Determine velocity pressure exposure coefficient, K_h, from Table 27.3-1.

           For Exposure D at a height of 68 feet at the top of the parapet, $K_h = 1.33$ by linear interpolation.

        vii. Determine velocity pressure, q_p, evaluated at the top of the parapet by Equation 27.3-1.

           $q_p = 0.00256 K_h K_{zt} K_d V^2$

           $= 0.00256 \times 1.33 \times 1.0 \times 0.85 \times 170^2 = 83.6$ psf

     b. Determine combined net pressure coefficient, (GC_{pn}), for the parapets.

        In accordance with 27.4.5, $(GC_{pn}) = 1.5$ for the windward parapet and $(GC_{pn}) = -1.0$ for the leeward parapet.

c. Determine combined net design pressure, p_p, on the parapet by Equation 27.4-4.

$p_p = q_p(GC_{pn})$

$= 83.6 \times 1.5 = 125.5$ psf on windward parapet

$= 83.6 \times (-1.0) = -83.6$ psf on leeward parapet

The loads on the windward and leeward parapets can be obtained by multiplying the pressures by the height of the parapet:

Windward parapet: $F = 125.5 \times 4.5 = 564.8$ plf

Leeward parapet: $F = 83.6 \times 4.5 = 376.2$ plf

2. Determine whether the building is rigid or flexible in accordance with 26.9.2.

In lieu of determining the natural frequency, n_1, of the building from a dynamic analysis, Equation 26.9-3 is used to compute an approximate natural frequency, n_a, for concrete moment-resisting frames:

$$n_a = \frac{43.5}{h^{0.9}} = \frac{43.5}{63.5^{0.9}} = 1.04 \text{ Hz}$$

Since $n_a > 1.0$ Hz, the building is defined as a rigid building.

3. Determine gust-effect factor (G or G_f) from Flowchart 5.1.

According to 26.9.4, gust-effect factor, G, for rigid buildings may be taken as 0.85 or can be calculated by Equation 26.9-6. For simplicity, use $G = 0.85$.

4. Determine velocity pressure: q_z for windward walls along the height of the building, and q_h for leeward walls, side walls and roof using Flowchart 5.2.

Most of the quantities needed to compute q_z and q_h were determined in this step above.

The velocity exposure coefficients K_z and K_h are summarized in Table 5.34.

Table 5.34 Velocity Pressure Exposure Coefficient, K_z

Height above ground level, z (feet)	K_z
63.5	1.32
60	1.31
50	1.27
40	1.22
30	1.16
25	1.12
20	1.08
15	1.03

Velocity pressures q_z and q_h are determined by Equation 27.3-1:

$q_z = 0.00256 K_z K_{zt} K_d V^2 = 0.00256 \times K_z \times 1.0 \times 0.85 \times 170^2 = 62.89 K_z$ psf

A summary of the velocity pressures is given in Table 5.35.

Table 5.35 Velocity Pressure, q_z

Height above ground level, z (feet)	K_z	q_z (psf)
63.5	1.32	83.0
60	1.31	82.4
50	1.27	79.9
40	1.22	76.7
30	1.16	73.0
25	1.12	70.4
20	1.08	67.9
15	1.03	64.8

5. Determine pressure coefficients, C_p, for the walls and roof from Figure 27.4-1.

   For wind in the E-W direction:

   Windward wall: $C_p = 0.8$ for use with q_z

   Leeward wall ($L/B = 328.75/75.33 = 4.4$): $C_p = -0.2$ for use with q_h

   Side wall: $C_p = -0.7$ for use with q_h

   Roof (normal to ridge with $\theta < 10$ degrees and parallel to ridge for all θ with $h/L = 63.5/328.75 = 0.19 < 0.5$):

   $C_p = -0.9, -0.18$ from windward edge to $h = 63.5$ feet for use with q_h

   $C_p = -0.5, -0.18$ from 63.5 feet to $2h = 127.0$ feet for use with q_h

   $C_p = -0.3, -0.18$ from 127.0 feet to 328.75 feet for use with q_h

   Note: The smaller uplift pressures on the roof due to $C_p = -0.18$ may govern the design when combined with roof live load or snow loads. This pressure is not shown in this example but in general must be considered.

   For wind in the N-S direction:

   Windward wall: $C_p = 0.8$ for use with q_z

   Leeward wall ($L/B = 75.33/328.75 = 0.23$): $C_p = -0.5$ for use with q_h

   Side wall: $C_p = -0.7$ for use with q_h

   Roof (normal to ridge with $\theta < 10$ degrees and parallel to ridge for all θ with $h/L = 63.5/75.33 = 0.84$):

   $C_p = -1.0, -0.18$ from windward edge to $h/2 = 31.75$ feet for use with q_h

   $C_p = -0.76, -0.18$ from 31.75 feet to $h = 63.5$ feet for use with q_h

   $C_p = -0.64, -0.18$ from 63.5 feet to 75.33 feet for use with q_h

   Note: The pressure coefficient $C_p = 1.3$ from the windward edge to $h/2$. This coefficient may be reduced based on the area over which it is applicable = $(63.5/2) \times 328.75 = 10{,}438$ square feet > 1,000 square feet. Corresponding reduction factor = 0.8 from Figure 27.4-1. Thus, $C_p = 0.8 \times (-1.3) = -1.04$ was used in the linear interpolation to determine C_p for $h/L = 0.84$.

Wind Loads   271

6. Determine q_i for the walls and roof using Flowchart 5.2.

   In accordance with 27.4.1, $q_i = q_h = 83.0$ psf for windward walls, side walls, leeward walls and roofs of enclosed buildings.

7. Determine internal pressure coefficients, (GC_{pi}), from Table 26.11-1.

   For an enclosed building, $(GC_{pi}) = +0.18, -0.18$.

8. Determine design wind pressures p_z and p_h by Equation 27.4-1.

   Windward walls:

   $$p_z = q_z GC_p - q_h(GC_{pi})$$

   $= (q_z \times 0.85 \times 0.8) - 83.0(\pm 0.18) = (0.68q_z \mp 14.9)$ psf (external ± internal pressure)

   Leeward wall, side walls and roof:

   $$p_h = q_h GC_p - q_h(GC_{pi})$$

   $= (83.0 \times 0.85 \times C_p) - 83.0(\pm 0.18) = (70.6C_p \mp 14.9)$ psf (external ± internal pressure)

   A summary of the maximum design wind pressures in the E-W and N-S directions is given in Tables 5.36 and 5.37, respectively.

   Illustrated in Figures 5.37 and 5.38 are the external design wind pressures in the E-W and N-S directions, respectively. Included in the figures are the loads on the windward and leeward parapets, which add to the overall wind loads in the direction of analysis. When considering horizontal wind forces on the MWFRS, it is clear that the effects from the internal pressure cancel out. On the roof, the effects from internal pressure add directly to those from the external pressure.

Table 5.36 Design Wind Pressures, p, in the E-W Direction

Location	Height above ground level, z (feet)	q (psf)	External pressure qGC_p (psf)	Internal pressure q_hGC_{pi} (psf)	Net pressure, p (psf) +(GC_{pi})	Net pressure, p (psf) −(GC_{pi})
Windward	63.5	83.0	56.4	±14.9	41.5	71.3
Windward	60	82.4	56.0	±14.9	41.1	70.9
Windward	50	79.9	54.3	±14.9	39.4	69.2
Windward	40	76.7	52.2	±14.9	37.3	67.1
Windward	30	73.0	49.6	±14.9	34.7	64.5
Windward	25	70.4	47.9	±14.9	33.0	62.8
Windward	20	67.9	46.2	±14.9	31.3	61.1
Windward	15	64.8	44.1	±14.9	29.2	59.0
Leeward	All	83.0	−14.1	±14.9	−29.0	−0.8
Side	All	83.0	−49.4	±14.9	−64.3	−34.5
Roof	63.5	83.0	−63.5*	±14.9	−78.4	−48.6
Roof	63.5	83.0	−35.3†	±14.9	−50.2	−20.4
Roof	63.5	83.0	−21.2‡	±14.9	−36.1	−6.3

* from windward edge to 63.5 feet
† from 63.5 to 127.0 feet
‡ from 127.0 to 328.75 feet

Table 5.37 Design Wind Pressures, p, in the N-S Direction

Location	Height above ground level, z (feet)	q (psf)	External pressure qGC_p (psf)	Internal pressure q_hGC_{pi} (psf)	Net pressure, p (psf) +(GC_{pi})	Net pressure, p (psf) −(GC_{pi})
Windward	63.5	83.0	56.4	±14.9	41.5	71.3
Windward	60	82.4	56.0	±14.9	41.1	70.9
Windward	50	79.9	54.3	±14.9	39.4	69.2
Windward	40	76.7	52.2	±14.9	37.3	67.1
Windward	30	73.0	49.6	±14.9	34.7	64.5
Windward	25	70.4	47.9	±14.9	33.0	62.8
Windward	20	67.9	46.2	±14.9	31.3	61.1
Windward	15	64.8	44.1	±14.9	29.2	59.0
Leeward	All	83.0	−35.3	±14.9	−50.2	−20.4
Side	All	83.0	−49.4	±14.9	−64.3	−34.5
Roof	63.5	83.0	−70.6*	±14.9	−85.5	−55.7
Roof	63.5	83.0	−53.7†	±14.9	−68.6	−38.8
Roof	63.5	83.0	−45.2‡	±14.9	−60.1	−30.3

* from windward edge to 31.75 feet
† from 31.75 to 63.5 feet
‡ from 63.5 to 75.33 feet

Figure 5.37 Design Wind Pressures in the E-W Direction, Example 5.7

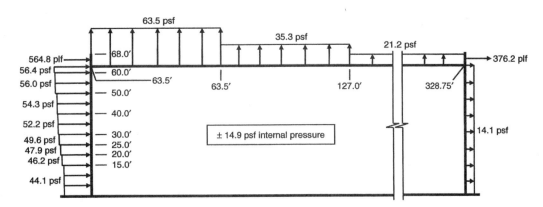

Figure 5.38 Design Wind Pressures in the N-S Direction, Example 5.7

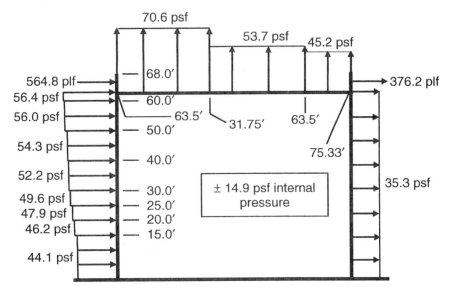

The MWFRS of buildings whose wind loads have been determined by Chapter 27 must be designed for the wind load cases defined in Figure 27.4-8 (27.4.6). In Case 1, the full design wind pressures act on the projected area perpendicular to each principal axis of the structure. These pressures are assumed to act separately along each principal axis. The wind pressures on the windward and leeward walls given in Tables 5.36 and 5.37 fall under Case 1.

In Case 2, 75 percent of the design wind pressures on the windward and leeward walls are applied on the projected area perpendicular to each principal axis of the building along with a torsional moment. The wind pressures and torsional moments, both of which vary over the height of the building, are applied separately for each principal axis.

As an example of the calculations that need to be performed over the height of the building, the wind pressures and torsional moment at the mean roof height for Case 2 are as follows (see Figure 5.13):

For E-W wind: $0.75p_{wx}$ = 0.75 × 56.4 = 42.3 psf (windward wall)

$\quad\quad\quad\quad\quad 0.75p_{lx}$ = 0.75 × 14.1 = 10.6 psf (leeward wall)

$\quad\quad\quad\quad\quad M_T = 0.75(p_{wx} + p_{lx})B_x e_x$

$\quad\quad\quad\quad\quad\quad\quad = 0.75(56.4 + 14.1) \times 75.33 \times (\pm 0.15 \times 75.33)$

$\quad\quad\quad\quad\quad\quad\quad = \pm 45{,}007$ ft-lb/ft

For N-S wind: $0.75p_{wy}$ = 0.75 × 56.4 = 42.3 psf (windward wall)

$\quad\quad\quad\quad\quad 0.75p_{ly}$ = 0.75 × 35.3 = 26.5 psf (leeward wall)

$\quad\quad\quad\quad\quad M_T = 0.75(p_{wy} + p_{ly})B_y e_y$

$\quad\quad\quad\quad\quad\quad\quad = 0.75(56.4 + 35.3) \times 328.75 \times (\pm 0.15 \times 328.75)$

$\quad\quad\quad\quad\quad\quad\quad = \pm 1{,}114{,}945$ ft-lb/ft

In Case 3, 75 percent of the wind pressures of Case 1 are applied to the building simultaneously. This accounts for wind along the diagonal of the building. In Case 4, 75 percent of the wind pressures and torsional moments defined in Case 2 act simultaneously on the building.

Figure 5.39 illustrates Load Cases 1 through 4 for MWFRS wind pressures acting on the projected area at the mean roof height. Note that internal pressures are always equal and opposite to each other and therefore not included. Similar loading diagrams can be obtained at other locations below the mean roof height.

Figure 5.39 Load Cases 1 through 4 at the Mean Roof Height, Example 5.7

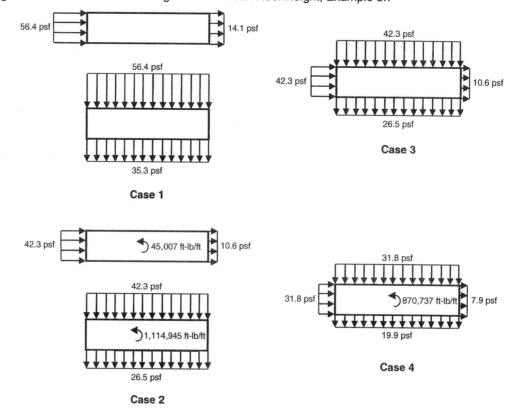

Minimum design wind loading prescribed in 27.1.5 must be considered as a load case in addition to those load cases described above (see Figure C27.4-1 and Figure 5.10).

5.8.8 Example 5.8 – Six-story Hotel Located on an Escarpment Using Chapter 27, Part 1 (MWFRS)

For the six-story hotel illustrated in Figure 5.36, determine design wind pressures on the MWFRS in both directions using Chapter 27, Part 1 (Directional Procedure for buildings of all heights) assuming that the structure is located on an escarpment.

DESIGN DATA

Location:	Miami, FL
Surface Roughness:	D (adjacent to water surface)
Topography:	2-D escarpment (see Figure 5.40)
Occupancy:	Residential building where less than 300 people congregate in one area

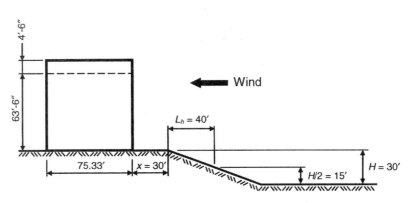

Figure 5.40 Six-story Hotel on Escarpment, Example 5.8

SOLUTION

- **Step 1:** Check if the building meets all of the conditions and limitations of 27.1.2 and 27.1.3 so that this procedure can be used to determine the wind pressures on the MWFRS.

  It was shown in Step 1 of Example 5.7 that Part 1 of Chapter 27 can be used to determine the design wind pressures on the MWFRS.

- **Step 2:** Determine the enclosure classification of the building.

  It is assumed in this example that the building is enclosed (see Step 3).

- **Step 3:** Use Flowchart 5.3 to determine the design wind pressures, p, on the MWFRS.

  1. Determine the design wind pressure effects of the parapet on the MWFRS.

     a. Determine the velocity pressure, q_p, at the top of the parapet from Flowchart 5.2.

        i. Determine the risk category of the building using IBC Table 1604.5.

           It was shown in Step 3 of Example 5.7 that this residential building falls under Risk Category II.

ii. Determine basic wind speed, V, for the applicable risk category from IBC Figure 1609A, B or C, or ASCE/SEI Figure 26.5-1A, B or C.

It was determined in Step 3 of Example 5.7 that V = 170 mph for Miami, FL.

iii. Determine wind directionality factor, K_d, from Table 26.6-1.

For the MWFRS of a building structure, $K_d = 0.85$.

iv. Determine exposure category.

In the design data, the surface roughness is given as D. Assume that Exposure D is applicable in all directions.

v. Determine topographic factor, K_{zt}.

Check if all five conditions of 26.8.1 are satisfied:

- Assume that the topography is such that conditions 1 and 2 are satisfied.
- Condition 3 is satisfied since the building is located near the crest of the escarpment.
- $H/L_h = 30/40 = 0.75 > 0.2$, so condition 4 is satisfied.
- $H = 30$ feet > 15 feet for Exposure D, so condition 5 is satisfied.

Since all five conditions of 26.8.1 are satisfied, wind speed-up effects at the escarpment must be considered in the design, and K_{zt} must be determined by Equation 26.8-1:

$$K_{zt} = (1 + K_1 K_2 K_3)^2$$

where the multipliers K_1, K_2 and K_3 are given in Figure 26.8-1 for Exposure C. Also given in the figure are parameters and equations that can be used to determine the multipliers for any exposure category.

It was determined above that $H/L_h = 0.75$. According to Note 2 in Figure 26.8-1, where $H/L_h > 0.5$, use $H/L_h = 0.5$ when evaluating K_1 and substitute $2H$ for L_h when evaluating K_2 and K_3.

From Figure 26.8-1, $K_1/(H/L_h) = 0.95$ for a 2-D escarpment for Exposure D. Thus, $K_1 = 0.95 \times 0.5 = 0.48$. This multiplier is related to the shape of the topographic feature and the maximum wind speed-up near the crest (see Section 5.2.5 of this publication).

The multiplier K_2 accounts for the reduction in speed-up with distance upwind or downwind of the crest. Since $x/L_h = x/2H = 30/60 = 0.5$, $K_2 = [1 - (0.5/4)] = 0.88$ for a 2-D escarpment from Figure 26.8-1.

The multiplier K_3 accounts for the reduction in speed-up with height z above the local ground surface. Even though the velocity pressure, q_p, is evaluated at the top of the parapet, the multiplier K_3 is conservatively determined at the height, z, corresponding to the centroid of the parapet, which is equal to $63.5 + (4.5/2) = 65.75$ feet.

Thus, $z/2H = 65.75/60 = 1.1$. Using the equation from Figure 26.8-1 for a 2-D escarpment yields the following (note: the values of K_3 can also be determined by linearly interpolating the values given in the table in Figure 26.8-1):

$$K_3 = e^{-\gamma z/L_h} = e^{-\gamma z/2H} = e^{-2.5 \times 1.1} = 0.07$$

Therefore, $K_{zt} = [1 + (0.48 \times 0.88 \times 0.07)]^2 = 1.06$

vi. Determine velocity pressure exposure coefficient, K_h, from Table 27.3-1.

For Exposure D at a height of 68 feet at the top of the parapet, $K_h = 1.33$ by linear interpolation.

vii. Determine velocity pressure, q_p, evaluated at the top of the parapet by Equation 27.3-1.

$q_p = 0.00256 K_h K_{zt} K_d V^2$

$= 0.00256 \times 1.33 \times 1.06 \times 0.85 \times 170^2 = 88.7$ psf

This velocity pressure is 6 percent greater than that determined in Example 5.7 where the building is not on an escarpment.

b. Determine combined net pressure coefficient, (GC_{pn}), for the parapets.

In accordance with 27.4.5, $(GC_{pn}) = 1.5$ for the windward parapet and $(GC_{pn}) = -1.0$ for the leeward parapet.

c. Determine combined net design pressure, p_p, on the parapet by Equation 27.4-4.

$p_p = q_p (GC_{pn})$

$= 88.7 \times 1.5 = 133.1$ psf on windward parapet

$= 88.7 \times (-1.0) = -88.7$ psf on leeward parapet

The forces on the windward and leeward parapets can be obtained by multiplying the pressures by the height of the parapet:

Windward parapet: $F = 133.1 \times 4.5 = 599.0$ plf

Leeward parapet: $F = 88.7 \times 4.5 = 399.2$ plf

As expected, these forces are 6 percent greater than those determined in Example 5.7 at this elevation.

2. Determine whether the building is rigid or flexible in accordance with 26.9.2.

It was determined in Step 3 of Example 5.7 that the building is rigid.

3. Determine gust-effect factor (G or G_f) from Flowchart 5.1.

According to 26.9.4, gust-effect factor, G, for rigid buildings may be taken as 0.85 or can be calculated by Equation 26.9-6. For simplicity, use $G = 0.85$.

4. Determine velocity pressure: q_z for windward walls along the height of the building, and q_h for leeward walls, side walls and roof using Flowchart 5.2.

Velocity pressures q_z and q_h are determined by Equation 27.3-1:

$q_z = 0.00256 K_z K_{zt} K_d V^2 = 0.00256 \times K_z \times K_{zt} \times 0.85 \times 170^2 = 62.89 K_z K_{zt}$ psf

The velocity exposure coefficient, K_z, and the topographic factor, K_{zt}, vary with height above the local ground surface. Values of K_z were determined in Example 5.7 (see Table 5.34) and are repeated in Table 5.38 for convenience.

From above,

$$K_{zt} = [1 + (0.48 \times 0.88 \times K_3)]^2 = [1 + 0.42K_3]^2$$

Values of K_{zt} are given in Table 5.38 as a function of $z/2H = z/60$ where z is taken midway between the height range (it is unconservative to use the top height of the range when determining K_3). Also given in the table is a summary of the velocity pressures, q_z, over the height of the building.

Table 5.38 Velocity Pressure, q_z

Height above ground level, z (feet)	K_z	$z/2H^*$	K_3	K_{zt}	q_z (psf)
63.5	1.32	1.03	0.08	1.06	88.0
60	1.31	0.92	0.10	1.09	89.8
50	1.27	0.75	0.15	1.13	90.3
40	1.22	0.58	0.24	1.21	92.8
30	1.16	0.46	0.32	1.28	93.4
25	1.12	0.38	0.39	1.35	95.1
20	1.08	0.29	0.48	1.45	98.5
15	1.03	0.13	0.72	1.70	110.1

*z is taken midway between the height range

As an example, determine K_{zt} at a height $z = 60$ feet above the local ground level. The multiplier K_3 is determined using a height, z, taken midway between the range of 60 feet and 50 feet, that is, $z = 55$ feet. Thus, $z/2H = 55/60 = 0.92$, and, by linear interpolation from Figure 26.8-1, $K_3 = 0.10$ for a 2-D escarpment (note: K_3 may also be computed from the equation given in Figure 26.8-1: $K_3 = e^{-\gamma z/L_h} = e^{-(2.5 \times 0.92)} = 0.10$). Then,

$$K_{zt} = [1 + (0.42 \times 0.10)]^2 = 1.09$$

5. Determine pressure coefficients, C_p, for the walls and roof from Figure 27.4-1.

   The pressure coefficients are the same as those determined in Step 3 of Example 5.7.

6. Determine q_i for the walls and roof using Flowchart 5.2.

   In accordance with 27.4.1, $q_i = q_h = 88.0$ psf for windward walls, side walls, leeward walls and roofs of enclosed buildings.

7. Determine internal pressure coefficients, (GC_{pi}), from Table 26.11-1.

   For an enclosed building, $(GC_{pi}) = +0.18, -0.18$.

8. Determine design wind pressures p_z and p_h by Equation 27.4-1.

Windward walls:

$$p_z = q_z GC_p - q_h(GC_{pi})$$

$$= (q_z \times 0.85 \times 0.8) - 88.0(\pm 0.18) = (0.68 q_z \mp 15.8) \text{ psf (external} \pm \text{internal pressure)}$$

Leeward wall, side walls and roof:

$$p_h = q_h GC_p - q_h(GC_{pi})$$

$$= (88.0 \times 0.85 \times C_p) - 88.0(\pm 0.18) = (74.8 C_p \mp 15.8) \text{ psf (external} \pm \text{internal pressure)}$$

A summary of the maximum design wind pressures in the E-W and N-S directions is given in Tables 5.39 and 5.40, respectively.

The percent increase in the external pressure on the windward wall of the building due to the escarpment is summarized in Table 5.41.

Table 5.39 Design Wind Pressures, p, in the E-W Direction

Location	Height above ground level, z (feet)	q (psf)	External pressure qGC_p (psf)	Internal pressure $q_h(GC_{pi})$ (psf)	Net pressure, p (psf) $+(GC_{pi})$	Net pressure, p (psf) $-(GC_{pi})$
Windward	63.5	88.0	59.8	±15.8	44.0	75.6
	60	89.8	61.1	±15.8	45.3	76.9
	50	90.3	61.4	±15.8	45.6	77.2
	40	92.8	63.1	±15.8	47.3	78.9
	30	93.4	63.5	±15.8	47.7	79.3
	25	95.1	64.7	±15.8	48.9	80.5
	20	98.5	67.0	±15.8	51.2	82.8
	15	110.1	74.9	±15.8	59.1	90.7
Leeward	All	88.0	−15.0	±15.8	−30.8	0.8
Side	All	88.0	−52.4	±15.8	−68.2	−36.6
Roof	63.5	88.0	−67.3*	±15.8	−83.1	−51.5
	63.5	88.0	−37.4†	±15.8	−53.2	−21.6
	63.5	88.0	−22.4‡	±15.8	−38.2	−6.6

* from windward edge to 63.5 feet
† from 63.5 to 127.0 feet
‡ from 127.0 to 328.75 feet

Table 5.40 Design Wind Pressures, p, in the N-S Direction

Location	Height above ground level, z (feet)	q (psf)	External pressure qGC_p (psf)	Internal pressure $q_h(GC_{pi})$ (psf)	Net pressure, p (psf) $+(GC_{pi})$	Net pressure, p (psf) $-(GC_{pi})$
Windward	63.5	88.0	59.8	±15.8	44.0	75.6
	60	89.8	61.1	±15.8	45.3	76.9
	50	90.3	61.4	±15.8	45.6	77.2
	40	92.8	63.1	±15.8	47.3	78.9
	30	93.4	63.5	±15.8	47.7	79.3
	25	95.1	64.7	±15.8	48.9	80.5
	20	98.5	67.0	±15.8	51.2	82.8
	15	110.1	74.9	±15.8	59.1	90.7
Leeward	All	88.0	−37.4	±15.8	−53.2	−21.6
Side	All	88.0	−52.4	±15.8	−68.2	−36.6
Roof	63.5	88.0	−74.8*	±15.8	−90.6	−59.0
	63.5	88.0	−56.9†	±15.8	−72.7	−41.1
	63.5	88.0	−47.9‡	±15.8	−63.7	−32.1

* from windward edge to 31.75 feet
† from 31.75 to 63.5 feet
‡ from 63.5 to 75.33 feet

Table 5.41 Comparison of External Design Wind Pressures on the Windward Wall with and without an Escarpment

Height above ground level, z (feet)	External pressure, qGC_p (psf) Without Escarpment	External pressure, qGC_p (psf) With Escarpment	Percent Increase
63.5	56.4	59.8	6
60	56.0	61.1	9
50	54.3	61.4	13
40	52.2	63.1	21
30	49.6	63.5	28
25	47.9	64.7	35
20	46.2	67.0	45
15	44.1	74.9	70

The external pressures on the leeward wall, side wall and roof as well as the internal pressure increase by 6 percent, since these pressures depend on the velocity pressure at the roof height, q_h.

Load Cases 1 through 4, depicted in Figure 27.4-8, must be investigated for the windward and leeward pressures, similar to that shown in Example 5.7.

Wind Loads 281

5.8.9 Example 5.9 – Six-story Hotel Using Alternate All-heights Method (MWFRS)

For the six-story hotel illustrated in Figure 5.36, determine design wind pressures on the MWFRS in both directions using the Alternate All-heights Method of IBC 1609.6.

SOLUTION

- **Step 1:** Check if the provisions of IBC 1609.6 can be used to determine the design wind pressures on this building.

   The provisions of IBC 1609.6 may be used to determine design wind pressures on this regularly-shaped building provided the conditions of IBC 1609.6.1 are satisfied:

   1. The height of the building is 63 feet 6 inches, which is less than 75 feet, and the height-to-least-width ratio = 63.5/75.33 = 0.84 < 4. Also, it was shown in Example 5.7 that the fundamental frequency $n_1 > 1$ Hz in both directions.

   2. As was discussed in Example 5.7, this building is not sensitive to dynamic effects.

   3. This building is not located on a site where channeling effects or buffeting in the wake of upwind obstructions need to be considered.

   4. The building is a simple diaphragm building as defined in 26.2, since windward and leeward wind loads are transmitted through the reinforced concrete floor and roof slabs (diaphragms) to the reinforced concrete moment frames (MWFRS), and there are no separations in the MWFRS.

   5. The fifth condition is not applicable in this example.

The provisions of the Alternate All-heights Method of IBC 1609.6 can be used to determine the design wind pressures on the MWFRS.

- **Step 2:** Use Flowchart 5.15 to determine the net design wind pressures, p_{net}, on the MWFRS.

   1. Determine the risk category of the building using IBC Table 1604.5.

      It was shown in Step 3 of Example 5.7 that this residential building falls under Risk Category II.

   2. Determine basic wind speed, V, for the applicable risk category from IBC Figure 1609A, B or C, or ASCE/SEI Figure 26.5-1A, B or C.

      It was determined in Step 3 of Example 5.7 that $V = 170$ mph for Miami, FL.

   3. Determine exposure category.

      In the design data of Example 5.7, the surface roughness is given as D. Assume that Exposure D is applicable in all directions.

4. Determine topographic factor, K_{zt}.

   As noted in the design data of Example 5.7, the building is not situated on a hill, ridge or escarpment. Thus, topographic factor $K_{zt} = 1.0$ (26.8.2).

5. Determine velocity pressure exposure coefficients, K_z, from Table 27.3-1.

   Values of K_z are summarized in Table 5.42.

Table 5.42 Velocity Pressure Exposure Coefficient, K_z

Height above ground level, z (feet)	K_z
68	1.33
63.5	1.32
60	1.31
50	1.27
40	1.22
30	1.16
25	1.12
20	1.08
15	1.03

6. Determine net pressure coefficients, C_{net}, for the walls and roof from IBC Table 1609.6.2 assuming the building is enclosed.

   Windward wall: $C_{net} = 0.43$ for positive internal pressure

   $C_{net} = 0.73$ for negative internal pressure

   Leeward wall: $C_{net} = -0.51$ for positive internal pressure

   $C_{net} = -0.21$ for negative internal pressure

   Side walls: $C_{net} = -0.66$ for positive internal pressure

   $C_{net} = -0.35$ for negative internal pressure

   Leeward roof (wind perpendicular to ridge):

   $C_{net} = -0.66$ for positive internal pressure

   $C_{net} = -0.35$ for negative internal pressure

   Roof: $C_{net} = -1.09$ for positive internal pressure

   $C_{net} = -0.79$ for negative internal pressure

   Parapet: $C_{net} = 1.28$ for windward

   $C_{net} = -0.85$ for leeward

   These net pressures are applicable for wind in both the N-S and E-W directions.

7. Determine net design wind pressures, p_{net}, by IBC Equation 16-35.

$$p_{net} = 0.00256 V^2 K_z C_{net} K_{zt}$$

Windward walls: $p_{net} = 74.0 K_z C_{net}$

Leeward walls, side walls and roof: $p_{net} = 74.0 \times 1.32 \times C_{net} = 97.7 C_{net}$

Parapet: $p_{net} = 74.0 \times 1.33 \times C_{net} = 98.4 C_{net}$

A summary of the maximum net design wind pressures is given in Table 5.43.

Table 5.43 Net Design Wind Pressures, p_{net}

Location	Height above ground level, z (feet)	K_z	C_{net} +Internal Pressure	C_{net} −Internal Pressure	Net design pressure, p_{net} (psf) +Internal Pressure	Net design pressure, p_{net} (psf) −Internal Pressure
Windward	63.5	1.32	0.43	0.73	42.0	71.3
	60	1.31	0.43	0.73	41.7	70.8
	50	1.27	0.43	0.73	40.4	68.6
	40	1.22	0.43	0.73	38.8	65.9
	30	1.16	0.43	0.73	36.9	62.7
	25	1.12	0.43	0.73	35.6	60.5
	20	1.08	0.43	0.73	34.4	58.3
	15	1.03	0.43	0.73	32.8	55.6
Leeward	All	1.33	−0.51	−0.21	−49.8	−20.5
Side	All	1.33	−0.66	−0.35	−64.5	−34.2
Roof	60	1.33	−1.09	−0.79	−106.5	−77.2

Illustrated in Figure 5.41 are the net design wind pressures in the N-S direction with positive internal pressure (wind pressures in the E-W direction are the same as in the N-S direction). Included in the figure are the following forces on the windward and leeward parapets, which add to the overall wind forces in the direction of analysis:

On the windward parapet: $p_{net} = 98.4 \times 1.28 = 126.0$ psf

$$F = 126.0 \times 4.5 = 567.0 \text{ plf}$$

On the leeward parapet: $p_{net} = 98.4 \times (-0.85) = -83.6$ psf

$$F = -83.6 \times 4.5 = -376.4 \text{ plf}$$

Figure 5.41 Net Design Wind Pressures in the N-S Direction, Example 5.9

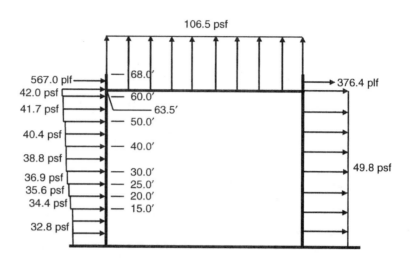

The MWFRS of buildings whose wind loads have been determined by IBC 1609.6 must be designed for the wind load cases identified in ASCE/SEI Figure 27.4-8 (IBC 1609.6.4.1). In Case 1, the full design wind pressures act on the projected area perpendicular to each principal axis of the structure. These pressures are assumed to act separately along each principal axis. The wind pressures on the windward and leeward walls depicted in Figures 5.41 fall under Case 1.

The calculations for the additional load cases that need to be considered in this example are similar to those in Example 5.7.

5.8.10 Example 5.10 – Fifteen-story Office Building Using Chapter 27, Part 1 (MWFRS)

For the 15-story office building depicted in Figure 5.42, determine design wind pressures on the MWFRS in both directions using Chapter 27, Part 1 (Directional Procedure for buildings of all heights).

Figure 5.42 Fifteen-story Office Building, Example 5.10

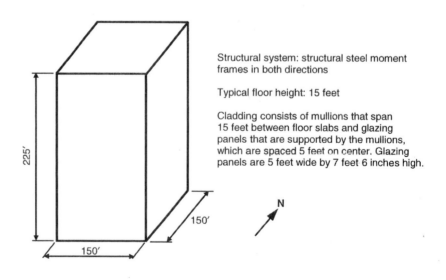

Structural system: structural steel moment frames in both directions

Typical floor height: 15 feet

Cladding consists of mullions that span 15 feet between floor slabs and glazing panels that are supported by the mullions, which are spaced 5 feet on center. Glazing panels are 5 feet wide by 7 feet 6 inches high.

Design Data

Location:	Chicago, IL
Surface Roughness:	B (suburban area with numerous closely spaced obstructions having the size of single-family dwellings and larger)
Topography:	Not situated on a hill, ridge or escarpment
Occupancy:	Business occupancy where less than 300 people congregate in one area

Solution

- **Step 1:** Check if the building meets all of the conditions and limitations of 27.1.2 and 27.1.3 so that this procedure can be used to determine the wind pressures on the MWFRS.

  The building is regularly-shaped, that is, it does not have any unusual geometric irregularities in spatial form. Also, the building does not have response characteristics that make it subject to across-wind loading or other similar effects, and it is not sited at a location where channeling effects or buffeting in the wake of upwind obstructions need to be considered.

Thus, Part 1 of Chapter 27 may be used to determine the design wind pressures on the MWFRS.

- **Step 2:** Determine the enclosure classification of the building.

  For illustrative purposes, it is assumed in this example that the building is partially enclosed. (Note: For office buildings of this type, it is common to assume that the building is enclosed. Where buildings have operable windows or where it is anticipated that debris may compromise some of the windows during a windstorm, it may be more appropriate to assume that the building is partially enclosed.)

- **Step 3:** Use Flowchart 5.3 to determine the design wind pressures, p, on the MWFRS.

  1. Determine whether the building is rigid or flexible in accordance with 26.9.2.

     A flexible building is defined in 26.2 as one in which the fundamental natural frequency of the building, n_1, is less than 1 Hz.

     In lieu of obtaining the natural frequency of the building from a dynamic analysis, check if the approximate lower-bound natural frequencies given in 26.9.3 can be used for this building (26.9.2.1):

     - Building height = 225 feet < 300 feet
     - Building height = 225 feet < $4L_{eff}$ = 4 × 150 = 600 feet in both directions

Since both of these limitations are satisfied, Equation 26.9-2 can be used to determine an approximate value of n_a for steel moment-resisting-frame systems:

$$n_a = \frac{22.2}{h^{0.8}} = \frac{22.2}{225^{0.8}} = 0.3 \text{ Hz} < 1.0 \text{ Hz}$$

Since $n_a < 1.0$ Hz, the building is defined as a flexible building.

2. Determine the gust-effect factor, G, using Flowchart 5.1.

   The gust-effect factor, G_f, for flexible buildings is determined by Equation 26.9-10 in 26.9.5.

   a. Determine g_Q and g_v

   $$g_Q = g_v = 3.4$$

   b. Determine g_R

   $$g_R = \sqrt{2\ln(3{,}600 n_1)} + \frac{0.577}{\sqrt{2\ln(3{,}600\, n_1)}}$$

   $$= \sqrt{2\ln(3{,}600 \times 0.3)} + \frac{0.577}{\sqrt{2\ln(3{,}600 \times 0.3)}} = 3.9 \qquad \text{Eq. 26.9-11}$$

   c. Determine $I_{\bar{z}}$

   $$\bar{z} = 0.6h = 0.6 \times 225$$

   $$= 135 \text{ feet} > z_{\min} = 30 \text{ feet} \qquad \text{Table 26.9-1 for Exposure B}$$

   $$I_{\bar{z}} = c\left(\frac{33}{\bar{z}}\right)^{1/6}$$

   $$= 0.30\left(\frac{33}{135}\right)^{1/6} = 0.24 \qquad \text{Eq. 26.9-7 and Table 26.9-1 for Exposure B}$$

   d. Determine Q

   $$L_{\bar{z}} = \ell\left(\frac{\bar{z}}{33}\right)^{\bar{\epsilon}}$$

   $$= 320\left(\frac{135}{33}\right)^{1/3} = 511.8 \text{ feet} \qquad \text{Eq. 26.9-9 and Table 26.9-1 for Exposure B}$$

   $$Q = \sqrt{\frac{1}{1 + 0.63\left(\frac{B+h}{L_{\bar{z}}}\right)^{0.63}}}$$

   $$= \sqrt{\frac{1}{1 + 0.63\left(\frac{150 + 225}{511.8}\right)^{0.63}}} = 0.81 \qquad \text{Eq. 26.9-8}$$

e. Determine R

Due to the nature of its occupancy, this office building falls under Risk Category II. For Risk Category II, use Figure 1609A or Figure 26.5-1A. From either of these figures, $V = 115$ mph for Chicago, IL.

$$\overline{V}_{\bar{z}} = \bar{b}\left(\frac{\bar{z}}{33}\right)^{\alpha}\left(\frac{88}{60}\right)V$$

$$= 0.45\left(\frac{135}{33}\right)^{1/4}\left(\frac{88}{60}\right) \times 115 \qquad \text{Eq. 26.9-16 and Table 26.9-1 for Exposure B}$$

$$= 107.9 \text{ feet/sec}$$

$$N_1 = \frac{n_1 L_{\bar{z}}}{\overline{V}_{\bar{z}}}$$

$$= \frac{0.3 \times 511.8}{107.9} = 1.4 \qquad \text{Eq. 26.9-14}$$

$$R_n = \frac{7.47 N_1}{(1 + 10.3 N_1)^{5/3}}$$

$$= \frac{7.47 \times 1.4}{[1 + (10.3 \times 1.4)]^{5/3}} = 0.11 \qquad \text{Eq. 26.9-13}$$

$$\eta_h = \frac{4.6 n_1 h}{\overline{V}_{\bar{z}}}$$

$$= \frac{4.6 \times 0.3 \times 225}{107.9} = 2.9$$

$$R_h = \frac{1}{\eta_h} - \frac{1}{2\eta_h^2}(1 - e^{-2\eta_h})$$

$$= \frac{1}{2.9} - \frac{1}{2 \times 2.9^2}(1 - e^{-2 \times 2.9}) = 0.29 \qquad \text{Eq. 26.9-15a}$$

$$\eta_B = \frac{4.6 n_1 B}{\overline{V}_{\bar{z}}}$$

$$= \frac{4.6 \times 0.3 \times 150}{107.9} = 1.9$$

$$R_B = \frac{1}{\eta_B} - \frac{1}{2\eta_B^2}(1 - e^{-2\eta_B})$$

$$= \frac{1}{1.9} - \frac{1}{2 \times 1.9^2}(1 - e^{-2 \times 1.9}) = 0.39 \qquad \text{Eq. 26.9-15a}$$

$$\eta_L = \frac{15.4 n_1 L}{\overline{V}_{\bar{z}}}$$

$$= \frac{15.4 \times 0.3 \times 150}{107.9} = 6.4$$

$$R_L = \frac{1}{\eta_L} - \frac{1}{2\eta_L^2}(1 - e^{-2\eta_L})$$

$$= \frac{1}{6.4} - \frac{1}{2 \times 6.4^2}(1 - e^{-2 \times 6.4}) = 0.14 \qquad \text{Eq. 26.9-15a}$$

Assume damping ratio $\beta = 0.01$ (see C26.9 for suggested value for steel buildings).

$$R = \sqrt{\frac{1}{\beta} R_n R_h R_B (0.53 + 0.47 R_L)}$$

$$= \sqrt{\frac{1}{0.01} \times 0.11 \times 0.29 \times 0.39 [0.53 + (0.47 \times 0.14)]} \qquad \text{Eq. 26.9-12}$$

$$= 0.86$$

f. Determine G_f

$$G_f = 0.925 \left(\frac{1 + 1.7 I_z \sqrt{g_Q^2 Q^2 + g_R^2 R^2}}{1 + 1.7 g_v I_z} \right)$$

$$= 0.925 \left(\frac{1 + (1.7 \times 0.24)\sqrt{(3.4^2 \times 0.81^2) + (3.9^2 \times 0.86^2)}}{1 + (1.7 \times 3.4 \times 0.24)} \right) \qquad \text{Eq. 26.9-10}$$

$$= 1.07$$

3. Determine velocity pressure: q_z for windward walls along the height of the building, and q_h for leeward walls, side walls and roof using Flowchart 5.2.

   a. Determine the risk category of the building using IBC Table 1604.5.

      As noted above, this office building falls under Risk Category II.

   b. Determine basic wind speed, V, for the applicable risk category from IBC Figure 1609A, B or C, or ASCE/SEI Figure 26.5-1A, B or C.

      As noted above, $V = 115$ mph for Chicago, IL.

   c. Determine wind directionality factor, K_d, from Table 26.6-1.

      For the MWFRS of a building structure, $K_d = 0.85$.

   d. Determine exposure category.

      In the design data, the surface roughness is given as B. Assume that Exposure B is applicable in all directions.

   e. Determine topographic factor, K_{zt}.

      As noted in the design data, the building is not situated on a hill, ridge or escarpment. Thus, topographic factor $K_{zt} = 1.0$ (26.8.2).

   f. Determine velocity pressure exposure coefficients, K_z and K_h, from Table 27.3-1.

      Values of K_z and K_h for Exposure B are summarized in Table 5.44.

Table 5.44 Velocity Pressure Exposure Coefficient, K_z, for MWFRS

Height above ground level, z (feet)	K_z
225	1.24
200	1.20
180	1.17
160	1.13
140	1.09
120	1.04
100	0.99
90	0.96
80	0.93
70	0.89
60	0.85
50	0.81
40	0.76
30	0.70
25	0.66
20	0.62
15	0.57

g. Determine velocity pressure, q_z and q_h, by Equation 27.3-1.

$$q_z = 0.00256 K_z K_{zt} K_d V^2 = 0.00256 \times K_z \times 1.0 \times 0.85 \times 115^2 = 28.78 K_z \text{ psf}$$

A summary of the velocity pressures is given in Table 5.45.

Table 5.45 Velocity Pressure, q_z, for MWFRS

Height above ground level, z (feet)	K_z	q_z (psf)
225	1.24	35.7
200	1.20	34.5
180	1.17	33.7
160	1.13	32.5
140	1.09	31.4
120	1.04	29.9
100	0.99	28.5
90	0.96	27.6
80	0.93	26.8
70	0.89	25.6
60	0.85	24.5
50	0.81	23.3
40	0.76	21.9
30	0.70	20.2
25	0.66	19.0
20	0.62	17.8
15	0.57	16.4

4. Determine pressure coefficients, C_p, for the walls and roof from Figure 27.4-1.

The pressure coefficients will be the same in both the N-S and E-W directions, since the building is square in plan.

Windward wall: $C_p = 0.8$ for use with q_z

Leeward wall ($L/B = 150/150 = 1.0$): $C_p = -0.5$ for use with q_h

Side wall: $C_p = -0.7$ for use with q_h

Roof (normal to ridge with $\theta < 10$ degrees and parallel to ridge for all θ with $h/L = 225/150 = 1.5 > 1.0$):

$C_p = -1.04, -0.18$ from windward edge to $h/2 = 112.5$ feet for use with q_h

$C_p = -0.7, -0.18$ from 112.5 feet to 150 feet for use with q_h

Note: The smaller uplift pressures on the roof due to $C_p = -0.18$ may govern the design when combined with roof live load or snow loads. This pressure is not shown in this example but in general must be considered. Also, $C_p = -1.3$, which is the pressure coefficient from the windward edge of the roof to $h/2$, may be reduced based on area over which it is applicable = $(225/2) \times 150 = 16{,}875$ square feet > 1,000 square feet. Based on this area, the reduction factor = 0.8 from Figure 27.4-1. Thus, $C_p = 0.8 \times (-1.3) = -1.04$.

5. Determine q_i for the walls and roof using Flowchart 5.2.

   In accordance with 27.4.1, $q_i = q_h = 35.7$ psf for windward walls, side walls, leeward walls and roofs of partially enclosed buildings.

6. Determine internal pressure coefficients, (GC_{pi}), from Table 26.11-1.

   For a partially enclosed building, $(GC_{pi}) = +0.55, -0.55$.

7. Determine design wind pressures, p_z and p_h, by Equation 27.4-2.

   Windward walls:

   $p_z = q_z G_f C_p - q_h (GC_{pi})$

   $= (q_z \times 1.07 \times 0.8) - 35.7(\pm 0.55)$

   $= (0.86 q_z \mp 19.6)$ psf (external $\pm$ internal pressure)

   Leeward wall, side walls and roof:

   $p_h = q_h G_f C_p - q_h (GC_{pi})$

   $= (35.7 \times 1.07 \times C_p) - 35.7(\pm 0.55)$

   $= (38.2 C_p \mp 19.6)$ psf (external $\pm$ internal pressure)

A summary of the maximum design wind pressures that are valid in both the N-S and E-W directions is given in Table 5.46.

Illustrated in Figure 5.43 are the external design wind pressures in the N-S or E-W directions. When considering horizontal wind forces on the MWFRS, it is clear that the effects from the internal pressure cancel out. On the roof, the effects from internal pressure add directly to those from the external pressure.

The MWFRS of buildings whose wind loads have been determined by Chapter 27 must be designed for the wind load cases defined in Figure 27.4-8 (27.4.6).

In Case 1, the full design wind pressures act on the projected area perpendicular to each principal axis of the structure. These pressures are assumed to act separately along each principal axis. The wind pressures on the windward and leeward walls depicted in Figure 5.43 fall under Case 1.

Table 5.46 Design Wind Pressures, p, for MWFRS

Location	Height above ground level, z (feet)	q (psf)	External pressure qGC_p (psf)	Internal pressure $q_h(GC_{pi})$ (psf)
Windward	225	35.7	30.7	±19.6
	200	34.5	29.7	±19.6
	180	33.7	29.0	±19.6
	160	32.5	28.0	±19.6
	140	31.4	27.0	±19.6
	120	29.9	25.7	±19.6
	100	28.5	24.5	±19.6
	90	27.6	23.7	±19.6
	80	26.8	23.1	±19.6
	70	25.6	22.0	±19.6
	60	24.5	21.1	±19.6
	50	23.3	20.0	±19.6
	40	21.9	18.8	±19.6
	30	20.2	17.4	±19.6
	25	19.0	16.3	±19.6
	20	17.8	15.3	±19.6
	15	16.4	14.1	±19.6
Leeward	All	35.7	−19.1	±19.6
Side	All	35.7	−26.7	±19.6
Roof	225	35.7	−39.7*	±19.6
	225	35.7	−26.7†	±19.6

* from windward edge to 112.5 feet
† from 112.5 to 150.0 feet

Figure 5.43
Design Wind Pressures in the N-S or E-W Directions, Example 5.10

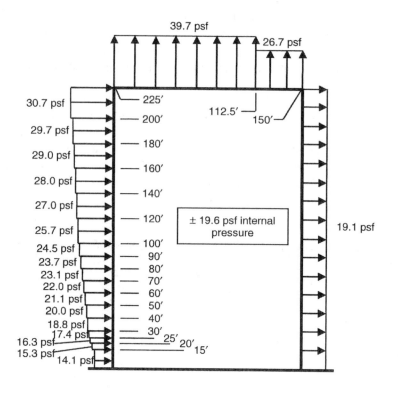

In Case 2, 75 percent of the design wind pressures on the windward and leeward walls are applied on the projected area perpendicular to each principal axis of the building along with a torsional moment. The wind pressures and torsional moments, both of which vary over the height of the building, are applied separately for each principal axis.

As an example of the calculations that need to be performed over the height of the building, the wind pressures and torsional moment at the mean roof height for Case 2 are as follows (see Figure 5.13):

$0.75p_{wx} = 0.75 \times 30.7 = 23.0$ psf (windward wall)

$0.75p_{lx} = 0.75 \times 19.1 = 14.3$ psf (leeward wall)

For flexible buildings, the eccentricity, e, that is used to determine the torsional moment, M_T, is given by Equation 27.4-5. Assuming that the elastic shear center and the center of mass coincide (i.e., $e_R = 0$),

$$e = \frac{e_Q + 1.7I_{\bar{z}}\sqrt{(g_Q Q e_Q)^2 + (g_R R e_R)^2}}{1 + 1.7I_{\bar{z}}\sqrt{(g_Q Q)^2 + (g_R R)^2}}$$

$$= \frac{(0.15 \times 150) + (1.7 \times 0.24)\sqrt{[3.4 \times 0.81 \times (0.15 \times 150)]^2 + 0}}{1 + (1.7 \times 0.24)\sqrt{(3.4 \times 0.81)^2 + (3.9 \times 0.86)^2}} = 17.3 \text{ feet}$$

The eccentricity determined by Equation 27.4-5 is less than that for a rigid building, which is equal to $0.15 \times 150 = 22.5$ feet (see Figure 27.4-8). For conservatism, an eccentricity of 22.5 feet is used in this example.

$$M_T = 0.75(p_{wx} + p_{lx})B_x e_x$$

$$= 0.75(30.7 + 19.1) \times 150 \times (\pm 0.15 \times 150)$$

$$= \pm 126{,}056 \text{ ft-lb/ft}$$

In Case 3, 75 percent of the wind pressures of Case 1 are applied to the building simultaneously. This accounts for wind along the diagonal of the building. In Case 4, 75 percent of the wind pressures and torsional moments defined in Case 2 act simultaneously on the building.

Figure 5.44 illustrates Load Cases 1 through 4 for MWFRS wind pressures acting on the projected area at the mean roof height. Similar loading diagrams can be obtained at other locations below the mean roof height.

Figure 5.44 Load Cases 1 through 4 at the Mean Roof Height, Example 5.10

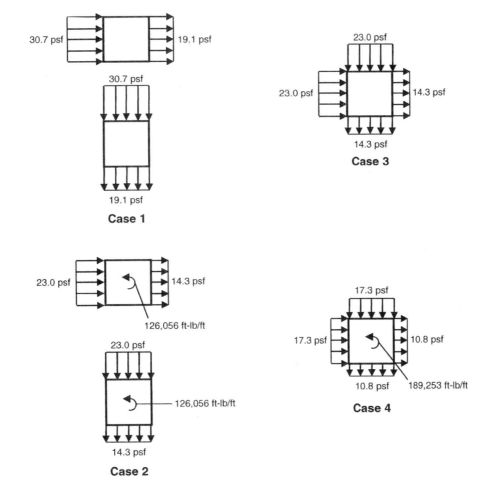

The minimum design wind loading prescribed in 27.1.5 must be considered as a load case in addition to those load cases described above (see Figure C27.4-1 and Figure 5.10).

The above wind pressure calculations assume a uniform design pressure over the incremental heights above ground level, which are given in Table 27.3-1. Alternatively, wind pressures can be computed at each floor level, and a uniform pressure is assumed between mid-story heights above and below the floor level under consideration.

Shown in Figure 5.45 are the wind pressures computed at each floor level for this example building. It can be shown that the base shears in Figures 5.43 and 5.45 are virtually the same.

Figure 5.45
Design Wind Pressures Computed at the Floor Levels in the N-S or E-W Directions, Example 5.10

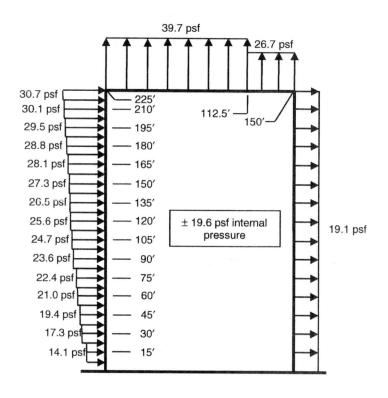

5.8.11 Example 5.11 – Agricultural Building Using Chapter 27, Part 1 (MWFRS)

For the agricultural building depicted in Figure 5.46, determine the design wind pressures on the MWFRS in both directions using Chapter 27, Part 1 (Directional Procedure for buildings of all heights).

DESIGN DATA	
Location:	Ames, IA
Surface Roughness:	C (open terrain with scattered obstructions having heights less than 30 feet)
Topography:	Not situated on a hill, ridge or escarpment
Occupancy:	Utility and miscellaneous occupancy

Figure 5.46 Agricultural Building, Example 5.11

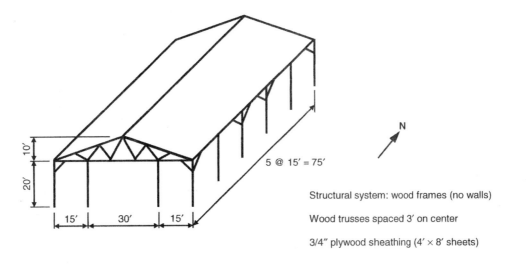

Structural system: wood frames (no walls)

Wood trusses spaced 3' on center

3/4" plywood sheathing (4' × 8' sheets)

SOLUTION

- **Step 1:** Check if the building meets all of the conditions and limitations of 27.1.2 and 27.1.3 so that this procedure can be used to determine the wind pressures on the MWFRS.

  The building is regularly-shaped, that is, it does not have any unusual geometric irregularities in spatial form. Also, the building does not have response characteristics that make it subject to across-wind loading or other similar effects, and it is not sited at a location where channeling effects or buffeting in the wake of upwind obstructions need to be considered.

Thus, Part 1 of Chapter 27 may be used to determine the design wind pressures on the MWFRS.

- **Step 2:** Determine the enclosure classification of the building.

  Since the building does not have any walls, it is classified as open.

- **Step 3:** Use Flowchart 5.3 to determine the design wind pressures, p, on the MWFRS.

  1. Determine the velocity pressure, q_h, using Flowchart 5.2.

     a. Determine the risk category of the structure using IBC Table 1604.5.

        This agricultural building falls under Risk Category I.

     b. Determine basic wind speed, V, for the applicable risk category from IBC Figure 1609A, B or C, or ASCE/SEI Figure 26.5-1A, B or C.

        For Risk Category I, use Figure 1609C or Figure 26.5-1C. From either of these figures, $V = 105$ mph for Ames, IA.

     c. Determine wind directionality factor, K_d, from Table 26.6-1.

        For the MWFRS of a building structure, $K_d = 0.85$.

     d. Determine exposure category.

        In the design data, the surface roughness is given as C. Assume that Exposure C is applicable in all directions.

     e. Determine topographic factor, K_{zt}.

        As noted in the design data, the building is not situated on a hill, ridge or escarpment. Thus, topographic factor $K_{zt} = 1.0$ (26.8.2).

     f. Determine velocity pressure exposure coefficient, K_h, from Table 27.3-1.

        $$\text{Mean roof height} = \frac{20 + 30}{2} = 25 \text{ feet}$$

        For Exposure C at a height of 25 feet, $K_h = 0.94$.

     g. Determine velocity pressure, q_h, by Equation 27.3-1.

        $q_h = 0.00256 K_h K_{zt} K_d V^2 = 0.00256 \times 0.94 \times 1.0 \times 0.85 \times 105^2 = 22.6$ psf

  2. Determine the gust-effect factor, G, using Flowchart 5.1.

     Assuming the building is rigid, gust-effect factor, G, for rigid buildings may be taken as 0.85 or can be calculated by Equation 26.9-6 (26.9.4). For simplicity, use $G = 0.85$.

  3. Determine net pressure coefficient, C_N, from Figure 27.4-5 for open buildings with pitched roofs.

     a. Wind in the E-W direction ($\gamma = 0°, 180°$)

        Figure 27.4-5 is used to determine the net pressure coefficients, C_{NW} and C_{NL}, on the windward and leeward portions of the roof surface for wind in the E-W direction. These net pressure coefficients include contributions from both the top and bottom surfaces of the roof (see Note 1 in Figure 27.4-5).

        The wind pressures on the roof depend on the level of wind flow restriction below the roof. Clear wind flow implies that little (less than or equal to 50 percent) or no portion of the cross-section below the roof is blocked by goods or materials (see Note 2 in Figure 27.4-5). Obstructed wind flow means that a significant portion (more than 50 percent) of the cross-section is blocked. Since the usage of the space below the roof is not known, wind pressures will be determined for both situations.

A summary of the net pressure coefficients is given in Table 5.47. The roof angle in this example is equal to approximately 18.4 degrees, and the values of C_{NW} and C_{NL} were obtained by linear interpolation (see Note 3 in Figure 27.4-5).

Table 5.47 Windward and Leeward Net Pressure Coefficients, C_{NW} and C_{NL}, for Wind in the E-W Direction

Load Case	Clear Wind Flow		Obstructed Wind Flow	
	C_{NW}	C_{NL}	C_{NW}	C_{NL}
A	1.10	−0.17	−1.20	−1.09
B	0.01	−0.95	−0.69	−1.65

b. Wind in the N-S direction ($\gamma = 90°$)

Figure 27.4-7 is used to determine the net pressure coefficients, C_N, at various distances from the windward edge of the roof.

Net pressure coefficients are given in Table 5.48.

4. Determine the net design pressures, p, by Equation 27.4-3.

$$p = q_h G C_N = 22.6 \times 0.85 \times C_N = 19.2 C_N \text{ psf}$$

A summary of the net design wind pressures is given in Table 5.49 for wind in the E-W direction and in Table 5.50 for wind in the N-S direction. These pressures act perpendicular to the roof surface.

Table 5.48 Net Pressure Coefficients, C_N, for Wind in the N-S Direction

Horizontal Distance from Windward Edge	Load Case	Clear Wind Flow C_N	Obstructed Wind Flow C_N
$\leq h = 25'$	A	−0.8	−1.2
	B	0.8	0.5
$> h = 25', \leq 2h = 50'$	A	−0.6	−0.9
	B	0.5	0.5
$\geq 2h = 50'$	A	−0.3	−0.6
	B	0.3	0.3

Table 5.49 Net Design Wind Pressures (psf) for Wind in the E-W Direction

Load Case	Clear Wind Flow		Obstructed Wind Flow	
	Windward	Leeward	Windward	Leeward
A	21.1	−3.3	−23.0	−20.9
B	0.2	−18.2	−13.3	−31.7

Table 5.50 Net Design Wind Pressures (psf) for Wind in the N-S Direction

Horizontal Distance from Windward Edge	Load Case	Clear Wind Flow C_N	Obstructed Wind Flow C_N
$\leq h = 25'$	A	−15.4	−23.0
	B	15.4	9.6
$> h = 25', \leq 2h = 50'$	A	−11.5	−17.3
	B	9.6	9.6
$\geq 2h = 50'$	A	−5.8	−11.5
	B	5.8	5.8

The minimum design wind loading prescribed in 27.1.5 must be considered as a load case in addition to Load Cases A and B described above (see Figure C27.4-1 and Figure 5.10).

5.8.12 Example 5.12 – Freestanding Masonry Wall Using Chapter 29

Determine the design wind forces on the architectural freestanding masonry screen wall depicted in Figure 5.47 using Chapter 29.

DESIGN DATA

Location: Sacramento, CA

Surface roughness: B

Topography: Not situated on a hill, ridge or escarpment

Figure 5.47
Freestanding Masonry Wall, Example 5.12

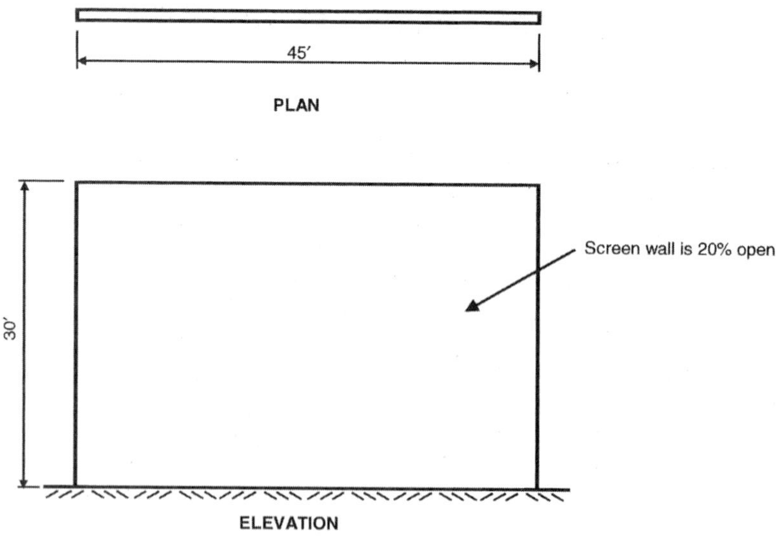

SOLUTION

Use Flowchart 5.7 to determine the design wind force on the freestanding wall.

1. Determine the velocity pressure, q_h, from Flowchart 5.2.

a. Determine the risk category of the structure using IBC Table 1604.5.

This wall represents a low risk to human life in the event of failure, so it falls under Risk Category I.

b. Determine basic wind speed, V, for the applicable risk category from IBC Figure 1609A, B or C, or ASCE/SEI Figure 26.5-1A, B or C.

For Risk Category I, use Figure 1609C or Figure 26.5-1C. From either of these figures, $V = 100$ mph for Sacramento, CA.

c. Determine wind directionality factor, K_d, from Table 26.6-1.

For a solid freestanding wall, $K_d = 0.85$.

d. Determine exposure category.

In the design data, the surface roughness is given as B. Assume that Exposure B is applicable in all directions.

e. Determine topographic factor, K_{zt}.

As noted in the design data, the wall is not situated on a hill, ridge or escarpment. Thus, topographic factor $K_{zt} = 1.0$ (26.8.2).

f. Determine velocity pressure exposure coefficient, K_h, from Table 29.3-1.

For Exposure B at a height of 30 feet, $K_h = 0.70$.

g. Determine velocity pressure, q_h, by Equation 29.3-1.

$q_h = 0.00256 K_h K_{zt} K_d V^2 = 0.00256 \times 0.70 \times 1.0 \times 0.85 \times 100^2 = 15.2$ psf

2. Determine the gust-effect factor, G, using Flowchart 5.1.

Assuming the wall is rigid, gust-effect factor, G, for rigid structures may be taken as 0.85 or can be calculated by Equation 26.9-6 (26.9.4). For simplicity, use $G = 0.85$.

3. Determine the net force coefficient, C_f, from Figure 29.4-1.

In general, Cases A, B and C must be considered for freestanding walls. A different loading condition is considered in each case.

In this example, the aspect ratio $B/s = 45/30 = 1.5 < 2$. Therefore, according to Note 3 in Figure 29.4-1, only Cases A and B need to be considered.

The net force coefficient $C_f = 1.43$ (by linear interpolation) for Cases A and B with $s/h = 1$ and $B/s = 1.5$ (see Note 5 in Figure 29.4-1).

According to Note 2 in Figure 29.4-1, force coefficients for solid freestanding walls with openings may be multiplied by the reduction factor $[1 - (1 - \varepsilon)^{1.5}]$ where ε is equal to the ratio of the solid area to gross area of the wall. In this example, $\varepsilon = 0.8$, and the reduction factor is equal to 0.91. Therefore, the net force coefficient $C_f = 0.91 \times 1.43 = 1.30$.

4. Determine design wind force, F, by Equation 29.4-1.

$F = q_h G C_f A_s = 15.2 \times 0.85 \times 1.30 \times (30 \times 45) = 22{,}675$ lbs

In Case A with $s/h = 1$, this force acts at a distance equal to 5 percent of the height of the wall above the geometric center of the wall, that is, the resultant force is located at $(30/2) + (0.05 \times 30) = 16.5$ feet above the ground level (see Figure 29.4-1).

In Case B, the resultant force is located 16.5 feet above the ground level and $(45/2) - (0.2 \times 45) = 13.5$ feet from the edge of the wall (see Figure 29.4-1).

5.8.13 Example 5.13 – Rooftop Equipment Using Chapter 29

Determine the design wind forces on the rooftop equipment of the six-story hotel depicted in Figure 5.36 using Chapter 29 (see Example 5.7 for additional design data).

SOLUTION

Use Flowchart 5.7 to determine the design wind force on the rooftop equipment.

1. Determine the velocity pressure, q_z, evaluated at height z of the centroid of area, A_f, of the rooftop unit.

   The distance from the ground level to the centroid of the rooftop unit = $63.5 + 0.5 + (6/2)$ = 67 feet.

   From Table 29.3-1, $K_z = 1.33$ (by linear interpolation) for Exposure D at a height of 67 feet above the ground level.

   Velocity pressures, q_z, is determined by Equation 29.3-1:

   $q_z = 0.00256 K_z K_{zt} K_d V^2 = 0.00256 \times 1.33 \times 1.0 \times 0.90 \times 170^2 = 88.6$ psf

   where $K_d = 0.90$ for square-shaped rooftop equipment (see Table 26.6-1).

2. Determine gust-effect factor, G, from Flowchart 5.1.

   From Example 5.7, G is equal to 0.85.

3. Determine force coefficient, C_f, from Figure 29.5-1 for rooftop equipment.

   $h = 67$ feet

   Least horizontal dimension, D, of rooftop unit = 7 feet (note: using the least dimension of the rooftop unit results in the largest force coefficient).

   $h/D = 67/7 = 9.6$

   From Figure 29.5-1, $C_f = 1.5$ by linear interpolation for a square cross-section with wind normal to the face. Conservatively use this for both faces.

4. Determine design wind force, F, by Equation 29.5-1 ($h > 60$ feet).

   $F = q_z G C_f A_f$

   $= 88.6 \times 0.85 \times 1.5 \times (6 \times 7) / 1{,}000 = 4.8$ kips on the smaller face

= 88.6 × 0.85 × 1.5 × (6 × 16) / 1,000 = 10.9 kips on the larger face

These forces act perpendicular to the respective faces of the equipment.

If the mean roof height of the building were less than or equal to 60 feet, Equations 29.5-2 and 29.5-3 would be used to determine the lateral force, F_h, and vertical uplift force, F_v, on the rooftop equipment, respectively.

Assuming that the mean roof height is less than or equal to 60 feet, the values of (GC_r) are determined as follows (see Section 5.3.4 and Figure 5.19 of this publication).

- For lateral forces:

    For the smaller face, $A_f = 6 \times 7 = 42$ square feet

    For the larger face, $A_f = 6 \times 16 = 96$ square feet

    $0.1Bh = 0.1 \times 75.33 \times 63.5 = 478.4$ square feet $> A_f$

    Thus, use $(GC_r) = 1.9$ in Equation 29.5-2.

- For vertical uplift forces:

    $A_r = 16 \times 7 = 112$ square feet

    $0.1BL = 0.1 \times 75.33 \times 328.75 = 2,476.5$ square feet $> A_r$

    Thus, use $(GC_r) = 1.5$ in Equation 29.5-3.

5.8.14 Example 5.14 – Warehouse Building Using Chapter 30, Part 1 (C&C)

For the one-story warehouse illustrated in Figure 5.23, determine design wind pressures on (1) a solid precast wall panel and (2) an open-web joist purlin using Part 1 of Chapter 30.

SOLUTION

Part 1: Determine design wind pressures on a solid precast wall panel

- **Step 1:** Check if the building meets all of the conditions and limitations of 30.1.2 and 30.1.3 and the additional conditions of 30.4 so that this procedure can be used to determine the wind pressures on the C&C.

    The building is regularly-shaped, that is, it does not have any unusual geometric irregularities in spatial form. Also, the building does not have response characteristics that make it subject to across-wind loading or other similar effects, and it is not sited at a location where channeling effects or buffeting in the wake of upwind obstructions need to be considered.

    Also, the mean roof height $h = 20$ feet < 60 feet, and the building is also a low-rise building as defined in 26.2.

    Thus, Part 1 of Chapter 30 may be used to determine the design wind pressures on the C&C.

- **Step 2:** Use Flowchart 5.8 to determine the net design wind pressures, p, on the C&C.

1. Determine velocity pressure, q_h, using Flowchart 5.2 with K_h determined by Table 30.3-1.

   Since the values of K_h in Tables 27.3-1 and 30.3-1 are identical for Exposure C, the value of q_h calculated in Example 5.1 is also applicable in this example; thus, with all other applicable quantities being the same, $q_h = 25.9$ psf.

2. Determine external pressure coefficients, (GC_p), from Figure 30.4-1 for Zones 4 and 5.

   Pressure coefficients for Zones 4 and 5 can be determined from Figure 30.4-1 based on the effective wind area.

   The effective wind area is defined as the span length multiplied by an effective width that need not be less than one-third the span length: $20 \times (20/3) = 133.3$ square feet (note: the smallest span length corresponding to the east and west walls is used, since this results in larger pressures).

   The pressure coefficients from the figure are summarized in Table 5.51.

Table 5.51 External Pressure Coefficients, (GC_p), for Precast Walls

Zone	(GC_p)	
	Positive	Negative
4	0.80	−0.90
5	0.80	−1.00

Note 5 in Figure 30.4-1 states that values of (GC_p) for walls are to be reduced by 10 percent when the roof angle is less than or equal to 10 degrees. Since this is applicable in this example, modified values of (GC_p) based on Note 5 are provided in Table 5.52.

Determine distance, a, from Note 9 in Figure 28.4-1:

$$a = 0.1 \text{ (least horizontal dimension)} = 0.1 \times 148 = 14.8 \text{ feet}$$

$$= 0.4h = 0.4 \times 20 = 8 \text{ feet (governs)}$$

Minimum $a = 0.04$ (least horizontal dimension) $= 0.04 \times 148 = 5.9$ feet

$$= 3 \text{ feet}$$

Table 5.52 Modified External Pressure Coefficients, (GC_p), for Precast Walls

Zone	(GC_p)	
	Positive	Negative
4	0.72	−0.81
5	0.72	−0.90

3. Determine internal pressure coefficients, (GC_{pi}), from Table 26.11-1.

   For an enclosed building, $(GC_{pi}) = +0.18, -0.18$

4. Determine design wind pressure, p, by Equation 30.4-1 on Zones 4 and 5.

$$p = q_h[(GC_p) - (GC_{pi})] = 25.9[(GC_p) - (\pm 0.18)]$$

Calculation of design wind pressures is illustrated for Zone 4:

For positive (GC_p): $p = 25.9[0.72 - (-0.18)] = 23.3$ psf

For negative (GC_p): $p = 25.9[-0.81 - (+0.18)] = -25.6$ psf

The maximum design wind pressures for positive and negative internal pressures are summarized in Table 5.53. These pressures act perpendicular to the face of the precast walls.

Table 5.53 Design Wind Pressures, p, on Precast Walls

Zone	(GC_p)	Design Pressure, p (psf)
4	0.72	23.3
4	−0.81	−25.6
5	0.72	23.3
5	−0.90	−28.0

In Zones 4 and 5, the computed positive and negative pressures are greater than the minimum values prescribed in 30.2.2 of +16 psf and −16 psf, respectively.

Part 2: Determine design wind pressures on an open-web joist purlin

- **Step 1:** Check if the building meets all of the conditions and limitations of 30.1.2 and 30.1.3 and the additional conditions of 30.4 so that this procedure can be used to determine the wind pressures on the C&C.

  As shown in Part 1 of this example, Part 1 of Chapter 30 can be used to determine the design wind pressures on the C&C of this example building.

- **Step 2:** Use Flowchart 5.8 to determine the net design wind pressures, p, on the C&C.

  1. Determine velocity pressure, q_h, using Flowchart 5.2 with K_h determined by Table 30.3-1.

     From Part 1 of this example, $q_h = 25.9$ psf.

  2. Determine external pressure coefficients, (GC_p), from Figure 30.4-2A for Zones 1, 2 and 3.

     Pressure coefficients for Zones 1, 2 and 3 can be determined from Figure 30.4-2A for gable roofs with a roof slope less than or equal to 7 degrees based on the effective wind area.

     Effective wind area = larger of $37 \times 8 = 296$ square feet or $37 \times (37/3) = 456.3$ square feet (governs).

     The pressure coefficients from the figure are summarized in Table 5.54.

Table 5.54 External Pressure Coefficients, (GC_p), for Open-web Joist Purlins

Zone	(GC_p)	
	Positive	Negative
1	0.20	−0.90
2	0.20	−1.10
3	0.20	−1.10

3. Determine internal pressure coefficients, (GC_{pi}), from Table 26.11-1.

   For an enclosed building, $(GC_{pi}) = +0.18, -0.18$

4. Determine design wind pressure, p, by Equation 30.4-1 on Zones 1, 2 and 3.

$$p = q_h[(GC_p) - (GC_{pi})] = 25.9[(GC_p) - (\pm 0.18)]$$

The maximum design wind pressures for positive and negative internal pressures are summarized in Table 5.55.

Table 5.55 Design Wind Pressures, p, on Open-web Joist Purlins

Zone	(GC_p)	Design Pressure, p (psf)
1	0.20	9.8
	−0.90	−28.0
2 and 3	0.20	9.8
	−1.10	−33.2

The pressures in Table 5.55 are applied normal to the open-web joist purlins and act over the tributary area of each purlin, which is equal to 37 × 8 = 296 square feet. If the tributary area were greater than 700 square feet, the purlins could have been designed using the provisions for MWFRSs (30.2.3).

The positive pressures on Zones 1, 2 and 3 must be increased to the minimum value of 16 psf per 30.2.2.

Figure 5.48 contains the loading diagrams for typical open-web joist purlins located within Zones 1, 2 and 3 on the roof.

Figure 5.48 Open-web Joist Purlin Loading Diagrams, Example 5.14

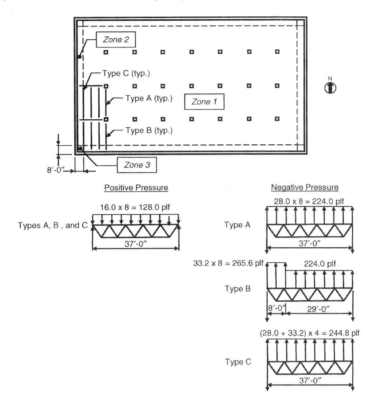

5.8.15 Example 5.15 – Warehouse Building Using Chapter 30, Part 2 (C&C)

For the one-story warehouse illustrated in Figure 5.23, determine design wind pressures on (1) a solid precast wall panel and (2) an open-web joist purlin using Part 2 of Chapter 30.

SOLUTION

Part 1: Determine design wind pressures on a solid precast wall panel

- **Step 1:** Check if the building meets all of the conditions and limitations of 30.1.2 and 30.1.3 and the additional conditions of 30.5 so that this procedure can be used to determine the wind pressures on the C&C.

  The building is regularly-shaped, that is, it does not have any unusual geometric irregularities in spatial form. Also, the building does not have response characteristics that make it subject to across-wind loading or other similar effects, and it is not sited at a location where channeling effects or buffeting in the wake of upwind obstructions need to be considered.

  The following conditions of 30.5.1 must also be satisfied:

  1. Mean roof height $h = 20$ feet < 60 feet
  2. The building is enclosed
  3. The building has a roof slope of 2.4 degrees, which is less than 45 degrees

306 Chapter 5

Thus, Part 2 of Chapter 30 may be used to determine the design wind pressures on the C&C.

- **Step 2:** Use Flowchart 5.9 to determine the net design wind pressures, p_{net}, on the C&C.

    1. Determine the risk category of the building using IBC Table 1604.5.

       Due to the nature of its occupancy, this warehouse building falls under Risk Category II.

    2. Determine basic wind speed, V, for the applicable risk category from IBC Figure 1609A, B or C, or ASCE/SEI Figure 26.5-1A, B or C.

       For Risk Category II, use Figure 1609A or Figure 26.5-1A. From either of these figures, $V = 115$ mph for St. Louis, MO.

    3. Determine exposure category.

       In the design data of Example 5.1, the surface roughness is given as C. It is assumed that Exposures B and D are not applicable, so Exposure C applies (see 26.7.3).

    4. Determine adjustment factor for height and exposure, λ, from Figure 30.5-1.

       For Exposure C and a mean roof height of 20 feet, $\lambda = 1.29$.

    5. Determine topographic factor, K_{zt}.

       As noted in the design data of Example 5.1, the building is not situated on a hill, ridge or escarpment. Thus, topographic factor $K_{zt} = 1.0$ (26.8.2).

    6. Determine net design wind pressures, p_{net30}, from Figure 30.5-1 for Zones 4 and 5, which are the interior and end zones of walls, respectively.

       Wind pressures, p_{net30}, can be read directly from Figure 30.5-1 for $V = 115$ mph and an effective wind area.

       The effective wind area is defined as the span length multiplied by an effective width that need not be less than one-third the span length: $20 \times (20/3) = 133.3$ square feet (note: the smallest span length corresponding to the east and west walls is used, since this results in larger pressures).

       According to Note 4 in Figure 30.5-1, tabulated pressures may be interpolated for effective wind areas between those given, or the value associated with the lower effective wind area may be used. The latter of these two options is utilized in this example. The pressures, p_{net30}, in Table 5.56 are obtained from Figure 30.5-1 for $V = 115$ mph and an effective wind area of 100 square feet, and they are based on Exposure B, $h = 30$ feet, and $K_{zt} = 1.0$.

Table 5.56 Wind Pressures, p_{net30}, on Precast Walls

Zone	p_{net30} (psf)	
4	20.2	−22.2
5	20.2	−24.7

7. Determine net design wind pressures $p_{net} = \lambda K_{zt} p_{net30}$ by Equation 30.5-1 for Zones 4 and 5.

$$p_{net} = 1.29 \times 1.0 \times p_{net30} = 1.29\, p_{net30}$$

The pressures in Table 5.57 represent the net (sum of internal and external) pressures that are applied normal to the precast walls.

Table 5.57 Wind Pressures, p_{net}, on Precast Walls

Zone	p_{net} (psf)	
4	26.1	−28.6
5	26.1	−31.9

Determine distance, a, from Note 9 in Figure 28.4-1:

$$a = 0.1 \text{ (least horizontal dimension)} = 0.1 \times 148 = 14.8 \text{ feet}$$

$$= 0.4h = 0.4 \times 20 = 8 \text{ feet (governs)}$$

Minimum $a = 0.04$ (least horizontal dimension) $= 0.04 \times 148 = 5.9$ feet

$$= 3 \text{ feet}$$

In Zones 4 and 5, the computed positive and negative (absolute) pressures are greater than the minimum values prescribed in 30.2.2 of +16 psf and −16 psf, respectively.

Part 2: Determine design wind pressures on an open-web joist purlin

- **Step 1:** Check if the building meets all of the conditions and limitations of 30.1.2 and 30.1.3 and the additional conditions of 30.5.1 so that this procedure can be used to determine the wind pressures on the C&C.

  As shown in Part 1 of this example, Part 2 of Chapter 30 can be used to determine the design wind pressures on the C&C of this example building.

- **Step 2:** Use Flowchart 5.9 to determine the net design wind pressures, p_{net}, on the C&C.

  1. The risk category, the basic wind speed (V), the exposure category, the adjustment factor for height and exposure (λ), and the topographic factor (K_{zt}) have all been determined previously (Part 1) and are used in calculating the wind pressures on the open-web joist purlins, which are subject to C&C pressures.

  2. Determine net design wind pressures, p_{net30}, from Figure 30.5-1 for Zones 1, 2 and 3, which are the interior, end and corner zones of the roof, respectively.

Effective wind area is equal to the larger value of the purlin tributary area = 37 × 8 = 296 square feet, or the span length multiplied by an effective width that need not be less than one-third the span length = 37 × (37/3) = 456.3 square feet (governs).

The pressures, p_{net30}, in Table 5.58 are from Figure 30.5-1 for V = 115 mph, a roof angle between 0 and 7 degrees, and an effective wind area of 100 square feet. (Note: where actual effective wind areas are greater than 100 square feet, the tabulated pressure values associated with an effective wind area of 100 square feet are applicable.) These pressures are based on Exposure B and h = 30 feet.

Table 5.58 Wind Pressures, p_{net30}, on Open-web Joist Purlins

Zone	p_{net30} (psf)	
1	7.7	−21.8
2	7.7	−25.8
3	7.7	−25.8

3. Determine net design wind pressures $p_{net} = \lambda K_{zt} p_{net30}$ by Equation 30.5-1 for Zones 1, 2 and 3.

$$p_{net} = 1.29 \times 1.0 \times p_{net30} = 1.29 p_{net30}$$

The pressures in Table 5.59 represent the net (sum of internal and external) pressures that are applied normal to the open-web joist purlins and that act over the tributary area of each purlin, which is equal to 37 × 8 = 296 square feet.

The width of the end and corner zones (Zones 2 and 3): a = 8 feet (see Part 1 of this example). The positive net design pressures in Zones 1, 2 and 3 must be increased to the minimum value of 16 psf in accordance with 30.2.2.

Table 5.59 Wind Pressures, p_{net}, on Open-web Joist Purlins

Zone	p_{net} (psf)	
1	9.9	−28.1
2	9.9	−33.3
3	9.9	−33.3

The pressures determined by Part 2 of Chapter 30 are essentially the same as those obtained from Part 1. As such, the loading diagrams for the joists in this example are the same as those depicted in Figure 5.48.

5.8.16 Example 5.16 – Warehouse Building Using Chapter 30, Part 4 (C&C)

For the one-story warehouse illustrated in Figure 5.23, determine design wind pressures on (1) a solid precast wall panel and (2) an open-web joist purlin using Part 4 of Chapter 30.

SOLUTION

Part 1: Determine design wind pressures on a solid precast wall panel

- **Step 1:** Check if the building meets all of the conditions and limitations of 30.1.2 and 30.1.3 and the additional conditions of 30.7 so that this procedure can be used to determine the wind pressures on the C&C.

The building is regularly-shaped, that is, it does not have any unusual geometric irregularities in spatial form. Also, the building does not have response characteristics that make it subject to across-wind loading or other similar effects, and it is not sited at a location where channeling effects or buffeting in the wake of upwind obstructions need to be considered.

The following conditions of 30.7 must also be satisfied:

1. Mean roof height h = 20 feet < 160 feet

2. The building is enclosed

3. The building has a gable roof (slope = 2.4 degrees)

Thus, Part 4 of Chapter 30 may be used to determine the design wind pressures on the C&C.

- **Step 2:** Use Flowchart 5.11 to determine the net design wind pressures, p, on the C&C.

    1. Determine the risk category of the building using IBC Table 1604.5.

        Due to the nature of its occupancy, this warehouse building falls under Risk Category II.

    2. Determine basic wind speed, V, for the applicable risk category from IBC Figure 1609A, B or C, or ASCE/SEI Figure 26.5-1A, B or C.

        For Risk Category II, use Figure 1609A or Figure 26.5-1A. From either of these figures, V = 115 mph for St. Louis, MO.

    3. Determine exposure category.

        In the design data of Example 5.1, the surface roughness is given as C. It is assumed that Exposures B and D are not applicable, so Exposure C applies (see 26.7.3).

    4. Determine topographic factor, K_{zt}.

        As noted in the design data of Example 5.1, the building is not situated on a hill, ridge or escarpment. Thus, topographic factor K_{zt} = 1.0 (26.8.2).

    5. Determine design wind pressures, p_{table}, from Table 30.7-2 for Zones 4 and 5, which are the interior and end zones of walls, respectively, on the building.

        The pressures, p_{table}, in Table 5.60 are obtained from Figure 30.7-2 for Exposure C with V = 115 mph, h = 20 feet and an effective wind area of 10 square feet.

Table 5.60 Wind Pressures, p_{table}, on Precast Walls

Zone	p_{table} (psf)	
4	28.0	−28.0
5	28.0	−51.4

6. Determine net design wind pressures, p, by Equation 30.7-1 for Zones 4 and 5.

$$p = p_{table}(\text{EAF})(\text{RF})K_{zt}$$

Since the building is located in Exposure C, the exposure adjustment factor (EAF) = 1.0.

The effective area reduction factor (RF) is determined from Table 30.7-2 based on the effective wind area, which in this case is equal to $20 \times (20/3) = 133.3$ square feet. Table 5.61 contains values of (RF) for an effective wind area of 100 square feet, which is conservative. The letters in parentheses after each value correspond to those given in the reduction factor graph and table that are a part of Table 30.7-2.

Table 5.61 Effective Area Reduction Factor (RF) for Precast Walls

Zone	Sign Pressure	
	Plus	Minus
4	0.82 (D)	−0.88 (C)
5	0.82 (D)	−0.76 (E)

The net design wind pressure is

$$p = p_{table} \times 1.0 \times (\text{RF}) \times 1.0 = p_{table} \times (\text{RF})$$

A summary of the net design wind pressures on a precast wall panel is given in Table 5.62.

Table 5.62 Net Design Wind Pressures, p, on Precast Walls

Zone	p (psf)	
4	23.0	−24.6
5	23.0	−39.1

The width of the corner zones (Zone 5) is equal to $a = 8$ feet (see Part 1 of Example 5.14).

Part 2: Determine design wind pressures on an open-web joist purlin

- **Step 1:** Check if the building meets all of the conditions and limitations of 30.1.2 and 30.1.3 and the additional conditions of 30.7 so that this procedure can be used to determine the wind pressures on the C&C.

As shown in Part 1 of this example, Part 4 of Chapter 30 can be used to determine the design wind pressures on the C&C of this example building.

- **Step 2:** Use Flowchart 5.11 to determine the net design wind pressures, p, on the C&C.

1. The risk category, the basic wind speed (V), the exposure category, the exposure adjustment factor (EAF) and the topographic factor (K_{zt}) have all been determined in Part 1 of this example and are used in calculating the wind pressures on the open-web joist purlins.

2. Determine design wind pressures, p_{table}, from Table 30.7-2 for Zones 1, 2 and 3, which are located on the roof of the building.

The pressures, p_{table}, in Table 5.63 are obtained from Figure 30.7-2 for Exposure C with $V = 115$ mph, $h = 20$ feet and an effective wind area of 10 square feet.

Table 5.63 Wind Pressures, p_{table}, on Open-web Joist Purlins

Zone	p_{table} (psf)	
1	17.6	−30.6
2	17.6	−51.4
3	17.6	−77.3

3. Determine net design wind pressures, p, by Equation 30.7-1 for Zones 1, 2 and 3.

$$p = p_{table}(\text{EAF})(\text{RF})K_{zt}$$

Since the building is located in Exposure C, the exposure adjustment factor (EAF) = 1.0.

The effective area reduction factor (RF) is determined from Table 30.7-2 based on the effective wind area, which in this case is equal to the larger of the purlin tributary area = $37 \times 8 = 296$ square feet or the span length multiplied by an effective width that need not be less than one-third the span length = $37 \times (37/3) = 456.3$ square feet (governs). Table 5.64 contains values of (RF) for an effective wind area of 450 square feet.

Table 5.64 Effective Area Reduction Factor (RF) for Open-web Joist Purlins

Zone	Sign Pressure	
	Plus	Minus
1	0.91(B)	−0.91(B)
2	0.91(B)	−0.81(C)
3	0.91(B)	−0.81(C)

The net design wind pressure is

$$p = p_{table} \times 1.0 \times (\text{RF}) \times 1.0 = p_{table} \times (\text{RF})$$

A summary of the net design wind pressures on an open-web joist purlin is given in Table 5.65.

Table 5.65 Net Design Wind Pressures, p, on Open-web Joist Purlins

Zone	p (psf)	
1	16.0	−27.9
2	16.0	−41.6
3	16.0	−62.6

The width of the corner zones (Zone 3) is equal to $2a = 16$ feet.

5.8.17 Example 5.17 – Warehouse Building Using Alternate All-heights Method (C&C)

For the one-story warehouse illustrated in Figure 5.23, determine design wind pressures on (1) a solid precast wall panel and (2) an open-web joist purlin using the Alternate All-heights Method of IBC 1609.6.

312 Chapter 5

Solution

Part 1: Determine design wind pressures on solid precast wall panel

- **Step 1:** Check if the provisions of IBC 1609.6 can be used to determine the design wind pressures on this building.

  It was shown in Step 1 of Example 5.5 that all of the conditions of 1609.6.1 are satisfied for this building.

The provisions of the Alternate All-heights Method of IBC 1609.6 can be used to determine the design wind pressures on the C&C.

- **Step 2:** Use Flowchart 5.15 to determine the net design wind pressures, p_{net}, on the C&C.

  1. Determine the risk category of the building using IBC Table 1604.5.

     Due to the nature of its occupancy, this warehouse building falls under Risk Category II.

  2. Determine basic wind speed, V, for the applicable risk category from IBC Figure 1609A, B or C, or ASCE/SEI Figure 26.5-1A, B or C.

     For Risk Category II, use Figure 1609A or Figure 26.5-1A. From either of these figures, $V = 115$ mph for St. Louis, MO.

  3. Determine exposure category.

     In the design data of Example 5.1, the surface roughness is given as C. It is assumed that Exposures B and D are not applicable, so Exposure C applies (see 26.7.3).

  4. Determine topographic factor, K_{zt}.

     As noted in the design data of Example 5.1, the building is not situated on a hill, ridge or escarpment. Thus, topographic factor $K_{zt} = 1.0$ (26.8.2).

  5. Determine velocity pressure exposure coefficients, K_z, from Table 27.3-1.

     Values of K_z are summarized in Table 5.66.

Table 5.66 Velocity Pressure Exposure Coefficient, K_z

Height above ground level, z (feet)	K_z
20	0.90
15	0.85

  6. Determine net pressure coefficients, C_{net}, for Zones 4 and 5 in ASCE/SEI Figure 30.4-1 from IBC Table 1609.6.2.

     The effective wind area is defined as the span length multiplied by an effective width that need not be less than one-third the span length: $20 \times (20/3) = 133.3$ square feet.

(Note: the smallest span length corresponding to the east and west walls is used, since this results in larger pressures.)

The net pressure coefficients from IBC Table 1609.6.2 for C&C (walls) not in areas of discontinuity (Item 4, $h \leq 60$ feet, Zone 4) and in areas of discontinuity (Item 5, $h \leq 60$ feet, Zone 5) are summarized in Table 5.67. (Note: linear interpolation was used to determine the values of C_{net} in Table 5.67; see Note a in IBC Table 1609.6.2.)

Table 5.67 Net Pressure Coefficients, C_{net}, for Precast Walls

Zone	C_{net}	
	Positive	Negative
4	0.94	−1.03
5	0.94	−1.21

Determine distance, a, from Note 9 in Figure 28.4-1:

$$a = 0.1 \text{ (least horizontal dimension)} = 0.1 \times 148 = 14.8 \text{ feet}$$

$$= 0.4h = 0.4 \times 20 = 8 \text{ feet (governs)}$$

Minimum $a = 0.04$ (least horizontal dimension) $= 0.04 \times 148 = 5.9$ feet

$$= 3 \text{ feet}$$

7. Determine net design wind pressures, p_{net}, by IBC Equation 16-35.

$$p_{net} = 0.00256 V^2 K_z C_{net} K_{zt} = 0.00256 \times 115^2 \times 0.90 \times C_{net} \times 1.0 = 30.5 C_{net}$$

where $K_z = K_h = 0.9$ from Table 5.66 is used for all walls.

A summary of the design wind pressures on the precast walls is given in Table 5.68. These pressures act perpendicular to the face of the precast walls.

Table 5.68 Design Wind Pressures on Precast Walls

Zone	C_{net}	Design Pressure, p_{net} (psf)
4	0.94	28.7
	−1.03	−31.4
5	0.94	28.7
	−1.21	−36.9

In Zones 4 and 5, the computed positive and negative pressures are greater than the minimum values prescribed in IBC 1609.6.3 of +16 psf and −16 psf, respectively.

Part 2: Determine design wind pressures on an open-web joist purlin

- **Step 1:** Check if the provisions of IBC 1609.6 can be used to determine the design wind pressures on this building.

As shown in Part 1 of this example, the Alternate All-heights Method of IBC 1609.6 can be used to determine the design wind pressures on the C&C of this example building.

- **Step 2:** Use Flowchart 5.15 to determine the net design wind pressures, p_{net}, on the C&C.

  1–5. These items are the same as those shown in Part 1 of this example.

  6. Determine net pressure coefficients, C_{net}, for Zones 1, 2 and 3 in ASCE/SEI Figure 30.4-2B from IBC Table 1609.6.2.

     Effective wind area = larger of $37 \times 8 = 296$ square feet or $37 \times (37/3) = 456.3$ square feet (governs).

     The net pressure coefficients from IBC Table 1609.6.2 for C&C (roofs) not in areas of discontinuity (Item 2, gable roof with flat < slope < 6:12, Zone 1) and in areas of discontinuity (Item 3, gable roof with flat < slope < 6:12, Zones 2 and 3) are summarized in Table 5.69.

Table 5.69 Net Pressure Coefficients, C_{net}, for Open-web Joist Purlins

Zone	C_{net}	
	Positive	Negative
1	0.41	−0.92
2	0.41	−1.17
3	0.41	−1.85

  7. Determine net design wind pressures, p_{net}, by IBC Equation 16-35.

  $$p_{net} = 0.00256 V^2 K_z C_{net} K_{zt} = 0.00256 \times 115^2 \times 0.90 \times C_{net} \times 1.0 = 30.5 C_{net}$$

  where $K_z = K_h = 0.9$ from Table 5.66.

  A summary of the design wind pressures on the open-web joist purlins is given in Table 5.70.

Table 5.70 Design Wind Pressures on Open-web Joist Purlins

Zone	C_{net}	Design Pressure, p_{net} (psf)
1	0.41	12.5
	−0.92	−28.1
2	0.41	12.5
	−1.17	−35.7
3	0.41	12.5
	−1.85	−56.4

The pressures in Table 5.70 are applied normal to the open-web joist purlins and act over the tributary area of each purlin, which is equal to $37 \times 8 = 296$ square feet. If the tributary area were greater than 700 square feet, the purlins could have been designed using the provisions for MWFRSs (30.2.3).

The positive pressures on Zones 1, 2 and 3 must be increased to the minimum value of 16 psf in accordance with IBC 1609.6.3.

5.8.18 Example 5.18 – Residential Building Using Chapter 30, Part 1 (C&C)

For the three-story residential building illustrated in Figure 5.34, determine design wind pressures on (1) a typical wall stud in the third story, (2) a typical roof truss and (3) a typical roof sheathing panel using Part 1 of Chapter 30.

SOLUTION

Part 1: Determine design wind pressures on a typical wall stud in the third story

- **Step 1:** Check if the building meets all of the conditions and limitations of 30.1.2 and 30.1.3 and the additional conditions of 30.4 so that this procedure can be used to determine the wind pressures on the C&C.

  The building is regularly-shaped, that is, it does not have any unusual geometric irregularities in spatial form. Also, the building does not have response characteristics that make it subject to across-wind loading or other similar effects, and it is not sited at a location where channeling effects or buffeting in the wake of upwind obstructions need to be considered.

  Also, the mean roof height $h = 20$ feet < 60 feet.

  Thus, Part 1 of Chapter 30 may be used to determine the design wind pressures on the C&C.

- **Step 2:** Use Flowchart 5.8 to determine the net design wind pressures, p, on the C&C.

  1. Determine velocity pressure, q_h, using Flowchart 5.2 with K_h determined by Table 30.3-1.

     Since the values of K_z in Tables 27.3-1 and 30.3-1 are identical for Exposure B at and above a height of 30 feet, the value of q_h calculated in Example 5.6 is also applicable in this example; thus, with all other applicable quantities being the same, $q_h = 19.8$ psf.

  2. Determine external pressure coefficients, (GC_p), from Figure 30.4-1 for Zones 4 and 5.

     Pressure coefficients for Zones 4 and 5 can be determined from Figure 30.4-1 based on the effective wind area.

     The effective wind area is the larger of the tributary area of a wall stud and the span length multiplied by an effective width that need not be less than one-third the span length.

     Effective wind area = larger of $10 \times (16/12) = 13.3$ square feet or $10 \times (10/3) = 33.3$ square feet (governs).

     The pressure coefficients from the figure are summarized in Table 5.71.

Table 5.71 External Pressure Coefficients, (GC_p), for Wall Studs

Zone	(GC_p)	
	Positive	Negative
4	0.9	−1.0
5	0.9	−1.2

Note 5 in Figure 30.4-1 states that values of (GC_p) for walls are to be reduced by 10 percent when the roof angle is less than or equal to 10 degrees. Modified values of (GC_p) based on Note 5 are provided in Table 5.72.

Table 5.72 Modified External Pressure Coefficients, (GC_p), for Wall Studs

Zone	(GC_p)	
	Positive	Negative
4	0.81	−0.9
5	0.81	−1.1

Determine distance, a, from Note 9 in Figure 28.4-1:

$$a = 0.1 \text{ (least horizontal dimension)} = 0.1 \times 72 = 7.2 \text{ feet (governs)}$$

$$= 0.4h = 0.4 \times 38 = 15.2 \text{ feet}$$

Minimum $a = 0.04$ (least horizontal dimension) $= 0.04 \times 72 = 2.9$ feet

$$= 3 \text{ feet}$$

3. Determine internal pressure coefficients, (GC_{pi}), from Table 26.11-1.

   For an enclosed building, $(GC_{pi}) = +0.18, -0.18$.

4. Determine design wind pressure, p, by Equation 30.4-1 on Zones 4 and 5.

$$p = q_h[(GC_p) - (GC_{pi})] = 19.8[(GC_p) - (\pm 0.18)]$$

Calculation of design wind pressures is illustrated for Zone 4:

For positive (GC_p): $p = 19.8[0.81 - (-0.18)] = 19.6$ psf

For negative (GC_p): $p = 19.8[-0.90 - (+0.18)] = -21.4$ psf

The maximum design wind pressures for positive and negative internal pressures are summarized in Table 5.73.

Table 5.73 Design Wind Pressures, p, on Wall Studs

Zone	(GC_p)	Design Pressure, p (psf)
4	0.81	19.6
	−0.90	−21.4
5	0.81	19.6
	−1.10	−25.3

The pressures in Table 5.73 are applied normal to the wall studs and act over the tributary area of each stud.

In Zones 4 and 5, the computed positive and negative pressures are greater than the minimum values prescribed in 30.2.2 of +16 psf and −16 psf, respectively.

Part 2: Determine design wind pressures on roof trusses

1. Determine velocity pressure, q_h, using Flowchart 5.2 with K_h determined by Table 30.3-1.

   Velocity pressure, q_h, was determined in Part 1 of this example and is equal to 19.8 psf.

2. Determine external pressure coefficients, (GC_p), from appropriate figures for roof trusses spanning in the N-S and E-W directions.

   a. Trusses spanning in the N-S direction

   Pressure coefficients for Zones 1, 2 and 3 can be determined from Figure 30.4-2B ($7° < \theta = 23° < 27°$) based on the effective wind area. Included are the pressure coefficients for the overhanging portions of the trusses.

   Effective wind area = larger of $56.5 \times 2 = 113$ square feet or $56.5 \times (56.5/3) = 1,064.1$ square feet (governs).

   The pressure coefficients from Figure 30.4-2B are summarized in Table 5.74. According to Note 5 in the figure, values of (GC_p) for roof overhangs include pressure contributions from both the upper and lower surfaces of the overhang.

Table 5.74 External Pressure Coefficients, (GC_p), for Roof Trusses Spanning in the N-S Direction

Zone	(GC_p)	
	Positive	Negative
1	0.3	−0.8
2	0.3	−1.2
2 (overhang)	---	−2.2
3	0.3	−2.0
3 (overhang)	---	−2.5

   b. Trusses spanning in the E-W direction

   Pressure coefficients for Zones 1, 2 and 3 can be determined from Figure 30.4-2C ($27° < \theta = 44.4° < 45°$) based on the effective wind area.

   Included are the pressure coefficients for the overhanging portions of the trusses.

   Effective wind area = larger of $24.5 \times 2 = 49$ square feet or $24.5 \times (24.5/3) = 200$ square feet (governs).

   The pressure coefficients from Figure 30.4-2C are summarized in Table 5.75. According to Note 5 in the figure, values of (GC_p) for roof overhangs include pressure contributions from both the upper and lower surfaces of the overhang.

Table 5.75 External Pressure Coefficients, (GC_p), for Roof Trusses Spanning in the E-W Direction

Zone	(GC_p)	
	Positive	Negative
1	0.8	−0.8
2	0.8	−1.0
2 (overhang)	---	−1.8
3	0.8	−1.0
3 (overhang)	---	−1.8

3. Determine internal pressure coefficients, (GC_{pi}), from Table 26.11-1.

   For an enclosed building, $(GC_{pi}) = +0.18, -0.18$.

4. Determine design wind pressure, p, by Equation 30.4-1 on Zones 1, 2 and 3.

   $$p = q_h[(GC_p) - (GC_{pi})] = 19.8[(GC_p) - (\pm 0.18)]$$

   a. Trusses spanning in the N-S direction

      The maximum design wind pressures for positive and negative internal pressures are summarized in Table 5.76.

Table 5.76 Design Wind Pressures, p, on Roof Trusses Spanning in the N-S Direction

Zone	(GC_p)	Design Pressure, p (psf)
1	0.3	9.5
	−0.8	−19.4
2	0.3	9.5
	−1.2	−27.3
2 (overhang)	−2.2	−43.6*
3	0.3	9.5
	−2.0	−43.2
3 (overhang)	−2.5	−49.5*

* Net overhang pressure = $q_h(GC_p)$

   b. Trusses spanning in the E-W direction

      The maximum design wind pressures for positive and negative internal pressures are summarized in Table 5.77.

Table 5.77 Design Wind Pressures, p, on Roof Trusses Spanning in the E-W Direction

Zone	(GC_p)	Design Pressure, p (psf)
1	0.8	19.4
	−0.8	−19.4
2 and 3	0.8	19.4
	−1.0	−23.4
2 and 3 (overhang)	−1.8	−35.6*

* Net overhang pressure = $q_h(GC_p)$

Figure 5.49 contains the loading diagrams for typical trusses located within various zones of the roof. The minimum pressure of +16.0 psf is used in Zones 1, 2 and 3 for the roof trusses spanning in the N-S direction (see 30.2.2).

Figure 5.49 Roof Truss Loading Diagrams, Example 5.18

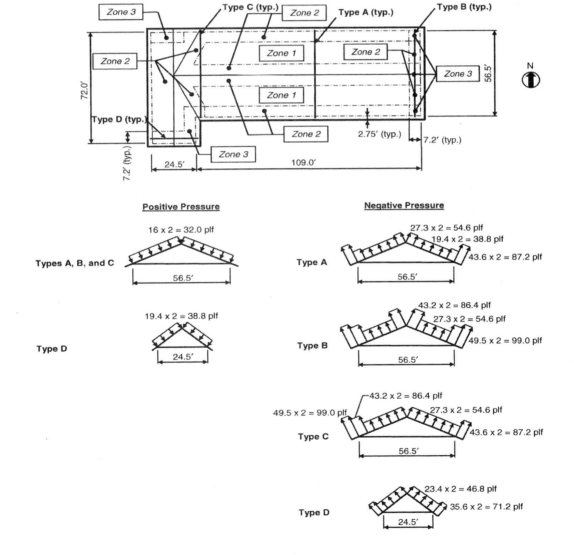

Part 3: Determine design wind pressures on roof sheathing panel

1. Determine velocity pressure, q_h, using Flowchart 5.2 with K_h determined by Table 30.3-1.

   Velocity pressure, q_h, was determined in Part 1 of this example and is equal to 19.8 psf.

2. Determine external pressure coefficients, (GC_p), from appropriate figures for roof panels on the east and west wings.

   a. Roof panels on the east wing

   Pressure coefficients for Zones 1, 2 and 3 can be determined from Figure 30.4-2B ($7° < \theta = 23° < 27°$) based on the effective wind area assuming a 4 by 2 foot panel size. Included are the pressure coefficients for the overhanging portions of panels.

   Effective wind area = larger of $2 \times 4 = 8$ square feet (governs) or $2 \times (2/3) = 1.33$ square feet.

   The pressure coefficients from Figure 30.4-2B are summarized in Table 5.78. According to Note 5 in the figure, values of (GC_p) for roof overhangs include pressure contributions from both the upper and lower surfaces of the overhang.

   Table 5.78 External Pressure Coefficients, (GC_p), for Roof Panels on the East Wing

Zone	(GC_p) Positive	(GC_p) Negative
1	0.5	−0.9
2	0.5	−1.7
2 (overhang)	---	−2.2
3	0.5	−2.6
3 (overhang)	---	−3.7

   b. Roof panels on the west wing

   Pressure coefficients for Zones 1, 2 and 3 can be determined from Figure 30.4-2C ($27° < \theta = 44.4° < 45°$) based on the effective wind area. Included are the pressure coefficients for the overhanging portions of the trusses.

   Effective wind area = larger of $2 \times 4 = 8$ square feet (governs) or $2 \times (2/3) = 1.33$ square feet.

   The pressure coefficients from Figure 30.4-2C are summarized in Table 5.79. According to Note 5 in the figure, values of (GC_p) for roof overhangs include pressure contributions from both the upper and lower surfaces of the overhang.

Table 5.79 External Pressure Coefficients, (GC_p), for Roof Panels on the West Wing

Zone	(GC_p) Positive	(GC_p) Negative
1	0.9	−1.0
2	0.9	−1.2
2 (overhang)	---	−2.0
3	0.9	−1.2
3 (overhang)	---	−2.0

3. Determine internal pressure coefficients, (GC_{pi}), from Table 26.11-1.

   For an enclosed building, $(GC_{pi}) = +0.18, -0.18$.

4. Determine design wind pressure, p, by Equation 30.4-1 on Zones 1, 2 and 3.

$$p = q_h[(GC_p) - (GC_{pi})] = 19.8[(GC_p) - (\pm 0.18)]$$

The maximum design wind pressures for positive and negative internal pressures are summarized in Tables 5.80 and 5.81 for roof panels on the east wing and the west wing, respectively. These pressures are applied normal to the roof panels and act over the tributary area of each panel.

The positive pressures in Zones 1, 2 and 3 for roof panels on the east wing must be increased to the minimum value of 16 psf in accordance with 30.2.2.

Table 5.80 Design Wind Pressures, p, on Roof Panels on the East Wing

Zone	(GC_p)	Design Pressure, p (psf)
1	0.5	13.5
	−0.9	−21.4
2	0.5	13.5
	−1.7	−37.2
2 (overhang)	−2.2	−43.6*
3	0.5	13.5
	−2.6	−55.0
3 (overhang)	−3.7	−73.3*

* Net overhang pressure = $q_h(GC_p)$

Table 5.81 Design Wind Pressures, p, on Roof Panels on the West Wing

Zone	(GC_p)	Design Pressure, p (psf)
1	0.9	21.4
1	−1.0	−23.4
2 and 3	0.9	21.4
2 and 3	−1.2	−27.3
2 and 3 (overhang)	−2.0	−39.6*

* Net overhang pressure = $q_h(GC_p)$

5.8.19 Example 5.19 – Fifteen-story Office Building Using Chapter 30, Part 3 (C&C)

For the 15-story office building depicted in Figure 5.42, determine design wind pressures on the C&C using Part 3 of Chapter 30.

SOLUTION

- **Step 1:** Check if the building meets all of the conditions and limitations of 30.1.2 and 30.1.3 and the additional conditions of 30.6 so that this procedure can be used to determine the wind pressures on the C&C.

  The building is regularly-shaped, that is, it does not have any unusual geometric irregularities in spatial form. Also, the building does not have response characteristics that make it subject to across-wind loading or other similar effects, and it is not sited at a location where channeling effects or buffeting in the wake of upwind obstructions need to be considered.

  Also, the mean roof height h = 225 feet > 60 feet.

  Thus, Part 3 of Chapter 30 may be used to determine the design wind pressures on the C&C.

- **Step 2:** Determine the enclosure classification of the building.

  For illustrative purposes, it is assumed in Example 5.10 that the building is partially enclosed.

- **Step 3:** Use Flowchart 5.10 to determine the design wind pressures, p, on the C&C.

  1. Determine velocity pressure: q_z for windward walls along the height of the building, and q_h for leeward walls, side walls and roof using Flowchart 5.2.

     a. Determine the risk category of the building using IBC Table 1604.5.

        As noted in Example 5.10, this office building falls under Risk Category II.

     b. Determine basic wind speed, V, for the applicable risk category from IBC Figure 1609A, B or C, or ASCE/SEI Figure 26.5-1A, B or C.

        It was determined in Example 5.10 that V = 115 mph for Chicago, IL.

     c. Determine wind directionality factor, K_d, from Table 26.6-1.

        For the C&C of a building structure, K_d = 0.85.

d. Determine exposure category.

In the design data of Example 5.10, the surface roughness is given as B. Assume that Exposure B is applicable in all directions.

e. Determine topographic factor, K_{zt}.

As noted in the design data of Example 5.10, the building is not situated on a hill, ridge or escarpment. Thus, topographic factor $K_{zt} = 1.0$ (26.8.2).

f. Determine velocity pressure exposure coefficients, K_z and K_h, from Table 30.3-1.

Values of K_z and K_h for Exposure B are summarized in Table 5.82.

Table 5.82 Velocity Pressure Exposure Coefficient, K_z, for C&C

Height above ground level, z (feet)	K_z
225	1.24
200	1.20
180	1.17
160	1.13
140	1.09
120	1.04
100	0.99
90	0.96
80	0.93
70	0.89
60	0.85
50	0.81
40	0.76
30	0.70
25	0.70
20	0.70
15	0.70

g. Determine velocity pressure, q_z and q_h, by Equation 30.3-1.

$$q_z = 0.00256 K_z K_{zt} K_d V^2 = 0.00256 \times K_z \times 1.0 \times 0.85 \times 115^2 = 28.78 K_z \text{ psf}$$

A summary of the velocity pressures is given in Table 5.83.

2. Determine external pressure coefficients, (GC_p), for zones on the walls from Figure 30.6-1.

Pressure coefficients for Zones 4 and 5 can be determined from Figure 30.6-1 based on the effective wind area.

In general, the effective wind area is the larger of the tributary area and the span length multiplied by an effective width that need not be less than one-third the span length.

Effective wind area for the mullions = larger of $15 \times 5 = 75$ square feet or $15 \times (15/3) = 75$ square feet (see Figure 5.20).

Effective wind area for the glazing panels = larger of $7.5 \times 5 = 37.5$ square feet (governs) or $5 \times (5/3) = 8.3$ square feet.

Table 5.83 Velocity Pressure, q_z, for C&C

Height above ground level, z (feet)	K_z	q_z (psf)
225	1.24	35.7
200	1.20	34.5
180	1.17	33.7
160	1.13	32.5
140	1.09	31.4
120	1.04	29.9
100	0.99	28.5
90	0.96	27.6
80	0.93	26.8
70	0.89	25.6
60	0.85	24.5
50	0.81	23.3
40	0.76	21.9
30	0.70	20.2
25	0.70	20.2
20	0.70	20.2
15	0.70	20.2

The pressure coefficients from Figure 30.6-1 for the C&C are summarized in Table 5.84.

Table 5.84 External Pressure Coefficients, (GC_p), for C&C

Zone	(GC_p)			
	Mullions		Glazing Panels	
	Positive	Negative	Positive	Negative
4	0.78	−0.82	0.84	−0.86
5	0.78	−1.47	0.84	−1.64

The width of the end zone (Zone 5) $a = 0.1$ (least horizontal dimension) $= 0.1 \times 150 = 15$ feet, which is greater than 3 feet (see Note 8 in Figure 30.6-1).

3. Determine q_i for the walls and roof using Flowchart 5.2.

In accordance with 27.4.1, $q_i = q_h = 35.7$ psf for windward walls, side walls, leeward walls and roofs of partially enclosed buildings.

4. Determine internal pressure coefficients, (GC_{pi}), from Table 26.11-1.

For a partially enclosed building, $(GC_{pi}) = +0.55, -0.55$.

5. Determine design wind pressures, p, by Equation 30.6-1 in Zones 4 and 5.

$p = q(GC_p) - q_h(GC_{pi})$

$ = q(GC_p) - 35.7(\pm 0.55)$

$ = [q(GC_p) \mp 19.6]$ psf (external ± internal pressure)

Wind Loads   325

where $q = q_z$ for positive external pressure, and $q = q_h$ for negative external pressure (30.6.2). Note that $q_i = q_h$ for negative internal pressure in partially enclosed buildings. Also, $q_i = q_h$ may be conservatively used for positive internal pressure.

The maximum design wind pressures for positive and negative internal pressures are summarized in Table 5.85. The maximum positive pressure, which varies with height, is obtained with negative internal pressure. Similarly, the maximum negative pressure, which is a constant over the height of the building, is obtained with positive internal pressure.

The pressures in Table 5.85 are applied normal to the C&C and act over the respective tributary areas. The computed positive and negative pressures are greater than the minimum values prescribed in 30.2.2 of +16 psf and −16 psf, respectively.

Table 5.85 Design Wind Pressures, p, on C&C

Height above ground level, z (feet)	Design Pressure, p (psf)							
	Mullions				Glazing Panels			
	Zone 4		Zone 5		Zone 4		Zone 5	
	Positive	Negative	Positive	Negative	Positive	Negative	Positive	Negative
225	47.4	−48.9	47.4	−72.1	49.6	−50.3	49.6	−78.1
200	46.5	−48.9	46.5	−72.1	48.6	−50.3	48.6	−78.1
180	45.9	−48.9	45.9	−72.1	47.9	−50.3	47.9	−78.1
160	45.0	−48.9	45.0	−72.1	46.9	−50.3	46.9	−78.1
140	44.1	−48.9	44.1	−72.1	46.0	−50.3	46.0	−78.1
120	42.9	−48.9	42.9	−72.1	44.7	−50.3	44.7	−78.1
100	41.8	−48.9	41.8	−72.1	43.5	−50.3	43.5	−78.1
90	41.2	−48.9	41.2	−72.1	42.8	−50.3	42.8	−78.1
80	40.5	−48.9	40.5	−72.1	42.1	−50.3	42.1	−78.1
70	39.6	−48.9	39.6	−72.1	41.1	−50.3	41.1	−78.1
60	38.7	−48.9	38.7	−72.1	40.1	−50.3	40.1	−78.1
50	37.8	−48.9	37.8	−72.1	39.2	−50.3	39.2	−78.1
40	36.7	−48.9	36.7	−72.1	38.0	−50.3	38.0	−78.1
30	35.3	−48.9	35.3	−72.1	36.5	−50.3	36.5	−78.1
25	35.3	−48.9	35.3	−72.1	36.5	−50.3	36.5	−78.1
20	35.3	−48.9	35.3	−72.1	36.5	−50.3	36.5	−78.1
15	35.3	−48.9	35.3	−72.1	36.5	−50.3	36.5	−78.1

5.8.20 Example 5.20 – Agricultural Building Using Chapter 30, Part 5 (C&C)

For the agricultural building depicted in Figure 5.46, determine the design wind pressures on the roof trusses using Chapter 30, Part 5.

SOLUTION

- **Step 1:** Check if the building meets all of the conditions and limitations of 30.1.2 and 30.1.3 so that this procedure can be used to determine the wind pressures on the C&C.

The building is regularly-shaped, that is, it does not have any unusual geometric irregularities in spatial form. Also, the building does not have response characteristics that make it subject to across-wind loading or other similar effects, and it is not sited at a location where channeling effects or buffeting in the wake of upwind obstructions need to be considered.

Thus, Part 5 of Chapter 30 may be used to determine the design wind pressures on the C&C.

- **Step 2:** Determine the enclosure classification of the building.

  Since the building does not have any walls, it is classified as open.

- **Step 3:** Use Flowchart 5.12 to determine the design wind pressures, p, on the C&C.

  1. Determine velocity pressure, q_h, using Flowchart 5.2.

     a. Determine the risk category of the building using IBC Table 1604.5.

        As noted in Example 5.11, this agricultural building falls under Risk Category I.

     b. Determine basic wind speed, V, for the applicable risk category from IBC Figure 1609A, B or C, or ASCE/SEI Figure 26.5-1A, B or C.

        It was determined in Example 5.11 that $V = 105$ mph for Ames, IA.

     c. Determine wind directionality factor, K_d, from Table 26.6-1.

        For the C&C of a building structure, $K_d = 0.85$.

     d. Determine exposure category.

        In the design data of Example 5.11, the surface roughness is given as C. Assume that Exposure C is applicable in all directions.

     e. Determine topographic factor, K_{zt}.

        As noted in the design data of Example 5.11, the building is not situated on a hill, ridge or escarpment. Thus, topographic factor $K_{zt} = 1.0$ (26.8.2).

     f. Determine velocity pressure exposure coefficient, K_h, from Table 30.3-1.

        $$\text{Mean roof height} = \frac{20 + 30}{2} = 25 \text{ feet}$$

        From Table 30.3-1, $K_h = 0.94$ for Exposure C.

     g. Determine velocity pressure, q_h, by Equation 30.3-1.

        $$q_h = 0.00256 K_h K_{zt} K_d V^2 = 0.00256 \times 0.94 \times 1.0 \times 0.85 \times 105^2 = 22.6 \text{ psf}$$

  2. Determine gust-effect factor, G, from Flowchart 5.1.

     Assuming the building is rigid (see Example 5.11), gust-effect factor, G, for rigid buildings may be taken as 0.85 or can be calculated by Equation 26.9-6 (26.9.4). For simplicity, use $G = 0.85$.

  3. Determine net pressure coefficient, C_N, from Figure 30.8-2 for open buildings with pitched roofs.

     Figure 30.8-2 is used to determine the net pressure coefficients, C_N, for Zones 1, 2 and 3 on the roof. In this example, $h/L = 25/60 = 0.42$, which is between the limits of 0.25 and 1.0.

     The magnitude of the net pressure coefficient depends on the effective wind area, which is the larger of the tributary area and the span length multiplied by an effective width that need not be less than one-third the span length.

Effective wind area = larger of 60 × 3 = 180 square feet or 60 × (60/3) = 1,200 square feet (governs).

Determine distance, a, from Note 9 in Figure 28.4-1:

$$a = 0.1 \text{ (least horizontal dimension)} = 0.1 \times 60 = 6.0 \text{ feet (governs)}$$

$$= 0.4h = 0.4 \times 25 = 10.0 \text{ feet}$$

Minimum $a = 0.04$ (least horizontal dimension) $= 0.04 \times 60 = 2.4$ feet

$$= 3 \text{ feet}$$

The effective wind area is greater than $4.0a^2 = 4.0 \times 6.0^2 = 144$ square feet; this information is also needed to determine the net pressure coefficients.

Like in the case of the MWFRS, the magnitude of C_N depends on whether the wind flow is clear or obstructed. Both situations are examined in this example.

For clear wind flow, $C_N = 1.15, -1.06$ in Zones 1, 2 and 3. Linear interpolation was used to determine these values for a roof angle of 18.4 degrees and an effective wind area > $4.0a^2$ (see Note 3 in Figure 30.8-2).

For obstructed wind flow, $C_N = 0.50, -1.51$ in Zones 1, 2 and 3 by linear interpolation.

4. Determine net design wind pressure, p, by Equation 30.8-1.

$$p = q_h G C_N = 22.6 \times 0.85 \times C_N = 19.2 C_N \text{ psf}$$

For clear wind flow: $p = 22.1$ psf, -20.4 psf in Zones 1, 2 and 3

For obstructed wind flow: $p = 9.6$ psf, -29.0 psf in Zones 1, 2 and 3

In the case of obstructed wind flow, the net positive pressure must be increased to 16 psf to satisfy the minimum requirements of 30.2.2.

Illustrated in Figure 5.50 are the loading diagrams on a typical interior roof truss for clear wind flow. Similar loading diagrams can be obtained for obstructed wind flow.

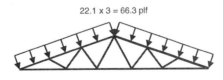

Figure 5.50 Roof Truss Loading Diagrams for Clear Wind Flow, Example 5.20

5.9 References

5.1. Simiu, E. and Scanlan, R. H. 1996. *Wind Effects on Structures: Fundamentals and Applications to Design*, 3rd Ed. John Wiley & Sons, Inc., New York, NY.

5.2. Davenport, A. G. 1960. "Rationale for Determining Design Wind Velocities." *Journal of the Structural Division*, American Society of Civil Engineers, 88: 39–68.

5.3. Smith, B. S. and Coull, A. 1991. *Tall Building Structures: Analysis and Design*, John Wiley & Sons, Inc., New York, NY.

5.4. Applied Technology Council. ATC Windspeed by Location, www.atcouncil.org/windspeed/.

5.5. Federal Emergency Management Agency. 2011. *Taking Shelter From the Storm: Building a Safe Room For Your Home or Small Business*, FEMA 320, 3rd Ed. Washington, DC.

5.6. European Committee for Standardization. 1995. *Eurocode 1: Basis of Design and Actions on Structures*, Part 2–4: Actions on Structures—Wind Actions. Brussels, Belgium.

5.7. American National Standards Institute. 1982. *Minimum Design Loads for Buildings and Other Structures*, ANSI A58.1-1982. New York, NY.

5.8. Swiss Society of Engineers and Architects (SIA). 1956. *Normen fur die Belastungsannahmen, die Inbetriebnahme und die Uberwachung der Bauten*, SIA Technische Normen No. 160. Zurich, Switzerland.

5.9. American Society of Civil Engineers. 1961. Wind Forces on Structures. *Trans. ASCE*, 126(2): 1124–1198.

5.10. Davenport, A. G., Surry, D., and Stathopoulos, T. 1978. *Wind Loads on Low-Rise Buildings*, Final Report on Phase III, BLWT-SS4. University of Western Ontario, London, Ontario, Canada.

5.11. International Association for Shell and Spatial Structures. 1991. *Structural Standards for Steel Antenna Towers and Antenna Supporting Structures*, ANSI/EIA/TIA-222-E-1991.

5.10 Problems

5.1. Given the one-story warehouse described in Example 5.1, determine design wind pressures on the MWFRS in both directions using Part 1 of Chapter 27 assuming the roof slope is equal to (a) 1 inch/foot and (b) 3 inch/foot.

5.2. Given the one-story warehouse described in Example 5.1, determine design wind pressures on (1) a solid precast wall panel and (2) an open-web joist purlin using Part 1 of Chapter 30 assuming the roof slope is equal to (a) 1 inch/foot and (b) 3 inch/foot.

5.3. The plan of a two-story building is shown in Figure 5.51. Determine the design wind pressures on the MWFRS in the E-W direction by (a) Part 1 of Chapter 27 and (b) Part 2 of Chapter 27 using the provided design data in Table 5.86. The first floor is located 20 feet above ground and the roof, which has an angle less than 5 degrees from the horizontal, is located 32 feet above ground. Use the floor heights as the reference heights and assume the building is rigid.

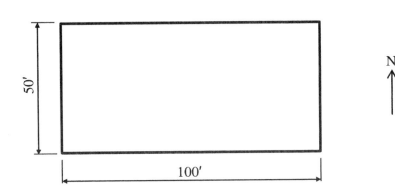

Figure 5.51 Plan of Two-story Building in Problem 5.3

Table 5.86 Design Data for Problem 5.3

Wind velocity, V	115 mph
Exposure category	B
Enclosure classification	Enclosed
Occupancy	Residential
Topography	No hills or escarpments

5.4. Given the plan dimensions in Figure 5.51 and the design data in Table 5.86, determine the design wind pressures on the MWFRS in the E-W direction by (a) Part 1 of Chapter 27 and (b) Part 2 of Chapter 27 assuming that the building has five stories, each with a story height of 15 feet. The roof angle is less than 5 degrees from the horizontal. Use the floor heights as the reference heights and assume the building is rigid.

5.5. Given the data in Problem 5.3, determine the design wind pressures on the MWFRS in the E-W direction by (a) Part 1 of Chapter 28 and (b) Part 2 of Chapter 28. Assume continuous roof and floor diaphragms.

5.6. A building with a mean roof height of 20 feet is located at the crest of a two-dimensional ridge that has the following properties: $H = 100$ feet, $L_h = 40$ feet and $x = 0$. Determine the topographic factor, K_{zt}, assuming Exposure C and conditions 1 and 2 of 26.8.1 are satisfied.

5.7. A three-story police station is depicted in Figure 5.52. Determine the design wind pressures on the walls and roof in both the N-S and E-W directions using Part 1 of Chapter 27 and the design data in Table 5.87. The CMU walls around the perimeter of the building have a nominal thickness of 8 inches.

Figure 5.52
Building in Problem 5.7

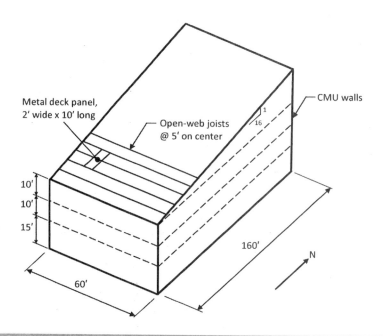

Table 5.87 Design Data for Problem 5.7

Wind velocity, V	130 mph
Exposure category	C
Enclosure classification	Enclosed
Topography	No hills or escarpments

5.8. Given the building in Problem 5.7, determine the design wind pressures on (a) a CMU wall located in the first story, (b) an open-web joist and (c) a metal deck panel using Part 4 of Chapter 30.

5.9. Given the building in Problem 5.7, assume that there is a roof overhang on the south elevation that is 5 feet in plan. Determine the design wind pressure on the roof overhang using (a) Part 1 of Chapter 27 and (b) Part 4 of Chapter 30.

5.10. Given the building in Problem 5.7, assume that there is one 3 foot-6 inch by 7 foot-0 inch opening and one 20 foot-0 inch by 8 foot-0 inch opening in each of the east and west walls in the first story. Also in the first story are two 3 foot-6 inch by 7 foot-0 inch openings in the south wall. There are no openings in the north wall or in the roof. Determine the enclosure classification of the building.

5.11. Given the building in Problem 5.7, assume that there is a mechanical unit located at the geometric center of the roof. The unit has base dimensions of 15 by 15 feet, and its height is 8 feet. Determine the wind forces on the mechanical unit.

5.12. A reinforced concrete chimney with an average outside diameter of 6 feet has a height 60 feet above the ground surface, which is relatively flat. A preliminary analysis indicates that the approximate fundamental frequency of the chimney is 0.3 Hz. Assuming that the response characteristics are predominantly along-wind, determine the design wind forces on the chimney given a wind velocity of 115 mph, Exposure C and a damping ratio of 2 percent.

CHAPTER 6

Earthquake Loads

6.1 Introduction

6.1.1 Nature of Earthquake Ground Motion

The earth's crust consists of a number of large and fairly stable rock slabs, which are referred to as *tectonic plates* (see Figure 6.1). For the most part, the plates are prevented from moving relative to each other due to friction at their boundaries. Over time, stress builds up along these boundaries, and an earthquake is generated when energy is released by the following mechanisms:

1. The built-up stress along a plate boundary exceeds the frictional force that locks the plates together resulting in slip along the fault lines.

2. The stress at an internal location in the plate exceeds the strength of the rock causing the rock to fracture.

Figure 6.1
Major Tectonic Plates (Courtesy of the U.S. Geological Survey)

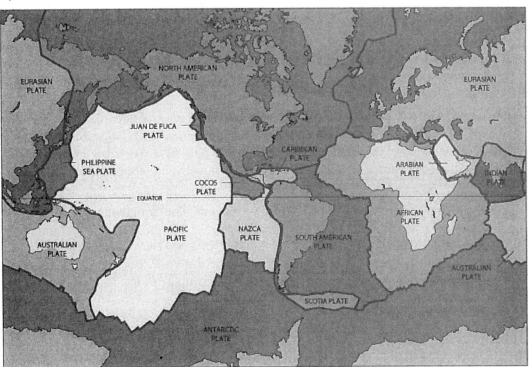

Faults are locations where earthquakes have occurred previously and are generally found at the plate boundaries although they can occur within the interior of a plate. A fault is considered to be active if there is evidence that it has moved within the past 10,000 years. New faults are created as the stress patterns in the earth's crust change over time.

Slippage can occur at the surface of the earth or many feet beneath it. *Ground fault ruptures* are created when the slippage extends to the surface. In such cases, abrupt vertical and lateral offsets occur; surface offsets can range from a few inches to a dozen feet depending on the magnitude of the earthquake. Because it is very difficult to design structures to accommodate such movement, building over known faults is avoided wherever possible.

The energy that is released during an earthquake radiates out in the form of random vibrations in all directions from the area where slippage has occurred. *Ground shaking* is felt on the surface due to these vibrations. As the ground displaces, a building will move and undergo a series of oscillations. Depending on the size of the earthquake, ground shaking can last from a few seconds to several minutes.

Deformations are produced in the structure due to ground shaking, and both structural and nonstructural members must be designed for the ensuing effects. Structures can also be subjected to other effects caused by ground shaking. Landslides and soil liquefaction are forms of ground failure that can result in significant damage or complete failure of a building. Tsunamis are seismically-induced water waves that can have a devastating impact on structures. An earthquake-induced fire is another example of the various types of damaging events that can occur to a building or structure following an earthquake.

6.1.2 Seismic Hazard Analysis

Magnitude and *intensity* measure different characteristics of earthquakes. In general, magnitude is a measure of the energy released at the source of the earthquake and is determined using seismographs. Intensity measures the strength of ground shaking produced by an earthquake at a certain location.

The Modified Mercalli Intensity (MMI) scale is a qualitative measure of earthquake intensity. Descriptions of effects on people, structures, and the natural environment are provided for twelve intensity levels, which range from earthquakes that are essentially not felt (MMI I) to earthquakes that cause total destruction (MMI XII).

The MMI scale is not directly useful for the design of structures because it is not based on quantifiable effects due to ground shaking. A mathematical relationship that can be used to quantify earthquake shaking is an *acceleration response spectrum*, which gives the peak accelerations that structures would experience if subjected to a specific earthquake motion.

Depicted in Figure 6.2 is a generic representation of an acceleration response spectrum that was obtained from recordings of an actual seismic event. The horizontal axis is the period, T, of a structure, which is related to its dynamic properties (methods on how to determine the period of a structure are given in Section 6.3.8 of this publication). The vertical axis is the acceleration a structure will experience based on its period. The *spectral acceleration*, S_a, is derived from a response spectrum and is usually expressed as a percentage of the acceleration due to gravity, g.

Figure 6.2
Acceleration Response Spectrum

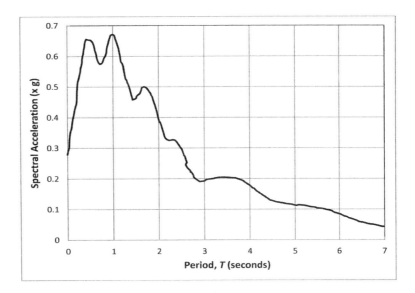

It is evident from the figure that structures with longer periods (for example, flexible, high-rise buildings) will experience smaller accelerations than structures with shorter periods (such as stiff, low-rise buildings).

In general, a unique acceleration response spectrum is obtained at a specific site for a specific earthquake. The following factors have a direct impact on the shape of the spectrum:

- Magnitude of earthquake
- Depth of earthquake
- Distance of earthquake from the site
- Soil type

A design response spectrum is given in the IBC and ASCE/SEI 7 to facilitate the design of structures for earthquake motions (see Section 6.2.1 of this publication for more information). In essence, the design response spectrum is a smoothed and normalized approximation for many different ground motions. Adjustments have been made at the extremes of the spectrum (that is, at the smaller and larger period portions of the spectrum) to include characteristics of larger structures.

The U.S. Geological Survey (USGS) has performed national seismic hazard analyses to determine spectral response accelerations for different recurrence intervals. The spectral response acceleration at short periods, S_S, the spectral acceleration at 1-second periods, S_1, and the long-period transition period, T_L, have been determined for the entire United States and can be used to derive a generalized response spectrum. Additional information on these accelerations is given in the next section.

The three aforementioned acceleration parameters can be plotted as a function of *annual frequency of exceedance*, which is the number of times in any one year that, on average, a specific site can experience ground shaking that is greater than or equal to a specific acceleration. Such plots are commonly referred to as *hazard curves* where the annual frequency of exceedance is on the vertical axis and the acceleration is on the horizontal axis. The *average return period* is the inverse of the annual frequency of exceedance and is the number of years likely to elapse between two events, each producing ground shaking that is at least the value indicated for a specific acceleration on the horizontal axis. For example, a particular hazard curve would provide the following information: ground shaking that produces a spectral response acceleration at short periods, S_S, equal to $2.1g$ has an annual frequency of exceedance of 10^{-3} per year, or an average return period of $1/10^{-3} = 1,000$ years. Hazard curves are used in establishing the risk associated with collapse of a structure (see Section 6.1.3 of this publication).

Methods on how to calculate the effects of earthquakes on structural and nonstructural components of a structure are covered in subsequent sections of this chapter. In these methods, such effects are determined using static earthquake loads. The purpose of these loads is to produce the same deformations that would occur in the structure (when multiplied by a deflection amplification factor) if it were subjected to the design-basis ground motion.

The following section provides information on the reference earthquake shaking that is stipulated in the IBC and ASCE/SEI 7; this reference shaking is used to determine earthquake effects on structures.

6.1.3 Reference Earthquake Shaking

Any structure will collapse if it subjected to sufficiently strong ground motion. However, it is virtually impossible to predict for a particular earthquake the response spectrum and the accompanying ground accelerations that can cause collapse to occur. Variability in material strengths and structural member sizes and quality of workmanship are a few of the many other variable factors that make collapse prediction essentially unattainable.

A structural *fragility function* can be used to determine an estimate of the earthquake-generated acceleration that will cause collapse of a structure. In particular, a fragility curve can be developed that will provide the probabilities of a structure collapsing as a function of the spectral response acceleration. For example, a building may have a 10 percent chance of collapsing if subjected to an acceleration of $0.5g$ and a 50 percent chance of collapsing if subjected to an acceleration of $0.9g$. More information on fragility functions for structural collapse can be found in *NEHRP Recommended Seismic Provisions for New Buildings and Other Structures* (Reference 6.1).

The risk associated with collapse is a function of the fragility of a structure and the seismic hazard at the site, the latter of which is represented by a hazard curve (see the discussion in Section 6.1.2 of this publication on hazard). The mathematical combination of these two functions gives the probability that a structure will collapse in any given year or number of years. The *NEHRP Recommended Seismic Provisions* (Reference 6.1), which forms the basis of the earthquake provisions in the IBC and ASCE/SEI 7, establishes the probability at 1 percent or less in 50 years that a structure with ordinary occupancy will collapse due to an earthquake. See Section 6.1.4 of this publication for additional information.

Ground motion that is based on uniform risk, as described above, is referred to as *probabilistic ground motion* (note: previous reference to earthquake shaking in the reference documents was based on uniform hazard). At most locations within the United States, this shaking corresponds to a mean recurrence interval of approximately 2,500 years. A recurrence interval that is somewhat shorter is expected at sites where earthquakes occur relatively frequently while a recurrence level that is longer is anticipated at sites where earthquakes rarely occur. An iterative process was used to establish the shaking parameters (S_S and S_1), and the details of this process can be found in Reference 6.1.

The probabilistic ground motion can be severe in regions of the United States with major active faults that are capable of producing large earthquakes; at certain locations, values of the parameters determined by this approach far exceed those that correspond to the strongest ground motion that has ever been recorded in those areas. Thus, a *deterministic* estimate of the ground shaking is used rather than the risk-based shaking in those regions. Estimates of the ground motion parameters S_S and S_1 are determined using the maximum earthquake that can be generated by the nearby faults. In particular, acceleration values are established using the near-source 84th-percentile 5-percent damped ground motions in the direction of maximum horizontal response. It is important to note that the use of deterministic ground shaking implies a somewhat higher level of collapse risk than the 1 percent probability of collapse in 50 years that is associated with the probabilistic ground shaking.

The *risk-targeted maximum considered earthquake shaking* (MCE$_R$) is the shaking from the probabilistic ground motion except at locations where the deterministic ground motion governs as described above. This reference earthquake shaking level provides a small probability (approximately 10 percent or less) that structures with ordinary occupancies will collapse when subjected to such shaking. Methods on how to determine MCE$_R$ shaking are given in Section 6.2.1 of this publication.

6.1.4 Overview of Code Requirements

According to IBC 1613.1, the effects of earthquake motion on structures and their components are to be determined in accordance with ASCE/SEI 7, excluding Chapter 14 (Material Specific Seismic Design and Detailing Requirements) and Appendix 11A (Quality Assurance Provisions). These chapters from ASCE/SEI 7 have been excluded because the IBC includes quality assurance provisions in Chapter 17 and structural material provisions in Chapters 19 through 23.

The earthquake provisions contained in these documents are based on the requirements set forth in the *NEHRP Recommended Seismic Provisions* (Reference 6.1). This resource document contains the fundamental seismic requirements, which are presented in a format that can be readily adopted into a code. The history on the creation and subsequent development of the *NEHRP Recommended Seismic Provisions* can be found in Reference 6.2.

A summary of the chapters of ASCE/SEI 7 that contain earthquake load provisions that are referenced by the 2012 IBC is provided in Table 6.1. The primary focus of the discussion in this chapter is on Chapters 11, 12, 13, 15, 20, 21 and 22.

Table 6.1 Summary of Chapters in ASCE/SEI 7 That Are Referenced by the 2012 *IBC* for Earthquake Load Provisions

Chapter	Title
11	Seismic Design Criteria
12	Seismic Design Requirements for Building Structures
13	Seismic Design Requirements for Nonstructural Components
15	Seismic Design Requirements for Nonbuilding Structures
16	Seismic Response History Procedures
17	Seismic Design Requirements for Seismically Isolated Structures
18	Seismic Design Requirements for Structures with Damping Systems
19	Soil-Structure Interaction for Seismic Design
20	Site Classification Procedure for Seismic Design
21	Site-Specific Ground Motion Procedures for Seismic Design
22	Seismic Ground Motion Long-Period Transition and Risk Coefficient Maps
23	Seismic Design Reference Documents

IBC 1613.1 lists four exemptions to seismic design requirements presented in this section:

1. Detached one- and two-family dwellings that are assigned to Seismic Design Category (SDC) A, B or C (that is, $S_{DS} < 0.5$ and $S_{D1} < 0.2$), or located where $S_S < 0.4$ (definitions of

SDC and spectral response accelerations S_S, S_{DS} and S_{D1} are given in subsequent sections of this publication);

2. Conventional light-frame wood construction that conforms to IBC 2308 (limitations for conventional light-frame wood construction are given in IBC 2308.2);

3. Agricultural storage structures where human occupancy is incidental; and

4. Structures that are covered under other regulations, such as vehicular bridges, electrical transmission towers, hydraulic structures, buried utility lines and nuclear reactors.

The first and third exemptions reflect the low seismic risks and the low risk to loss of human life, respectively, associated with these occupancies. The second exemption recognizes that wood-frame seismic design requirements of IBC 2308 substantially meet the intent of conventional construction (wood-frame) provisions included in the *NEHRP Recommended Seismic Provisions*. The types of structures listed under the fourth exemption have unique design and performance issues and are not covered in the IBC and ASCE/SEI 7 because the requirements in those documents were developed for buildings and building-like structures.

ASCE/SEI 11.1.2 contains essentially the same exemptions as the IBC with two major differences. One difference occurs in the second exemption, which is stated in ASCE/SEI 11.1.2 as follows: detached one- and two-family wood-frame dwellings not included in Exception 1 that are less than or equal to two stories, satisfying the limitations and constructed in accordance with the *International Residential Code®* (IRC®). This exemption recognizes that the wood-frame seismic design requirements of the IRC substantially meet the intent of conventional construction (wood-frame) provisions included in the *NEHRP Recommended Seismic Provisions*. The second difference is that ASCE/SEI 7 includes a fifth exemption, which is for piers and wharves that are not accessible to the general public.

The seismic requirements of the IBC need not be applied to structures that meet at least one of these four exemptions. Flowchart 6.1 in Section 6.6 of this publication can be used to determine where the seismic conditions of ASCE/SEI 7 must be considered and need not be considered.

Seismic requirements for existing buildings are contained in IBC Chapter 34, Existing Structures. Covered in that chapter are provisions related to additions, alterations, repairs or change of occupancy of structures.

The requirements of IBC 3403.4 and 3404.4 must be satisfied where additions and alterations are made to an existing building, respectively. Also, where an existing seismic-force-resisting system is a type that can be designated as ordinary, the seismic parameters for that system shall be those that are specified in Chapter 12 of ASCE/SEI 7 for an ordinary system unless it can be demonstrated that the existing seismic-force-resisting system can provide a performance level equivalent to a detailed, intermediate or special system (IBC 3401.4.3).

Requirements for buildings that have sustained damage to the vertical elements of the lateral-force-resisting system are given in IBC 3405.2. IBC 3408 provides guidance with respect to the impact a change of occupancy has on an existing building. Two important exceptions to the general requirement that the structure must conform to the seismic requirements of a higher risk category are given in IBC 3408.4.

Similar existing building provisions are given in Appendix 11B of ASCE/SEI 7.

6.2 Seismic Design Criteria

6.2.1 Seismic Ground Motion Values

Mapped Acceleration Parameters

IBC Figures 1613.3.1(1) and 1613.3.1(2) and ASCE/SEI Figures 22-1 and 22-2 contain contour maps of the conterminous United States giving S_S and S_1, which are the mapped risk-targeted maximum considered earthquake (MCE_R) spectral response accelerations at periods of 0.2 second and 1.0 second, respectively, for a Site Class B soil profile and 5-percent damping (definitions of site classes are given in the next section). IBC Figures 1613.3.1(3) through 1613.3.1(6) and ASCE/SEI Figures 22-3 through 22-6 contain similar contour maps for Alaska, Hawaii, Puerto Rico and the United States Virgin Islands. The mapped spectral accelerations are the smaller of the probabilistic risk-based and deterministic ground motion values obtained at a particular site (see Section 6.1.3 of this publication). It is important to note that these maps signify the spectral response accelerations in the maximum direction, which are larger than the geometric mean spectral response accelerations developed by the USGS by factors of 1.1 for the short period and 1.3 for the 1-second period accelerations.

The short-period acceleration, S_S, has been determined at a period of 0.2 second because it was concluded that 0.2 second was reasonably representative of the shortest effective period of buildings and structures that are designed using these requirements. The 1-second acceleration, S_1, is used because spectral response acceleration at periods other than 1 second typically can be derived from the acceleration at 1 second. As noted in Section 6.1.2 of this publication, these two acceleration parameters are sufficient to define an entire response spectrum for the range of periods that are applicable for most buildings and structures.

In lieu of the maps, MCE_R spectral response accelerations can be obtained from the U.S. Seismic "DesignMaps" Web Application, which can be accessed on the United States Geological Survey (USGS) website (https://geohazards.usgs.gov/secure/designmaps/us/) or the Structural Engineering Institute (SEI) website (http://content.seinstitute.org). Accelerations are output for a specific latitude-longitude or site address, which is input by the user.

Where $S_S \leq 0.15$ and $S_1 \leq 0.04$, the structure is permitted to be assigned to Seismic Design Category A (IBC 1613.3.1 and ASCE/SEI 11.4.1). These areas are considered to have very low seismic risk based solely on the mapped ground motions.

Site Class

Six site classes are defined in Table 20.3-1 (see Table 6.2). A site is to be classified as one of these six based on one of three soil properties measured over the top 100 feet of the site:

- Average shear wave velocity at small shear strains, $\overline{v}_s$
- Average field standard penetration, $\overline{N}$, or average standard penetration resistance for cohesionless soil layers, $\overline{N}_{ch}$
- Average undrained shear strength, $\overline{s}_u$

Table 6.2 Site Classification in Accordance with ASCE/SEI 20.3

Site Class	$\bar{v}_s$ (feet/sec)	$\bar{N}$ or $\bar{N}_{ch}$	$\bar{s}_u$ (psf)
A – Hard rock	> 5000	NA	NA
B – Rock	2500 to 5000	NA	NA
C – Very dense soil and soft rock	1200 to 2500	> 50	> 2000
D – Stiff soil	600 to 1200	15 to 50	1000 to 2000
E – Soft clay soil	< 600	< 15	< 1000
E – Soft clay soil	Any profile with more than 10 feet of soil with the following characteristics: •Plasticity index $PI > 20$ •Moisture content $w \geq 40\%$ •Undrained shear strength $\bar{s}_u < 500$ psf		
F – Soils requiring site response analysis in accordance with ASCE/SEI 21.1	See ASCE 20.3.1		

As noted above, the properties of the soil in the 100 feet below the ground surface must be known in order to determine the site class. This generally requires a geotechnical investigation that includes drilling borings into the soil and collecting soil samples at various depths. Steps for classifying a site are given in ASCE/SEI 20.3. Methods of determining the site class where the soil is not homogeneous over the top 100 feet are provided in ASCE/SEI 20.4.

Site Class A is hard rock, which is typically found in the eastern United States. Site Class B or C is a softer rock, including various volcanic deposits, sandstones, shales and granites; these types are commonly found in western parts of the country. Very dense sands and gravels as wells as very stiff clay deposits usually qualify as Site Class C. The most common site class throughout the United States is Site Class D, which consists of sites with relatively stiff soils, including mixtures of silty clays, silts, and sands. Sites that are located along rivers or other waterways that are underlain by deep soft clay deposits are classified as Site Class E. Site Class F indicates soil so poor that a site response analysis is required to determine site coefficients; these are sites where soils are subject to liquefaction or other ground instabilities. Site-specific ground motion procedures for seismic design are given in Chapter 21 of ASCE/SEI 7 (see below).

Under certain conditions, it may not be possible to drill the entire 100-foot depth to acquire soil properties (for example, a layer of hard rock may be encountered 25 feet below the ground surface). At such locations or in cases where soil property measurements to a depth of 100 feet are not feasible, the registered design professional that is responsible for preparing the geotechnical report may estimate soil properties from geological conditions. When soil properties are not known in sufficient detail to determine the site class in accordance with code provisions, Site Class D must be used, unless the authority having jurisdiction requires or has reason to believe that the site would be more properly classified as Site Class E or F.

It is important to note that it can be beneficial to perform a geotechnical investigation instead of automatically assuming Site Class D. A site that can be classified as Site Class A, B or C will usually result in a more economical structure since any of those site classes result in earthquake shaking that is less intense than that associated with Site Class D (see the following discussion).

Flowchart 6.2 in Section 6.6 of this publication can be used to determine the site class of a site in accordance with Chapter 20 of ASCE/SEI 7.

Site Coefficients and Risk-targeted MCE$_R$ Spectral Response Acceleration Parameters

Once the mapped MCE$_R$ spectral accelerations and site class have been established, the risk-targeted MCE$_R$ spectral response acceleration for short periods, S_{MS}, and at a 1-second period, S_{M1}, adjusted for site class effects are determined by IBC Equations 16-37 and 16-38, respectively, or ASCE/SEI Equations 11.4-1 and 11.4-2, respectively:

$$S_{MS} = F_a S_S \tag{6.1}$$

$$S_{M1} = F_v S_1 \tag{6.2}$$

where F_a = short-period site coefficient determined from IBC Table 1613.3.3(1) or ASCE/SEI Table 11.4-1 and F_v = long-period site coefficient determined from IBC Table 1613.3.3(2) or ASCE/SEI Table 11.4-2. As expected the values of F_a and F_v are equal to 1.0 for Site Class B irrespective of the seismicity at the site. For site classes other than Site Class B, an adjustment to the mapped spectral response accelerations is necessary.

Typically, ground motion is amplified in softer soils (Site Classes C through E) and attenuated in stiffer soils (Site Class A). This can be observed in the tables where the magnitudes of F_a and F_v increase going from Site Class A to F for a given mapped ground motion acceleration. The only exception to this occurs for short periods where $S_S \geq 1$ and the Site Class changes from D to E. Very soft soils are not capable of amplifying the short-period components of subsurface rock motion; in fact, deamplification occurs in such cases.

Design Spectral Response Acceleration Parameters

Design spectral response accelerations at short periods, S_{DS}, and at a 1-second period, S_{D1}, are determined by IBC Equations 16-39 and 16-40, respectively, or ASCE/SEI Equations 11.4-3 and 11.4-4, respectively:

$$S_{DS} = \frac{2}{3} S_{MS} \tag{6.3}$$

$$S_{D1} = \frac{2}{3} S_{M1} \tag{6.4}$$

It is shown later that these accelerations are used in determining, among other things, the magnitude of the effects on a structure created by a seismic event.

In previous editions of ASCE/SEI 7, the lower bound margin against collapse that was inherent in buildings designed by the seismic provisions in that standard was judged, based on experience, to correspond to a factor of about 1.5 in ground motions, which corresponds to design spectral response accelerations equal to approximately 1/1.5 = 2/3 of the soil-modified design accelerations. As noted above, the uncertainty in this margin is accounted for in ASCE/SEI 7 in the collapse fragility defined in ASCE/SEI 21.2.1.2. Regardless of this, the design earthquake ground motion is based on 2/3 of the MCE$_R$ ground motion (Equations 6.3 and 6.4) for consistency with previous editions of ASCE/SEI 7.

Flowchart 6.3 in Section 6.6 of this publication provides a step-by-step procedure on how to determine the design spectral accelerations for a site in accordance with ASCE/SEI 11.4.

Design Response Spectrum

Illustrated in Figure 6.3 is the generalized form of the design acceleration response spectrum based on the design spectral response acceleration parameters discussed previously. This spectrum is to be used wherever required by the provisions in ASCE/SEI 7; it should not be used in cases where site-specific ground motion procedures are required (requirements for the response spectrum where a site response analysis is performed or required are given in ASCE/SEI 7 Chapter 21; see the discussion below).

Figure 6.3
Design Response Spectrum

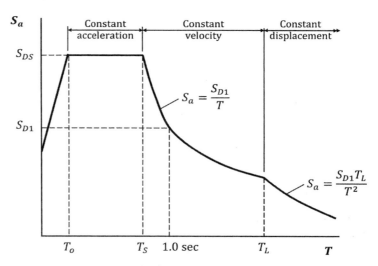

For periods up to the period T_o, which is equal to $0.2S_{D1}/S_{DS}$, the design spectral response acceleration, S_a, is determined by ASCE/SEI Equation 11.4-5:

$$S_a = S_{DS}\left(0.4 + 0.6\frac{T}{T_o}\right) \tag{6.5}$$

The constant acceleration portion of the spectrum is between the periods T_o and T_S, where $T_S = S_{D1}/S_{DS}$. In this region, S_a is a constant and is equal to S_{DS}.

The constant velocity region is between T_S and T_L where the response acceleration is proportional to the inverse of the period, T. In this region, S_a is determined by ASCE/SEI Equation 11.4-6:

$$S_a = \frac{S_{D1}}{T} \tag{6.6}$$

Mapped values of the long-period transition period, T_L, are given in ASCE/SEI Figures 22-12 through 22-16. These maps were also prepared by the USGS.

The constant displacement region occurs where the period is greater than T_L, and S_a is determined by ASCE/SEI Equation 11.4-7:

$$S_a = \frac{S_{D1}T_L}{T^2} \tag{6.7}$$

In cases where a risk-targeted MCE_R response spectrum is required in the provisions, it is obtained by multiplying the values determined for the design response spectrum by 1.5 (ASCE/SEI 11.4.6).

Site-specific Ground Motion Procedures

Overview

The site-specific ground motion procedures given in ASCE/SEI Chapter 21 may be used to determine ground motions for any structure and must be used for any structure that is located on a site that has been classified as Site Class F, except in cases where the exception in ASCE/SEI 20.3.1 is applicable.

In short, the purpose of these procedures is to determine ground motions for local seismic and site conditions with a higher degree of confidence than is possible using the general procedure in the preceding discussion.

Site-specific procedures for computing earthquake ground motion include the following:

1. **Site response analyses.** Equivalent linear and nonlinear analytical methods are generally used based on the requirements in ASCE/SEI 21.1.

2. **Probabilistic seismic hazard analysis.** Requirements are given in ASCE/SEI 21.2.1.

3. **Deterministic seismic hazard analysis.** Requirements are given in ASCE/SEI 21.2.2.

Risk-targeted MCE_R Ground Motion Hazard Analysis

Probabilistic MCE_R Ground Motion. Methods on how to determine probabilistic MCE_R ground motions are given in ASCE/SEI 21.2.1. Two methods are provided on how to determine the spectral response acceleration, S_a. In Method 1, values of S_a are obtained for a given period, T, by multiplying the risk coefficient, C_R, by the spectral response acceleration obtained from a 5-percent damped acceleration response spectrum having a 2-percent probability of exceedance within a 50-year period. ASCE/SEI Figures 22-17 and 22-18 contain values of C_{RS} and C_{RI} that are used in determining C_R:

- For $T \leq 0.2$ second: $C_R = C_{RS}$
- For $T \geq 1.0$ second: $C_R = C_{RI}$
- For 0.2 second $< T < 1.0$ second: C_R is to be determined by linear interpolation between C_{RS} and C_{RI}

In Method 2, values of S_a are computed from iterative integration of a site-specific hazard curve with a lognormal probability density function representing the collapse fragility (that is, probability of collapse). A 1-percent probability of collapse within a 50-year period for a collapse fragility with specific attributes must be achieved (see ASCE/SEI 21.2.1.2). This method is essentially the same as that which was used to determine the risk-targeted MCE_R ground motion parameters discussed in Section 6.1.3 of this publication.

Deterministic MCE_R Ground Motion. Deterministic MCE_R ground motions meeting the requirements of ASCE/SEI 21.2.2 must also be obtained at each period. Values of S_a are calculated as an 84th-percentile 5-percent damped spectral response acceleration in the direction of maximum horizontal response computed at that period. The largest such acceleration calculated for the characteristic earthquakes on all known active faults within the region shall be used. Results from this analysis must be greater than or equal to the values given in the response spectrum of ASCE/SEI Figure 21.2-1 (see Figure 6.4). The site coefficients F_a and F_v are determined using ASCE/SEI Tables 11.4-1 and 11.4-2, respectively.

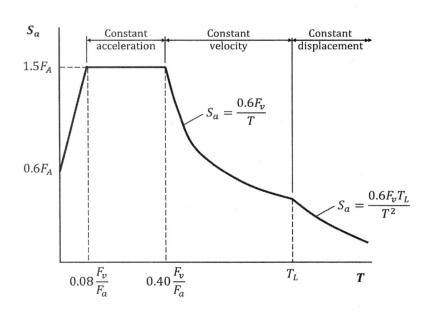

Figure 6.4 Deterministic Lower Limit on MCE_R Response Spectrum

Site-specific MCE_R

Once the values of S_a are obtained from the probabilistic and deterministic methods described above, the site-specific MCE_R spectral response acceleration, S_{aM}, is set equal to the lesser of the values obtained from these two methods.

Design Response Spectrum

The design spectral response acceleration is equal to two-thirds of the site-specific MCE_R spectral response acceleration, S_{aM} (see ASCE/SEI Equation 21.3-1). The design spectral response acceleration at any period shall not be taken less than 80 percent of S_a, determined in accordance with ASCE/SEI 11.4.5. For sites classified as Site Class F requiring site response analysis in accordance with ASCE/SEI 11.4.7, the design spectral response acceleration at any period shall not be taken as less than 80 percent of S_a, determined for Site Class E in accordance with ASCE/SEI 11.4.5.

Design Acceleration Parameters

The following requirements on how to determine the design accelerations S_{DS} and S_{D1} from the site-specific response spectrum are given in ASCE/SEI 21.4:

- $S_{DS} = S_a$ at $T = 0.2$ second and must be $\geq 0.9 S_a$ at any T larger than 0.2 second
- $S_{D1} =$ greater of (S_a at $T = 1.0$ second and $2 S_a$ at $T = 2.0$ seconds)

Values of S_{DS} and S_{D1} must be taken greater than or equal to 80 percent of the values determined by ASCE/SEI 11.4.4.

Also, the soil-modified accelerations are determined as follows:

- $S_{MS} = 1.5 S_{DS} \geq 0.8 S_{MS}$ determined from ASCE/SEI 11.4.3
- $S_{M1} = 1.5 S_{D1} \geq 0.8 S_{M1}$ determined from ASCE/SEI 11.4.3

When the site-specific ground motion procedure is used in conjunction with the Equivalent Lateral Force Procedure in ASCE/SEI 12.8, the site-specific spectral acceleration, S_a, shall be permitted to replace S_{D1}/T in Equation 12.8-3 and $S_{D1}T_L/T^2$ in Equation 12.8-4. Additionally, (1) the value of S_{DS} determined in accordance with ASCE/SEI 21.4 is permitted to be used in

Equations 12.8-2, 12.8-5, 15.4-1 and 15.4-3, and (2) the mapped value of S_1 is to be used in Equations 12.8-6, 15.4-2 and 15.4-4.

Maximum Considered Earthquake Geometric Mean (MCE_G) Peak Ground Acceleration

Where required by the provisions in ASCE/SEI 7, the site specific MCE_G peak ground acceleration, PGA_M, is the lesser of the probabilistic geometric mean peak ground acceleration of ASCE/SEI 21.5.1 (which is taken as the geometric mean peak ground acceleration with a 2-percent probability of exceedance within a 50-year period) and the deterministic geometric mean peak ground acceleration of ASCE/SEI 21.5.2 (which is taken as the largest 84th-percentile geometric mean peak ground acceleration for characteristic earthquakes on all known active faults within the site region). Note that these accelerations have not been adjusted for targeted risk. The governing value must be taken greater than or equal to 80 percent of the value determined by Equation 11.8-1.

It is shown later that the MCE_G peak ground acceleration adjusted for site effects, PGA_M, is used for evaluation of liquefaction, lateral spreading, seismic settlements and other soil-related issues.

6.2.2 Importance Factor and Risk Category

Risk categories are defined in IBC Table 1604.5 and ASCE/SEI Table 1.5-1. An in-depth discussion on risk categories is provided in Chapter 1 of this publication.

An importance factor, I_e, is assigned to a building or structure in accordance with ASCE/SEI Table 1.5-2 based on its risk category. Larger values of I_e are assigned to more important risk categories, such as assembly and essential facilities, to increase the likelihood that such structures would suffer less damage and continue to function during and following a design earthquake. The risk category of a structure is also used in determining the Seismic Design Category (see Section 6.2.3 of this publication).

Provisions pertaining to operational access to a Risk Category IV structure through an adjacent structure are given in ASCE/SEI 11.5.2. In such cases, the adjacent structure must conform to the requirements for Risk Category IV structures. Additionally, where operational access is less than 10 feet from an interior lot line or another structure on the same lot, protection from potential falling debris from adjacent structures shall be provided by the owner of the Risk Category IV structure.

6.2.3 Seismic Design Category

All buildings and structures must be assigned to a *Seismic Design Category* (SDC) in accordance with IBC 1613.3.5 or ASCE/SEI 11.6. In general, the SDC is a function of the risk category and the design spectral accelerations at the site.

Structures are categorized according to the seismic risk they could pose. Six SDCs are defined ranging from A (minimal seismic risk) to F (highest seismic risk). As the SDC of a structure increases, so do the strength and detailing requirements.

The SDC of a structure is assigned as follows where S_1 is greater than or equal to 0.75:

- Structures classified as Risk Category I, II or III are assigned to SDC E
- Structures classified as Risk Category IV are assigned to SDC F

Where S_1 is less than 0.75, the SDC is determined twice: first as a function of S_{DS} by IBC Table 1613.3.5(1) or ASCE/SEI Table 11.6-1 and second as a function of S_{D1} by IBC Table 1613.3.5(2) or ASCE/SEI Table 11.6-2. The more severe SDC of the two governs.

The SDC may be determined by IBC Table 1613.3.5(1) or ASCE/SEI Table 11.6-1 based solely on S_{DS} in cases where S_1 is less than 0.75 provided that all of the following conditions listed under IBC 1613.3.5.1 or ASCE/SEI 11.6 are satisfied:

1. In each of the two orthogonal directions, the approximate fundamental period of the structure, T_a, determined in accordance with ASCE/SEI 12.8.2.1 is less than $0.8T_S$.

2. In each of the two orthogonal directions, the fundamental period of the structure used to calculate the story drift is less than T_S.

3. ASCE/SEI Equation 12.8-2 is used to determine the seismic response coefficient, C_s.

4. The diaphragms are rigid as defined in ASCE/SEI 12.3.1 or, for diaphragms that are flexible, the distances between vertical elements of the seismic-force-resisting system do not exceed 40 feet.

This exception should always be considered—especially when determining the SDC of low-rise buildings that are stiff (that is, buildings with small structural periods)—because it makes it possible for a building to be assigned to a lower SDC. As noted above, a lower SDC generally means less stringent design and detailing requirements, which translates into cost savings.

The SDC is a trigger mechanism for many seismic requirements, including the following:

- Permissible seismic-force-resisting systems
- Limitations on building height
- Consideration of structural irregularities
- The need for additional special inspections, structural testing and structural observation for seismic resistance

For structures that can be designed in accordance with the alternate simplified design procedure in ASCE/SEI 12.14, it is permitted to determine the SDC based on ASCE/SEI Table 11.6-1 alone. The provisions for this method are given in Section 6.3.14 of this publication.

Flowchart 6.4 in Section 6.6 of this publication can be used to determine the SDC of a building or structure.

6.2.4 Design Requirements for SDC A

Structures assigned to SDC A need only comply with the general structural integrity requirements of 1.4 (note: from this point on in this chapter, referenced section, table and figure numbers are from ASCE/SEI 7-10 unless noted otherwise). To ensure general structural integrity, the lateral-force-resisting system must be proportioned to resist a lateral force at each floor level equal to 1 percent of the total dead load at that floor level, as depicted in Figure 6.5. These forces have the potential to be larger than the governing wind forces for low-rise structures that are heavy (such as those made of reinforced concrete).

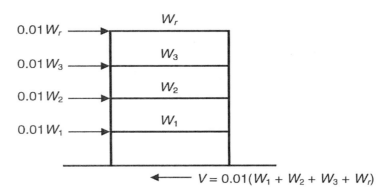

Figure 6.5 Design Seismic Force Distribution for Structures Assigned to SDC A

According to 1.4.3, the lateral forces are to be applied independently in each of two orthogonal directions.

Requirements for load path connections, connection to supports and anchorage of structural walls are given in 1.4.2, 1.4.4 and 1.4.5, respectively. See Section 2.6 of this publication for more information on general structural integrity requirements.

Nonstructural components of a building or structure are automatically exempt from any seismic design requirements.

Flowchart 6.5 in Section 6.6 of this publication can be used to determine the design requirements for buildings or structures assigned to SDC A.

6.2.5   Geological Hazards and Geotechnical Investigation

Site Limitation for Seismic Design Category (SDC) E or F

Any structure that is assigned to SDC E or F must not be located where there is a potential for an active fault to cause rupture of the ground surface at that location (11.8.1). It was discussed in Section 6.1.1 of this publication that it is very difficult to design structures to accommodate large movement associated with ground fault ruptures; thus, building at locations where this is possible is not permitted.

Geotechnical Investigation Report Requirements for Seismic Design Categories C through F

For structures assigned to SDC C, D, E or F, the geotechnical report for a particular site must include an evaluation of the following potential geologic and seismic hazards:

- Slope instability
- Liquefaction
- Surface displacement due to faulting or seismically-induced lateral spreading or lateral flow

As mentioned in Section 6.1.1 of this publication, any one of these hazards can have an extremely adverse effect on a structure.

Recommendations for the type of foundation and other methods to mitigate any of the aforementioned hazards should also be included in the geotechnical report.

The exception in 11.8.2 states that a site-specific geotechnical report is not required at locations where evaluations of nearby sites with similar soil conditions have been performed previously; in such cases, the proposed site can utilize the nearby information provided the authority having jurisdiction approves it.

Additional Geotechnical Investigation Report Requirements for Seismic Design Categories D through F

For structures assigned to SDC D, E or F, the geotechnical report must include the following information in addition to that noted previously:

- The determination of dynamic seismic lateral earth pressures on basement and retaining walls due to design earthquake ground motions.

  Dynamic lateral pressures are considered to be an earthquake load, E, which is superimposed on the pre-existing lateral earth load, H, during ground shaking.

- The potential for liquefaction and soil strength loss evaluated for site peak ground acceleration, earthquake magnitude and source characteristics consistent with the maximum considered earthquake geometric mean (MCE_G) peak ground acceleration.

  Peak accelerations are to be determined using one of the following methods: (a) a site-specific study that takes into account soil amplification effects as prescribed in the site-specific ground motion procedures of 11.4.7 (see Section 6.2.1 of this publication), or (b) the peak ground acceleration, PGA_M, determined by Equation 11.8-1:

  $$PGA_M = F_{PGA}PGA \tag{6.8}$$

  where F_{PGA} is the site coefficient from Table 11.8-1 and PGA is the mapped MCE_G peak ground acceleration provided in Figures 22-7 through 22-11 for the conterminous United States, Alaska, Hawaii, Puerto Rico, the United States Virgin Islands, Guam and American Samoa for Site Class B.

  It is evident from Equation 6.8 that the peak ground acceleration is used when considering the effects of liquefaction and soil strength loss because such effects can be catastrophic to a structure. Unlike the uniform hazard and deterministic ground motion maps described in Section 6.2.1 of this publication, the geometric mean ground motion maps do not represent response in the maximum direction. Also, risk coefficients were not included in the development of these maps (that is why the "risk-targeted" prefix is not included when referring to these accelerations).

- Assessment of potential consequences of liquefaction and soil strength loss, including, but not limited to the following: (a) estimation of total and differential settlement, (b) lateral soil movement, (c) lateral soil loads on foundations, (d) reduction in soil bearing capacity and lateral soil reactions, (e) soil downdrag and reduction in axial and lateral soil reaction for pile foundations, (f) increases in soil lateral pressures on retaining walls, and (g) flotation of buried structures.

  Since the effects from liquefaction and soil strength loss can be disastrous, it is very important to have a clear understanding of how the effects will impact the structure and its foundation.

- Discussion of mitigation measures, such as, but not limited to, the following: (a) selection of appropriate foundation type and depth, (b) selection of appropriate structural systems to accommodate anticipated displacements and forces, (c) ground stabilization, or (d) any combinations of these measures and how they shall be considered in the design of the structure.

It is essential to have a clear plan on how to control the potential effects from liquefaction and soil strength loss. Appropriate foundation and lateral-force-resisting systems must be established in the early design stages to counteract any potential damaging effects.

6.3 Seismic Design Requirements for Building Structures

6.3.1 Structural Design Basis

Basic Requirements

Basic requirements for seismic analysis and design of building structures are contained in 12.1. In general, the structure must have complete lateral- and vertical-force-resisting systems that are capable of providing adequate strength, stiffness and energy-dissipation capacity when subjected to the design ground motion, which is assumed to occur along any horizontal direction of a structure. Without sufficient strength and stiffness, large displacements can occur, which could lead to local or overall instability, or both.

Design seismic forces and their distribution over the height of a structure are to be established in accordance with one of the procedures in 12.6. When multiplied by the appropriate deflection amplification factor, these forces essentially produce the same deformations that would occur in the structure as if it were subjected to the design-basis ground motion. A simplified design procedure may also be used, provided that all of the limitations of 12.14 are satisfied. These methods are covered in subsequent sections of this chapter.

Member Design, Connection Design and Deformation Limit

All structural members in a building or structure, including those that do not directly resist the effects from earthquakes, must be designed for the applicable shear forces, axial forces and bending moments determined in accordance with the provisions of ASCE/SEI 7. This includes the connections between the members.

Deformation limits are also prescribed and the structure must have adequate strength and stiffness such that these limits are not violated. As discussed previously, instability can occur if deflections become too large; so ensuring that all required limits are met is very important when designing a structure for the design-basis earthquake.

Continuous Load Path and Interconnection

A continuous load path with adequate strength and stiffness must be provided to transfer all forces from the point of application to the final point of resistance (usually the ground). If all of the components of a building are not adequately tied together, including nonstructural components, the individual members will move independently and can pull apart from one another, which could lead to partial or total collapse.

According to 12.1.3, all smaller portions of a structure shall be tied to the remainder of the structure with elements that have design strength capable of transmitting the greater of $0.133S_{DS}$ times the weight of the smaller portion or 5 percent of the portion's weight. This requirement is meant to provide a minimum amount of continuity in a structure so that it can act as a unit in resisting the effects from earthquakes.

Additional information on load paths can be found in Chapter 8 of this publication.

Connection to Supports

Connections must be designed to transfer horizontal forces through the structural system to the supporting elements, including the foundation. In particular, every connection between each beam, girder or truss to its supporting elements, including diaphragms, must have a minimum design strength of 5 percent of the dead load plus live load reaction on that member (12.1.4).

Foundation Design

Foundations must be designed to resist the forces caused by the ground motion and must be able to transfer these forces between the structure and the ground. Additionally, they must be able to accommodate the ground movements without inducing large displacements into the structure, especially at sites that are subject to liquefaction or lateral spreading.

Since large lateral displacements can occur during a seismic event, it is very important to tie together all of the individual foundation elements that support a structure. By doing so, this helps ensure that the structure is not torn apart by the differential ground displacements. A mat foundation beneath the entire footprint of a structure is a very suitable foundation system. Where spread footings or piers are utilized, reinforced concrete grade beams should be used to tie the individual foundation elements together so that they can move as a unit.

Design requirements for foundation design are contained in 12.13. Chapter 18 of the IBC contains comprehensive design and detailing requirements for a variety of foundation systems. The provisions are organized with respect to SDC.

6.3.2 Structural System Selection

Selection and Limitations

The basic seismic-force-resisting systems (SFRSs) are listed in Table 12.2-1. There are six general categories of SFRSs:

- **Bearing wall system.** A bearing wall system is a structural system where bearing walls support all or a major portion of the vertical loads. Some or all of the walls also provide resistance to the seismic forces.

- **Building frame system.** In a building frame system, an essentially complete space frame provides support for the vertical loads. Shear walls or braced frames provide resistance to the seismic forces.

- **Moment-resisting frame system.** A moment-resisting frame system is a structural system with an essentially complete space frame that supports the vertical loads. Seismic forces are resisted primarily by flexural action of the frame members through the joints. The entire space frame or selected portions of the frame may be designated as the SFRS. Requirements for deformation compatibility must be satisfied for structures assigned to SDC D, E or F (12.12.5).

- **Dual system.** In a dual system, an essentially complete space frame provides support for the vertical loads. Moment-resisting frames and shear walls or braced frames provide resistance to seismic forces in accordance with their rigidities (see 12.2.5.1). Additionally, the moment-resisting frames must act as a backup for the walls or braces; this is accomplished by requiring that the moment frames be capable of resisting at least 25 percent of the design seismic forces. Dual systems are also referred to as shear wall-frame interactive systems.

- **Cantilever column system.** In this system, the vertical forces and the seismic forces are resisted entirely by columns acting as cantilevers from their base. This system is usually used in one-story buildings or in the top story of a multistory building. Severe restrictions are placed on the use of this system since it has performed poorly in past earthquakes.

- **Steel systems not specifically designed for seismic resistance.** Structural steel buildings that do not specifically conform to any of the other types are designated as steel systems not specifically designed for seismic resistance. They are permitted only in areas of relatively low seismic risk. Note that these systems do not include cantilever column systems.

It is evident from Table 12.2-1 that all of the primary materials of construction are included under the general categories of SFRSs. Also, the SFRSs are categorized according to the quality and extent of the seismic-resistant detailing that is used in the design of a structure. "Special" systems provide superior seismic resistance, which is accomplished through extensive design and detailing of the structural members. An "ordinary" system generally has basic design and detailing requirements. "Intermediate" systems provide seismic resistance that is better than ordinary systems, but not as good as special systems. Design and detailing requirements for these three categories are given in the material standards for each type of material.

Table 12.2-1 also includes the following important quantities:

- *Response modification coefficient, R.*

  In general, this coefficient accounts for the ability of the SFRS to respond to ground shaking in a ductile manner without loss of load-carrying capacity. In other words, R is an approximate way of accounting for the effective damping and energy dissipation that can be mobilized during inelastic response to ground shaking. It represents the ratio of the forces that would develop under the ground motion specified in ASCE/SEI 7 if the structure had responded to the ground motion in a linear-elastic manner.

  A system that has no ability to respond in a ductile manner has an R-value equal to 1; from Table 12.2-1, the only such system is a cantilevered column system consisting of ordinary reinforced concrete moment frames. Systems that are capable of highly ductile response have an R-value equal to 8. A wide variety of systems possess this maximum value of R.

  It will be shown later that this coefficient is used to reduce the seismic base shear, V, on a structure. Although the required design strength is decreased as R increases, the level of design and detailing substantially increase with increasing R.

- *Overstrength factor, Ω_0.*

  This factor accounts for the fact that the actual seismic forces on some members of a structure can be significantly larger than those indicated by analysis using the prescribed design seismic forces. In general, these members cannot provide reliable inelastic response or energy dissipation.

  For most of the structural systems given in the table, Ω_0 ranges from 2 to 3.

- *Deflection amplification factor, C_d.*

  This factor is used to adjust the lateral displacements that are determined for a structure using the prescribed design seismic forces to the actual anticipated lateral displacements during the design earthquake.

  It can be seen from the table that values of C_d are equal to or slightly less than the corresponding R-values for a given SFRS. The more ductile a system is (that is, the greater the R-value), the greater the difference is between the values of R and C_d.

Also included in Table 12.2-1 are limitations with respect to SDC and height. Some SFRSs, such as special steel moment frames and special reinforced concrete moment frames, can be utilized in structures assigned to any SDC with no height limitations. Other less ductile systems can be utilized with no height limitations in some SDCs while others are permitted in structures up to certain heights. The least ductile systems are not permitted in the higher SDCs under any circumstances.

ASCE/SEI 12.2.1 permits the use of SFRSs that are not contained in Table 12.2-1 provided analytical and test data are submitted to the authority having jurisdiction for approval. In general, the supporting data must demonstrate that the lateral-force resistance and energy dissipation capacity of the proposed structural system is equivalent to the structural systems listed in Table 12.2-1 for equivalent values of R, Ω_0 and C_d.

Combination of Framing Systems in Different Directions

It is permitted to use different SFRSs along each of the two orthogonal directions of a structure provided the respective R, Ω_0 and C_d values are used. Depending on the type of SFRSs that are utilized, it is possible that one of the systems will have more restrictive limitations on use and height. If this occurs, the more restrictive limitations govern.

Combination of Framing Systems in the Same Direction

Overview

The more stringent system limitations of Table 12.2-1 apply where different SFRSs are used in the same direction. The intent of these requirements is to prevent concentration of inelastic behavior in the lower stories of a structure.

The following requirements must be satisfied for vertical and horizontal combinations of SFRSs in the same direction.

Vertical Combinations of SFRSs

Vertical combinations of SFRSs in the same direction are permitted provided one of the following sets of applicable requirements is satisfied:

1. Lower SFRS has a lower R-value than the upper SFRS.

    In this case, the R, Ω_0 and C_d values for the upper SFRS can be used to calculate the forces and drifts of the upper SFRS, and the R, Ω_0 and C_d values for the lower SFRS are to be used for the lower SFRS.

    Forces that are transferred from the upper SFRS to the lower SFRS must be increased by multiplying the forces by the ratio of larger R-value to the lower R-value.

2. Upper SFRS has a lower R-value than the lower SFRS.

    In this case, the R, Ω_0 and C_d values for the upper SFRS are to be used in the design of both SFRSs.

Three exceptions are given in this section and are applicable, in general, for conditions that do not affect the dynamic characteristics of the structure or that will not result in concentration of inelastic demand in critical areas of a structure:

1. Rooftop structures less than or equal to two stories in height that weigh less than 10 percent of the total structure weight;

2. Other supported structural systems with a weight less than or equal to 10 percent of the weight of the structure; and

3. Detached one- and two-family dwellings of light-frame construction.

Two-stage Analysis Procedure

In situations where a structure has a flexible upper portion supported by a rigid lower portion, a two-stage analysis procedure is permitted provided the following conditions are satisfied:

- The lower portion must have a stiffness that is at least 10 times greater than that of the upper portion.

- The period of the combined structure must be less than or equal to 1.1 times the period of the upper portion, the latter of which is determined assuming that the upper portion is a separate structure that is supported at the transition from the upper to the lower portion.

- The upper portion is to be designed as a separate structure using the appropriate values of R and the redundancy factor, ρ (see Section 6.3.3 of this publication for a discussion on ρ).

- The lower portion is to be designed as a separate structure using the appropriate values of R and the redundancy factor, ρ. The reactions from the upper portion are to be those that are determined from the analysis of the upper portion amplified by (R/ρ of the upper portion)/(R/ρ of the lower portion) ≥ 1.0. In essence, this ratio produces demands on the lower rigid portion that are proportionate with its inelastic capability.

- The upper portion is analyzed with the Equivalent Lateral Force Procedure or the modal response spectrum procedure, and the lower portion is analyzed with the Equivalent Lateral Force Procedure. Detailed information on these analysis procedures can be found in Sections 6.3.8 and 6.3.9 of this publication, respectively.

This two-stage procedure can be used for any structure that meets these conditions. It is often utilized where light-frame construction is built on a rigid concrete base.

Horizontal Combinations of SFRSs

For structures that utilize different SFRSs in the same horizontal direction, the least value of R for the different systems in that direction is to be used in design. This requirement reflects the expectation that the entire structure will undergo deformation in the direction of analysis that is controlled by the least ductile SFRS in that direction.

According to the exception in this section, Risk Category I or II buildings that are two stories or less in height and that use light-frame construction or flexible diaphragms are permitted to be designed using the least value of R for the different SFRSs in each independent line of resistance (12.2.3.3). Also, the value of R used for the design of the diaphragms in such structures must be less than or equal to the least value for any of the systems utilized in that same direction.

Combination Framing Detailing Requirements

Structural members that are common to different SFRSs in any direction must be designed by the detailing requirements of Chapter 12 using the highest value of R of the connected SFRSs. The intent of this requirement is to ensure that both SFRSs be designed to remain functional throughout the design event in order to preserve the integrity of the structure.

System Specific Requirements

Specific requirements for dual systems, cantilever column systems, inverted pendulum-type structures, special moment frames and other systems are given in 12.2.5. It is important that these requirements be satisfied for these SFRSs in addition to those specified in Table 12.2-1 and throughout Chapter 12.

6.3.3 Diaphragm Flexibility, Configuration Irregularities, and Redundancy

Diaphragm Flexibility

In general, diaphragms distribute the horizontal forces to the elements of the SFRS. The manner in which the forces get distributed depends on the relative flexibility of the diaphragm. In the case of flexible diaphragms, the lateral forces are distributed to the vertical elements of the SFRS using tributary areas. For diaphragms that are not flexible or that are rigid, the forces are distributed in proportion to the relative rigidities of the SFRSs in the direction of analysis based on the location of the center of mass and the center of rigidity.

The relative flexibility of diaphragms must be considered in the structural analysis along with the stiffness of the SFRS (12.3.1). Certain types of diaphragm construction can readily be idealized as flexible or rigid. Floor and roof systems that can be idealized as flexible are given in 12.3.1.1. In particular, diaphragms that are constructed of untopped steel decking or wood structural panels are permitted to be idealized as flexible if any of the following conditions in 12.3.1.1 are satisfied:

- In structures where the vertical elements of the SFRS are relatively stiff compared to the diaphragm stiffness, including (1) steel braced frames, (2) steel and concrete composite braced frames or (3) concrete, masonry, steel or steel and concrete composite shear walls;

- One- and two-family dwellings; or

- In structures of light-frame construction where all of the following conditions are met:
  1. Topping of concrete or similar materials is not placed over wood structural panel diaphragms except for nonstructural topping that is less than or equal to 1.5 inches.
  2. Each line of vertical elements in the SFRS complies with the allowable story drift given in Table 12.12-1.

Diaphragms that can be considered rigid are concrete slabs and concrete-filled metal deck with span-to-depth ratios less than or equal to 3. One further stipulation is that the structure must have no horizontal irregularities as defined in 12.3.2.1 (see the discussion on irregularities in the next section).

Floor/roof systems that do not satisfy the conditions of either an idealized flexible or rigid diaphragm are permitted to be idealized as flexible diaphragms where the computed maximum in-plane deflection of the diaphragm under lateral load is greater than two times the average story drift of adjoining vertical elements of the SFRS (12.3.1.3). This provision is illustrated in Figure 12.3-1 (see Figure 6.6).

Figure 6.6
Definition of Flexible Diaphragm (ASCE/SEI 12.3.1.3)

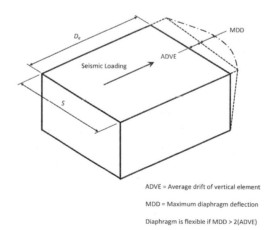

ADVE = Average drift of vertical element
MDD = Maximum diaphragm deflection
Diaphragm is flexible if MDD > 2(ADVE)

Flowchart 6.6 in Section 6.6 of this publication can be used to determine whether a diaphragm is flexible, rigid or semirigid.

Irregular and Regular Classifications

Overview

A regular structure can be defined as one that possesses a distribution of mass, stiffness and strength that results in essentially a uniform sway when subjected to ground shaking. In other words, the lateral displacement in each story on each side of the structure will be about the same.

The dissipation of earthquake energy is basically uniform throughout a regular structure; this results in relatively light and well-distributed damage. In contrast, structures that are irregular can suffer extreme damage at only one or a few locations, which can result in localized failure of structural members and can lead to a loss of the structure's ability to survive the ground shaking. Past earthquakes have revealed that irregular structures suffer greater damage than those that are regular.

Structures are classified as regular or irregular based on the criteria in 12.3.2. Structural configurations can be divided into horizontal and vertical types for purposes of defining irregularities. Both types of irregularities are covered below.

Most of the earthquake provisions that are given in ASCE/SEI 7 are applicable to regular structures; for example, the Equivalent Lateral Force Procedure using elastic analysis is not capable of accurately predicting the earthquake effects on certain types of irregular buildings (see Section 6.3.8 of this publication). Limitations and additional requirements must be satisfied for irregular structures depending on the SDC.

Horizontal Irregularity

Table 12.3-1 contains the types and descriptions of horizontal structural irregularities that must be considered as a function of SDC. A summary of these horizontal irregularities follows.

This table also contains the sections in ASCE/SEI 7 that have limitations and requirements that must be satisfied for the various SDCs that are applicable. These provisions are covered below.

Torsional irregularity (Type 1a). Torsional irregularity is defined to exist in structures with rigid or semirigid diaphragms where the maximum story drift at one end of the structure that is transverse to an axis is greater than 1.2 times the average of the story drifts at the two ends of the structure. Note that the drifts are to be computed, including accidental torsion, with the torsional amplification factor, A_x, set equal to 1.0 (see 12.8.4.3 and Section 6.3.8 of this publication). This provision is illustrated in Figure 6.7.

Figure 6.7
Horizontal Irregularity—Torsional and Extreme Torsional Irregularity (Type 1a and 1b)

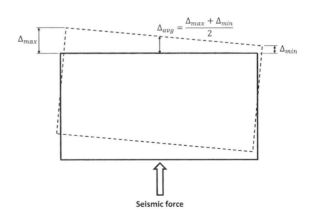

Torsional irregularity: $\Delta_{max} > 1.2\Delta_{avg}$

Extreme torsional irregularity: $\Delta_{max} > 1.4\Delta_{avg}$

The building illustrated in Figure 6.7 has a symmetrical geometric shape in plan without reentrant corners or wings; such configurations may still be classified as having a torsional irregularity due to the distribution of the SFRSs and their relationship to the center of mass in plan. Plan configurations with wings, such as L-, H- or Y-shapes, are very susceptible to this type of irregularity even if symmetrical in plan.

Extreme torsional irregularity (Type 1b). Extreme torsional irregularity is defined the same as that for torsional irregularity except the limit is 1.4 times the average of the story drifts at the two ends of the structure instead of 1.2 times the average of the story drifts at the two ends of the structure (see Figure 6.7). Extreme torsional irregularities should be avoided if at all possible.

Reentrant corner irregularity (Type 2). A reentrant corner irregularity is defined to exist where both plan projections of a structure beyond a reentrant corner are greater than 15 percent of the plan dimension of the structure in the given direction (see Figure 6.8).

Figure 6.8
Horizontal Irregularity—
Reentrant Corner Irregularity (Type 2)

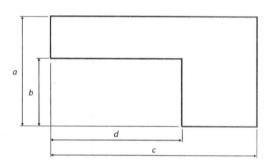

Reentrant corner irregularity: $b > 0.15a$ and $d > 0.15c$

It is evident that a concentration of forces can occur at a sufficiently large reentrant corner, which can lead to significant local damage.

Diaphragm discontinuity irregularity (Type 3). A diaphragm discontinuity irregularity is defined to exist where a diaphragm has an abrupt discontinuity or variation in stiffness. This includes diaphragms that have (a) a cutout or open area that is greater than 50 percent of the gross enclosed diaphragm area (see Figure 6.9) and (b) a change in effective diaphragm stiffness of more than 50 percent from one story to the next.

Figure 6.9
Horizontal Irregularity—
Diaphragm Discontinuity Irregularity (Type 3)

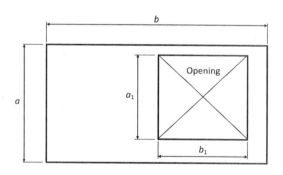

Diaphragm discontinuity irregularity: $a_1 b_1 > 0.5ab$

Significant variations in stiffness due to relatively large openings or abrupt changes in the type of construction material used for the diaphragm affects the distribution of seismic forces to the vertical components of the SFRS. Torsional forces can be created that must be considered in design; these types of torsional forces are not accounted for in the force distribution that is normally used for regular buildings. Typical opening sizes for elevators and stairways in buildings are generally not large enough to cause significant differences in diaphragm stiffness.

Out-of-plane offset irregularity (Type 4). Out-of-plane offset irregularity is defined to exist where there is a discontinuity in the path of lateral-force resistance. An out-of-plane offset of at least one of the vertical elements in the SFRS is an example of such a discontinuity. Illustrated in Figure 6.10 is an out-of-plane offset of a shear wall that is part of the SFRS (a similar such offset would occur if instead of a wall, a column or a brace was present at that location). This offset is the most critical discontinuity of its type since forces must be adequately transferred through horizontal members at the location of the offset.

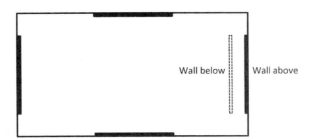

Figure 6.10
Horizontal Irregularity—Out-of-plane Offset Irregularity (Type 4)

Nonparallel system irregularity (Type 5). A nonparallel system irregularity is defined to exist where vertical lateral force-resisting elements are not parallel to the major orthogonal axes of the SFRS (see Figure 6.11).

Figure 6.11
Horizontal Irregularity— Nonparallel System Irregularity (Type 5)

The Equivalent Lateral Force Procedure that is covered in Section 6.3.8 of this publication cannot be used when this type of irregularity exists.

Vertical Irregularity

Table 12.3-2 contains the types and descriptions of vertical structural irregularities that must be considered as a function of SDC. A summary of these vertical irregularities follows.

This table also contains the sections in ASCE/SEI 7 that contain limitations and requirements that must be satisfied for the various SDCs that are applicable. These provisions are covered below.

Stiffness-soft story irregularity (Type 1a). A stiffness-soft story irregularity is defined to exist where there is a story in which the lateral stiffness is less than 70 percent of that in the story above or less than 80 percent of the average stiffness of the three stories above.

An example of a stiffness-soft story irregularity is illustrated in Figure 6.12. This type of irregularity commonly occurs in the lower level of a building that has a lobby with a taller story height than the typical floors above.

Figure 6.12
Vertical Irregularity—Stiffness-soft Story Irregularity (Type 1a and 1b)

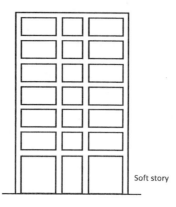

- Stiffness-soft story irregularity:
  Soft story stiffness < 0.7(story stiffness above)
  or
  < 0.8(average stiffness of 3 stories above)
- Stiffness-extreme soft story irregularity:
  Soft story stiffness < 0.6(story stiffness above)
  or
  < 0.7(average stiffness of 3 stories above)

Stiffness-extreme soft story irregularity (Type 1b). Similar to stiffness-soft story irregularity, a stiffness-extreme soft story irregularity is defined to exist where there is a story in which the lateral stiffness is less than 60 percent of that in the story above or less than 70 percent of the average stiffness of the three stories above (see Figure 6.12).

Weight (mass) irregularity (Type 2). Weight (mass) irregularity is defined to exist where the effective mass of any story is more than 150 percent of the effective mass of an adjacent story (see Figure 6.13). A roof that is lighter than the floor below need not be considered in the evaluation.

Figure 6.13 Vertical Irregularity—Weight (Mass) Irregularity (Type 2)

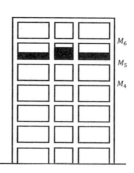

Weight (mass) irregularity: $M_5 > 1.5 M_4$ or $M_5 > 1.5 M_6$

Floor levels with heavy mechanical equipment or storage areas are examples of where this type of irregularity can commonly occur.

Vertical geometric irregularity (Type 3). A vertical geometric irregularity is defined to exist where the horizontal dimension of the SFRS in any story is more than 130 percent of that in an adjacent story (see Figure 6.14). This irregularity applies regardless of whether the larger dimension is above or below the smaller one.

Figure 6.14 Vertical Irregularity—Vertical Geometric Irregularity (Type 3)

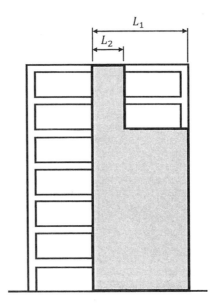

Vertical geometric irregularity: $L_1 > 1.3 L_2$

In-plane discontinuity in vertical lateral force-resisting element irregularity (Type 4). This type of irregularity is defined to exist where there is an in-plane offset of a vertical seismic-force-resisting element resulting in overturning demands on a supporting beam, column, truss or slab (see Figure 6.15).

Figure 6.15 Vertical Irregularity—In-plane Discontinuity in Vertical Lateral Force-resisting Element Irregularity (Type 4)

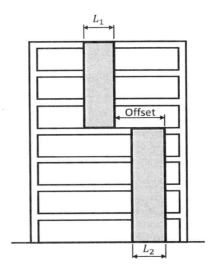

In-plane discontinuity irregularity:
Offset $> L_1$ or Offset $> L_2$

Discontinuity in lateral strength—weak story irregularity (Type 5a). This type of irregularity is defined to exist where a story lateral strength is less than 80 percent of that in the story above (see Figure 6.16). By definition, the story lateral strength is the total lateral strength of all of the seismic-resisting elements that share the story shear in the direction of analysis.

Figure 6.16 Vertical Irregularity—Discontinuity in Lateral Strength—Weak Story Irregularity (Type 5a and 5b)

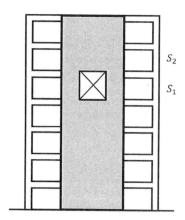

Lateral strength – weak story irregularity: $S_1 < 0.80 S_2$

Lateral strength – extreme weak story irregularity: $S_1 < 0.65 S_2$

Buildings with a weak story irregularity usually develop all of their inelastic behavior at the weak story along with the resulting damage, which could lead to a possible collapse.

Discontinuity in lateral strength—extreme weak story irregularity (Type 5b). This irregularity is defined the same as that of Type 5a except the limit is 65 percent instead of 80 percent (see Figure 6.16).

Two exceptions are given in 12.3.2.2 related to vertical irregularities:

1. Types 1a, 1b and 2 irregularities do not apply where no story drift ratio determined using the design seismic forces is greater than 130 percent of the story drift ratio of the next story above. Note that torsional effects need not be considered in the calculation of the story drifts when applying this exception. Also, the story drift ratio relationship for the top two stories are not required to be evaluated.

2. Types 1a, 1b and 2 irregularities need not be considered for one-story buildings assigned to any SDC and for two-story buildings assigned to SDCs B, C or D.

Limitations and Additional Requirements for Systems with Structural Irregularities

As noted previously, most of the earthquake provisions that are given in ASCE/SEI 7 are applicable to regular structures. The requirements in 12.3.3 are meant to account for detrimental effects that can be caused by irregularities and to prohibit or limit certain types of irregularities in areas of high seismic risk.

A summary of the limitations and additional requirements for systems with particular types of structural irregularities are given in Table 6.3.

The types of horizontal and vertical irregularities that are prohibited in SDCs D through F derive from poor performance in past earthquakes. These types of irregularities have the tendency to concentrate large inelastic demands in certain portions of a structure, which could lead to collapse. As such, these irregularities should be avoided wherever possible even in cases where they are permitted.

The height limitations for structures with extreme weak stories are applicable to SDC B and C only since extreme weak stories (vertical irregularity Type 5b) are prohibited in SDCs D, E and F (see 12.3.3.1 and Table 6.3).

Table 6.3 Limitations and Additional Requirements for Systems with Structural Irregularities (12.3.3)

	SDC	Irregularity Type	Limitations/Additional Requirements
Prohibited horizontal and vertical irregularities for SDCs D–F	D	Vertical irregularity Type 5b	Not permitted
	E or F	• Horizontal irregularity Type 1b • Vertical irregularities Type 1b, 5a or 5b	
Extreme weak stories	B or C	Vertical irregularity Type 5b	Height is limited to two stories or 30 feet.*
Elements supporting discontinuous walls or frames	B–F	Horizontal irregularity Type 4 or Vertical irregularity Type 4	Design supporting members and their connections to resist load combinations with overstrength factor of 12.4.3.
Increase in forces due to irregularities for SDCs D–F	D–F	• Horizontal irregularities Type 1a, 1b, 2, 3 or 4 • Vertical irregularity Type 4	Design forces determined by 12.10.1.1 shall be increased by 1.25 for (a) connections of diaphragms to vertical elements and to collectors and (b) for collectors and their connections, including connections of collectors to vertical elements of the SFRS.**

* The limit does not apply where the weak story is capable of resisting a total seismic force equal to Ω_0 times the design force prescribed in 12.8.
** Forces calculated using the seismic load effects including the overstrength factor of 12.4.3 need not be increased.

The main purpose of the requirement for elements supporting discontinuous walls or frames is to protect the supporting elements from overload caused by overstrength of a discontinued seismic-force-resisting element (columns, beams, slabs or trusses). The connection between the discontinuous element and the supporting member must be adequate to transfer the forces for which the discontinuous elements are required to be designed.

The increase in forces due to the irregularities listed for SDCs D, E and F are intended to account for the difference between the loads that are distributed using the Equivalent Lateral Force Procedure and the load distribution including the irregularities. The difference in load distribution is especially important for the connection of the diaphragm to the vertical elements of the SFRS. No additional increase is warranted where the load combinations with overstrength factor apply.

Redundancy

Redundancy has long been recognized as a desirable attribute to have in any structure. The redundancy factor, ρ, is a measure of the redundancy inherent in a structure. A higher degree of redundancy exists where there are multiple paths to resist the lateral forces. In essence, ρ has the effect of reducing the response modification coefficient, R, for less redundant structures, which, in turn, increases the seismic demand on the system. A redundancy factor equal to 1.0 means that a structure is sufficiently redundant, and no increase in the seismic load effects is warranted.

According to 12.3.4, the value ρ is either 1.0 or 1.3 and is to be determined in each of the two orthogonal directions for all structures. Conditions where ρ is equal to 1.0 are as follows:

1. Structures assigned to SDC B or C;

2. Drift calculations and P-delta effects;

3. Design of nonstructural components;

4. Design of nonbuilding structures that are not similar to buildings;

5. Design of collector elements, splices and their connections for which the seismic load effects including overstrength factor of 12.4.3 are used;

6. Design of members or connections where the seismic load effects including overstrength factor of 12.4.3 are required for design;

7. Diaphragm loads determined by Equation 12.10-1;

8. Structures with damping systems designed in accordance with Chapter 18; and

9. Design of structural walls for out-of-plane forces, including their anchorage.

For structures that are assigned to SDC D, E or F that do not satisfy any of the conditions listed above, ρ must be set equal to 1.3 unless one of the two conditions in 12.3.4.2 is met, in which case it is permitted to be set equal to 1.0.

In the first condition, the applicable requirements in Table 12.3-3 must be satisfied in each story in a structure that resists more than 35 percent of the base shear in the direction of analysis. The 35-percent value was based on unpublished parametric studies that indicated that stories with at least 35 percent of the base shear include all stories of low-rise buildings up to 6 stories and 87 percent of the stories of tall buildings. The intent is to exclude the uppermost stories and penthouses in a structure from the redundancy requirements.

The requirements under the first condition are organized with respect to the following lateral-force-resisting elements:

- Braced frames
- Moment frames
- Shear walls or wall piers with a height-to-length ratio greater than 1.0 (see Figure 12.3-2)
- Cantilever columns
- Other

A dual system is included under the "Other" types of elements and is considered to be inherently redundant.

In this approach, individual lateral-force-resting elements are removed to determine the effect on the remaining structure. If the removal of such elements, one by one, does not result in (1) more than a 33-percent reduction in story strength (or, in the case of cantilever columns, a 33-percent reduction in moment resistance) or (2) an extreme torsional irregularity, ρ may be taken as 1.0. This check determines whether an individual member plays a significant role in lateral-force resistance.

In the second condition, ρ is permitted to be taken as 1.0 for structures that are regular in plan at all levels provided the SFRSs consist of at least two bays of seismic-force-resisting perimeter framing on each side of the structure in each orthogonal direction at each story resisting more than 35 percent of the base shear. In the case of shear walls, the number of bays is taken equal to the length of the shear wall divided by the story height in all cases except where light-frame construction is used; in that case the number of bays is equal to two times the length of the shear wall divided by the story height.

As part of the parametric study that was mentioned previously, braced frame and moment frame systems were investigated to determine their sensitivity to the analytical redundancy criteria. The second condition is consistent with the results of this study.

Taking ρ equal to 1.3 is permitted without checking either of the two conditions discussed above. It is important to keep in mind that ρ has a direct impact on the magnitude of the horizontal seismic effects (see Section 6.3.4 of this publication) so automatically taking the default value of 1.3 may not result in the most economical solution.

6.3.4 Seismic Load Effects and Combinations

Overview

Unless otherwise exempted by ASCE/SEI 7, all structural members, including those that are not part of the SFRS, must be designed using the seismic load effects of 12.4. In general, these effects are the axial forces, shear forces and bending moments that are produced in the structural members due to the application of the horizontal and vertical forces on the structure (see 12.4.2).

Certain structural members must be designed to account for the seismic load effects including the overstrength factor, Ω_0 (see 12.4.3). The conditions under which these effects are applicable are covered below.

Seismic Load Effect

The seismic load effect, E, that is used in the load combinations defined in 2.3 and 2.4 consists of effects of horizontal (E_h) and vertical (E_v) seismic forces, which are determined by Equations 12.4-3 and 12.4-4, respectively:

$$E_h = \rho Q_E \tag{6.9}$$

$$E_v = 0.2 S_{DS} D \tag{6.10}$$

In Equation 6.9, Q_E are the effects (axial forces, shear forces and bending moments) on the structural members obtained from the structural analysis from either the base shear, V, which is distributed over the height of the structure, or the seismic force acting on a component of a structure, F_p (12.4). It is evident from Equation 6.10 that the effect from the vertical seismic forces, E_v, is based on an assumed effective vertical acceleration of $0.2S_{DS}$ times the acceleration due to gravity.

It is important to understand the proper sign convention to use in the load combinations that contain E. In the strength design load combination 5 in 2.3.2 or the allowable stress load combinations 5 and 6 in 2.4.1, E is determined by Equation 12.4-1:

$$E = E_h + E_v \tag{6.11}$$

In these load combinations, effects from the dead load, D, are additive to those due to seismic, E.

In the strength design load combination 7 in 2.3.2 or the allowable stress load combination 8 in 2.4.1, E is determined by Equation 12.4-2:

$$E = E_h - E_v \tag{6.12}$$

In these load combinations, the effects of E counteract those from D. As such, E_v is subtracted from E_h.

Two exceptions are given in 12.4.2.2 that permit E_v to be taken as zero:

1. In Equations 12.4-1, 12.4-2, 12.4-5 and 12.4-6 where $S_{DS} \leq 0.125$; and

2. In Equation 12.4-2 where determining demands on soil-structure interface of foundations.

Basic load combinations for strength design and allowable stress design are summarized in 12.4.2.3 and Table 6.4. These are applicable in the design of the members that have been designated as part of the SFRS. Additional information on these load combinations can be found in Chapter 2 of this publication.

Table 6.4 Seismic Load Combinations (12.4.2.3)

Combination No.	Combination
*Strength Design**	
5	$(1.2 + 0.2S_{DS})D + \rho Q_E + L + 0.2S$
7	$(0.9 - 0.2S_{DS})D + \rho Q_E$
*Allowable Stress Design***	
5	$(1.0 + 0.14S_{DS})D + 0.7\rho Q_E$
6b	$(1.0 + 0.105S_{DS})D + 0.525\rho Q_E + 0.75L + 0.75S$
8	$(0.6 - 0.14S_{DS})D + 0.7\rho Q_E$

* Notes for strength design load combinations:
  (1) Load factor on L in combination 5 is permitted to equal 0.5 for all occupancies in which L_o in Table 4-1 is less than or equal to 100 psf, with the exception of garages or areas occupied as places of public assembly.
  (2) Where fluid loads, F, are present, they shall be included with the same load factor as dead load, D, in combinations 1 through 5 and 7.
  (3) Where load H is present, it shall be included as follows:
      (a) Where the effect of H adds to the primary variable load effect, include H with a load factor of 1.6.
      (b) Where the effect of H resists the primary variable load effect, include H with a load factor of 0.9 where the load is permanent or a load factor of 0 for all other conditions.

** Notes for allowable stress design load combinations:
  (1) Where fluid loads, F, are present, they shall be included in combinations 1 through 6 and 8 with the same factor as that used for dead load, D.
  (2) Where load H is present, it shall be included as follows:
      (a) Where the effect of H adds to the primary variable load effect, include H with a load factor of 1.0.
      (b) Where the effect of H resists the primary variable load effect, include H with a load factor of 0.6 where the load is permanent or a load factor of 0 for all other conditions.

Seismic Load Effect Including Overstrength Factor

The seismic load effect including overstrength factor, E_m, that is used in the load combinations defined in 12.4.3.2 consists of effects of horizontal seismic forces including overstrength factor (E_{mh}) and vertical seismic load effect (E_v), the latter of which is determined by Equation 6.10. The horizontal seismic load effect with overstrength factor is determined by Equation 12.4-7:

$$E_{mh} = \Omega_0 Q_E \tag{6.13}$$

where Ω_0 is the overstrength factor given in Table 12.2-1 as a function of the SFRS (12.4.3).

In the strength design load combination 5 in 2.3.2 or the allowable stress load combinations 5 and 6b in 2.4.1, E shall be taken equal to E_m, which is determined by Equation 12.4-5:

$$E_m = E_{mh} + E_v \tag{6.14}$$

In the strength design load combination 7 in 2.3.2 or the allowable stress load combination 8 in 2.4.1, E shall be taken equal to E_m, which is determined by Equation 12.4-6:

$$E_m = E_{mh} - E_v \tag{6.15}$$

Note that the value of E_m used in these equations need not exceed the maximum force that can develop in the member as determined by a rational, plastic mechanism analysis or nonlinear response analysis utilizing realistic expected values of material strengths.

Basic combinations for strength design with overstrength factor and allowable stress design with overstrength factor are summarized in 12.4.3.2 and Table 6.5. It is important to note that these combinations pertain only to the following structural elements:

- Elements supporting discontinuous walls or frames in structures assigned to SDC B through F.
- Collector elements and their connections, including connections to vertical elements in structures assigned to SDC C through F.

See Chapter 2 of this publication for additional information on these load combinations.

Allowable stresses are permitted to be determined using an allowable stress increase of 1.2; this increase is not to be combined with any other increases in allowable stresses or load combination reductions permitted in any of the provisions.

Minimum Upward Force for Cantilevers (SDC D through F)

Minimum upward force provisions for horizontal cantilever members in structures assigned to SDC D, E or F are given in 12.4.4. In particular, such members are to be designed for a minimum net upward force equal to 0.2 times the dead load in addition to the applicable load combinations in 12.4.

Horizontal cantilever members are designed for an upward force that results from an effective vertical acceleration of 1.2 times the acceleration due to gravity. This provision is meant to provide minimum strength in the upward direction and to account for any dynamic amplification of vertical ground motion due to the vertical flexibility of the cantilever.

Table 6.5 Seismic Load Combinations with Overstrength Factor (12.4.3.2)

Combination No.	Combination
Strength Design*	
5	$(1.2 + 0.2S_{DS})D + \Omega_0 Q_E + L + 0.2S$
7	$(0.9 - 0.2S_{DS})D + \Omega_0 Q_E$
Allowable Stress Design**	
5	$(1.0 + 0.14S_{DS})D + 0.7\Omega_0 Q_E$
6b	$(1.0 + 0.105S_{DS})D + 0.525\Omega_0 Q_E + 0.75L + 0.75S$
8	$(0.6 - 0.14S_{DS})D + 0.7\Omega_0 Q_E$

* Notes for strength design load combinations:
  (1) Load factor on L in combination 5 is permitted to equal 0.5 for all occupancies in which L_o in Table 4-1 is less than or equal to 100 psf, with the exception of garages or areas occupied as places of public assembly.
  (2) Where fluid loads, F, are present, they shall be included with the same load factor as dead load, D, in combinations 1 through 5 and 7.
  (3) Where load H is present, it shall be included as follows:
    (a) Where the effect of H adds to the primary variable load effect, include H with a load factor of 1.6.
    (b) Where the effect of H resists the primary variable load effect, include H with a load factor of 0.9 where the load is permanent or a load factor of 0 for all other conditions.

** Notes for allowable stress design load combinations:
  (1) Where fluid loads, F, are present, they shall be included in combinations 1 through 6 and 8 with the same factor as that used for dead load, D.
  (2) Where load H is present, it shall be included as follows:
    (a) Where the effect of H adds to the primary variable load effect, include H with a load factor of 1.0.
    (b) Where the effect of H resists the primary variable load effect, include H with a load factor of 0.6 where the load is permanent or a load factor of 0 for all other conditions.

6.3.5 Direction of Loading

According to 12.5.1, seismic forces must be applied to the structure in directions that produce the most critical load effects on the structural members. The requirements are based on the SDC and are summarized in Table 6.6.

Earthquakes can produce ground motion in any direction and a structure must be designed to resist the maximum possible effects. The purpose of these requirements is to provide rational methods of analyses to capture critical loading effects.

Orthogonal effects on horizontal members such as beams and slabs are generally minimal while those on vertical elements (columns and walls) that are part of the SFRS can be substantial to the point of governing the design. The maximum effects, Q_E, obtained from the methods in Table 6.6 must be modified by ρ or Ω_0, whichever is applicable, and then combined with E_v accordingly.

Table 6.6 Direction of Loading Requirements (12.5)

SDC	Requirement
B	Design seismic forces applied independently in each of two orthogonal directions and orthogonal interaction effects are permitted to be neglected.

C	• Conform to requirements of SDC B
	• Structures with horizontal irregularity Type 5:
	  – *Orthogonal combination procedure*: Apply 100 percent of the seismic forces in one direction and 30 percent of the seismic forces in the perpendicular direction on the structure simultaneously where the forces are computed in accordance with 12.8 (Equivalent Lateral Force Procedure), 12.9 (modal response spectrum analysis) or 16.1 (linear response history procedure), or
	  – *Simultaneous application of orthogonal ground motion*: Apply orthogonal pairs of ground motion acceleration histories simultaneously to the structure using 16.1 (linear response history procedure) or 16.2 (nonlinear response history procedure).

D–F	• Conform to requirements of SDC C • Any column or wall that forms part of two or more intersecting SFRSs that is subjected to axial load due to seismic forces along either principal axis greater than or equal to 20 percent of the axial design strength of the column or wall must be designed for the most critical load effect due to application of seismic forces in any direction.*

* Either of the procedures of 12.5.3 (a) or (b) for SDC C are permitted to be used to satisfy this requirement.

6.3.6 Analysis Procedure Selection

Requirements on the type of procedure that can be used to analyze the structure for seismic load effects are given in 12.6 and are summarized in Table 12.6-1 (see Table 6.7). Permitted analytical procedures depend on the SDC, the risk category of the structure, characteristics of the structure (height and period) and the presence of any structural irregularities.

It is evident from the table that the higher analysis methods—modal response spectrum analysis and seismic response history procedures—can be used to analyze any structure. The equivalent lateral force procedure is not permitted for structures with certain types of irregularities because basic assumptions of that method are a gradually varying distribution of mass and stiffness along the height of the structure and an SFRS that has a negligible torsional response to ground shaking.

An upper limit of $3.5T_S$ is set for the period of a structure that can be analyzed using the equivalent lateral force procedure, which inherently assumes that the first mode of vibration is the most significant mode. In taller structures, higher modes of vibration are more significant; thus, the equivalent lateral force procedure may underestimate the design base shear and may not predict accurately the distribution of the seismic forces over the height of the structure in such cases.

Table 6.7 Permitted Analytical Procedures (12.6)

SDC	Structural Characteristics	Equivalent Lateral Force Procedure (12.8)	Modal Response Spectrum Analysis (12.9)	Seismic Response History Procedures (Chapter 16)
B, C	All structures	P	P	P
D, E, F	Risk Category I or II buildings not exceeding two stories above the base	P	P	P
	Structures of light-frame construction	P	P	P
	Structures with no structural irregularities and not exceeding 160 feet in structural height	P	P	P
	Structures exceeding 160 feet in structural height with no structural irregularities and with $T < 3.5 T_S$	P	P	P
	Structures not exceeding 160 feet in structural height and having only horizontal irregularities of Type 2, 3, 4 or 5 or vertical irregularities of Type 4, 5a or 5b	P	P	P
	All other structures	NP	P	P

P = Permitted, NP = Not permitted

Flowchart 6.7 in Section 6.6 of this publication can be used to determine the permitted analytical procedure for structures.

6.3.7 Modeling Criteria

Overview

Requirements pertaining to the construction of an adequate model for the purposes of seismic load analysis are given in 12.7. Included are provisions for foundation modeling, effective seismic weight, structural modeling and interaction effects.

Foundation Modeling

It is permitted to assume that the base of the structure is fixed for purposes of determining the seismic loads on a structure. However, if the flexibility of the foundation is considered, the requirements of 12.13.3 (foundation load-deformation characteristics) or Chapter 19 (soil structure interaction for seismic design) must be satisfied.

The base of a structure is defined as the level at which the horizontal seismic ground motions are considered to be imparted to the structure (11.2). Additional information on where the base occurs for a number of common situations can be found in C11.2.

Effective Seismic Weight

In general, a structure accelerates laterally during an earthquake, which leads to its mass producing inertial forces. These inertial forces accumulate over the height of the structure and produce the design base shear, V.

The definition of the effective seismic weight, W, that is used in determining the base shear, V, is given in 12.7.2. In addition to the dead load of the structure, the following loads must be included in W:

1. In areas used for storage, a minimum of 25 percent of the floor live load;

2. Where partitions must be included in accordance with 4.2.2, the actual partition weight or 10 psf of floor area, whichever is greater;

3. Total operating weight of permanent equipment;

4. Where the flat roof snow load, p_f, exceeds 30 psf, 20 percent of the uniform design snow load regardless of the roof slope; and

5. Weight of landscaping and other materials at roof gardens and similar areas.

It is clear from the list that only that portion of the weight that is physically tied to the structure is considered effective. Live loads due to human occupancy and loose furniture, to name a few, need not be considered effective.

The probability that a portion of live load in storage areas is present during an earthquake is not small, especially in those that are densely packed; that is why 25 percent of the floor live load is considered effective. An exception to this occurs where the inclusion of storage loads adds no more than 5 percent to the effective seismic weight: in such cases, it need not be included in the effective seismic weight. Also, floor live load in public garages and open parking structures need not be included.

The full snow load is not considered to be part of the effective seismic weight because it is unlikely that maximum snow load and maximum earthquake load will occur simultaneously.

Structural Modeling

A mathematical model of the structure that includes member stiffness and strength must be constructed in order to determine forces in the structural members and story drifts due to ground shaking.

Cracked section properties must be used when analyzing concrete and masonry structures. In steel moment frames, the contribution of panel zone deformations to overall story drift must be included. It is important to include these factors in the model in order to obtain more realistic values of lateral deformations.

A three-dimensional analysis is required for structures that have horizontal irregularity Types 1a, 1b, 4 or 5. In such cases, a minimum of three dynamic degrees of freedom (translation in two orthogonal directions and torsional rotation about the vertical axis) must be included at each level of the structure. Structures with flexible diaphragms and Type 4 horizontal structural irregularities need not be analyzed using a three-dimensional model.

Interaction Effects

Interaction effects between moment-resisting frames that make up the SFRS and rigid elements that are not part of the SFRS that enclose or adjoin the moment-resisting frame must be considered in the analysis in order to prevent any unexpected detrimental effects from occurring during an earthquake.

Consider the situation in Figure 6.17 where masonry is used to infill between the reinforced concrete columns in a moment-resisting frame. Assume that the masonry fits tightly against the columns. When subjected to ground shaking, hinges will form in the columns at the tops of the columns and at the top of the masonry rather than at the top and bottom of the columns as would be expected because the masonry is much stiffer than the columns. The shear forces in the columns increase by a factor of H/h. If this increase is not accounted for in design, an abrupt shear failure of the columns can occur, which could result in catastrophic failure of the structure.

Figure 6.17
Interaction Effects

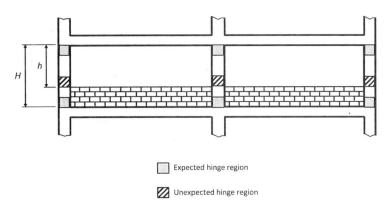

☐ Expected hinge region
▨ Unexpected hinge region

6.3.8 Equivalent Lateral Force Procedure

Overview

The Equivalent Lateral Force Procedure (ELFP) is a simplified method of determining the effects of ground motion on a structure. It is valid for essentially regular structures without certain types of significant discontinuities where the primary response to ground motion is in the horizontal direction without substantial torsion (that is, the first mode of vibration).

The effects of inelastic dynamic response are determined by using a linear static analysis: a design base shear, V, is determined, which is distributed over the height of the structure at each floor level. The structure is analyzed for these static forces, which are distributed to the members of the SFRS considering the flexibility of the diaphragms.

The provisions of the ELFP are contained in 12.8. This analysis procedure can be used for all structures assigned to SDC B and C as well as some types of structures assigned to SDC D, E and F (see Table 6.7).

Seismic Base Shear, V

The seismic base shear, V, is determined by multiplying the seismic response coefficient, C_S, by the effective seismic weight, W (see Equation 12.8-1):

$$V = C_S W \tag{6.16}$$

Equations for C_S are given in 12.8.1, which form the design response spectrum:

- Constant acceleration, $0 \leq T \leq T_S = S_{D1}/S_{DS}$ (Equation 12.8-2):

$$C_S = \frac{S_{DS}}{\left(\frac{R}{I_e}\right)} \tag{6.17}$$

- Constant velocity, $T_S < T \leq T_L$ (Equation 12.8-3):

$$C_S = \frac{S_{D1}}{T\left(\frac{R}{I_e}\right)} \tag{6.18}$$

- Constant displacement, $T > T_L$ (Equation 12.8-4):

$$C_S = \frac{S_{D1} T_L}{T^2\left(\frac{R}{I_e}\right)} \tag{6.19}$$

Note that the transition to the design acceleration, S_{DS}, in the segment from $T = 0$ to $T = T_O$ in the design spectrum of 11.4.5 (see Figure 6.3) is not used in the ELFP; for simplicity, the horizontal portion of the spectrum associated with constant acceleration extends from $T = 0$ to $T = T_S$.

In the constant velocity portion of the spectrum, V is inversely proportional to T. The constant displacement segment of the response spectrum is generally valid for tall and flexible structures.

A lower limit of C_S is defined in Equation 12.8-5:

$$C_S = 0.044 S_{DS} I_e \geq 0.01 \tag{6.20}$$

This equation provides a minimum base shear as a function of the short period design acceleration, S_{DS}, and the importance factor, I_e. In no case is V permitted to be less than 1 percent of the effective seismic weight, W.

For structures that are located where $S_1 \geq 0.6$, which applies to sites near active faults, Equation 12.8-6 provides an additional lower limit for C_S:

$$C_S = \frac{0.5 S_1}{\left(\dfrac{R}{I_e}\right)} \tag{6.21}$$

The design response spectrum in accordance with the ELFP is depicted in Figure 6.18.

Figure 6.18
Design Response Spectrum According to the Equivalent Lateral Force Procedure (12.8)

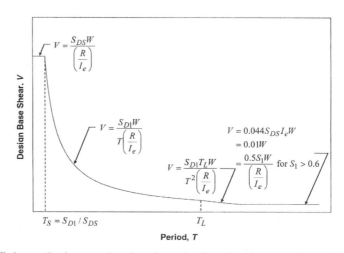

The seismic response coefficient, C_s, is permitted to be calculated using a value of $S_S = 1.5$ for regular structures that are five stories or less above the base and that have a period, T, that is less than or equal to 0.5 second. This cap on S_S is based on the performance of buildings in past earthquakes that were designed using the applicable provisions of ASCE/SEI 7.

Period Determination

The fundamental period of the structure, T, in the direction of analysis is to be determined based on the structural properties and deformational characteristics of the SFRS using fundamental principles of a dynamic analysis.

In the preliminary design stage, the information that is needed to determine T (such as member sizes) may not be known. Equations 12.8-7 through 12.8-10 in 12.8.2.1 can be used to determine an approximate fundamental period, T_a, for a variety of structure types (see Table 6.8). These equations provide a lower-bound value for the approximate period, which, in turn, provides an upper-bound value of V.

Note that the approximate period for masonry of concrete shear wall systems may be determined using the equation for all other structural systems in lieu of the equation given in Table 6.8.

If the fundamental period of a structure, T, is determined using a rational dynamic analysis, it must be taken less than or equal to the coefficient for upper limit on calculated period, C_u, in Table 12.8-1 times the appropriate approximate fundamental period, T_a. The purpose of setting a limit on the calculated period is so that an unusually low base shear is not obtained for overly flexible structures.

Table 6.8 Approximate Period, T_a

Structure Type		T_a
Moment-resisting frame systems in which the frames resist 100 percent of the required seismic forces and are not enclosed or adjoined by components that are more rigid and will prevent the frames from deflecting when subjected to seismic forces	Steel	$0.028h_n^{0.8}$
	Concrete	$0.016h_n^{0.9}$
Steel eccentrically braced frame in accordance with Table 12.2-1 lines B1 or D1		$0.030h_n^{0.75}$
Steel buckling-restrained braced frames		$0.030h_n^{0.75}$
Structures less than or equal to 12 stories in height where the SFRS consists entirely of concrete or steel moment-resisting frames where the average story height is greater than or equal to 10 feet		$0.1N$
Masonry or concrete shear wall structures		$\dfrac{0.0019h_n}{\sqrt{\dfrac{100}{A_B}\sum\limits_{i=1}^{x}\left(\dfrac{h_n}{h_i}\right)^2 \dfrac{A_i}{\left[1+0.83\left(\dfrac{h_i}{D_i}\right)^2\right]}}}$
All other structural systems		$0.020h_n^{0.75}$

Notes:
- h_n = vertical distance from the base to the highest level of the SFRS of a structure
- N = number of stories above the base of a structure
- A_B = area of base of structure (square feet)
- A_i = web area of shear wall i (square feet)
- D_i = length of shear wall i (feet)
- h_i = height of shear wall i (feet)
- x = number of shear walls in the structure effective in resisting lateral forces in the direction of analysis

The values of C_u in Table 12.8-1 remove the conservatism of the lower-bound equations that are used to determine T_a. As is evident from the values of C_u in the table, the limiting periods are larger in regions of lower seismicity since structures in these areas usually have longer periods (that is, are more flexible) than those in regions of higher seismicity.

Vertical Distribution of Seismic Forces

Once the seismic base shear, V, has been determined, it is distributed over the height of the building in accordance with Equations 12.8-11 and 12.8-12:

$$F_x = C_{vx}V \tag{6.22}$$

$$C_{vx} = \frac{w_x h_x^k}{\sum_{i=1}^{n} w_i h_i^k} \qquad (6.23)$$

In these equations, which are based on a simplified first mode shape, F_x is the lateral seismic force located at level x above the base of the structure; w_i and w_x are the portions of the total effective seismic weight, W, located or assigned to level i or x; and h_i and h_x are the heights from the base of the structure to level i or x. The exponent related to the structure period, k, is determined as follows:

$$k = \begin{cases} 1.0 \text{ for } T \leq 0.5 \text{ second} \\ 0.75 + 0.5T \text{ for } 0.5 \text{ second} < 2.5 \text{ seconds} \\ 2.0 \text{ for } T \geq 2.5 \text{ seconds} \end{cases} \qquad (6.24)$$

This parameter is intended to approximate the effects of higher modes, which are usually more dominant in structures with a longer fundamental period of vibration (that is, in structures that are more flexible). Note that it is permitted to take $k = 2.0$ for structures with a period between 0.5 and 2.5 seconds instead of using the equation given above for that period range.

For structures with a fundamental period less than or equal to 0.5 second, V is distributed linearly over the height, varying from zero at the base to a maximum value at the top (see Figure 6.19a). When T is greater than 2.5 seconds, a parabolic distribution is to be used (see Figure 6.19b). As noted above, for a period between these two values, a linear interpolation between a linear and parabolic distribution is permitted or a parabolic distribution may be utilized.

Figure 6.19
Vertical Distribution of Seismic Forces (12.8.3)

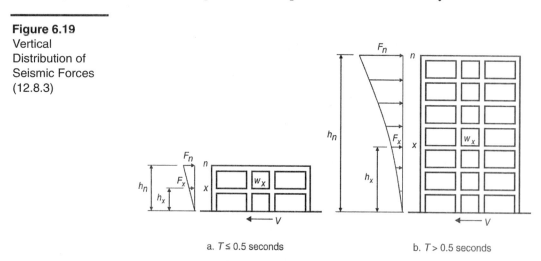

a. $T \leq 0.5$ seconds   b. $T > 0.5$ seconds

It is important to note that the lateral forces, F_x, determined using the equations above are not the inertial forces that occur in the structure at any particular time during an actual earthquake. Rather, they provide design story shears that are consistent with enveloped results obtained from more refined analyses.

Horizontal Distribution of Forces

The seismic design story shear, V_x, in story x is the sum of the lateral forces acting at the floor or roof level supported by that story and all of the floor levels above, including the roof. The story shear is distributed to the vertical elements of the SFRS in the story based on the lateral stiffness of the diaphragm.

For diaphragms that are not flexible, V_x is distributed based on the relative stiffness of the vertical resisting elements and the diaphragm. Inherent and accidental torsion must also be included in the overall distribution (12.8.4.1 and 12.8.4.2):

- The inherent torsional moment, M_t, is determined based on the distance between the center of mass (CM) and the center of rigidity (CR).

- The accidental torsional moment, M_{ta}, is determined based on the assumption that the CM is displaced each way from its actual location by a distance equal to 5 percent of the dimension of the structure perpendicular to the direction of analysis. This is meant to account for any uncertainties in the actual locations of the center of mass and center of rigidity (for example, due to tolerances in the constructed structure, that is, actual member sizes and actual locations of the members in the structure).

Illustrated in Figure 6.20 is a floor or roof level with a diaphragm that is not flexible. The story shear, V_x, at this level acts through the CM. The inherent torsional moment, M_t, is equal to V_x times the eccentricity, e_x, between the CR and the CM. The accidental torsional moment, M_{ta}, is equal to V_x times $0.05b$ for V_x acting in the direction shown. The total torsional moment is the sum of M_t and M_{ta}. Similar calculations can be performed to obtain the total torsional moment for V_x acting in the perpendicular direction.

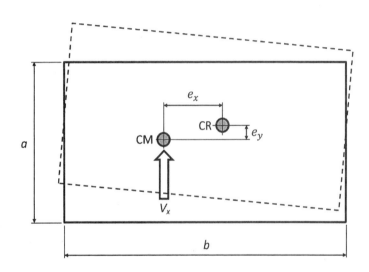

Figure 6.20
Inherent and Accidental Torsional Moments (12.8.4.1 and 12.8.4.2)

Inherent torsional moment: $M_t = V_x e_x$

Accidental torsional moment: $M_{ta} = V_x(0.05b)$

Total torsional moment: $M_t + M_{ta}$

In cases where earthquake forces are required to be applied concurrently in two orthogonal directions (see Section 6.3.5 of this publication), the 5-percent displacement of the CM should be applied along a single orthogonal axis, which produces the greatest effects in the structural members; simultaneous application of the 5-percent displacement along two axes is not required.

For flexible diaphragms, V_x is distributed to the vertical elements of the SFRS based on the mass that is tributary to the elements. Since the floor system usually has the greatest mass, V_x can be distributed based on the area of the diaphragm tributary to each line of resistance. Torsion need not be considered in the overall distribution.

Chapter 6

Where Type 1a or 1b torsional irregularity is present in structures assigned to SDC C, D, E or F, the accidental torsional moment is to be amplified in accordance with 12.8.4.3. In particular, M_{ta} at each level is to be multiplied by the torsional amplification factor, A_x, which is determined by Equation 12.8-14:

$$1.0 \leq A_x = \left(\frac{\delta_{max}}{1.2\delta_{avg}}\right)^2 \leq 3.0 \qquad (6.25)$$

In this equation, δ_{max} is the maximum displacement that occurs at level x in the structure computed assuming $A_x = 1$ and δ_{avg} is the average of the displacements at the extreme points of the structure at level x computed assuming $A_x = 1$ (see Figure 6.21). By setting $A_x = 1$ when determining the displacements, the need for iterations in the calculation of A_x is eliminated.

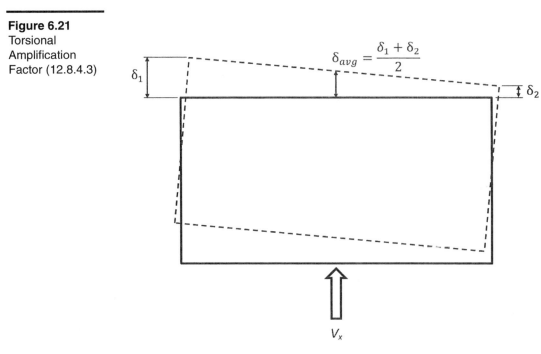

Figure 6.21
Torsional Amplification Factor (12.8.4.3)

Torsional amplification factor $A_x = \left(\dfrac{\delta_{max}}{1.2\delta_{avg}}\right)^2$

Overturning

The structure must be designed to resist the overturning effects caused by the seismic forces. In such cases, the critical load combinations are typically those where the effects from gravity and seismic loads counteract.

Story Drift Determination

Design story drift, Δ, is determined in accordance with 12.8.6 and is defined as the difference of the deflections, δ_x, at the center of mass of the diaphragms at the top and bottom of the story under consideration as shown in Figure 12.8-2 (see Figure 6.22).

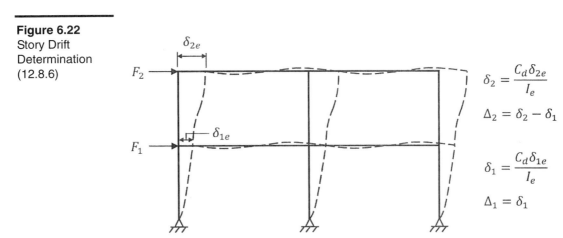

Figure 6.22
Story Drift Determination
(12.8.6)

$$\delta_2 = \frac{C_d \delta_{2e}}{I_e}$$

$$\Delta_2 = \delta_2 - \delta_1$$

$$\delta_1 = \frac{C_d \delta_{1e}}{I_e}$$

$$\Delta_1 = \delta_1$$

The horizontal displacements at each floor level due to the application of the design seismic forces, F_x, are designated δ_{xe} and are obtained from an elastic analysis of the structure. These elastic deflections are multiplied by the deflection amplification factor, C_d, and divided by the importance factor, I_e, to obtain the deflections, δ_x, which are estimates of the actual deflections that are likely to occur from the ground motion (see Equation 12.8-15):

$$\delta_x = \frac{C_d \delta_{xe}}{I_e} \tag{6.26}$$

According to the exception in 12.8.6.1, the minimum seismic base shear computed in accordance with Equation 12.8-5 need not be considered when determining the drifts, δ_{xe}. Also, it is permitted to determine the drifts, δ_{xe}, using seismic forces based on the computed fundamental period of the structure without the upper limit, $C_u T_a$, specified in 12.8.2 (12.8.6.2); this essentially allows the drift to be determined using forces that are consistent with the actual period of the structure that has been determined by a rational analysis, resulting in drifts that are not overly conservative.

In general, values of C_d are slightly less than the corresponding R-values for a given SFRS (see Table 12.2-1). The more ductile a system is (that is, the greater the R-value), the greater the difference is between the values of R and C_d. In cases where C_d is substantially less than R, the SFRS is considered to have damping greater than the nominal 5 percent of critical damping assumed in the analysis.

Illustrated in Figure 6.23 is the relationship between the lateral seismic force and the displacements. If a structure were to respond elastically to the ground motion, the deflection would be δ_E and the corresponding seismic force would be V_E, which is many times greater than the design force, V, that is determined by the provisions of 12.8. In particular, $V_E = RV$.

Figure 6.23
Inelastic Force-displacement Curve

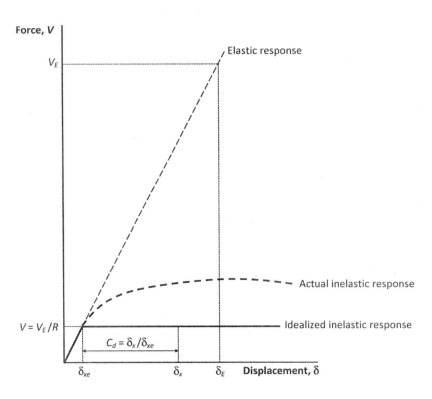

The actual inelastic response of the structure to ground motion is also depicted in Figure 6.23. When designed and detailed properly, the SFRS is expected to reach significant yield when subjected to seismic forces greater than the design seismic forces. Significant yield is defined as the point where the first plastic hinge forms in the structure. Additional plastic hinges form as the seismic force increases until a maximum displacement is reached. It can be shown that full yielding of properly designed and detailed regular structures that are redundant can occur at force levels that are two to four times the prescribed design force levels.

For structures that are idealized with a bilinear inelastic response with a fundamental period greater than T_S, it has been observed that maximum displacement is approximately equal to that of an elastic system with the same initial stiffness (see Figure 6.23). The maximum displacement of an inelastic system with a period smaller than T_S is likely to exceed that of the corresponding elastic system. As noted above, the design forces produce essentially fictitious displacements, δ_{xe}, which are smaller than the actual displacements since the design forces include R.

As can be seen from Figure 6.23, the actual inelastic response is different than the idealized one mainly due to over strength and the related increase in stiffness. Thus, the actual displacement of the system may be less than $R\delta_{xe}$. This difference is accounted for by multiplying the design elastic displacements, δ_{xe}, by the deflection amplification factor, C_d.

Since the design forces, F_x, that are used to determine δ_{xe} contain the importance factor, I_e, it is included in the denominator of Equation 6.26. It would be overly conservative to exclude this term from the denominator for structures with importance factors greater than 1.0.

As noted above, the deflections, δ_x, at each floor level are used in obtaining the design story drifts. Limits on the design story drifts are given in 12.12 and are covered in Section 6.3.12 of this publication.

P-delta Effects

Member forces and story drifts induced by P-delta effects must be considered in member design and in the evaluation of overall stability of a structure where such effects are significant (12.8.7). As a structure deflects horizontally, the vertical loads are displaced from their original position and additional effects are introduced into the structural members, which cause the structure to deflect even further. If sufficient strength and stiffness are not provided, the deflections will continue to increase, leading to overall instability of the structure. As expected, flexible structures are typically more prone to P-delta effects than rigid ones.

In lieu of a more refined analysis, P-delta effects are not required to be considered when the stability coefficient θ is less than or equal to 0.10 where θ is determined by Equation 12.8-16:

$$\theta = \frac{P_x \Delta I_e}{V_x h_{sx} C_d} \qquad (6.27)$$

In this equation P_x is the total vertical design load at and above level x:

$$P_x = \sum_{i=x}^{n} (P_D + P_L)_i \qquad (6.28)$$

where P_D and P_L are the dead load and live loads, respectively, and n is the total number of levels in the structure.

The other quantities in Equation 6.27 are defined as follows (see Figure 6.24):

- Δ = design story drift defined in 12.8.6 occurring simultaneously with V_x
- V_x = seismic force acting between levels x and $x - 1$
- h_{sx} = story height below level x

Figure 6.24
P-delta Effects
(12.8.7)

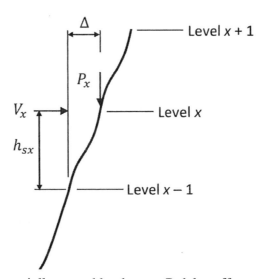

A structure is considered to be potentially unstable due to P-delta effects and must be redesigned where the stability coefficient, θ, is greater than θ_{max}, which is determined by Equation 12.8-17:

$$\theta_{max} = \frac{0.5}{\beta C_d} \leq 0.25 \qquad (6.29)$$

The quantity β is the ratio of the shear demand to the shear capacity for the story under consideration, which is permitted to be conservatively taken as 1.0. Equation 6.29 must be satisfied even where computer software is utilized to determine second-order effects. However, the value of θ is permitted to be divided by (1 + θ) before checking Equation 6.29.

Where P-delta effects must be considered ($0.10 < \theta \leq \theta_{max}$), the member forces and displacements must be increased accordingly using a rational analysis or by multiplying them by $1.0/(1 - \theta)$. Two types of rational analyses are presented in Reference 6.1.

Flowchart 6.8 in Section 6.6 of this publication provides a step-by-step procedure on how to determine the design seismic forces and their distribution based on the requirements of the ELFP.

6.3.9 Modal Response Spectrum Analysis

Overview

In the modal response spectrum analysis, a structure is broken down into a number of single-degree-of-freedom systems. Each system has its own mode shape and natural period of vibration.

For seismic design, the number of modes corresponds to the number of mass degrees of freedom of the structure. Even for a simple low-rise building, the number of mass degrees of freedom is immense. Fortunately, some simple idealizations can be made to significantly reduce the number of degrees of freedom while still preserving a meaningful solution. For example, in the case of planar structures with rigid diaphragms, the number of mass degrees of freedom can be taken as one per story (horizontal translation in the direction of analysis) with the mass of the story concentrated in the floor and roof systems.

Illustrated in Figure 6.25 are the three modes of vibration for a three-story building. For three-dimensional structures with rigid diaphragms, three mass degrees of freedom can be utilized per story (two horizontal translations and rotation about the vertical axis). The number of modes that is required for design is discussed below.

Figure 6.25
Modes of Vibration
of an Idealized
Planar Structure

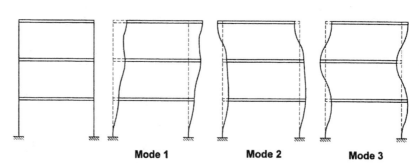

Details on how to determine modal periods, mode shapes and modal participation are given in numerous standard references on dynamic analysis and are not covered here (see for example, Reference 6.3).

A modal response spectral analysis is permitted for any structure assigned to SDC B through F and is required for regular and irregular structures where $T \geq 3.5T_S$ and certain types of irregular structures assigned to SDC D through F (see Table 12.6-1). The following requirements must be satisfied for this method of analysis.

Number of Modes

The basis of performing a modal response spectral analysis is to determine how the distribution of mass and stiffness of a structure affects the forces and displacements due to ground shaking. According to 12.9.1, at least 90 percent of the modal mass must participate in the response in order to obtain a distribution of forces and displacements that are sufficient for design.

In general, 90 percent modal mass participation is usually a small fraction of the total number of modes, even in the case of high-rise buildings. For example, it is possible that the first three modes of a twenty-story building may achieve 90 percent modal mass participation.

Modal Response Parameters

Once the required number of modes has been determined, the forces and deflections are calculated for each mode using the properties of each mode and either the general response spectrum defined in 11.4.5 or the site-specific response spectrum defined 21.2 (see Section 6.2.1 of this publication). Regardless of the spectrum that is used, the spectral ordinates must be divided by (R/I_e): division by R accounts for inelastic behavior and multiplication by I_e provides the additional strength needed for important structures. Also, the displacements that have been obtained using the response spectrum that has been modified by (R/I_e) must be multiplied by (C_d/I_e) in order to obtain the expected inelastic displacements.

At this stage, a base shear, V_t, has been determined for each mode, which has been distributed over the height of the structure. The structure has been analyzed for these forces, and displacements have been obtained at each level.

Combined Response Parameters

The results from the analysis of each mode are to be combined using one of the following methods:

- Square root of the sum of the squares (SRSS)
- Complete quadratic combination (CQC)
- Complete quadratic combination as modified by ASCE 4 (CQC-4) (Reference 6.4)
- Approved equivalent approach

ASCE/SEI 12.9.3 requires that either the CQC or CQC-4 methods be used where closely spaced modes have significant cross-correlation of translational and torsional response.

In general, any one of these combination methods is applied to one direction of analysis at a time. In cases where 12.5 requires consideration of orthogonal effects (see Section 6.3.5 and Table 6.6 of this publication), the results from one direction of loading may be added to 30 percent of the results from the loading in the orthogonal direction.

Scaling Design Values of Combined Response

Scaling of Forces

ASCE/SEI 12.9.4.1 stipulates that the forces must be scaled by a factor of $0.85(V/V_t)$ in cases where the combined base shear using modal analysis (V_t) is less than 85 percent of the base shear calculated using the ELFP (V). It is permitted to use a fundamental period of $T = C_u T_a$ in situations where the calculated T exceeds $C_u T_a$.

The following are reasons why V_t may be less than V:

- The calculated modal, T, may be longer than that used in calculating V.
- The response is not characterized by a single mode.
- The ELFP assumes 100 percent mass participation in the first mode, which is always an overestimate.

In essence, this scaling requirement provides a minimum base shear for design in cases where the computed T has been based on an incorrect (overly flexible) analytical model.

Scaling of Drifts

According to 12.9.4.2, drifts must be scaled by a factor of $0.85C_sW/V_t$ when V_t is less than $0.85C_sW$ where the governing seismic response coefficient, C_s, is the minimum value determined by Equation 12.8-6 for structures that are located where $S_1 \geq 0.6$. This provision is meant to provide a minimum drift for structures that are located in proximity to an active fault.

Horizontal Shear Distribution

Distribution of the horizontal forces to the SFRSs is to be in accordance with 12.8.4 (see Section 6.3.8 of this publication). In the case of modal analysis, two approaches can be used to include accidental torsion in the dynamic analysis model:

1. Perform static analyses with accidental torsion applied at each level of the structure and then add these results to those from the modal response spectrum analysis. Torsional amplification in accordance with 12.8.4.3 is required when this approach is used.

2. Offset the center of mass of each story 5 percent in each direction in a three-dimensional model of the structure. The effects of direct loading and accidental torsion are automatically combined in such an analysis. Amplification of accidental torsion is not required in this approach since the relocation of the centers of mass in a dynamic analysis changes the mode shapes and periods, which results in torsions that are larger than static accidental torsions.

P-delta Effects

The P-delta requirements of 12.8.7 are also applicable to modal response spectrum analysis. The base shear that is used to determine the story shears and the story drifts is determined in accordance with 12.8.6.

Soil Structure Interaction Effects

Reduction in forces associated with soil structure interaction is permitted provided the provisions of Chapter 19 are used. Any generally accepted procedures that are approved by the authority having jurisdiction are also permitted.

6.3.10 Diaphragms, Chords, and Collectors

Diaphragm Design

According to IBC 202, a diaphragm is a horizontal or sloped system that transmits lateral forces to the vertical elements of the SFRS. They are usually treated as deep beams that span between these elements. The roof and floor systems act as the web of the beam, which resists the design shear forces, while the chords act as the flanges, which resist the flexural tension and compression design forces. Horizontal seismic forces are transferred to the elements of the SFRS based on the flexibility of the diaphragm (see Section 6.3.3 of this publication).

Diaphragms also transfer gravity loads that are perpendicular to the diaphragm surface to floor or roof members (beams, joists, and columns). When diaphragms are properly attached to the top surface of horizontal framing members, they increase the flexural lateral stability of such members.

Illustrated in Figure 6.26 is the case of a diaphragm with an SFRS consisting of shear walls with collector beams. The horizontal seismic force is transferred through the web of the diaphragm to the shear walls, which act as supports for the diaphragm. Since the shear wall on the left does not extend the full depth of the diaphragm, collector beams are needed to collect the shear from the diaphragm and to transfer it to the wall. The chords are perpendicular to the seismic force at the top and bottom of the diaphragm and resist the tension and compression flexural forces that are induced in the diaphragm.

Figure 6.26
Diaphragm Force Distribution

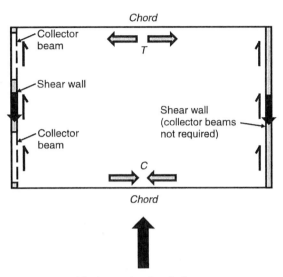

Depicted in Figure 6.27 is the same diaphragm system but now with an opening. In this case, sub-chord forces develop in the sub-diaphragms at the top and bottom of the opening. Collector elements on each side of the opening are required to collect the diaphragm shear into the sub-diaphragms.

Figure 6.27
Diaphragm Force Distribution with an Opening

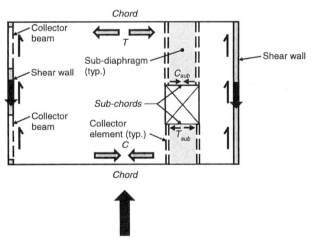

For both of the scenarios in Figures 6.26 and 6.27, it is important to keep in mind that seismic forces also act in the direction opposite to that shown; this must be considered in the design and detailing of the diaphragm and all other elements.

Diaphragm Design Forces

In structures assigned to SDC B and higher, floor and roof diaphragms must be designed to resist seismic forces from base shear determined from the structural analysis or the force, F_{px}, determined by Equation 12.10-1, whichever is greater:

$$0.4S_{DS}I_e w_{px} \geq F_{px} = \frac{\sum_{i=x}^{n} F_i}{\sum_{i=1}^{n} w_i} w_{px} \geq 0.2 S_{DS} I_e w_{px} \qquad (6.30)$$

In this equation, F_i is the design seismic force applied to level i, which is obtained from Equation 6.22, w_i is the weight tributary to level i, and w_{px} is the weight tributary to the diaphragm at level x.

In cases where there are offsets in the vertical SFRS or where there are changes in lateral stiffness of the vertical elements, the diaphragm must be designed for the force from Equation 6.30 plus the force due to the offset or change in lateral stiffness. Figure 6.28 illustrates the case of an offset in the vertical components of the SFRS.

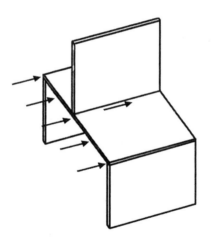

Figure 6.28
Example of Vertical Offsets in the Seismic-force-resisting System (12.10.1)

The redundancy factor, ρ, applies to the design of diaphragms in structures assigned to SDC D, E and F; it is equal to 1.0 for inertial forces calculated by Equation 6.30 (see Section 6.3.3 of this publication). Where the diaphragm transfers design forces from the vertical elements above the diaphragm to the vertical elements below the diaphragm, the redundancy factor, ρ, for the structure applies to these forces, thus completing the load path. Also, the requirements of 12.3.3.4 must be satisfied for structures having horizontal or vertical irregularities indicated in that section.

Collector Elements

The provisions of 12.10.2.1 apply to collector elements in structures assigned to SDC C and higher. As discussed previously, collectors, which are also commonly referred to as drag struts, are elements in a structure that are used to transfer diaphragm loads from the diaphragm to the elements of the SFRS where the lengths of the vertical elements in the SFRS are less than the length of the diaphragm at that location (see Figure 6.26).

In general, collector elements and their connections to the vertical elements must be designed to resist the maximum of the following:

1. Forces calculated using the seismic load effects including overstrength factor of 12.4.3 with seismic forces determined by the ELFP of 12.8 or the modal response spectrum analysis procedure of 12.9;

2. Forces calculated using the seismic load effects including overstrength factor of 12.4.3 with seismic forces determined by Equation 12.10-1 for diaphragms; or

3. Forces calculated using the seismic load combinations of 12.4.2.3 with seismic forces determined by Equation 12.10-2, which is the lower-limit diaphragm force.

The purpose of this overstrength requirement is to help keep inelastic behavior in the ductile elements of the SFRS and not in the collectors or their connections. It is essential that the collectors and their connections be able to perform as intended during a seismic event.

The maximum forces calculated by the three methods above need not exceed those that are calculated using the seismic load combinations of 12.4.2.3 with the seismic forces determined by Equation 12.10-3, which is the upper-limit diaphragm force. Also, in structures or portions of structures braced entirely by light-frame shear walls, collector elements and their connections need only be designed to resist forces using the seismic load combinations of 12.4.2.3 with the seismic forces determined in accordance with the diaphragm design forces of 12.10.1.1.

Additional information on the design and detailing of diaphragms, chords and collectors can be found in Reference 6.5.

6.3.11 Structural Walls and Their Anchorage

Design for Out-of-plane Forces

In addition to forces in the plane of the wall, structural walls and their anchorage in structures assigned to SDC B and higher must be designed for an out-of-plane force, F_p, equal to $0.4S_{DS}I_e$ times the weight of the structural wall or 0.10 times the weight of the structural wall, whichever is greater (12.11.1).

Since walls are often subjected to local deformations due to material shrinkage, temperature changes, and foundation settlement, structural wall elements and connections to supporting framing systems require some degree of ductility in order to accommodate these deformations while providing the required strength for combined gravity and seismic forces.

Nonstructural walls need not be designed for this requirement but must be designed in accordance with the seismic design for nonstructural components given in Chapter 13.

Anchorage of Concrete or Masonry Structural Walls

Wall Anchorage Forces

Concrete or masonry structural walls in structures assigned to SDC B and higher must be adequately anchored to the roof and floor members that provide lateral support for the walls. The anchorage of structural walls must provide a direct connection that is capable of resisting the following force:

$$F_p = 0.4S_{DS}k_aI_eW_p \geq 0.2k_aI_eW_p \qquad (6.31)$$

where

F_p = design force in the individual anchors

k_a = amplification factor for diaphragm flexibility = $1.0 + (L_f/100) \leq 2.0$

L_f = the span in feet of a flexible diaphragm that provides lateral support for the wall. The span is measured between vertical elements that provide lateral support to the diaphragm in the direction of analysis. The span L_f is equal to 0 for rigid diaphragms.

W_p = weight of the wall tributary to the anchor

The purpose of this requirement is to help prevent separation of concrete and masonry structural walls from the roof and floors. The force, F_p, applies only to the design of the anchorage or connection of the wall to the structure and not to the overall design of the wall. This force is to be applied for both tension (out-of-plane) and sliding (in-plane), where applicable (for example, where roof or floor framing is not perpendicular to the anchored walls).

The amplification factor, k_a, accounts for amplification of out-of-plane accelerations caused by diaphragm flexibility. This factor is equal to 1.0 for rigid diaphragms, which do not exhibit this type of amplification.

In a structure where all of the diaphragms are not flexible, the value of F_p determined by Equation 6.31 is permitted to be multiplied by the factor $(1 + 2z/h)/3$ where anchorage is not provided by the roof (that is, the wall cantilevers above its highest attachment to or near a higher level of the structure). The quantity z is the height of the anchor above the base of the structure and h is the height of the roof above the base. This reduction factor can be conservatively set equal to 1.0 to ensure that a smaller than appropriate anchorage force is not used in design.

Where the anchor spacing is greater than 4 feet along the length of the wall, the section of wall that spans between the anchors must be designed to resist the local out-of-plane bending caused by F_p.

Additional information on out-of-plane wall anchorage can be found in Reference 6.6.

Additional Requirements for Diaphragms in Structures Assigned to SDCs C through F

Additional requirements for diaphragms in structures assigned to SDC C through F are given in 12.11.2.2. The main purpose of these requirements is to ensure that a continuous load path exists that will survive the ground shaking.

6.3.12 Drift and Deformation

Story Drift Limit

Story drift must be controlled for both structural and nonstructural reasons. Limiting the drift helps limit inelastic strain in structural members and helps control overall structural stability by reducing P-delta effects. Drift control is also needed to restrict damage to nonstructural elements such as partitions, elevator and stair enclosures, and glass.

Once the design drifts have been determined in accordance with 12.8.6, they are compared to the allowable story drift, Δ_a, in Table 12.12-1. The drift limits depend on the risk category. Drift limits are generally more restrictive for Risk Categories III and IV and are meant to provide a higher level of performance for more important structures.

Drift limits also depend on the type of structure. For ordinary structures, the drift limit is 2 percent of the story height, which is approximately ten times the drift that is commonly permitted under wind loads. For low-rise structures (4 stories or less) where the interior walls, partitions, ceilings and exterior wall systems have been designed to accommodate story drifts, the drift limits are not as stringent as those for other types of structures.

Satisfying strength requirements may result in a structure that also satisfies story drift limits. Moment-resisting frames or tall, slender structures are often controlled by drift, and member sizes of the SFRS generally have to be increased to satisfy prescribed drift limits.

For structures assigned to SDC D, E or F with SFRSs consisting of only moment-resisting frames, the design story drift, Δ, must be less than or equal to Δ_a/ρ for any story where ρ is determined in accordance with 12.3.4.2 (see Section 6.3.3 of this publication). Given the inherent flexibility of moment-resisting systems, this provision essentially penalizes structures that utilize such systems in regions of high seismic risk that are not redundant.

Diaphragm Deflection

Elements that are attached to diaphragms must be able to maintain their structural integrity and to support any applicable loads as the diaphragm deflects in its own plane due to the seismic forces. Thus, the permissible in-plane deflection of a diaphragm is equal to the permissible deflections of the attached elements. Diaphragm deflections can be determined by a variety of methods based on the flexibility of the diaphragm.

Structural Separation

Portions of a structure or adjacent structures must have sufficient distance between them so that they can respond independently to ground motion without pounding into each other.

The maximum inelastic response displacement, δ_M, that occurs at the critical sections is used in determining separation distances (see Equation 12.12-1):

$$\delta_M = \frac{C_d \delta_{max}}{I_e} \quad (6.32)$$

where δ_{max} is the maximum elastic displacement at the critical location, which is determined based on the seismic story forces, F_x, and includes translational and torsional displacements of the structure. Where applicable, amplification of the accidental torsional deflections, as prescribed in 12.8.4.3, must also be included in the determination of δ_{max}.

Once δ_M has been calculated for each of the adjacent structures, the minimum separation distance between them, δ_{MT}, is determined by Equation 12.12-2:

$$\delta_{MT} = \sqrt{(\delta_{M1})^2 + (\delta_{M2})^2} \quad (6.33)$$

where δ_{M1} and δ_{M2} are the maximum inelastic response displacements of the adjacent structures at their adjacent edges.

Structures must be set back a minimum distance of δ_M from an adjoining property line that is not common to a public way.

Members Spanning between Structures

For members that are connected to adjacent structures or to seismically separated portions of structures, the gravity connections or supports of such members must be designed to accommodate the maximum anticipated relative displacements. These displacements are to be calculated based on all of the following:

1. The deflections at the locations of support are to be calculated using Equation 12.8-15 multiplied by $1.5R/C_d$.

2. Additional deflection due to diaphragm rotation (including the torsional amplification factor of 12.8.4.3 where applicable) must be included.

3. Diaphragm deformations must be included.

4. The two structures are assumed to be moving in opposite directions and the absolute sum of the displacements must be used.

Deformation Compatibility

Certain members or components in a structure are assigned to be part of the SFRS. This means that these elements must be designed for the combined effects due to gravity and seismic loads. During an earthquake, all of the structural members will be displaced regardless if they are part of the SFRS or not.

For structures assigned to SDC D through F, structural members that are not part of the SFRS must be designed for deformation compatibility. In particular, these members must be designed to adequately resist gravity load effects that are assigned to them when subjected to the design story drift, Δ, determined in accordance with 12.8.6. The story drifts essentially induce seismic load effects into these members that must be accounted for in design.

The stiffening effects of adjoining rigid structural and nonstructural elements must be considered in the design of the members that are not part of the SFRS. Interaction effects between moment-resisting frames that make up the SFRS and rigid elements that are not part of the SFRS that enclose or adjoin the moment-resisting frame are discussed in Section 6.3.7 of this publication.

6.3.13 Foundation Design

General requirements for the design of various types of foundation systems are given in 12.13 and Chapter 18 of the IBC. The special seismic requirements are driven by SDC.

For other than cantilever column and inverted pendulum systems, overturning effects at the soil-foundation interface are permitted to be reduced by 25 percent where the ELFP is used. A 10-percent reduction is permitted for foundations of structures designed in accordance with the modal analysis requirements of 12.9.

6.3.14 Simplified Alternative Structural Design Criteria for Simple Bearing Wall or Building Frame Systems

General

The simplified design procedure is entirely self-contained in 12.14. The provisions of 12.14 are intended to apply to a defined set of essentially regularly-configured structures where a reduction in requirements is deemed to be warranted. Only those SFRSs specifically listed in Table 12.14-1 are permitted to be used. Note that drift-controlled structural systems, such as moment-resisting frames, are not permitted.

The simplified method is permitted to be used in lieu of other analytical procedures in Chapter 12 for the analysis and design of simple bearing wall or building frame systems provided the 12 limitations of 12.14.1.1 are satisfied, which are summarized in Table 6.9. Even though there are 12 conditions to be met, it is important to note that the procedure is applicable to a wide range of relatively stiff, low-rise structures that fall under Risk Categories I and II and possess SFRSs that are arranged in a torsionally-resistant, regular layout.

Seismic Load Effects and Combinations

The seismic load effect, E, in the simplified method is defined the same as in the general requirements of 12.4 and consists of the effects of horizontal and vertical earthquake-induced forces, E_h and E_v, respectively (see Section 6.3.4 of this publication). However, in the simplified method, E_h is equal to Q_E, whereas in 12.4.2.1, E_h is equal to the redundancy factor, ρ, times Q_E. Thus, the redundancy factor, ρ, is equal to 1.0 in the simplified method. It is assumed that the structure possesses a reasonable level of redundancy (see the fifth limitation in Table 6.9).

Table 6.9 Limitations of the Simplified Seismic Procedure of ASCE/SEI 12.14

Number	Limitation
1	Risk Category I or II structures
2	Site Class A through D
3	Structure shall be less than or equal to 3 stories above grade
4	SFRS must be bearing wall or building frame
5	Two lines of walls or frames are required in each of two major axis directions
6	At least one line of walls or frames is required on each side of the center of mass in each direction
7	Equation 12.14-1 must be satisfied for structures with flexible diaphragms that have overhangs
8	For structures with diaphragms that are not flexible, (1) the distance between the center of rigidity and the center of mass parallel to each major axis must be less than or equal to 15 percent of the greatest width of the diaphragm parallel to that axis and (2) Equations 12.14-2A and 12.14-2B must be satisfied where applicable
9	Lines of walls or frames must be oriented at angles of no more than 15 degrees from the major orthogonal horizontal axes of the building
10	The simplified procedure must be used for each major orthogonal horizontal axis of the building
11	System irregularities caused by in-plane or out-of-plane offsets of lateral-force-resisting elements are not permitted
12	Lateral-load resistance of any story must be at least 80 percent of the story above

The seismic load combinations of 12.14 are the same as those in 12.4. In the simplified method, the system overstrength factor, Ω_0, is taken as 2.5, which is consistent with the values of Ω_0 for bearing wall and building frame systems in Table 12.2-1.

Seismic-force-resisting System

Parts of Table 12.2-1 are reproduced in Table 12.14-1 for the systems that are permitted to be designed by the simplified method. Contained in Table 12.14-1 are the sections of ASCE/SEI 7 pertaining to detailing requirements, the response modification coefficient, R, and system limitations for SDCs B through E. As noted above, the system overstrength factor, Ω_0, is taken as 2.5 for these systems, so it is not included in the table. Also, since drift calculations are not required (see 12.14.8.5 and the discussion below), the deflection amplification factor, C_d, is not in Table 12.14-1.

Combinations of framing systems are permitted horizontally and vertically. The value of R that is to be used must be the least value of any of the SFRSs in that direction.

Diaphragm Flexibility

Untopped metal deck, wood structural panels, and similar panelized construction are permitted to be considered flexible diaphragms (12.14.5), and lateral load is distributed to the vertical elements of the SFRS using tributary area (12.14.8.3.1).

For diaphragms that are not flexible, a simple rigidity analysis is required, which includes torsional moments resulting from eccentricity between the locations of center mass and center of rigidity (12.14.8.3.2). Analysis of accidental torsion and dynamic amplification of torsion is not required since the simplified method is applicable to regular structures with essentially a uniform distribution of lateral stiffness.

Application of Loading

Design seismic forces, which are prescribed in 12.14.8 (see discussion below), are permitted to be applied separately in each of the orthogonal directions. In other words, two separate analyses

386 Chapter 6

are acceptable, and the combination of load effects from the two directions need not be considered.

Design and Detailing Requirements

In order to achieve a continuous load path through a structure down to the foundation, all parts of the structure must be adequately interconnected. The requirements for interconnection in 12.14.7.1 are somewhat more stringent than those in 12.1.3, while the requirements for connections to supports are the same in 12.1.4 and 12.14.7.1. Foundation design requirements are the same in 12.14.7 and 12.1.5. Note that all of the design and detailing requirements are independent of the SDC.

Collector elements and their connections must be designed to resist the seismic load effects including a 2.5 overstrength factor. In structures that are braced entirely by light-frame shear walls, it is permitted to design collectors and their connections to resist the forces prescribed in 12.14.7.4 (see below).

Provisions for anchorage of structural walls to the roof and floors are given in 12.14.7.5. The equation to determine the design force in the individual anchors is the same as that in 12.11.2.1. The requirements in 12.14.7.6 for out-of-plane forces on exterior and interior structural walls are the same as those in 12.11.1.

Simplified Lateral Force Analysis Procedure

Seismic Base Shear

The seismic base shear, V, is determined by the equation in 12.14.8.1, which represents the horizontal short-period segment of the design response spectrum that is independent of the period of the structure (see Figure 6.18):

$$V = \frac{FS_{DS}}{R}W \qquad (6.34)$$

where $S_{DS} = 2F_a S_S /3$. In lieu of determining the acceleration-based site coefficient, F_a, in accordance with 11.4.3, which requires knowledge of the soil profile to a depth of 100 feet below the surface of the site, F_a is permitted to be taken as 1.0 for rock sites and 1.4 for soil sites. A site is considered to be a rock site where there is no more than 10 feet of soil between the rock surface and the bottom of the spread footing or mat foundation. By limiting applicability of the simplified design procedure to these sites, only a basic geotechnical investigation is needed; 100-foot-deep borings and seismic shear velocity tests are not necessary.

The response modification coefficient, R, is given in Table 12.14.-1, and W is the effective seismic weight (which is defined in ASCE/SEI 12.14.8.1 and is the same definition as in ASCE/SEI 12.7.2).

The factor F is related to the number of stories in the building and is equal to 1.0 for one-story buildings, 1.1 for two-story buildings, and 1.2 for three-story buildings. It is evident that the value of the seismic base shear is increased by 10 and 20 percent for two-story and three-story buildings, respectively. These increases primarily account for the method that is used for vertical distribution of the base shear, which is based on tributary weight, and have been shown by parametric studies to be adequate without being overly conservative (see below).

Vertical Distribution

The seismic force that is to be applied at each floor level, x, is determined by the equation in 12.14.8.2:

$$F_x = \frac{w_x}{W}V \qquad (6.35)$$

In this equation, w_x is equal to the portion of the effective seismic weight, W, at level x.

Horizontal Shear Distribution and Overturning

These requirements are the same as those for the ELFP. Included is the 25 percent reduction that can be applied when considering foundation overturning.

Drift Limits and Building Separation

According to 12.14.8.5, the simplified design procedure does not require a drift check because it is assumed that bearing wall and building frame systems within the prescribed height range will not require one (unlike moment frame systems, where drift is a major concern in design). Accordingly, calculations for P-delta effects need not be considered.

For requirements such as those for structural separations between buildings or the design of cladding, the allowable drift is to be taken as 1 percent of the building height.

Flowchart 6.9 in Section 6.6 of this publication provides a step-by-step procedure on how to determine the design seismic forces and their distribution based on the requirements of the simplified method.

6.4 Seismic Design Requirements for Nonstructural Components

6.4.1 General

Chapter 13 establishes minimum design criteria for nonstructural components that are permanently attached to structures and for their supports and attachments. Included are provisions for architectural components, mechanical and electrical components, and anchorage.

Nonstructural components that weigh greater than or equal to 25 percent of the effective seismic weight, W, of the structure must be designed as nonbuilding structures in accordance with 15.3.2 (13.1.1). Nonstructural components are assigned to the same SDC as the structure that they occupy or to which they are attached (13.1.2).

The component importance factor, I_p, is equal to 1.0 except when the following conditions apply where it is equal to 1.5 (13.1.3):

1. The component is required to function for life-safety purposes after an earthquake (including fire protection sprinkler systems and egress stairways).

2. The component conveys, supports or otherwise contains toxic, highly toxic or explosive substances in sufficient quantity to pose a hazard to the public if released.

3. The component is in or attached to a Risk Category IV structure and is needed for continued operation of the facility or its failure could impair the continued operation of the facility.

4. The component conveys, supports or otherwise contains hazardous substances and is attached to a structure or portion thereof that is classified as a hazardous occupancy.

A list of nonstructural components that are exempt from the requirements of this section is given in 13.1.4. It is assumed that these nonstructural components and systems can achieve the required performance goals either due to their inherent strength and stability, the lower level of earthquake demand, or both. Table 13.2-1 contains a summary of applicable requirements.

6.4.2 Seismic Demands on Nonstructural Components

Seismic Design Force

The horizontal seismic design force that is to be applied to the center of gravity of the component, F_p, is determined by Equation 13.3-1 with upper and lower limits determined by Equations 13.3-2 and 13.3-3, respectively:

$$0.3 S_{DS} I_p W_p \leq F_p = \frac{0.4 a_p S_{DS} W_p}{\left(\frac{R_p}{I_p}\right)}\left(1 + \frac{2z}{h}\right) \leq 1.6 S_{DS} I_p W_p \qquad (6.36)$$

where:

a_p = component amplification factor given in Table 13.5-1 for architectural components and Table 13.6-1 for mechanical and electrical components

S_{DS} = design spectral response acceleration at short periods determined in accordance with 11.4.4

W_p = component operating weight

R_p = component response modification factor given in Table 13.5-1 for architectural components and Table 13.6-1 for mechanical and electrical components

I_p = component importance factor (1.0 or 1.5)

z = height in structure of point of attachment of component with respect to the base (for items at or below the base, $z = 0$; the value of z/h need not exceed 1.0)

h = average roof height of structure with respect to the base

This equation represents a trapezoidal distribution of floor and roof accelerations within a structure where the accelerations vary linearly from the acceleration at the ground ($0.4 S_{DS}$ at $z = 0$) to the acceleration at the roof ($1.2 S_{DS}$ at $z = h$).

Component seismic forces, F_p, are to be applied independently in at least two orthogonal horizontal directions in combination with the service loads that are associated with the component. Equation 6.36 takes into consideration the dynamic and structural characteristics of nonstructural components, which typically means that component forces are larger than those for the design of the SFRS.

The component amplification factor, a_p, is a function of the fundamental periods of the structure and component. Amplification occurs when these two periods are closely matched. Except for long-period structures, the primary mode of vibration will have the most influence on a_p.

For vertically cantilevered systems, F_p shall be assumed to act in any direction and the component is to be designed for a vertical force equal to $\pm 0.2 S_{DS} W_p$ that acts concurrently with F_p.

In lieu of Equation 6.36, Equation 13.3-4 can be used to calculate F_p based on accelerations determined by a modal analysis in accordance with 12.9 with $R = 1.0$:

$$0.3 S_{DS} I_p W_p \leq F_p = \frac{a_i a_p S_{DS} W_p A_x}{\left(\frac{R_p}{I_p}\right)} \leq 1.6 S_{DS} I_p W_p \qquad (6.37)$$

In this equation, a_i is the acceleration at level i obtained from the modal analysis, and A_x is the torsional amplification factor determined by Equation 12.8-14 (see Section 6.3.8 of this publication).

Seismic Relative Displacements

The requirements for seismic relative displacements that are given in 13.3.2 are applicable in the design of cladding, stairways, windows, piping systems, sprinkler systems and other components that are connected either to (1) one structure at multiple levels or (2) multiple structures.

In the first case, the seismic relative displacement, D_{pI}, is equal to $(\delta_{xA} - \delta_{yA})I_e$ where δ_{xA} and δ_{yA} are the displacements corresponding to connection points located at height h_x and h_y in structure A, respectively. This relative displacement must be accommodated in the design of the component and connection points to the structure. The terms in parentheses need not be taken greater than $(h_x - h_y)\Delta_{aA}/h_{sx}$ where Δ_{aA} is the allowable structure drift for structure A in accordance with Table 12.12-1 and h_{sx} is the story height that is used in determining the drift.

Illustrated in Figure 6.29 is a glazing system that must be designed to accommodate the seismic relative displacements. In lieu of this requirement, 13.3.2.1 also describes a modal analysis procedure that can be used.

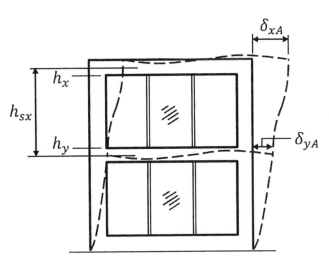

Figure 6.29
Seismic Relative Displacements within a Structure

In the second case, the seismic relative displacement, D_{pI}, is equal to $(|\delta_{xA}| + |\delta_{yB}|)I_e$ where δ_{xA} and δ_{yB} are the displacements corresponding to a connection point at height h_x in structure A and a connection point at height h_y in structure B, respectively. The terms in parentheses need not exceed $(h_x\Delta_{aA}/h_{sx}) + (h_y\Delta_{aB}/h_{sx})$ where Δ_{aA} and Δ_{aB} are the allowable story drifts given in Table 12.12-1 for structures A and B, respectively, and h_{sx} is the story height that is used in determining the allowable drifts. Illustrated in Figure 6.30 is a piping system that connects two structures, which must be designed to accommodate the seismic relative displacement for both cases.

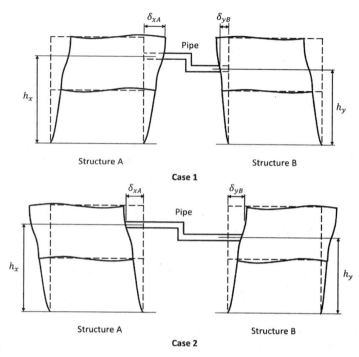

Figure 6.30
Seismic Relative Displacements between Structures

6.4.3 Nonstructural Component Anchorage

Components and their supports are to be attached or anchored to the structure in accordance with the requirements of 13.4. Forces in the attachments are to be determined using the prescribed forces and displacements specified in 13.3.1 and 13.3.2. Anchors in concrete or masonry elements must satisfy the provisions of 13.4.2.

6.4.4 Architectural Components

General design and detailing requirements for architectural components are contained in 13.5. All architectural components and attachments must be designed for the seismic forces defined in 13.3.1. Specific requirements are stipulated for:

- Exterior nonstructural wall elements and connections
- Glass
- Suspended ceilings
- Access floors
- Partitions
- Glass in glazed curtain walls, glazed storefronts and glazed partitions

For structures assigned to Risk Category I, II or III, the intent of these requirements is to reduce life-safety hazards and property damage posed by the loss of stability or integrity of architectural components. In the case of structures assigned to Risk Category IV, the additional intent is to reduce the potential disruption of essential functions due to failure of architectural components.

6.4.5 Mechanical and Electrical Components

The requirements of 13.6 are to be satisfied for mechanical and electrical components and their supports.

Equation 13.6-1 can be used to determine the fundamental period, T_p, of the mechanical or electrical component.

Requirements are provided for the following systems:

- Utility and service lines
- Ductwork
- Piping systems
- Boilers and pressure vessels
- Elevators and escalators

Flowchart 6.10 in Section 6.6 of this publication provides step-by-step procedures on how to determine design seismic forces on nonstructural components.

6.5 Seismic Design Requirements for Nonbuilding Structures

6.5.1 General

Nonbuilding structures supported by the ground or by other structures must be designed and detailed to resist the minimum seismic forces set forth in Chapter 15.

The selection of a structural analysis procedure for a nonbuilding structure is based on its similarity to buildings. Nonbuilding structures that are similar to buildings exhibit behavior similar to that of building structures; however, their function and performance are different. According to 15.1.3, structural analysis procedures for such buildings are to be selected in accordance with 12.6 and Table 12.6-1, which are applicable to building structures. Guidelines and recommendations on the use of these methods are given in C15.1.3. In short, the provisions for building structures need to be carefully examined before they are applied to nonbuilding structures.

Nonbuilding structures that are not similar to buildings exhibit behavior that is markedly different than that of building structures. Most of these types of structures have reference documents that address their unique structural performance and behavior. Such reference documents are permitted to be used to analyze the structure (15.1.3). In addition, the following procedures may be used: equivalent lateral force procedure (12.8), modal analysis procedure (12.9), linear response history analysis procedure (16.1) and nonlinear response history analysis procedure (16.2). In the case of nonbuilding structures similar to buildings, guidelines and recommendations on the proper analysis method to utilize for nonbuilding structures that are not similar to buildings are given in C15.1.3.

6.5.2 Reference Documents

As noted above, reference documents may be used to design nonbuilding structures for earthquake load effects. References that have seismic requirements based on the same force and displacement levels used in ASCE/SEI 7 are listed in Chapter 23 (15.2). The provisions in the reference documents are subject to the amendments given in 15.4.1. See C15.2 for additional references that cannot be referenced directly by ASCE/SEI 7.

6.5.3 Nonbuilding Structures Supported by Other Structures

Provisions are given in 15.3 for the nonbuilding structures in Table 15.4-2 (that is, nonbuilding structures that are not similar to buildings) that are supported by other structures and that are not part of the primary SFRS. The design method depends on the weight of the nonbuilding structure relative to the weight of the combined nonbuilding and supporting structure (see

15.3.1 and 15.3.2). Where the weight of the nonbuilding structure is less than 25 percent of the combined effective seismic weights of the nonbuilding structure and the supporting structure, the design seismic forces for the nonbuilding structure is to be determined in accordance with the nonstructural component requirements in Chapter 13 with R_p and a_p determined by 13.1.5. The appropriate requirements in Chapter 12 or 15.5 are to be used in the design of the supporting structure.

Where the weight of the nonbuilding structure is greater than or equal to 25 percent of the combined effective seismic weights of the nonbuilding structure and the supporting structure, an analysis that combines the structural characteristics of both structures must be performed, which is based on the fundamental period, T, of the nonbuilding structure:

1. Where $T < 0.06$ second:

    a. The nonbuilding system is considered a rigid element.

    b. The supporting structure shall be designed using the appropriate requirements of Chapter 12 or 15.5.

    c. The R-value of the combined system is permitted to be taken as the R-value of the supporting system.

    d. The nonbuilding structure shall be designed in accordance with Chapter 13 where R_p shall be taken equal to R of the nonbuilding structure given in Table 15.4-2 and a_p shall be taken as 1.0.

2. Where $T \geq 0.06$ second:

    a. The nonbuilding structure and supporting structure shall be modeled together in a combined model with appropriate stiffness and effective seismic weight distributions.

    b. The combined system shall be designed in accordance with 15.3 with the R-value of the combined system taken as the lesser of the R-value of the nonbuilding structure or the supporting structure.

6.5.4 Structural Design Requirements

Specific design requirements for nonbuilding structures are given in 15.4. As noted previously, provisions in referenced documents are amended by the requirements of this section.

For nonbuilding structures that are similar to buildings, the permitted structural systems, design values and limitations are given in Table 15.4-1. Similar information is provided in Table 15.4-2 for nonbuilding structures that are not similar to buildings. Requirements on the determination of the base shear, vertical distribution of seismic forces, importance factor and load combinations are contained in 15.4.1.

Additional provisions for rigid buildings, loads, fundamental period, drift limitations, deflection limits and other requirements can be found in 15.4.2 through 15.4.8.

6.5.5 Nonbuilding Structures Similar to Buildings

Additional requirements are given in 15.5 for pipe racks, steel storage racks, electrical power generating facilities, structural towers for tanks and vessels and piers and wharves.

6.5.6 Nonbuilding Structures Not Similar to Buildings

Additional requirements are given in 15.6 for earth-retaining structures, stacks and chimneys, amusement structures, special hydraulic structures, secondary containment systems and telecommunication towers.

6.5.7 Tanks and Vessels

Comprehensive seismic design requirements are given in 15.7 for tanks and vessels. As noted in C15.7, most, if not all, industry standards that contain seismic design requirements are based on earlier seismic codes. Many of the provisions of 15.7 show how to modify existing standards to get to the same force levels as ASCE/SEI 7.

Flowchart 6.11 in Section 6.6 of this publication provides a step-by-step procedure on how to determine design seismic forces on nonbuilding structures that are similar to buildings and that are not similar to buildings.

6.6 Flowcharts

A summary of the flowcharts provided in this chapter is given in Table 6.10. Included is a description of the content of each flowchart.

All referenced section numbers and equations in the flowcharts are from ASCE/SEI 7 unless noted otherwise.

Table 6.10 Summary of Flowcharts Provided in Chapter 6

Flowchart	Title	Description
Section 6.6.1 Seismic Design Criteria		
Flowchart 6.1	Consideration of Seismic Design Requirements	Summarizes conditions where the seismic requirements of ASCE/SEI 7 must be considered and need not be considered.
Flowchart 6.2	Site Classification Procedure for Seismic Design	Provides step-by-step procedure on how to determine the site class of a site in accordance with Chapter 20.
Flowchart 6.3	Seismic Ground Motion Values	Provides step-by-step procedure on how to determine the design spectral accelerations for a site in accordance with 11.4.
Flowchart 6.4	Seismic Design Category	Provides step-by-step procedure on how to determine the seismic design category of a structure.
Flowchart 6.5	Design Requirements for SDC A	Summarizes the seismic design requirements for structures assigned to SDC A.
Section 6.6.2 Seismic Design Requirements for Building Structures		
Flowchart 6.6	Diaphragm Flexibility	Provides methods on how to determine whether a diaphragm is flexible, rigid or semirigid.
Flowchart 6.7	Permitted Analytical Procedures	Summarizes analytical procedures that are permitted in determining design seismic forces for building structures.
Flowchart 6.8	Equivalent Lateral Force Procedure	Provides step-by-step procedure on how to determine the design seismic forces and their distribution based on the requirements of this procedure.
Flowchart 6.9	Alternate Simplified Design Procedure	Provides step-by-step procedure on how to determine the design seismic forces and their distribution based on the requirements of this procedure.
Section 6.6.3 Seismic Design Requirements for Nonstructural Components		
Flowchart 6.10	Seismic Demands on Nonstructural Components	Provides step-by-step procedure on how to determine design seismic forces on nonstructural components.
Section 6.6.4 Seismic Design Requirements for Nonbuilding Structures		
Flowchart 6.11	Seismic Design Requirements for Nonbuilding Structures	Provides step-by-step procedure on how to determine design seismic forces on nonbuilding structures that are similar to buildings and that are not similar to buildings.

6.6.1 Seismic Design Criteria

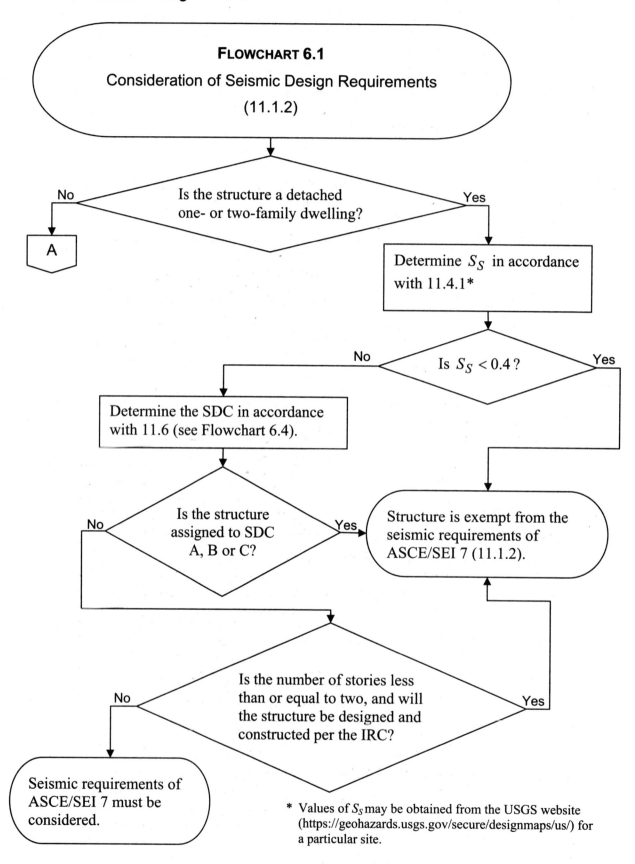

Earthquake Loads 395

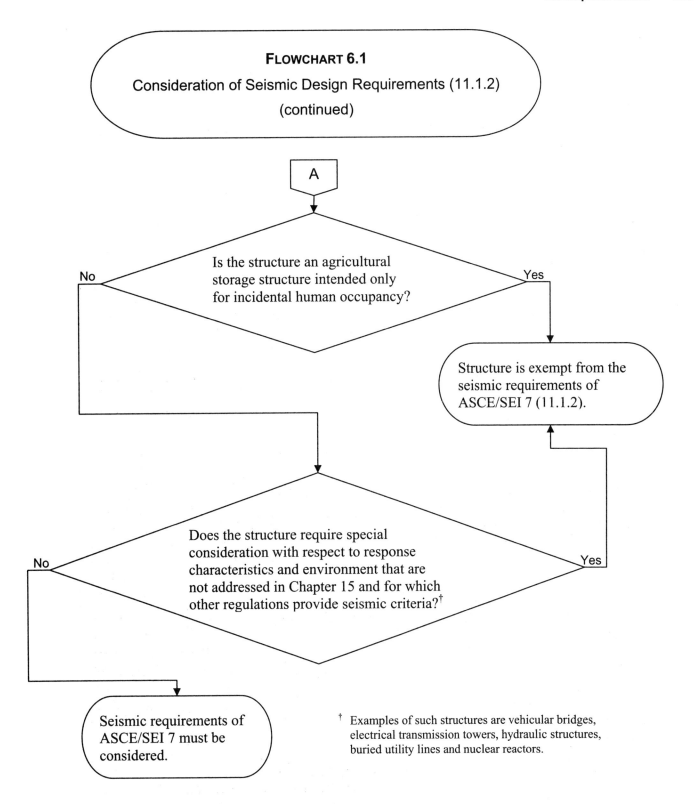

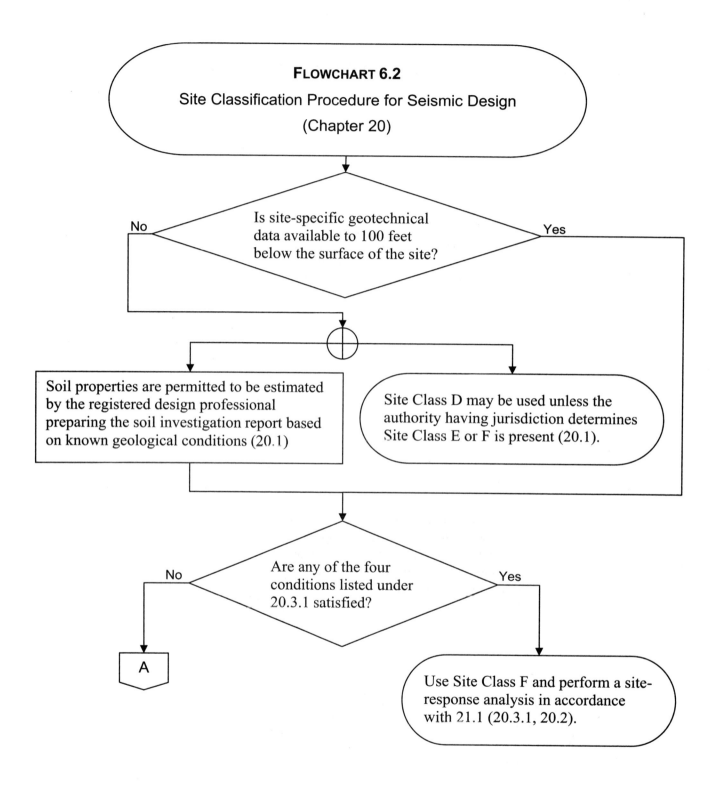

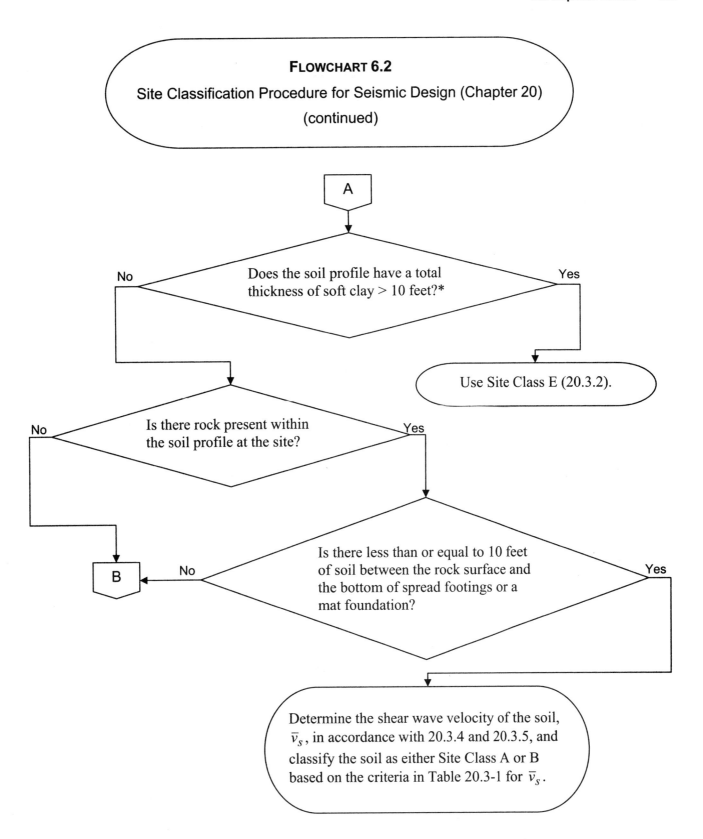

* A soft clay layer is defined by $s_u < 500$ psf, $w \geq 40$ percent, and $PI > 20$ (20.3.2).

398 Chapter 6

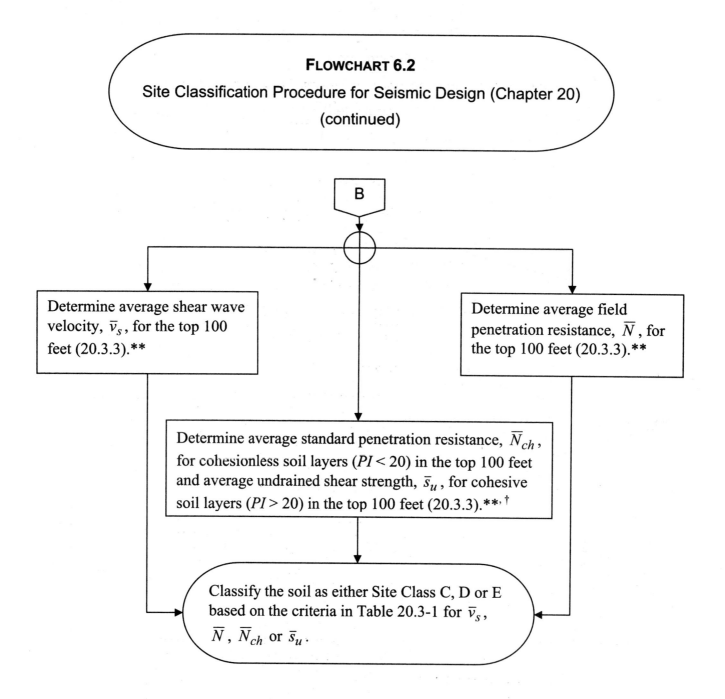

FLOWCHART 6.2

Site Classification Procedure for Seismic Design (Chapter 20)

(continued)

B

Determine average shear wave velocity, $\bar{v}_s$, for the top 100 feet (20.3.3).**

Determine average field penetration resistance, $\bar{N}$, for the top 100 feet (20.3.3).**

Determine average standard penetration resistance, $\bar{N}_{ch}$, for cohesionless soil layers ($PI < 20$) in the top 100 feet and average undrained shear strength, $\bar{s}_u$, for cohesive soil layers ($PI > 20$) in the top 100 feet (20.3.3).**,†

Classify the soil as either Site Class C, D or E based on the criteria in Table 20.3-1 for $\bar{v}_s$, $\bar{N}$, $\bar{N}_{ch}$ or $\bar{s}_u$.

** Values of $\bar{v}_s$, $\bar{N}$ and $\bar{s}_u$ are computed in accordance with 20.4 where soil profiles contain distinct soil and rock layers (20.3.3).

† Where the $\bar{N}_{ch}$ and $\bar{s}_u$ criteria differ, the site shall be assigned to the category with the softer soil [20.3.3(3)].

Earthquake Loads 399

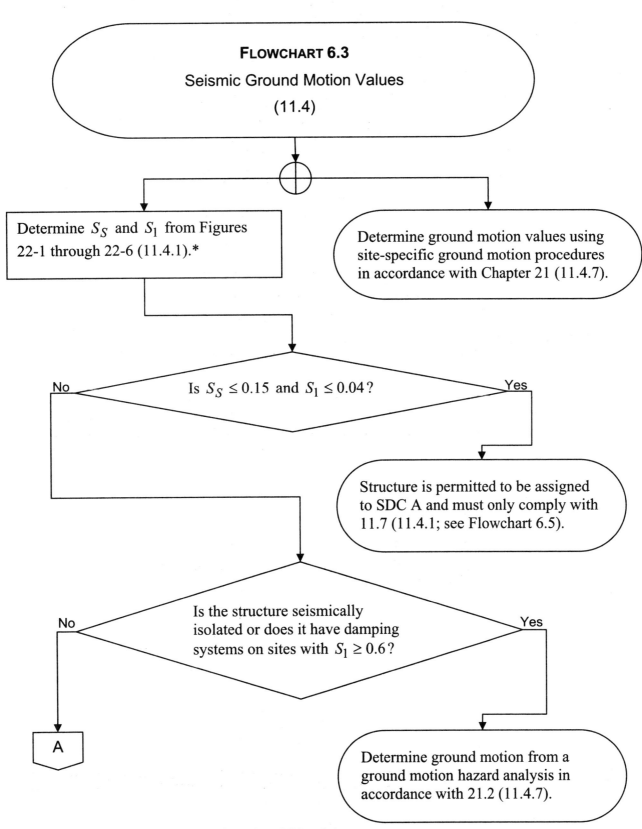

FLOWCHART 6.3
Seismic Ground Motion Values
(11.4)

* Values of S_S and S_1 may be obtained from the USGS website (https://geohazards.usgs.gov/secure/designmaps/us/) for a particular site.

FLOWCHART 6.3

Seismic Ground Motion Values (11.4)

(continued)

Determine the site class of the soil in accordance with 11.4.2 and Chapter 20 (11.4.2; see Flowchart 6.2).

Is the site classified as Site Class F?

Yes: Determine ground motion values using a site response analysis in accordance with 21.1 (11.4.7).**

No: Determine S_{MS} and S_{M1} by Eqs. 11.4-1 and 11.4-2, respectively:

$$S_{MS} = F_a S_S$$
$$S_{M1} = F_v S_1$$

where site coefficients F_a and F_v are given in Tables 11.4-1 and 11.4-2, respectively (11.4.3).†

Determine S_{DS} and S_{D1} by Eqs. 11.4-3 and 11.4-4, respectively:†

$$S_{DS} = 2S_{MS}/3$$
$$S_{D1} = 2S_{M1}/3 \quad (11.4.4)$$

** A site response analysis in accordance with 21.1 is required for structures on Site Class F sites unless the exception in 20.3.1(1) is satisfied for structures with periods $T \leq 0.5$ second.
† Where the simplified design procedure of 12.14 is used, only the values of F_a and S_{DS} must be determined in accordance with 12.14.8.1 (11.4.3, 11.4.4).

Earthquake Loads 401

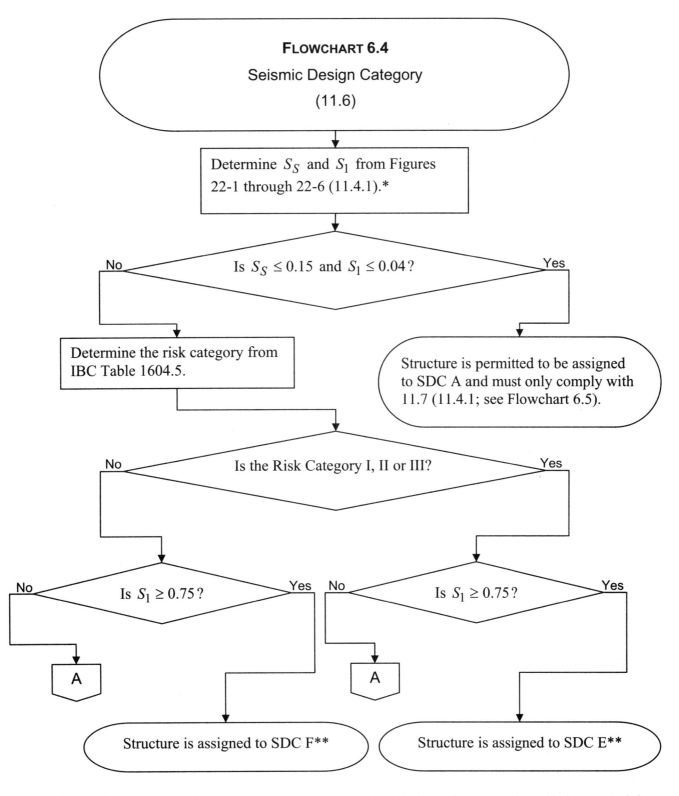

* Values of S_S and S_1 may be obtained from the USGS website (https://geohazards.usgs.gov/secure/designmaps/us/) for a particular site.
** A structure assigned to SDC E or F shall not be located where there is a known potential for an active fault to cause rupture of the ground surface at the structure (11.8).

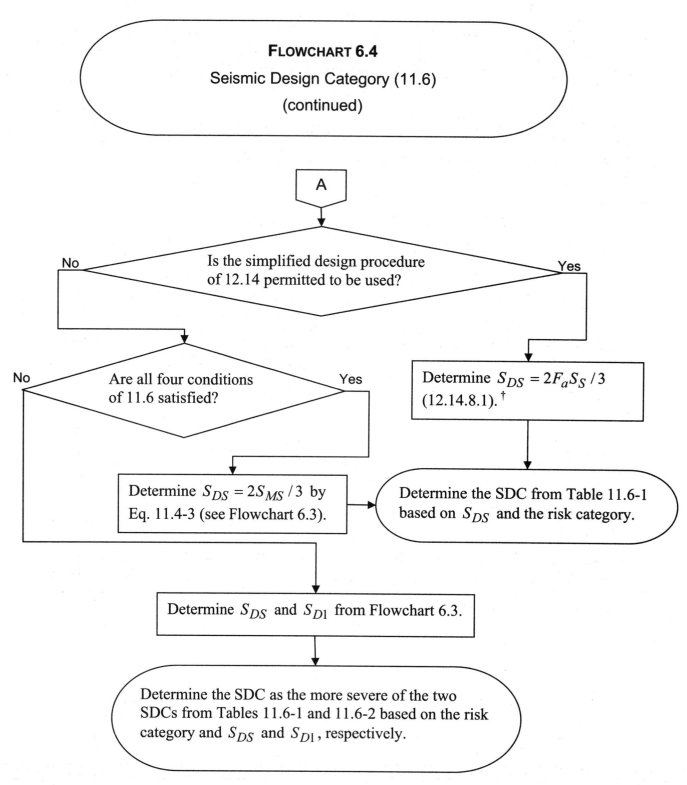

FLOWCHART 6.4

Seismic Design Category (11.6)

(continued)

† Short-period site coefficient, F_a, is permitted to be taken as 1.0 for rock sites, 1.4 for soil sites or may be determined in accordance with 11.4.3. Rock sites have no more than 10 feet of soil between the rock surface and the bottom of spread footing or mat foundation. Mapped spectral response acceleration, S_S, is determined in accordance with 11.4.1 and need not be taken larger than 1.5 (12.14.8.1).

FLOWCHART 6.5

Design Requirements for SDC A

(11.7)

↓

Determine the seismic force, F_x, applied at each floor level by Eq. 1.4-1: $F_x = 0.01W_x$ where W_x = portion of the total dead load of the structure located or assigned to level x (11.7, 1.4.3).*

↓

Provide load path connections and connection to supports in accordance with 1.4.2 and 1.4.4, respectively.

↓

Anchor any concrete or masonry walls to roof and floors in accordance with 1.4.5.

---

* These forces are applied simultaneously at all levels in one direction. The structure is analyzed for the effects of these forces applied independently in each of two orthogonal directions (1.4.3).

404   Chapter 6

6.6.2 Seismic Design Requirements for Building Structures

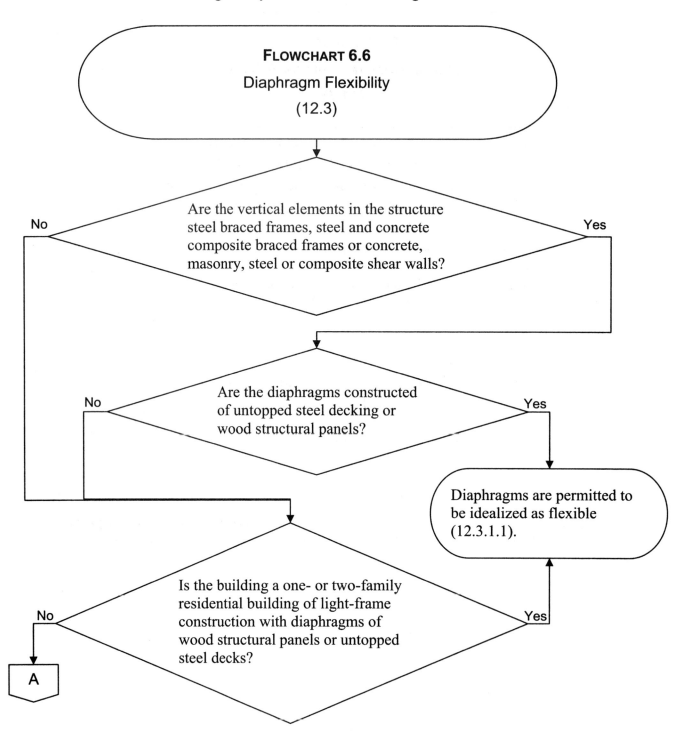

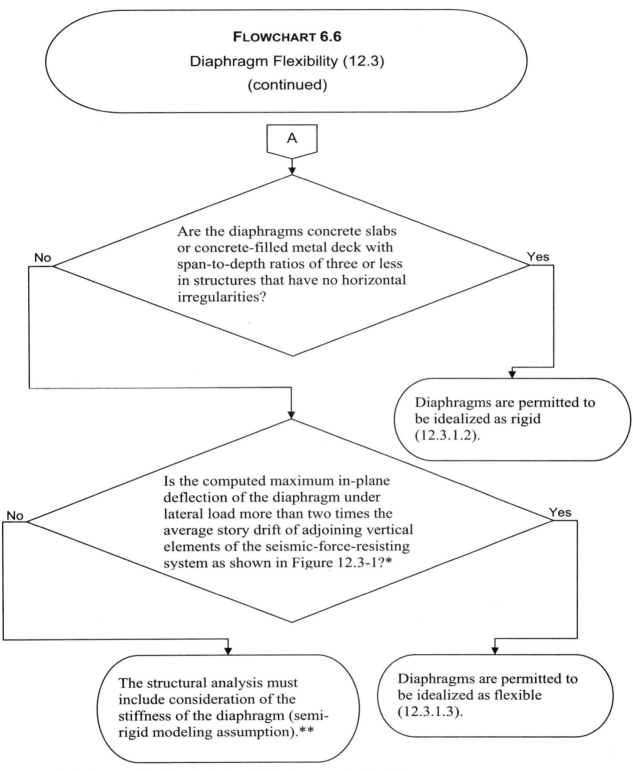

* The loading used for this calculation shall be that prescribed in 12.8.

** IBC 202 provides definitions for rigid and flexible diaphragms only; semirigid diaphragms are not defined in the IBC.

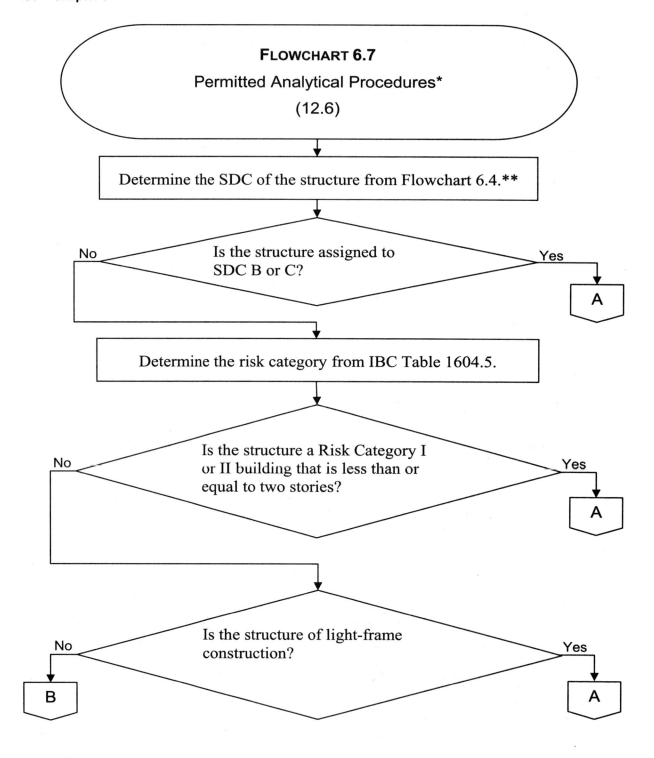

* The simplified alternative structural design method of 12.14 may be used for simple bearing wall or building frame systems that satisfy the 12 limitations in 12.14.1.1.

** This flowchart is applicable to buildings assigned to SDC B and higher. See Flowchart 6.5 for design requirements for SDC A.

Earthquake Loads 407

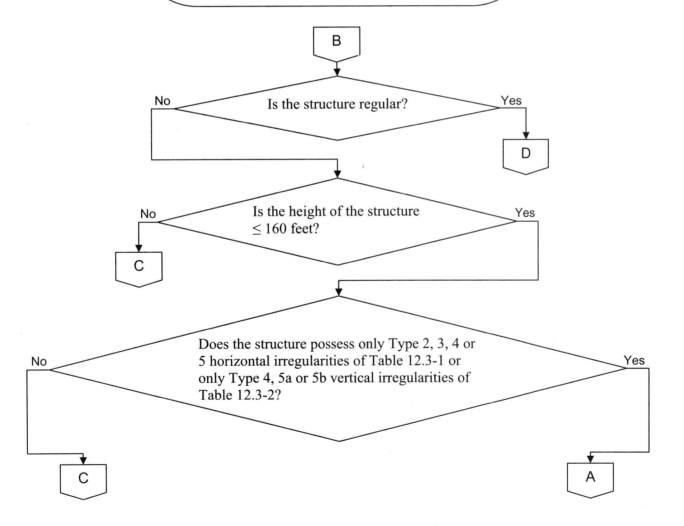

FLOWCHART 6.7
Permitted Analytical Procedures (12.6)
(continued)

408   Chapter 6

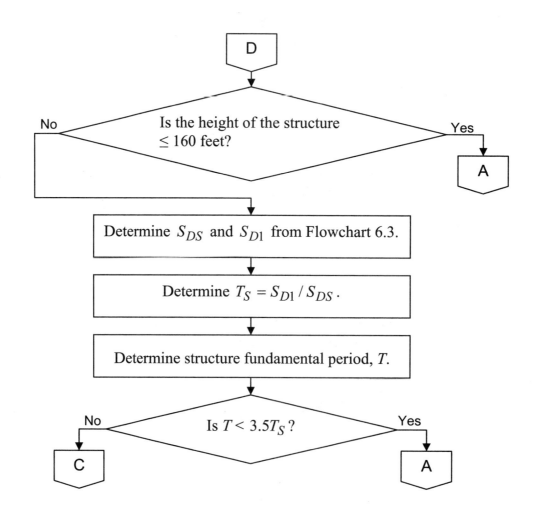

FLOWCHART 6.8
Equivalent Lateral Force Procedure
(12.8)

Determine S_S, S_1, S_{DS}, S_{D1} and the SDC from Flowcharts 6.3 and 6.4.

↓

Determine the response modification coefficient, R, from Table 12.2-1 for the appropriate structural system based on SDC.

↓

Determine the importance factor, I_e, from Table 11.5-1 based on the risk category.

↓

Left branch:

Determine the fundamental period of the structure, T, by a substantiated analysis that considers the structural properties and deformational characteristics of the structure.

↓

Determine the approximate fundamental period of the structure, T_a, by Eq. 12.8-7: $T_a = C_t h_n^x$ where values of approximate period parameters C_t and x are given in Table 12.8-2.*,**

↓

A

Right branch:

Determine the approximate fundamental period of the structure, T_a, by Eq. 12.8-7: $T_a = C_t h_n^x$ where values of approximate period parameters C_t and x are given in Table 12.8-2.*,**

↓

B

---

* h_n = height in feet above the base to the highest level of the structure.

** Alternate equations for T_a are given in 12.8.2.1 for concrete or steel moment resisting frames and masonry or concrete shear wall structures (see Table 6.8).

410 Chapter 6

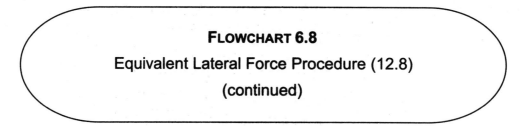

FLOWCHART 6.8
Equivalent Lateral Force Procedure (12.8)
(continued)

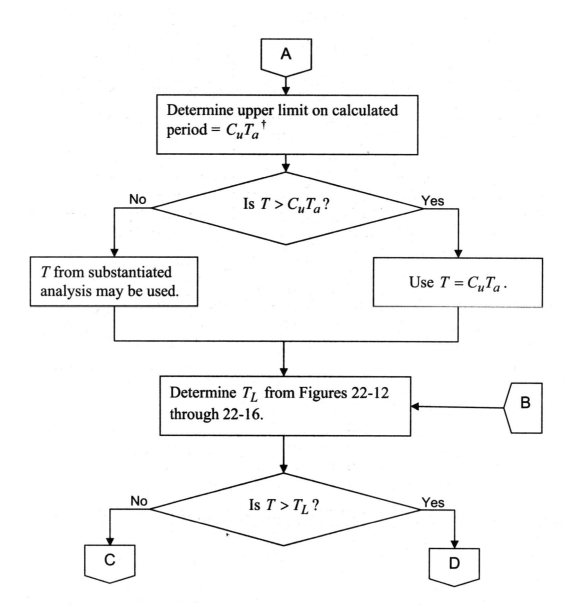

† C_u = coefficient for upper limit on calculated period given in Table 12.8-1.

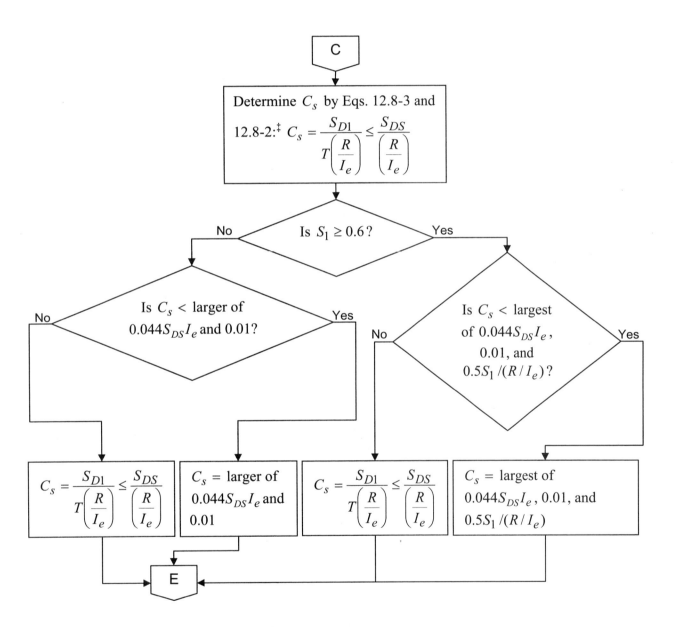

FLOWCHART 6.8

Equivalent Lateral Force Procedure (12.8)

(continued)

‡ For regular structures five stories or less in height and having a period, T, less than or equal to 0.5 second, C_s is permitted to be calculated using a value of 1.5 for S_S (12.8.1.3).

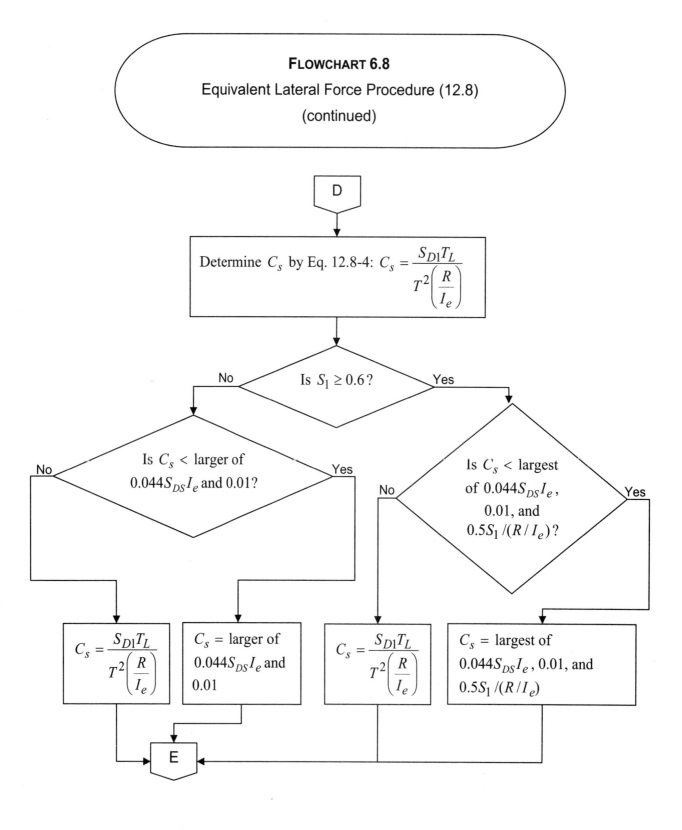

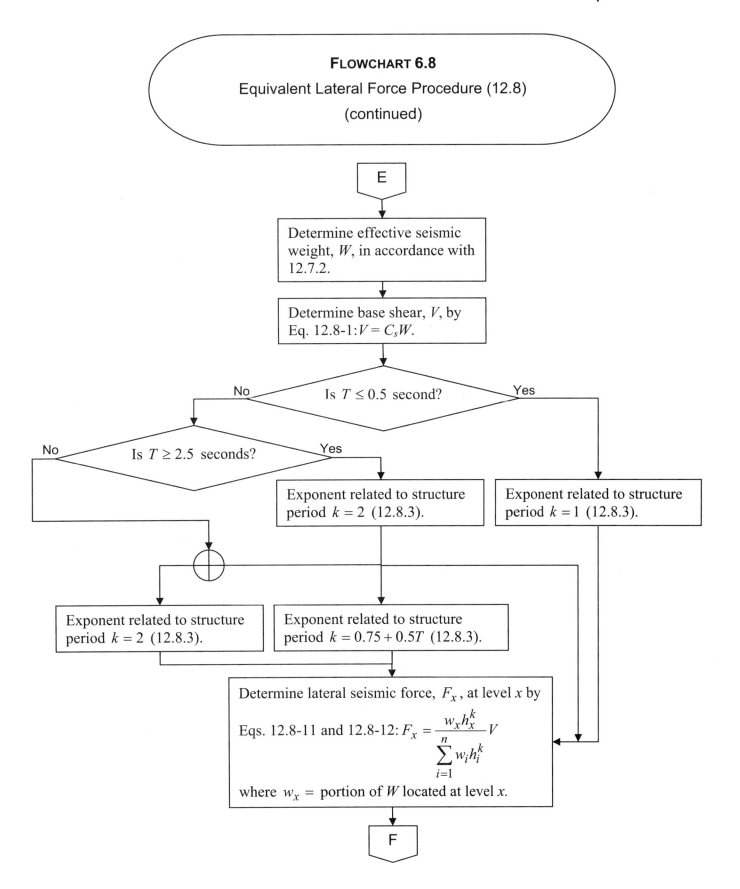

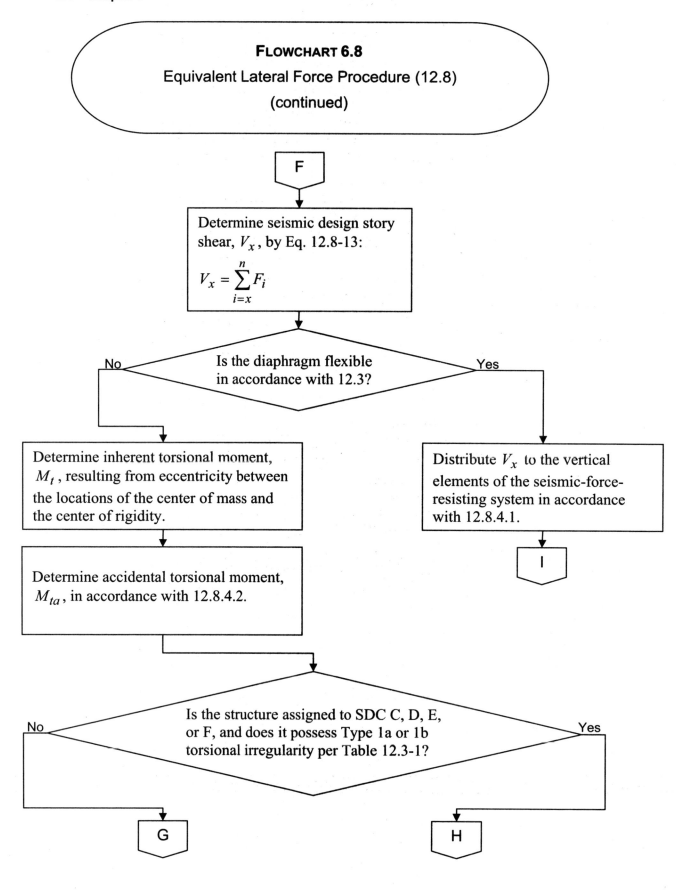

FLOWCHART 6.8

Equivalent Lateral Force Procedure (12.8)

(continued)

Distribute V_x to the vertical elements of the seismic-force-resisting system considering the relative lateral stiffness of the vertical resisting elements and the diaphragm, including $M_t + M_{ta}$.

Determine the torsional amplification factor A_x by Eq. 12.8-14: $A_x = \left(\dfrac{\delta_{max}}{1.2\delta_{avg}}\right)^2 \leq 3$

where δ_{max} and δ_{avg} are defined in 12.8.4.3.

Determine $M'_{ta} = A_x M_{ta}$.

Distribute V_x to the vertical elements of the seismic-force-resisting system considering the relative lateral stiffness of the vertical resisting elements and the diaphragm, including $M_t + M'_{ta}$.

 Design structure to resist overturning effects caused by the seismic forces, F_x (12.8.5).

Determine the deflection amplification factor, C_d, from Table 12.2-1.

Determine the deflection, δ_x, at levels x by Eq. 12.8-15: $\delta_x = \dfrac{C_d \delta_{xe}}{I_e}$ where δ_{xe} are the deflections at level x based on an elastic analysis of the structure subjected to the seismic forces, F_x.[††]

[††] It is permitted to determine the δ_{xe} using seismic design forces based on the computed fundamental period of the structure without the upper limit $C_u T_a$ specified in 12.8.2 (12.8.6.2).

FLOWCHART 6.8
Equivalent Lateral Force Procedure (12.8)
(continued)

J

Determine the design story drift, Δ, as the difference of the deflections, δ_x, at the center of mass at the top and bottom of the story under consideration (see Figures 12.8-2 and 6.22).

Check that the allowable story drifts, Δ_a, given in Table 12.12-1 are satisfied at each story.

Ensure that the other applicable drift and deformation requirements of 12.12 are satisfied.

Determine stability coefficient, θ, at each story by Eq. 12.8-16:
$$\theta = \frac{P_x \Delta I_e}{V_x h_{sx} C_d}$$
where P_x = total vertical design load at and above level x (P_x is determined using load factors no greater than 1.0), and h_{sx} = story height below level x.

Is $\theta \leq 0.10$?

No → Determine θ_{max} by Eq. 12.8-17:
$$\theta_{max} = \frac{0.5}{\beta C_d}$$
where β = ratio of shear demand to shear capacity for the story.‡‡

K

Yes → P-delta effects need not be considered on the structure (12.8.7).

‡‡ β can conservatively be taken as 1.0. Where P-delta effects are included in an automated analysis, the value of θ computed by Eq. 12.8-16 is permitted to be divided by $(1 + \theta)$ before checking Eq. 12.8-17 (12.8.7).

Earthquake Loads 417

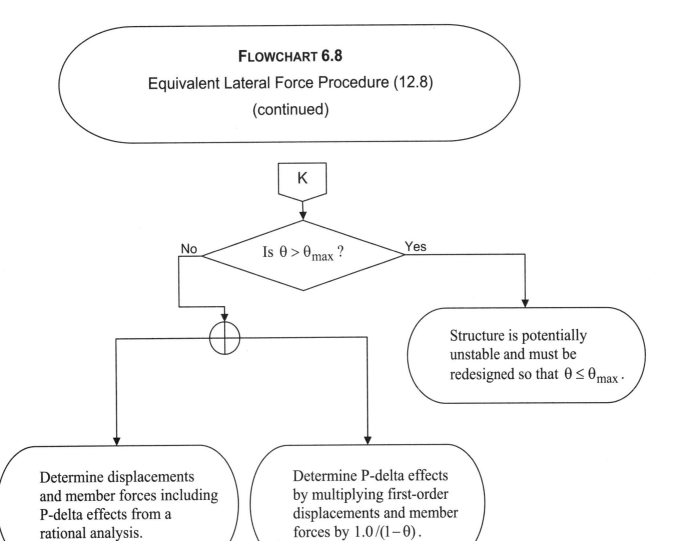

FLOWCHART 6.9
Alternate Simplified Design Procedure
(12.14)

↓

Are the 12 limitations of 12.14.1.1 satisfied?

- **No** → Use another analytical procedure in Chapter 12.
- **Yes** ↓

Determine S_S, S_{DS} and the SDC from Flowcharts 6.3 and 6.4.

↓

Determine the response modification coefficient, R, from Table 12.14-1 for the appropriate structural system based on SDC.

↓

Determine effective seismic weight, W, in accordance with 12.14.8.1.

↓

Determine base shear, V: $V = \dfrac{FS_{DS}W}{R}$ where

    $F = 1.0$ for one-story buildings
       $= 1.1$ for two-story buildings
       $= 1.2$ for three-story buildings

  $S_{DS} = 2F_a S_S / 3$

  $F_a = 1.0$ for rock sites
      $= 1.4$ for soil sites, or
      $=$ value determined in accordance with 11.4.3

↓

A

FLOWCHART 6.9
Alternate Simplified Design Procedure (12.14)
(continued)

A

Determine lateral seismic force, F_x, at level x:

$F_x = \dfrac{w_x}{W} V$ where w_x = portion of W located at level x.

Determine seismic design story shear, V_x:

$V_x = \sum\limits_{i=x}^{n} F_i$

Is the diaphragm flexible in accordance with 12.14.5?

No: Determine inherent torsional moment, M_t, resulting from eccentricity between the locations of the center of mass and the center of rigidity (12.14.8.3.2.1).

Distribute V_x to the vertical elements of the seismic-force-resisting system based on relative stiffness of the vertical elements and the diaphragm, including M_t.

Yes: Distribute V_x to the vertical elements of the seismic-force-resisting system based on tributary area (12.14.8.3.1).

B

FLOWCHART 6.9
Alternate Simplified Design Procedure (12.14)
(continued)

B

Design structure to resist overturning effects caused by the seismic forces, F_x (12.14.8.4).

Foundations of structures shall be designed for not less than 75 percent of the foundation overturning design moment.

Structural drift need not be calculated. If drift is required for other design requirements, it shall be taken as 1 percent of building height unless it is computed to be less (12.14.8.5).

6.6.3 Seismic Design Requirements for Nonstructural Components

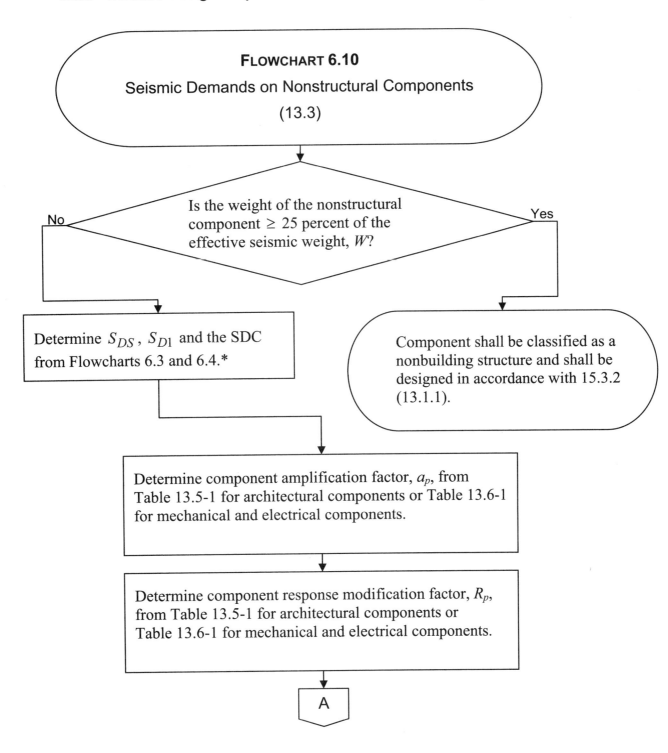

*Nonstructural components shall be assigned to the same SDC as the structure that they occupy or to which they are attached (13.1.2).

FLOWCHART 6.10

Seismic Demands on Nonstructural Components (13.3)

(continued)

Determine component importance factor, I_p, in accordance with 13.1.3.

Determine horizontal seismic design force, F_p, applied at component's center of gravity by Eqs. 13.3-1, 13.3-2 and 13.3-3:**

$$0.3 S_{DS} I_p W_p \leq F_p = \frac{0.4 a_p S_{DS} W_p}{\left(\dfrac{R_p}{I_p}\right)} \left(1 + 2\frac{z}{h}\right) \leq 1.6 S_{DS} I_p W_p$$

where W_p = component operating weight

z = height in structure of point of attachment of component with respect to the base.†

h = average roof height of structure with respect to the base

---

** F_p shall be applied independently in at least two orthogonal horizontal directions in combination with service loads. For vertically cantilevered systems, F_p shall be assumed to act in any horizontal direction, and the component shall be designed for a concurrent vertical force $\pm 0.2 S_{DS} W_p$. Redundancy factor, ρ, is permitted to be taken equal to 1.0, and the overstrength factor, Ω_0, does not apply (13.3.1).

† For items at or below the base, z shall be taken as zero. The value of z/h need not exceed 1.0.

6.6.4 Seismic Design Requirements for Nonbuilding Structures

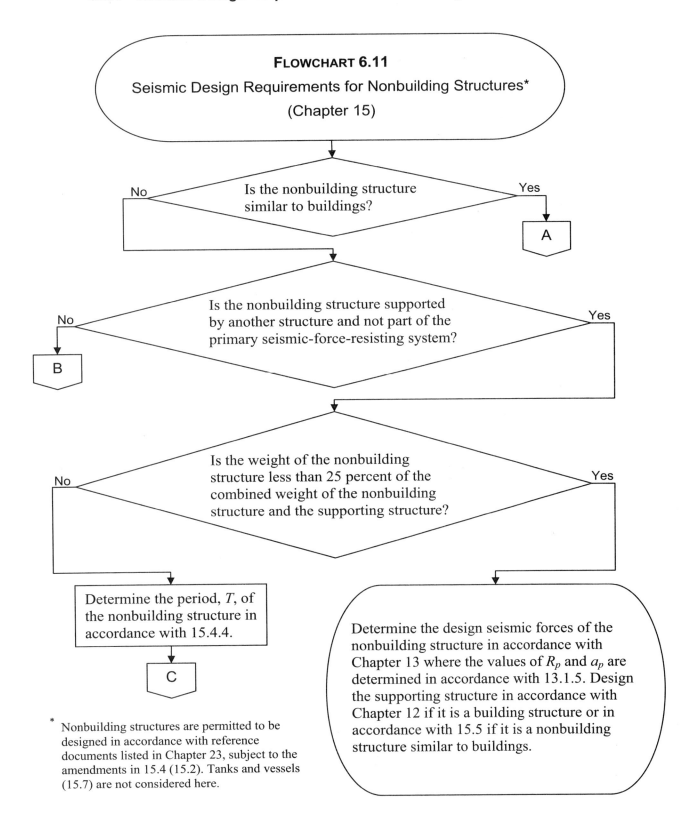

424 Chapter 6

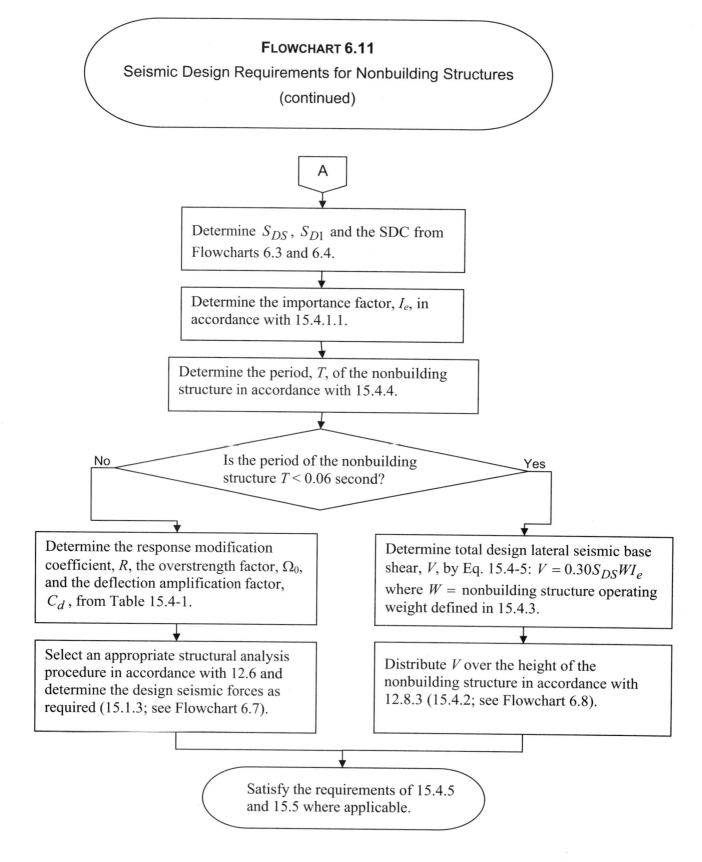

Earthquake Loads 425

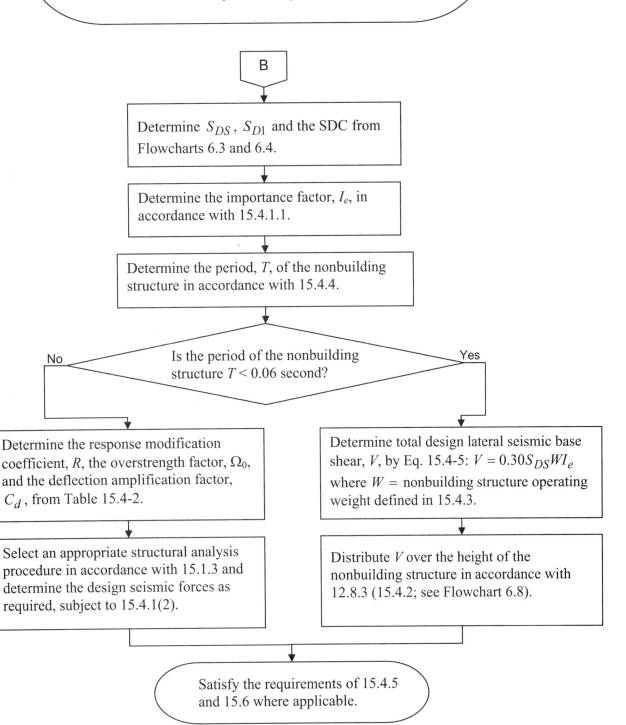

FLOWCHART 6.11
Seismic Design Requirements for Nonbuilding Structures
(continued)

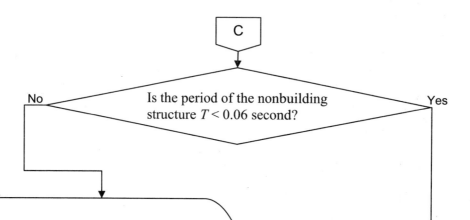

6.7 Examples

The following examples illustrate the IBC and ASCE/SEI 7 requirements for seismic design loads.

6.7.1 Example 6.1 – Residential Building, Seismic Design Category

A typical floor plan and elevation of a 12-story residential building is depicted in Figure 6.31. Given the design data below, determine the seismic design category (SDC).

Figure 6.31 Typical Floor Plan and Elevation of 12-story Residential Building, Example 6.1

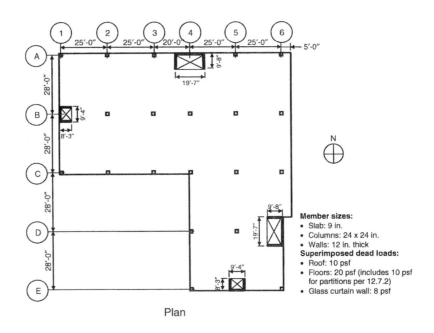

Member sizes:
- Slab: 9 in.
- Columns: 24 x 24 in.
- Walls: 12 in. thick

Superimposed dead loads:
- Roof: 10 psf
- Floors: 20 psf (includes 10 psf for partitions per 12.7.2)
- Glass curtain wall: 8 psf

Plan

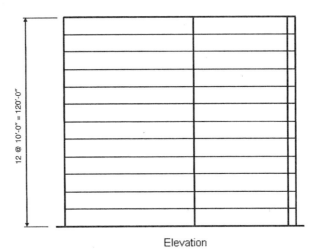

Elevation

DESIGN DATA

Location:	Charleston, SC (Latitude: 32.74°, Longitude: −79.93°)
Soil classification:	Site Class D
Occupancy:	Residential occupancy where less than 300 people congregate in one area
Material:	Cast-in-place, reinforced concrete
Structural system:	Building frame system

SOLUTION

Step 1: Determine the seismic ground motion values from Flowchart 6.3.

1. Determine the mapped accelerations S_S and S_1.

    In lieu of using Figures 22-1 and 22-2, the mapped accelerations are determined by inputting the latitude and longitude of the site into the USGS "DesignMaps" Web Application. The output is as follows: $S_S = 1.05$ and $S_1 = 0.33$.

2. Determine the site class of the soil.

    The site class of the soil is given in the design data as Site Class D.

3. Determine soil-modified accelerations S_{MS} and S_{M1}.

    Site coefficients F_a and F_v are determined from Tables 11.4-1 and 11.4-2, respectively:

    For Site Class D and $1.0 < S_S < 1.25$: $F_a = 1.08$ from linear interpolation

    For Site Class D and $0.3 < S_1 < 0.4$: $F_v = 1.73$ from linear interpolation

    Thus,

    $S_{MS} = 1.08 \times 1.05 = 1.13$

    $S_{M1} = 1.73 \times 0.33 = 0.57$

4. Determine design accelerations S_{DS} and S_{D1}.

    From Equations 11.4-3 and 11.4-4:

    $S_{DS} = \frac{2}{3} \times 1.13 = 0.75$

    $S_{D1} = \frac{2}{3} \times 0.57 = 0.38$

Step 2: Determine the SDC from Flowchart 6.4.

1. Determine if the building can be assigned to SDC A in accordance with 11.4.1.

    Since $S_S = 1.05 > 0.15$ and $S_1 = 0.33 > 0.04$, the building cannot be automatically assigned to SDC A.

2. Determine the risk category from IBC Table 1604.5.

    For a residential occupancy where less than 300 people congregate in one area, the Risk Category is II.

3. Since $S_1 < 0.75$, the building is not assigned to SDC E or F.

4. Check if all four conditions of 11.6 are satisfied.

    Check if the approximate period, T_a, is less than $0.8 T_S$.

    Use Equation 12.8-7 with approximate period parameters for "other structural systems:"

    $T_a = C_t h_n^x = 0.02(120)^{0.75} = 0.73$ second

    where C_t and x are given in Table 12.8-2.

    $T_S = S_{D1} / S_{DS} = 0.38 / 0.75 = 0.51$ second

    0.73 second $> 0.8 \times 0.51 = 0.41$ second

Since this condition is not satisfied, the SDC cannot be determined by Table 11.6-1 alone (11.6).

5. Determine the SDC from Tables 11.6-1 and 11.6-2.

    From Table 11.6-1, with $S_{DS} > 0.50$ and Risk Category II, the SDC is D.

    From Table 11.6-2, with $S_{D1} > 0.20$ and Risk Category II, the SDC is D.

Therefore, the SDC is D for this building.

6.7.2 Example 6.2 – Residential Building, Permitted Analytical Procedure

For the 12-story residential building in Example 6.1, determine the analytical procedure that can be used in calculating the seismic forces.

SOLUTION

Use Flowchart 6.7 to determine the permitted analytical procedure.

1. Determine the SDC from Flowchart 6.4.

    It was determined in Step 2 of Example 6.1 that the SDC is D.

2. Determine S_{DS} and S_{D1} from Flowchart 6.3.

    The design accelerations were determined in Step 1, Item 4 of Example 6.1: $S_{DS} = 0.75$ and $S_{D1} = 0.38$.

3. Determine T_S.

    $T_S = S_{D1} / S_{DS}$ was determined in Step 2, Item 4 of Example 6.1 as 0.51 second.

4. Determine the fundamental period of the building, T.

    It was determined in Step 2, Item 4 of Example 6.1 that $T = T_a = 0.73$ second.

5. Check if $T < 3.5T_S$.

    $T = 0.73$ second $< 3.5T_S = 3.5 \times 0.51 = 1.8$ seconds

6. Determine if the structure is regular or not.

    a. Determine if the structure has any horizontal structural irregularities in accordance with Table 12.3-1.

       i. Torsional irregularity

          In accordance with Table 12.3-1, Type 1a torsional irregularity and Type 1b extreme torsional irregularity for rigid or semirigid diaphragms exist where the ratio of the maximum story drift at one end of a structure to the average story drifts at two ends of the structure exceeds 1.2 and 1.4, respectively. The story drifts are to be determined using code-prescribed forces, including accidental torsion. In this example, the floors and roof are cast-in-place reinforced concrete slabs, which are considered to be rigid diaphragms (12.3.1.2).

At this point in the analysis, it is obviously not evident which method is required to be used to determine the prescribed seismic forces. In lieu of using a more complicated higher order analysis, the equivalent lateral force procedure may be used to determine the lateral seismic forces. These forces are applied to the building and the subsequent analysis yields the story drifts, Δ, which are used in determining whether Type 1a or 1b torsional irregularity exists. The results from the equivalent lateral force procedure will be needed if it is subsequently determined that a modal analysis is required (see 12.9.4).

Use Flowchart 6.8 to determine the lateral seismic forces from the equivalent lateral force procedure.

a. The design accelerations and the SDC have been determined in Example 6.1.

b. Determine the response modification coefficient, R, from Table 12.2-1.

   The walls in this building frame system must be special reinforced concrete shear walls, since the building is assigned to SDC D (system B4 in Table 12.2-1). In this case, $R = 6$. Note that the height of the building, which is 120 feet, is less than the limiting height of 160 feet for this type of system in SDC D. The increased building height limit of 12.2.5.4 is not considered in this example.

c. Determine the importance factor, I_e, from Table 1.5-2.

   For Risk Category II, $I_e = 1.0$.

d. Determine the period of the structure, T.

   It was determined in Step 2, Item 4 of Example 6.1 that the approximate period of the structure, T_a, which is permitted to be used in the equivalent lateral force procedure, is equal to 0.73 second.

e. Determine long-period transition period, T_L, from Figure 22-12.

   For Charleston, SC, $T_L = 8$ seconds $> T_a = 0.73$ second.

f. Determine seismic response coefficient, C_S.

   The seismic response coefficient, C_S, is determined by Equation 12.8-3:

   $$C_S = \frac{S_{D1}}{T(R/I_e)} = \frac{0.38}{0.73(6/1.0)} = 0.09$$

   The value of C_S need not exceed that from Equation 12.8-2:

   $$C_S = \frac{S_{DS}}{R/I_e} = \frac{0.75}{6/1.0} = 0.13$$

   Also, C_S must not be less than the larger of $0.044 S_{DS} I_e = 0.03$ (governs) and 0.01 (Equation 12.8-5).

   Thus, the value of C_S from Equation 12.8-3 governs.

g. Determine effective seismic weight, W, in accordance with 12.7.2.

   The member sizes and superimposed dead loads are given in Figure 6.31 and the effective weights at each floor level are given in Table 6.11. The total weight, W, is the summation of the effective dead loads at each level.

Table 6.11 Seismic Forces and Story Shears

Level	Story weight, w_x (kips)	Height, h_x (feet)	$w_x h_x^k$	Lateral force, F_x (kips)	Story shear, V_x (kips)
R	1,308	120	278,799	233	233
11	1,692	110	327,160	274	507
10	1,692	100	294,036	246	753
9	1,692	90	261,308	219	972
8	1,692	80	229,014	192	1,164
7	1,692	70	197,202	165	1,329
6	1,692	60	165,932	139	1,468
5	1,692	50	135,284	113	1,581
4	1,692	40	105,368	88	1,669
3	1,692	30	76,344	64	1,733
2	1,692	20	48,479	41	1,774
1	1,692	10	22,305	19	1,793
Σ	19,920		2,141,231	1,793	

h. Determine seismic base shear, V.

Seismic base shear is determined by Equation 12.8-1:

$V = C_S W = 0.09 \times 19,920 = 1,793$ kips

i. Determine exponent related to structure period, k.

Since 0.5 second < T = 0.73 second < 2.5 seconds, k is determined as follows:

$k = 0.75 + 0.5T = 1.12$

j. Determine lateral seismic force, F_x, at each level, x.

F_x is determined by Equations 12.8-11 and 12.8-12. A summary of the lateral forces, F_x, and the story shears, V_x, is given in Table 6.11.

Three-dimensional analyses were performed independently in the N-S and E-W directions for the seismic forces in Table 6.11 using a commercial computer program. In the model, rigid diaphragms were assigned at each floor level. The stiffness properties of the shear walls were input assuming cracked sections (12.7.3): $I_{eff} = 0.5 I_g$ where I_g is the gross moment of inertia of the section. In accordance with 12.8.4.2, the center of mass was displaced each way from its actual location a distance equal to 5 percent of the building dimension perpendicular to the applied forces to account for accidental torsion in seismic design.

A summary of the elastic displacements, δ_{xe}, at each end of the building in both the N-S and E-W directions due to the code-prescribed forces in Table 6.11 is given in Table 6.12 at all floor levels. Also provided in the table are the story drifts, Δ, at each end of the building in both directions.

According to Table 12.3-1, a torsional irregularity occurs where maximum story drift at one end of the structure is greater than 1.2 times the average of the story drifts at the two ends of the structure. The average story drift, Δ_{avg}, and the ratio of the maximum story drift to the average story drift, $\Delta_{max}/\Delta_{avg}$, are also provided in Table 6.12.

Table 6.12 Lateral Displacements and Story Drifts Due to Seismic Forces

Story	N-S Direction						E-W Direction					
	$(\delta_{xe})_1$ (in.)	Δ_1 (in.)	$(\delta_{xe})_2$ (in.)	Δ_2 (in.)	Δ_{avg} (in.)	$\dfrac{\Delta_{max}}{\Delta_{avg}}$	$(\delta_{xe})_1$ (in.)	Δ_1 (in.)	$(\delta_{xe})_2$ (in.)	Δ_2 (in.)	Δ_{avg} (in.)	$\dfrac{\Delta_{max}}{\Delta_{avg}}$
12	10.79	1.16	5.72	0.60	0.88	1.32	7.29	0.75	6.20	0.65	0.70	1.07
11	9.63	1.18	5.12	0.60	0.89	1.33	6.54	0.77	5.55	0.66	0.72	1.07
10	8.45	1.17	4.52	0.61	0.89	1.32	5.77	0.76	4.89	0.66	0.71	1.07
9	7.28	1.17	3.91	0.61	0.89	1.32	5.01	0.77	4.23	0.66	0.72	1.07
8	6.11	1.13	3.30	0.59	0.86	1.31	4.24	0.76	3.57	0.65	0.71	1.07
7	4.98	1.09	2.71	0.58	0.84	1.30	3.48	0.74	2.92	0.62	0.68	1.09
6	3.89	1.02	2.13	0.54	0.78	1.31	2.74	0.69	2.30	0.58	0.64	1.08
5	2.87	0.90	1.59	0.49	0.70	1.29	2.05	0.63	1.72	0.53	0.58	1.09
4	1.97	0.78	1.10	0.42	0.60	1.30	1.42	0.56	1.19	0.46	0.51	1.10
3	1.19	0.61	0.68	0.34	0.48	1.27	0.86	0.43	0.73	0.40	0.42	1.04
2	0.58	0.41	0.34	0.24	0.33	1.24	0.43	0.31	0.37	0.26	0.29	1.07
1	0.17	0.17	0.10	0.10	0.14	1.21	0.12	0.12	0.11	0.11	0.12	1.00

For example, at the 12th story in the N-S direction:

$\Delta_1 = 10.79 - 9.63 = 1.16$ inches

$\Delta_2 = 5.72 - 5.12 = 0.60$ inch

$$\Delta_{avg} = \frac{1.16 + 0.60}{2} = 0.88 \text{ inch}$$

$$\frac{\Delta_{max}}{\Delta_{avg}} = \frac{1.16}{0.88} = 1.32 > 1.2$$

Therefore, a Type 1a torsional irregularity exists at all floor levels in the N-S direction.

A Type 1b extreme torsional irregularity does not exist since the ratio of maximum drift to average drift is less than 1.4 at all floor levels (see Table 12.3-1 and Figure 6.7).

According to 12.8.4.3, where torsional irregularity exists at floor level x, the accidental torsional moments, M_{ta}, must be increased by the torsional amplification factor, A_x, given by Equation 12.8-14:

$$A_x = \left(\frac{\delta_{max}}{1.2\delta_{avg}}\right)^2$$

For example, at the 12th story in the N-S direction:

$$A_{12} = \left[\frac{10.79}{1.2\left(\dfrac{10.79 + 5.72}{2}\right)}\right]^2 = 1.19 > 1.0$$

ii. Reentrant corner irregularity

   According to Table 12.3-1 and Figure 6.8, a reentrant corner irregularity exists where both plan projections of the structure beyond a reentrant corner are greater than 15 percent of the plan dimension in a given direction:

   Projection $b = 56$ feet $> 0.15a = 0.15 \times 112 = 16.8$ feet

   Projection $d = 70$ feet $> 0.15c = 0.15 \times 120 = 18.0$ feet

   Therefore, a Type 2 reentrant corner irregularity exists.

iii. Diaphragm discontinuity irregularity

   This type of irregularity does not exist, since the area of any of the openings is much less than 50 percent of the area of the diaphragm. Also, the diaphragm has the same effective stiffness on all of the floor levels.

iv. Out-of-plane offsets irregularity

   There are no out-of-plane offsets of the shear walls, so this irregularity does not exist.

v. Nonparallel systems irregularity

   This discontinuity does not exist, since all of the shear walls are parallel to a major orthogonal axis of the building.

b. Determine if the structure has any vertical structural irregularities in accordance with Table 12.3-2.

   By inspection, none of the vertical irregularities defined in Table 12.3-2 exist for this building (also see Figures 6.12 through 6.16).

In summary, the building is not regular and has the following horizontal irregularities: Type 1a torsional irregularity and Type 2 reentrant corner irregularity.

7. Determine the permitted analytical procedure from Table 12.6-1.

The structure has a height of 120 feet (< 160 feet) and is irregular with $T < 3.5T_S$. If the structure had only a Type 2 reentrant corner irregularity, the equivalent lateral force procedure could be used to analyze the structure. However, since a Type 1a torsional irregularity also exists, the equivalent lateral force procedure is not permitted; either a modal response spectrum analysis (12.9) or a seismic response history procedure (Chapter 16) must be utilized.

The reference sections in Table 12.3-1 must also be satisfied for these types of irregularities in structures assigned to SDC D. Design forces shall be increased 25 percent for connections of diaphragms to vertical elements and to collectors and for connection of collectors to the vertical elements (12.3.3.4). Members that are not part of the SFRS must satisfy the deformational compatibility requirements of 12.12.4.

6.7.3 Example 6.3 – Office Building, Seismic Design Category

Typical floor plans and elevations of a seven-story office building are depicted in Figure 6.32. Given the design data below, determine the seismic design category (SDC).

Figure 6-32a
Typical Floor Plans and Elevations of Seven-story Office Building, Example 6.3

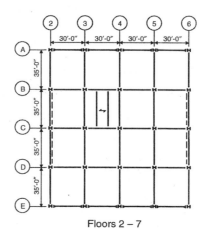

Floors 2 – 7

Member sizes:
- Metal deck: 3 in. deck + 2.5 in. lightweight concrete (39 psf)

Superimposed dead loads:
- Roof: 10 psf
- Floors: 20 psf (includes 10 psf for partitions per 12.7.2).
- Glass curtain wall: 8 psf

– – – Braces
▶ Moment connections

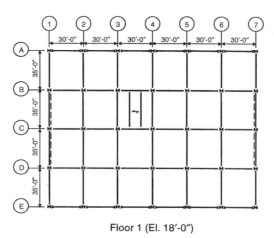

Floor 1 (El. 18'-0")

Figure 6-32b
Typical Floor Plans and Elevations of Seven-story Office Building, Example 6.3 (continued)

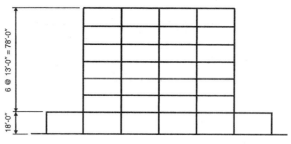

North or South Elevation

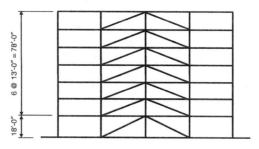

East or West Elevation

Design Data

Location:	Memphis, TN (Latitude: 35.13°, Longitude: –90.05°)
Soil classification:	Site Class D
Occupancy:	Office occupancy where less than 300 people congregate in one area
Material:	Structural steel
Structural system:	Moment-resisting frame and building frame systems

Solution

Step 1: Determine the seismic ground motion values from Flowchart 6.3.

1. Determine the mapped accelerations S_S and S_1.

   In lieu of using Figures 22-1 and 22-2, the mapped accelerations are determined by inputting the latitude and longitude of the site into the USGS "DesignMaps" Web Application. The output is as follows: $S_S = 0.99$ and $S_1 = 0.34$.

2. Determine the site class of the soil.

   The site class of the soil is given in the design data as Site Class D.

3. Determine soil-modified accelerations S_{MS} and S_{M1}.

   Site coefficients, F_a and F_v, are determined from Tables 11.4-1 and 11.4-2, respectively:

   For Site Class D and $0.75 < S_S < 1.0$: $F_a = 1.11$ from linear interpolation

   For Site Class D and $0.3 < S_1 < 0.4$: $F_v = 1.72$ from linear interpolation

   Thus,

   $S_{MS} = 1.11 \times 0.99 = 1.10$

   $S_{M1} = 1.72 \times 0.34 = 0.59$

4. Determine design accelerations S_{DS} and S_{D1}.

   From Equations 11.4-3 and 11.4-4:

   $S_{DS} = \frac{2}{3} \times 1.10 = 0.73$

   $S_{D1} = \frac{2}{3} \times 0.59 = 0.39$

Step 2: Determine the SDC from Flowchart 6.4.

1. Determine if the building can be assigned to SDC A in accordance with 11.4.1.

   Since $S_S = 0.99 > 0.15$ and $S_1 = 0.34 > 0.04$, the building cannot be automatically assigned to SDC A.

2. Determine the risk category from IBC Table 1604.5.

   For a business occupancy where less than 300 people congregate in one area, the Risk Category is II.

3. Since $S_1 < 0.75$, the building is not assigned to SDC E or F.

4. Check if all four conditions of 11.6 are satisfied.

   Check if the approximate period, T_a, is less than $0.8T_S$.

   In the N-S direction, the concentrically braced steel frames fall under "other structural systems" in Table 12.8-2. Using Equation 12.8-7, T_a is determined as follows:

   $$T_a = C_t h_n^x = 0.02(96)^{0.75} = 0.61 \text{ second}$$

   In the E-W direction, values of C_t and x for steel moment-resisting frames are used from Table 12.8-2:

   $$T_a = C_t h_n^x = 0.028(96)^{0.8} = 1.1 \text{ seconds}$$

   $$T_S = S_{D1}/S_{DS} = 0.39/0.73 = 0.53 \text{ second}$$

   $0.8 \times 0.53 = 0.43$ second is less than the approximate periods in both the N-S and E-W directions.

   Since this condition is not satisfied, the SDC cannot be determined by Table 11.6-1 alone (11.6).

5. Determine the SDC from Tables 11.6-1 and 11.6-2.

   From Table 11.6-1, with $S_{DS} > 0.50$ and Risk Category II, the SDC is D.

   From Table 11.6-2, with $S_{D1} > 0.20$ and Risk Category II, the SDC is D.

Therefore, the SDC is D for this building.

6.7.4 Example 6.4 – Office Building, Permitted Analytical Procedure

For the seven-story office building in Example 6.3, determine the analytical procedure that can be used in calculating the seismic forces.

SOLUTION

Use Flowchart 6.7 to determine the permitted analytical procedure.

1. Determine the SDC from Flowchart 6.4.

   It was determined in Step 2 of Example 6.3 that the SDC is D.

2. Determine S_{DS} and S_{D1} from Flowchart 6.3.

   The design accelerations were determined in Step 1, Item 4 of Example 6.3: $S_{DS} = 0.73$ and $S_{D1} = 0.39$.

3. Determine T_S.

   It was determined in Step 2, Item 4 of Example 6.3 that $T_S = S_{D1}/S_{DS} = 0.53$ second.

4. Determine the fundamental period of the building, T.

   The periods, T, were determined in Step 2, Item 4 of Example 6.3 as 0.61 second in the N-S direction and 1.1 seconds in the E-W direction.

5. Check if $T < 3.5T_S$.

   $3.5T_S = 3.5 \times 0.53 = 1.9$ seconds, which is greater than the periods in both directions.

6. Determine if the structure is regular or not.

a. Determine if the structure has any horizontal structural irregularities in accordance with Table 12.3-1.

   i. Torsional irregularity

   In accordance with Table 12.3-1, Type 1a torsional irregularity and Type 1b extreme torsional irregularity for rigid or semirigid diaphragms exist where the ratio of the maximum story drift at one end of a structure to the average story drifts at two ends of the structure exceeds 1.2 and 1.4, respectively. The story drifts are to be determined using code-prescribed forces, including accidental torsion. In this example, the floor and roof are metal deck with concrete, which is considered to be a rigid diaphragm (12.3.1.2).

   At this point in the analysis, it is obviously not evident which method is required to be used to determine the prescribed seismic forces. In lieu of using a more complicated higher order analysis, the equivalent lateral force procedure may be used to determine the lateral seismic forces. These forces are applied to the building and the subsequent analysis yields the story drifts, Δ, which are used in determining whether Type 1a or 1b torsional irregularity exists. The results from the equivalent lateral force procedure will be needed if it is subsequently determined that a modal analysis is required (see 12.9.4).

   Use Flowchart 6.8 to determine the lateral seismic forces from the equivalent lateral force procedure:

   a. The design accelerations and the SDC have been determined in Example 6.3.

   b. Determine the response modification coefficient, R, from Table 12.2-1.

      In the N-S direction, special steel concentrically braced frames are required, since the building is assigned to SDC D (system B2 in Table 12.2-1). In this case, $R = 6$. Note that the height of the building, which is 96 feet, is less than the limiting height of 160 feet for this type of system in SDC D. The increased building height limit of 12.2.5.4 is not considered in this example.

      In the E-W direction, special steel moment frames are required (system C1 in Table 12.2-1). In this case, $R = 8$, and there is no height limit.

   c. Determine the importance factor, I_e, from Table 1.5-2.

      For Risk Category II, $I_e = 1.0$.

   d. Determine the period of the structure, T.

      It was determined in Step 2, Item 4 of Example 6.3 that the approximate period of the structure, T_a, which is permitted to be used in the equivalent lateral force procedure, is 0.61 second in the N-S direction and 1.1 seconds in the E-W direction.

   e. Determine long-period transition period, T_L, from Figure 22-12.

      For Memphis, TN, $T_L = 12$ seconds, which is greater than the periods in both directions.

   f. Determine seismic response coefficients, C_S, in both directions.

      - N-S direction:

      The seismic response coefficient, C_S, is determined by Equation 12.8-3:

      $$C_S = \frac{S_{D1}}{T(R/I_e)} = \frac{0.39}{0.61(6/1.0)} = 0.11$$

The value of C_S need not exceed that from Equation 12.8-2:

$$C_S = \frac{S_{DS}}{R/I_e} = \frac{0.73}{6/1.0} = 0.12$$

Also, C_S must not be less than the larger of $0.044S_{DS}I_e = 0.03$ (governs) and 0.01 (Equation 12.8-5).

Thus, the value of C_S from Equation 12.8-3 governs in the N-S direction.

- E-W direction:

$$C_S = \frac{S_{D1}}{T(R/I_e)} = \frac{0.39}{1.1(8/1.0)} = 0.04$$

The value of C_S need not exceed that from Equation 12.8-2:

$$C_S = \frac{S_{DS}}{R/I_e} = \frac{0.73}{8/1.0} = 0.09$$

Also, C_S must not be less than the larger of $0.044S_{DS}I_e = 0.03$ (governs) and 0.01 (Equation 12.8-5).

Thus, the value of C_S from Equation 12.8-3 governs in the E-W direction.

g. Determine effective seismic weight, W, in accordance with 12.7.2.

The member sizes and superimposed dead loads are given in Figure 6.32 and the effective weights at each floor level are given in Tables 6.13 and 6.14. The total weight, W, is the summation of the effective dead loads at each level.

h. Determine seismic base shear, V.

Seismic base shear is determined by Equation 12.8-1 in both the N-S and E-W directions.

- N-S direction: $V = C_S W = 0.11 \times 9{,}960 = 1{,}096$ kips
- E-W direction: $V = C_S W = 0.04 \times 9{,}960 = 398$ kips

i. Determine exponent related to structure period, k, in both directions.

Since 0.5 second $< T <$ 2.5 seconds in both directions, k is determined as follows:

- N-S direction: $k = 0.75 + 0.5T = 1.06$
- E-W direction: $k = 0.75 + 0.5T = 1.30$

Table 6.13 Seismic Forces and Story Shears in the N-S Direction

Level	Story weight, w_x (kips)	Height, h_x (ft)	$w_x h_x^k$	Lateral force, F_x (kips)	Story shear, V_x (kips)
R	1,018	96	128,515	209	209
6	1,381	83	149,423	242	451
5	1,381	70	124,738	202	653
4	1,381	57	100,328	163	816
3	1,381	44	76,252	124	940
2	1,381	31	52,606	85	1,025
1	2,037	18	43,609	71	1,096
Σ	9,960		675,471	1,096	

Table 6.14 Seismic Forces and Story Shears in the E-W Direction

Level	Story weight, w_x (kips)	Height, h_x (ft)	$w_x h_x^k$	Lateral force, F_x (kips)	Story shear, V_x (kips)
R	1,018	96	384,327	84	84
6	1,381	83	431,514	94	178
5	1,381	70	345,797	76	254
4	1,381	57	264,747	58	312
3	1,381	44	189,096	41	353
2	1,381	31	119,940	26	379
1	2,037	18	87,266	19	398
Σ	9,960		1,822,687	398	

j. Determine lateral seismic force, F_x, at each level, x.

Lateral forces, F_x, are determined by Equations 12.8-11 and 12.8-12.

A summary of the lateral forces, F_x, and the story shears, V_x, are given in Tables 6.13 and 6.14 for the N-S and E-W directions, respectively.

Three-dimensional analyses were performed independently in the N-S and E-W directions for the seismic forces in Tables 6.13 and 6.14 using a commercial computer program. In the model, rigid diaphragms were assigned at each floor level. In accordance with 12.8.4.2, the center of mass was displaced each way from its actual location a distance equal to 5 percent of the building dimension perpendicular to the applied forces to account for accidental torsion in seismic design.

A summary of the elastic displacements, δ_{xe}, at each end of the building in both the N-S and E-W directions due to the code-prescribed forces in Tables 6.13 and 6.14 is given in Table 6.15 at all floor levels.

Table 6.15 Lateral Displacements and Story Drifts Due to Seismic Forces

Story	N-S Direction						E-W Direction					
	$(\delta_{xe})_1$ (in.)	Δ_1 (in.)	$(\delta_{xe})_2$ (in.)	Δ_2 (in.)	Δ_{avg} (in.)	$\frac{\Delta_{max}}{\Delta_{avg}}$	$(\delta_{xe})_1$ (in.)	Δ_1 (in.)	$(\delta_{xe})_2$ (in.)	Δ_2 (in.)	Δ_{avg} (in.)	$\frac{\Delta_{max}}{\Delta_{avg}}$
7	1.59	0.14	1.23	0.12	0.13	1.08	5.36	0.36	5.22	0.34	0.35	1.03
6	1.45	0.17	1.11	0.15	0.16	1.06	5.00	0.49	4.88	0.47	0.48	1.02
5	1.28	0.22	0.96	0.20	0.21	1.05	4.51	0.71	4.41	0.71	0.71	1.00
4	1.06	0.21	0.76	0.19	0.20	1.05	3.80	1.10	3.70	1.06	1.08	1.02
3	0.85	0.22	0.57	0.17	0.19	1.16	2.70	1.19	2.64	1.16	1.17	1.02
2	0.63	0.17	0.40	0.22	0.19	1.16	1.51	0.89	1.48	0.87	0.88	1.01
1	0.46	0.46	0.28	0.28	0.37	1.24	0.62	0.62	0.61	0.61	0.61	1.02

According to Table 12.3-1, a torsional irregularity occurs where maximum story drift at one end of the structure is greater than 1.2 times the average of the story drifts at the two ends of the structure. The average story drift, Δ_{avg}, and the ratio of the maximum story drift to the average story drift, $\Delta_{max}/\Delta_{avg}$, are also provided in Table 6.15.

For example, at the first story in the N-S direction:

$\Delta_1 = 0.46$ inch

$\Delta_2 = 0.18$ inch

$\Delta_{avg} = \dfrac{0.46 + 0.28}{2} = 0.37$ inch

$\dfrac{\Delta_{max}}{\Delta_{avg}} = \dfrac{0.46}{0.37} = 1.24 > 1.2$

Therefore, a Type 1a torsional irregularity exists at the first story in the N-S direction.

According to 12.8.4.3, where torsional irregularity exists at floor level x, the accidental torsional moments, M_{ta}, must be increased by the torsional amplification factor, A_x, given by Equation 12.8-14:

$$A_x = \left(\dfrac{\delta_{max}}{1.2\delta_{avg}}\right)^2$$

At the first story in the N-S direction:

$$A_1 = \left[\dfrac{0.46}{1.2\left(\dfrac{0.46 + 0.28}{2}\right)}\right]^2 = 1.07 > 1.0$$

Assuming that the center of mass and center of rigidity are at the same location in this building, the accidental torsional moment at the first story is:

$(M_{ta})_1 = A_1 F_1 e = 1.07 \times 71 \times (0.05 \times 180) = 684$ ft-kips

ii. Reentrant corner irregularity

   By inspection, this irregularity does not exist.

iii. Diaphragm discontinuity irregularity

   This irregularity does not exist in this building when opening sizes for typical elevators and stairs are present.

iv. Out-of-plane offsets irregularity

   In the first story, the SFRS consists of special steel concentrically braced frames along column lines 1 and 7. Above the first floor, there is a 30-foot offset of the braced frames, which occur along column lines 2 and 6.

   Therefore, a Type 4 out-of-plane offset irregularity exists.

   Note that the forces from the braced frames along column lines 2 and 6 must be transferred through the structure to the braced frames along column lines 1 and 7, respectively.

v. Nonparallel systems irregularity

   This discontinuity does not exist, since all of the braced frames and moment-resisting frames are parallel to a major orthogonal axis of the building.

b. Determine if the structure has any vertical structural irregularities in accordance with Table 12.3-2.
   i. Stiffness–Soft Story Irregularity

   A soft story is defined in Table 12.3-2 based on the relative lateral stiffness of stories in a building. In general, it is not practical to determine story stiffness. Instead, this type of irregularity can be evaluated using drift ratios due to the code-prescribed lateral forces. Note that story displacements based on the code-prescribed lateral forces can be used to evaluate soft stories when the story heights are equal.

   A soft story exists when one of the following conditions is satisfied:

   Soft story irregularity: $0.7\dfrac{\delta_{1e}}{h_1} > \dfrac{\delta_{2e} - \delta_{1e}}{h_2}$

   or $0.8\dfrac{\delta_{1e}}{h_1} > \dfrac{1}{3}\left[\dfrac{\delta_{2e} - \delta_{1e}}{h_2} + \dfrac{\delta_{3e} - \delta_{2e}}{h_3} + \dfrac{\delta_{4e} - \delta_{3e}}{h_4}\right]$

   Extreme soft story irregularity: $0.6\dfrac{\delta_{1e}}{h_1} > \dfrac{\delta_{2e} - \delta_{1e}}{h_2}$

   or $0.7\dfrac{\delta_{1e}}{h_1} > \dfrac{1}{3}\left[\dfrac{\delta_{2e} - \delta_{1e}}{h_2} + \dfrac{\delta_{3e} - \delta_{2e}}{h_3} + \dfrac{\delta_{4e} - \delta_{3e}}{h_4}\right]$

   Check for a soft story in the first story:

   In the N-S direction, the displacements of the center of mass at the first, second, third and fourth floors are $\delta_{1e} = 0.31$ inch, $\delta_{2e} = 0.51$ inch, $\delta_{3e} = 0.69$ inch, and $\delta_{4e} = 0.91$ inch.

   $0.7\dfrac{\delta_{1e}}{h_1} = 0.7\dfrac{0.31}{18 \times 12} = 0.0010 < \dfrac{\delta_{2e} - \delta_{1e}}{h_2} = \dfrac{0.51 - 0.31}{13 \times 12} = 0.0013$

   $0.8\dfrac{\delta_{1e}}{h_1} = 0.8\dfrac{0.31}{18 \times 12} = 0.0011 < \dfrac{1}{3}\left[\dfrac{\delta_{2e} - \delta_{1e}}{h_2} + \dfrac{\delta_{3e} - \delta_{2e}}{h_3} + \dfrac{\delta_{4e} - \delta_{3e}}{h_4}\right]$

   $= \dfrac{1}{3}\left[\dfrac{0.51 - 0.31}{13 \times 12} + \dfrac{0.69 - 0.51}{13 \times 12} + \dfrac{0.91 - 0.69}{13 \times 12}\right] = 0.0013$

   In the E-W direction, the displacements of the center of mass at the first, second, third and fourth floors are $\delta_{1e} = 0.61$ inch, $\delta_{2e} = 1.49$ inches, $\delta_{3e} = 2.68$ inches, and $\delta_{4e} = 3.75$ inches.

   $0.7\dfrac{\delta_{1e}}{h_1} = 0.7\dfrac{0.61}{18 \times 12} = 0.0020 < \dfrac{\delta_{2e} - \delta_{1e}}{h_2} = \dfrac{1.49 - 0.61}{13 \times 12} = 0.0056$

   $0.8\dfrac{\delta_{1e}}{h_1} = 0.8\dfrac{0.61}{18 \times 12} = 0.0023 < \dfrac{1}{3}\left[\dfrac{\delta_{2e} - \delta_{1e}}{h_2} + \dfrac{\delta_{3e} - \delta_{2e}}{h_3} + \dfrac{\delta_{4e} - \delta_{3e}}{h_4}\right]$

   $= \dfrac{1}{3}\left[\dfrac{1.49 - 0.61}{13 \times 12} + \dfrac{2.68 - 1.49}{13 \times 12} + \dfrac{3.75 - 2.68}{13 \times 12}\right] = 0.0067$

   Therefore, a soft story irregularity does not exist in the first story.

   ii. Weight (mass) irregularity.

Check the weight ratio of the first and second stories: 2,037/1,381 = 1.48 < 1.50. Thus, this irregularity is not present.

iii. Vertical geometric irregularity.

A vertical geometric irregularity is considered to exist where the horizontal dimension of the SFRS in any story is 1.3 times that in an adjacent story.

In this case, the setbacks at the first floor level must be investigated for the moment-resisting frames along column lines A and E:

Width of floor 1/Width of floor 2 = 180/120 = 1.5 > 1.3

Thus, a Type 3 vertical geometric irregularity exists.

iv. In-plane discontinuity in vertical lateral force-resisting element irregularity.

There are no in-plane offsets of this type, so this irregularity does not exist.

v. Discontinuity in lateral strength-weak story irregularity.

This type of irregularity exists where a story lateral strength is less than 80 percent of that in the story above. The story strength is considered to be the total strength of all seismic-resisting elements that share the story shear for the direction under consideration.

<u>E-W Direction</u>

Determine whether a weak story exists in the first story in the E-W direction. In this case, the story strength is equal to the sum of the column shears in the moment-resisting frames in that story when the member moment capacity is developed by lateral loading. It is assumed in this example that the same column and beam sections are used in the moment-resisting frames in the first and second stories.

Assume the following nominal flexural strengths (note: the assumed nominal flexural strengths of the columns and beams are based on preliminary member sizes and are provided for illustration purposes only):

Columns: M_{nc} = 550 ft-kips

Beams: M_{nb} = 525 ft-kips

- First story shear strength

  Corner columns A1/E1 and A7/E7 are checked for strong column-weak beam considerations:

  $2M_{nc} = 2 \times 550 = 1{,}100$ ft-kips $> M_{nb} = 525$ ft-kips

  The maximum shear force that can develop in each exterior column is based on the moment capacity of the beam (525/2 = 263 ft-kips), since it is less than the moment capacity of the column (550 ft-kips) at the top of the column (note: at the bottom of the column, it is assumed that the full moment capacity of the column can be developed):

  $$V_1 = V_7 = \frac{263 + 550}{18} = 45 \text{ kips}$$

  Interior columns A2 through A6/E2 through E6 are checked for strong column-weak beam considerations:

  $2M_{nc} = 2 \times 550 = 1{,}100$ ft-kips $> 2M_{nb} = 1{,}050$ ft-kips

The maximum shear force that can develop in each interior column is based on the moment capacity of the beam (525 ft-kips), since it is less than the moment capacity of the column (550 ft-kips) at the top of the column:

$$V_2 = V_3 = V_4 = V_5 = V_6 = \frac{525 + 550}{18} = 60 \text{ kips}$$

Total first story strength = $2(V_1 + V_2 + V_3 + V_4 + V_5 + V_6 + V_7)$ = 780 kips

- Second story shear strength

$$V_1 = V_7 = \frac{263 + 263}{13} = 41 \text{ kips}$$

$$V_2 = V_3 = V_4 = V_5 = V_6 = \frac{525 + 525}{13} = 81 \text{ kips}$$

Total second story strength = $2(V_1 + V_2 + V_3 + V_4 + V_5 + V_6 + V_7)$ = 974 kips

780 kips > 0.80 × 974 = 779 kips

Therefore, a weak story irregularity does not exist in the first story in the E-W direction.

<u>N-S Direction</u>

Assuming the same beam, column and brace sizes in the first and second floors, the shear strengths of these floors are essentially the same, and no weak story irregularity exists in the N-S direction.

In summary, the building is not regular and has the following irregularities: horizontal Type 1a torsional irregularity, horizontal Type 4 out-of-plane offsets irregularity, and vertical Type 3 vertical geometric irregularity.

7. Determine the permitted analytical procedure from Table 12.6-1.

The structure has a height of 96 feet (< 160 feet) and is irregular with $T < 3.5T_S$. If the structure had only a Type 4 out-of-plane offsets irregularity, the equivalent lateral force procedure could be used to analyze the structure. However, since a Type 1a torsional irregularity and a Type 3 vertical geometric irregularity also exist, the equivalent lateral force procedure is not permitted; either a modal response spectrum analysis (12.9) or a seismic response history procedure (Chapter 16) must be utilized.

The reference sections in Tables 12.3-1 and 12.3-2 must be satisfied for these types of irregularities in structures assigned to SDC D. Columns B2, C2, D2, B6, C6 and D6 must be designed to resist the load combinations with overstrength factor of 12.4.3.2, since they support the discontinued braced frames along column lines 2 and 6 (12.3.3.2). Design forces shall be increased 25 percent for connections of diaphragms to vertical elements and to collectors and for connection of collectors to the vertical elements (12.3.3.4). Members that are not part of the SFRS must satisfy the deformational compatibility requirements of 12.12.4.

6.7.5 Example 6.5 – Office Building, Allowable Story Drift

For the seven-story office building in Example 6.3, check the story drift limits in both the N-S and E-W directions. For illustration purposes, use the lateral deflections determined by the equivalent lateral force procedure.

SOLUTION

1. Drift limits in N-S direction.

   To check drift limits, the deflections determined by Equation 12.8-15 must be used:

   $$\delta_x = \frac{C_d \delta_{xe}}{I_e}$$

   The maximum story displacements, δ_{xe}, in the N-S direction are summarized in Table 6.15. For special steel concentrically braced frames, the deflection amplification factor, C_d, is equal to 5 from Table 12.2-1.

   A summary of the displacements at each floor level in the N-S direction is given in Table 6.16.

   Table 6.16 Lateral Displacements and Story Drifts Due to Seismic Forces in the N-S Direction

Story	δ_{xe} (in.)	δ_x (in.)	Δ (in.)
7	1.59	7.95	0.70
6	1.45	7.25	0.85
5	1.28	6.40	1.10
4	1.06	5.30	1.05
3	0.85	4.25	1.10
2	0.63	3.15	0.85
1	0.46	2.30	2.30

   The interstory drifts, Δ, computed from the δ_x are also given in the table. The drift at story level x is determined by subtracting the design earthquake displacement at the bottom of the story from the design earthquake displacement at the top of the story:

   $$\Delta = \delta_x - \delta_{x-1}$$

   The design story drifts, Δ, shall not exceed the allowable story drift, Δ_a, given in Table 12.12-1. For Risk Category II and "all other structures," $\Delta_a = 0.020 h_{sx}$ where h_{sx} is the story height below level x.

   For the 18-foot story height, $\Delta_a = 0.020 \times 18 \times 12 = 4.32$ inches > 2.30 inches

   For the 13-foot story heights, $\Delta_a = 0.020 \times 13 \times 12 = 3.12$ inches, which is greater than the values of Δ at floor levels 2 through 7.

   Thus, drift limits are satisfied in the N-S direction.

2. Drift limits in the E-W direction.

   The maximum story displacements, δ_{xe}, in the E-W direction are summarized in Table 6.15. For special steel moment frames, the deflection amplification factor, C_d, is equal to 5.5 from Table 12.2-1.

A summary of the displacements at each floor level in the E-W direction is given in Table 6.17. The interstory drifts, Δ, computed from the δ_x are also given in the table.

Table 6.17 Lateral Displacements and Story Drifts Due to Seismic Forces in the E-W Direction

Story	δ_{xe} (in.)	δ_x (in.)	Δ (in.)
7	5.36	29.48	1.98
6	5.00	27.50	2.69
5	4.51	24.81	3.91
4	3.80	20.90	6.05
3	2.70	14.85	6.54
2	1.51	8.31	4.90
1	0.62	3.41	3.41

In accordance with 12.12.1.1, design story drifts, Δ, must not exceed Δ_a/ρ for SFRSs comprised solely of moment frames in structures assigned to SDC D, E or F where ρ is the redundancy factor determined in accordance with 12.3.4.2.

In lieu of checking the conditions in 12.3.4.2 to determine if ρ can be taken equal to 1.0, assume ρ is equal to 1.3. Therefore, for the 18-foot story height, the allowable story drift is $\Delta_a/\rho = 0.020 \times 18 \times 12/1.3 = 3.32$ inches, which is less than the design story drift of 3.41 inches at the first story.

For the 13-foot story heights, $\Delta_a/\rho = 0.020 \times 13 \times 12/1.3 = 2.40$ inches, which is less than the design story drifts at stories 2 through 6.

Thus, drift limits are not satisfied in the E-W direction. Note that even if ρ were equal to 1.0, drift limits would not be satisfied at floors 2 through 5. Increasing member sizes may not be sufficient to reduce the design drift; including additional members in the SFRS may be needed to control drift. This, in turn, may help reduce the torsional effects.

6.7.6 Example 6.6 – Office Building, P-delta Effects

For the seven-story office building in Example 6.3, determine the P-delta effects in both the N-S and E-W directions.

For illustration purposes, use the lateral deflections determined by the equivalent lateral force procedure.

Assume a 10-psf live load on the roof and a 50-psf live load on the floors.

SOLUTION

1. P-delta effects in the N-S direction.

    In lieu of automatically considering P-delta effects in a computer analysis, the following procedure can be used to determine whether P-delta effects need to be considered in accordance with 12.8.7.

    P-delta effects need not be considered when the stability coefficient, θ, determined by Equation 12.8-16 is less than or equal to 0.10:

    $$\theta = \frac{P_x \Delta I_e}{V_x h_{sx} C_d}$$

446 Chapter 6

where P_x = total unfactored vertical design load at and above level x
Δ = design story drift occurring simultaneously with V_x
I_e = importance factor
V_x = seismic shear force acting between level x and $x-1$
h_{sx} = story height below level x
C_d = deflection amplification factor in Table 12.2-1

The stability coefficient, θ, must not exceed θ_{max} determined by Equation 12.8-17:

$$\theta_{max} = \frac{0.5}{\beta C_d} \leq 0.25$$

where β is the ratio of shear demand to shear capacity between level x and $x-1$, which may be taken equal to 1.0 when it is not calculated.

The P-delta calculations for the N-S direction are shown in Table 6.18. It is clear that P-delta effects need not be considered at any of the floor levels. Note that θ_{max} is equal to 0.1000 in the N-S direction using $\beta = 1.0$.

2. P-delta effects in the E-W direction.

The P-delta calculations for the E-W direction are shown in Table 6.19. Note that θ_{max} is equal to 0.0909 in the E-W direction using $\beta = 1.0$, and, since θ is greater than θ_{max} at levels 2 through 4, the structure is potentially unstable and needs to be redesigned.

It was determined in Example 6.4 that the shear capacity of the second floor is equal to 780 kips. Thus, $\beta = 474/780 = 0.61$, and $\theta_{max} = 0.5/(0.61 \times 5.5) = 0.15$. Assuming the same shear capacities at levels 3 and 4, $\theta_{max} = 0.16$ at level 3 and $\theta_{max} = 0.18$ at level 4. Therefore, the structure is still potentially unstable.

Table 6.18 P-delta Effects in the N-S Direction

Level	h_{sx} (ft)	P_x (kips)	V_x (kips)	Δ (in.)	θ
7	13	1,186	209	0.70	0.0051
6	13	3,407	451	0.85	0.0082
5	13	5,628	653	1.10	0.0122
4	13	7,849	816	1.05	0.0129
3	13	10,070	940	1.10	0.0151
2	13	12,291	1,025	0.85	0.0131
1	18	15,588	1,096	2.30	0.0303

Table 6.19 P-delta Effects in the E-W Direction

Level	h_{sx} (ft)	P_x (kips)	V_x (kips)	Δ (in.)	θ
7	13	1,186	105	1.98	0.0261
6	13	3,407	223	2.69	0.0479
5	13	5,628	317	3.91	0.0809
4	13	7,849	389	6.05	0.1423
3	13	10,070	441	6.54	0.1741
2	13	12,291	474	4.90	0.1481
1	18	15,588	498	3.41	0.0898

6.7.7 Example 6.7 – Health Care Facility, Diaphragm Design Forces

Determine the diaphragm design forces for the three-story health care facility depicted in Figure 6.33 given the design data below.

DESIGN DATA

Location:	St. Louis, MO (Latitude: 38.63°, Longitude: –90.20°)
Soil classification:	Site Class C
Occupancy:	Health care facility without surgery or emergency treatment facilities
Material:	Cast-in-place concrete
Structural system:	Moment-resisting frames in both directions

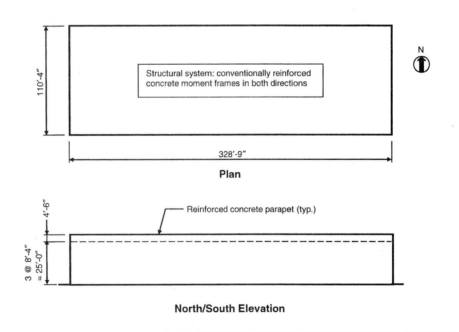

Figure 6.33 Plan and Elevation of Health Care Facility, Example 6.7

SOLUTION

In order to determine the diaphragm design forces in accordance with 12.10.1.1, the design seismic forces must be determined at each floor level.

Assuming that the building is regular, the equivalent lateral force procedure can be used to determine the design seismic forces (see Table 12.6-1).

Step 1: Determine the seismic ground motion values from Flowchart 6.3.

1. Determine the mapped accelerations S_S and S_1.

   In lieu of using Figures 22-1 and 22-2, the mapped accelerations are determined by inputting the latitude and longitude of the site into the USGS "DesignMaps" Web Application. The output is as follows: $S_S = 0.43$ and $S_1 = 0.17$.

2. Determine the site class of the soil.

   The site class of the soil is given in the design data as Site Class C.

3. Determine soil-modified accelerations S_{MS} and S_{M1}.

   Site coefficients, F_a and F_v, are determined from Tables 11.4-1 and 11.4-2, respectively:

   For Site Class C and $0.25 < S_S < 0.50$: $F_a = 1.20$

   For Site Class C and $0.1 < S_1 < 0.2$: $F_v = 1.63$ from linear interpolation

   Thus,

   $S_{MS} = 1.20 \times 0.43 = 0.52$

   $S_{M1} = 1.63 \times 0.17 = 0.28$

4. Determine design accelerations S_{DS} and S_{D1}.

   From Equations 11.4-3 and 11.4-4:

   $$S_{DS} = \frac{2}{3} \times 0.52 = 0.35$$

   $$S_{D1} = \frac{2}{3} \times 0.28 = 0.19$$

Step 2: Determine the SDC from Flowchart 6.4.

1. Determine if the building can be assigned to SDC A in accordance with 11.4.1.

   Since $S_S = 0.43 > 0.15$ and $S_1 = 0.17 > 0.04$, the building cannot be automatically assigned to SDC A.

2. Determine the risk category from IBC Table 1604.5.

   For the health care facility described in the design data, the Risk Category is III.

3. Since $S_1 < 0.75$, the building is not assigned to SDC E or F.

4. Check if all four conditions of 11.6 are satisfied.

   - Check if the approximate period, T_a, is less than $0.8T_S$.

     From Equation 12.8-7 for a concrete moment-resisting frame:

     $T_a = C_t h_n^x = 0.016(25.0)^{0.9} = 0.29$ second

     where C_t and x are given in Table 12.8-2.

     $T_S = S_{D1}/S_{DS} = 0.19/0.35 = 0.54$ second

     0.29 second $< 0.8 \times 0.54 = 0.43$ second

Earthquake Loads 449

- The fundamental period used to calculate the design drift is taken as 0.29 second, which is less than $T_S = 0.54$ second.

- Equation 12.8-2 will be used to determine the seismic response coefficient, C_S.

- Since the roof and the floors are cast-in-place concrete, the diaphragms are considered to be rigid.

Since all four conditions are satisfied, the SDC can be determined by Table 11.6-1 alone (11.6).

5. Determine the SDC from Table 11.6-1.

   From Table 11.6-1, with $0.33 < S_{DS} = 0.35 < 0.50$ and Risk Category III, the SDC is C.

   Note that if Table 11.6-2 were also used, the SDC would also be C.

Step 3: Determine the design seismic forces of the equivalent lateral force procedure from Flowchart 6.8.

1. The design accelerations and the SDC have been determined in Steps 1 and 2 above.

2. Determine the response modification coefficient, R, from Table 12.2-1.

   The moment-resisting frames in this building must be intermediate reinforced concrete moment frames, since the building is assigned to SDC C (system C6 in Table 12.2-1). In this case, $R = 5$. There is no height limit for this system in SDC C.

3. Determine the importance factor, I_e, from Table 1.5-2.

   For Risk Category III, $I_e = 1.25$.

4. Determine the period of the structure, T.

   It was determined in Step 2, Item 4 above that the approximate period of the structure, T_a, which is permitted to be used in the equivalent lateral force procedure, is equal to 0.29 second.

5. Determine long-period transition period, T_L, from Figure 22-12.

   For St. Louis, MO, $T_L = 12$ seconds $> T_a = 0.29$ second

6. Determine seismic response coefficient, C_S.

   The value of C_S must be determined by Equation 12.8-2 (see Step 2, Item 4):

   $$C_S = \frac{S_{DS}}{R/I_e} = \frac{0.35}{5/1.25} = 0.09$$

7. Determine effective seismic weight, W, in accordance with 12.7.2.

   The effective weights at each floor level are given in Table 6.20. The total weight, W, is the summation of the effective dead loads at each level.

Table 6.20 Seismic Forces and Story Shears

Level	Story weight, w_x (kips)	Height, h_x (ft)	$w_x h_x^k$	Lateral force, F_x (kips)	Story shear, V_x (kips)
3	4,958	25.00	123,950	685	685
2	5,681	16.67	94,702	523	1,208
1	5,681	8.33	47,323	261	1,469
Σ	16,320		265,975	1,469	

8. Determine seismic base shear, V.

   Seismic base shear is determined by Equation 12.8-1:

   $V = C_s W = 0.09 \times 16{,}320 = 1{,}469$ kips

9. Determine exponent related to structure period, k.

   Since $T < 0.5$ second, $k = 1.0$.

10. Determine lateral seismic force, F_x, at each level, x.

    F_x is determined by Equations 12.8-11 and 12.8-12. A summary of the lateral forces, F_x, and the story shears, V_x, is given in Table 6.20.

Step 4: Determine the diaphragm design seismic forces using Equation 12.10-1.

$$\text{Diaphragm design force } F_{px} = \frac{\sum_{i=x}^{n} F_i}{\sum_{i=x}^{n} w_i} w_{px}$$

where w_i = weight tributary to level i, and w_{px} = weight tributary to the diaphragm at level x.

Minimum $F_{px} = 0.2 S_{DS} I_e w_{px} = 0.2 \times 0.35 \times 1.25 \times w_{px} = 0.0875 w_{px}$

Maximum $F_{px} = 0.4 S_{DS} I_e w_{px} = 0.1750 w_{px}$

Assuming that the exterior walls are primarily glass, which weigh significantly less than the diaphragm weight at each level, the weight that is tributary to each diaphragm is identical to the weight of the structure at that level (i.e., $w_{px} = w_x$).

A summary of the diaphragm forces is given in Table 6.21.

Table 6.21 Design Seismic Diaphragm Forces

Level	w_x (kips)	Σw_x (kips)	F_x (kips)	ΣF_x (kips)	$\Sigma F_x / \Sigma w_x$	w_{px} (kips)	F_{px} (kips)
3	4,958	4,958	685	685	0.1382	4,958	685
2	5,681	10,639	523	1,208	0.1135	5,681	645
1	5,681	16,320	261	1,469	0.0900	5,681	511

6.7.8 Example 6.8 – Health Care Facility, Nonstructural Component

Determine the design seismic force on the parapet of the health care facility in Example 6.7.

SOLUTION

Use Flowchart 6.10 to determine the seismic force on the parapet.

1. Determine S_{DS}, S_{D1} and the SDC.

   The design accelerations and the SDC are determined in Example 6.7.

   The parapet is assigned to SDC C, which is the same SDC as the building to which it is attached (13.1.2).

2. Determine the component amplification factor, a_p, and the component response modification factor, R_p, from Table 13.5-1 for architectural components.

   Assuming that the parapet is not braced, $a_p = 2.5$ and $R_p = 2.5$ from Table 13.5-1.

3. Determine component importance factor, I_p, in accordance with 13.1.3.

   Since the parapet does not meet any of the four criteria that require $I_p = 1.5$, then $I_p = 1.0$.

4. Determine the horizontal seismic design force, F_p, by Equation 13.3-1.

$$F_p = \frac{0.4 a_p S_{DS} W_p}{\left(\frac{R_p}{I_p}\right)}\left(1 + 2\frac{z}{h}\right)$$

Assuming that the thickness of the parapet is 8 inches and that normal weight concrete is utilized, $W_p = 8 \times 150/12 = 100$ psf.

Since the parapet is attached to the top of the structure, $z/h = 1$.

Thus,

$$F_p = \frac{0.4 \times 2.5 \times 0.35 \times 100}{\left(\frac{2.5}{1.0}\right)}(1 + 2) = 42 \text{ psf}$$

Minimum $F_p = 0.3 S_{DS} I_p W_p = 10.5$ psf

Maximum $F_p = 1.6 S_{DS} I_p W_p = 56.0$ psf

The 42 psf seismic load is applied to the parapet as shown in Figure 6.34.

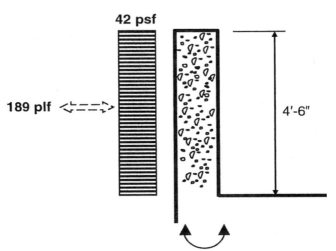

Figure 6.34 Design Seismic Force on Parapet, Example 6.8

6.7.9 Example 6.9 – Residential Building, Vertical Combination of Structural Systems

Determine the design seismic forces on the residential building depicted in Figure 6.35 given the design data below.

Figure 6.35
First Floor Plan and Elevation of Residential Building, Example 6.9

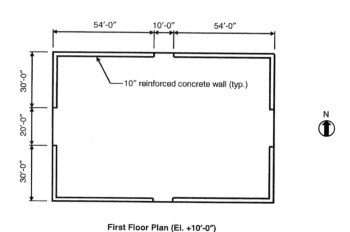

First Floor Plan (El. +10'-0")

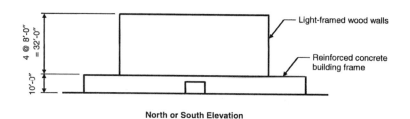

North or South Elevation

DESIGN DATA

Location:	Philadelphia, PA (Latitude: 39.92°, Longitude: –75.23°)
Soil classification:	Site Class D
Occupancy:	Residential occupancy where less than 300 people congregate in one area
Structural systems:	Light-frame wood bearing walls with shear panels rated for shear resistance and cast-in-place reinforced concrete building frame system with ordinary reinforced concrete shear walls

SOLUTION

Step 1: Determine the seismic ground motion values from Flowchart 6.3.

Using the USGS "DesignMaps" Web Application, $S_S = 0.20$ and $S_1 = 0.06$.

Using Tables 11.4-1 and 11.4-2, the soil-modified accelerations are $S_{MS} = 0.32$ and $S_{M1} = 0.14$.

Design accelerations: $S_{DS} = \frac{2}{3} \times 0.32 = 0.21$ and $S_{D1} = \frac{2}{3} \times 0.14 = 0.09$

Step 2: Determine the SDC from Flowchart 6.4.

From IBC Table 1604.5, the Risk Category is II.

From Table 11.6-1 with $0.167 < S_{DS} < 0.33$ and Risk Category II, the SDC is B.

From Table 11.6-2 with $0.067 < S_{D1} < 0.133$ and Risk Category II, the SDC is B.

Therefore, the SDC is B for this building.

Step 3: Determine the response modification coefficients, R, in accordance with 12.2.3 for vertical combinations of structural systems.

The vertical combination of structural systems in this building is a flexible wood frame upper portion above a rigid concrete lower portion. Thus, a two-stage equivalent lateral force procedure is permitted to be used provided the design of the structure complies with the four criteria listed in 12.2.3.2:

a. The stiffness of the lower portion must be at least 10 times the stiffness of the upper portion.

   It can be shown that the stiffness of the lower portion of this structure is more than 10 times that of the upper portion. **O.K.**

b. The period of the entire structure shall not be greater than 1.1 times the period of the upper portion considered as a separate structure fixed at the base.

   Determine the period of the upper portion by Equation 12.8-7 using the approximate period coefficients in Table 12.8-2 for "all other structural systems":

   $$T_a = C_t h_n^x = 0.02(32)^{0.75} = 0.27 \text{ second}$$

   Determine the period of the lower portion in the N-S direction by Equations 12.8-9 and 12.8-10 for concrete shear wall structures:

   $$T_a = \frac{0.0019}{\sqrt{C_W}} h_n$$

   $$C_W = \frac{100}{A_B} \sum_{i=1}^{x} \left(\frac{h_n}{h_i}\right)^2 \frac{A_i}{\left[1 + 0.83\left(\frac{h_i}{D_i}\right)^2\right]}$$

   where: A_B = area of base of structure = $118 \times 80 = 9{,}440$ square feet

   h_n = height of lower portion of building = 10 feet

   h_i = height of shear wall i = 10 feet

   A_i = area of shear wall i = $\frac{10}{12} \times 30 = 25$ square feet

   D_i = length of shear wall i = 30 feet

Thus,

$$C_W = \frac{4 \times 100}{9{,}440} \frac{25}{\left[1 + 0.83\left(\frac{10}{30}\right)^2\right]} = 0.97$$

$$T_a = \frac{0.0019}{\sqrt{0.97}} \times 10 = 0.02 \text{ second}$$

The period of the lower portion in the E-W direction is equal to 0.01 second.

The period of the combined structure, which was obtained from a commercial computer program, is approximately 0.28 second.

0.28 second < 1.1 × 0.27 = 0.30 second **O.K.**

   c. The flexible upper portion shall be designed as a separate structure using the appropriate values of R and ρ.

     From Table 12.2-1 for a bearing wall system with light-frame walls with wood structural panels rated for shear resistance (system A15), $R = 6.5$ with no height limitation for SDC B.

     For SDC B, $\rho = 1.0$ (12.3.4.1).

   d. The rigid lower portion shall be designed as a separate structure using the appropriate values of R and ρ. Amplified reactions from the upper portion are applied to the lower portion where the amplification factor is equal to the ratio of R/ρ of the upper portion divided by R/ρ of the lower portion and must be greater than or equal to one.

     From Table 12.2-1 for a building frame system with ordinary reinforced concrete shear walls (system B5), $R = 5$ with no height limitation for SDC B.

     For SDC B, $\rho = 1.0$ (12.3.4.1).

     Amplification factor = (6.5/1)/(5/1) = 1.3.

Therefore, a two-stage equivalent lateral force procedure is permitted to be used.

Step 4: Determine the design seismic forces on the upper and lower portions of the structure using the equivalent lateral force procedure.

   1. Use Flowchart 6.8 to determine the lateral seismic forces on the flexible upper portion of the structure.

     a. The design accelerations and the SDC have been determined in Steps 1 and 2 above, respectively.

     b. Determine the response modification coefficient, R, from Table 12.2-1.

       The response modification coefficient was determined in Step 3 as 6.5.

     c. Determine the importance factor, I_e, from Table 1.5-2.

       For Risk Category II, $I_e = 1.0$.

     d. Determine the period of the structure, T.

       It was determined in Step 3 that the approximate period of the structure $T_a = 0.27$ second.

     e. Determine long-period transition period, T_L, from Figure 22-12.

       For Philadelphia, PA, $T_L = 6$ seconds $> T_a = 0.27$ second.

     f. Determine seismic response coefficient, C_S.

The seismic response coefficient, C_S, is determined by Equation 12.8-3:

$$C_S = \frac{S_{D1}}{T(R/I_e)} = \frac{0.09}{0.27(6.5/1.0)} = 0.05$$

The value of C_S need not exceed that from Equation 12.8-2:

$$C_S = \frac{S_{DS}}{R/I_e} = \frac{0.21}{6.5/1.0} = 0.03$$

Also, C_S must not be less than the larger of $0.044 S_{DS} I_e = 0.01$ and 0.01 (Equation 12.8-5).

Thus, the value of C_S from Equation 12.8-2 governs.

g. Determine effective seismic weight, W, in accordance with 12.7.2.

The effective weights at each floor level are given in Table 6.22. The total weight W is the summation of the effective dead loads at each level.

Table 6.22 Seismic Forces and Story Shears on Flexible Upper Portion

Level	Story weight, w_x (kips)	Height, h_x (ft)	$w_x h_x^k$	Lateral force, F_x (kips)	Story shear, V_x (kips)
4	185	32	5,920	9	9
3	192	24	4,608	7	16
2	192	16	3,072	5	21
1	192	8	1,536	2	23
Σ	761		15,136	23	

h. Determine seismic base shear, V.

Seismic base shear is determined by Equation 12.8-1:

$V = C_S W = 0.03 \times 761 = 23$ kips

i. Determine exponent related to structure period, k.

Since $T = 0.27$ second < 0.5 second, $k = 1.0$.

j. Determine lateral seismic force, F_x, at each level, x.

F_x is determined by Equations 12.8-11 and 12.8-12. A summary of the lateral forces, F_x, and the story shears, V_x, is given in Table 6.22.

2. Use Flowchart 6.8 to determine the lateral seismic forces on the rigid lower portion of the structure in the N-S direction.

   a. The design accelerations and the SDC have been determined in Steps 1 and 2 above, respectively.

   b. Determine the response modification coefficient, R, from Table 12.2-1.

   It was determined in Step 3 that the response modification coefficient $R = 5$.

   c. Determine the importance factor, I_e, from Table 1.5-2.

   For Risk Category II, $I_e = 1.0$.

   d. Determine the period of the structure, T.

   It was determined in Step 3 that the approximate period of the structure $T_a = 0.02$ second in the N-S direction.

456 Chapter 6

e. Determine long-period transition period, T_L, from Figure 22-12.

For Philadelphia, PA, $T_L = 6$ seconds $> T_a = 0.02$ second.

f. Determine seismic response coefficient, C_S.

The seismic response coefficient, C_S, is determined by Equation 12.8-3:

$$C_S = \frac{S_{D1}}{T(R/I_e)} = \frac{0.09}{0.02(5/1.0)} = 0.9$$

The value of C_S need not exceed that from Equation 12.8-2:

$$C_S = \frac{S_{DS}}{R/I_e} = \frac{0.21}{5/1.0} = 0.04$$

Also, C_S must not be less than the larger of $0.044 S_{DS} I_e = 0.01$ and 0.01 (Equation 12.8-5).

Thus, the value of C_S from Equation 12.8-2 governs.

g. Determine effective seismic weight, W, in accordance with 12.7.2.

The effective weight, W, is equal to 1,458 kips for the lower portion.

h. Determine seismic base shear, V.

Seismic base shear is determined by Equation 12.8-1:

$$V = C_S W = 0.04 \times 1{,}458 = 58 \text{ kips}$$

i. Determine total lateral seismic forces on the lower portion.

For the one-story lower portion, the total seismic force is equal to the lateral force due to the base shear of the lower portion plus the amplified seismic force from the upper portion:

$$V_{total} = 58 + (1.3 \times 23) = 88 \text{ kips}$$

Since the base shear is independent of the period, the total seismic force in the E-W direction is also equal to 88 kips.

The distribution of the lateral seismic forces in the upper and lower parts of the structure is shown in Figure 6.36.

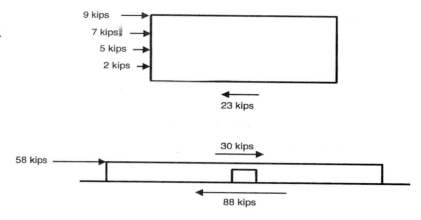

Figure 6.36
Distribution of Lateral Seismic Forces in the Upper and Lower Portions of the Structure, Example 6.9

6.7.10 Example 6.10 – Warehouse Building, Design of Roof Diaphragm, Collectors, and Wall Panels

For the one-story warehouse illustrated in Figure 6.37, determine (1) design seismic forces on the diaphragm, including diaphragm shear forces in both directions, (2) design seismic forces on the steel collector beam in the N-S direction, and (3) out-of-plane design seismic forces on the precast concrete wall panels, given the design data below.

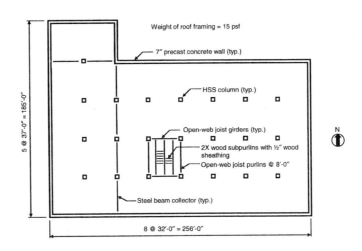

Figure 6.37
Plan and Elevation of Warehouse Building, Example 6.10

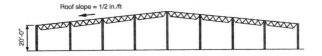

DESIGN DATA

Location:	San Francisco, CA (Latitude: 37.75°, Longitude: –122.43°)
Soil classification:	Site Class D
Occupancy:	Less than 300 people congregate in one area and the building is not used to store hazardous or toxic materials
Structural system:	Building frame system with intermediate precast concrete shear walls

SOLUTION

Part 1: Determine design seismic forces on the diaphragm

Step 1: Determine the seismic ground motion values from Flowchart 6.3.

Using the USGS "DesignMaps" Web Application, $S_S = 1.57$ and $S_1 = 0.72$.

Using Tables 11.4-1 and 11.4-2, the soil-modified accelerations are $S_{MS} = 1.57$ and $S_{M1} = 1.08$.

Design accelerations: $S_{DS} = \frac{2}{3} \times 1.57 = 1.05$ and $S_{D1} = \frac{2}{3} \times 1.08 = 0.72$

458 Chapter 6

Step 2: Determine the SDC from Flowchart 6.4.

From IBC Table 1604.5, the Risk Category is II.

Since $S_1 < 0.75$, use Tables 11.6-1 and 11.6-2 to determine the SDC. For Risk Category II, the SDC is D from both tables; thus, the SDC is D for this building (11.6).

Step 3: Use Flowchart 6.8 to determine the seismic base shear using the equivalent lateral force procedure.

1. Check if equivalent lateral force procedure can be used (see Flowchart 6.7).

   The building has a Type 2 reentrant corner irregularity, since 37.0 feet > 0.15 × 185.0 = 27.8 feet and 192.0 feet > 0.15 × 256.0 = 38.4 feet (see Table 12.3-1 and Figure 6.8).

   In accordance with Table 12.6-1, the equivalent lateral force procedure is permitted to be used for this irregular structure with a Type 2 reentrant corner since the height of the structure is less than 160 feet.

2. Use Flowchart 6.8 to determine the lateral seismic forces in the N-S and E-W directions.

   a. The design accelerations and the SDC have been determined in Steps 1 and 2 above, respectively.

   b. Determine the response modification coefficient, R, from Table 12.2-1.

   For a building frame system with intermediate precast concrete shear walls (system B8), $R = 5$. Note that the average building height of 22.67 feet is less than the 45-foot height limit for this system assigned to SDC D (see Footnote k in Table 12.2-1).

   c. Determine the importance factor, I_e, from Table 1.5-2.

   For Risk Category II, $I_e = 1.0$.

   d. Determine the period of the structure, T.
      i. Determine the period in the N-S direction by Equations 12.8-9 and 12.8-10 for concrete shear wall structures:

   $$T_a = \frac{0.0019}{\sqrt{C_W}} h_n$$

   $$C_W = \frac{100}{A_B} \sum_{i=1}^{x} \left(\frac{h_n}{h_i}\right)^2 \frac{A_i}{\left[1 + 0.83\left(\frac{h_i}{D_i}\right)^2\right]}$$

   where: A_B = area of base of structure = 40,256 square feet

   h_n = average height of building = 22.67 feet

   h_i = average height of shear wall i = 22.67 feet

   A_i = area of shear wall i: $A_1 = \frac{7}{12} \times 185 = 107.9$ square feet,

   $A_2 = \frac{7}{12} \times 37 = 21.6$ square feet, $A_3 = \frac{7}{12} \times 148 = 86.3$ square feet

   D_i = length of shear wall i: $D_1 = 185$ feet, $D_2 = 37$ feet, $D_3 = 148$ feet

Thus,

$$C_W = \frac{100}{40,256}\left\{\frac{107.9}{\left[1+0.83\left(\frac{22.67}{185}\right)^2\right]} + \frac{21.6}{\left[1+0.83\left(\frac{22.67}{37}\right)^2\right]} + \frac{86.3}{\left[1+0.83\left(\frac{22.67}{148}\right)^2\right]}\right\}$$

$$= \frac{100}{40,256}(106.6 + 16.5 + 84.7) = 0.52$$

$$T_a = \frac{0.0019}{\sqrt{0.52}} \times 22.67 = 0.06 \text{ second}$$

ii. Determine the period in the E-W direction.

$$C_W = \frac{100}{40,256}\left\{\frac{37.3}{\left[1+0.83\left(\frac{22.67}{64}\right)^2\right]} + \frac{112.0}{\left[1+0.83\left(\frac{22.67}{192}\right)^2\right]} + \frac{149.3}{\left[1+0.83\left(\frac{22.67}{256}\right)^2\right]}\right\}$$

$$= \frac{100}{40,256}(33.8 + 110.7 + 148.3) = 0.73$$

$$T_a = \frac{0.0019}{\sqrt{0.73}} \times 22.67 = 0.05 \text{ second}$$

e. Determine long-period transition period, T_L, from Figure 22-12.

For San Francisco, CA, $T_L = 12$ seconds $> T_a$ in both directions.

f. Determine seismic response coefficient, C_S.

The value of C_S from Equation 12.8-2 is:

$$C_S = \frac{S_{DS}}{R/I_e} = \frac{1.05}{5/1.0} = 0.21$$

Also, C_S must not be less than the larger of $0.044S_{DS}I_e = 0.05$ and 0.01 (Equation 12.8-5) or the value obtained by Equation 12.8-6 (governs), since $S_1 > 0.6$:

$$C_S = \frac{0.5S_1}{\left(\frac{R}{I_e}\right)} = \frac{0.5 \times 0.72}{\left(\frac{5}{1}\right)} = 0.07$$

Thus, the value of C_S from Equation 12.8-2 governs.

g. Determine effective seismic weight, W, in accordance with 12.7.2.

The effective weight, W, is equal to the weight of the roof framing plus the weight of the walls tributary to the roof where it is conservatively assumed that there are no openings in the walls:

Weight of roof framing = $0.015 \times 40,256 = 604$ kips

Weight of walls = $\frac{7}{12} \times 0.15 \times \frac{22.67}{2} \times [2(256 + 185)] = 875$ kips

$W = 604 + 875 = 1,479$ kips

460 Chapter 6

h. Determine seismic base shear, V.

Seismic base shear is determined by Equation 12.8-1 and is the same in both the N-S and E-W directions, since the governing C_s is independent of the period:

$$V = C_s W = 0.21 \times 1{,}479 = 311 \text{ kips}$$

3. Determine the design seismic forces in the diaphragm in both directions by Equation 12.10-1.

$$\text{Diaphragm design force } F_{px} = \frac{\sum_{i=x}^{n} F_i}{\sum_{i=x}^{n} w_i} w_{px}$$

where w_i = weight tributary to level i, and w_{px} = weight tributary to the diaphragm at level x.

Since this is a one-story building, Equation 12.10-1 reduces to $F_{px} = 0.21 w_{px}$

Minimum $F_{px} = 0.2 S_{DS} I_e w_{px} = 0.2 \times 1.05 \times 1.0 \times w_{px} = 0.21 w_{px}$

Maximum $F_{px} = 0.4 S_{DS} I_e w_{px} = 0.42 w_{px}$

The wood sheathing is permitted to be idealized as a flexible diaphragm in accordance with 12.3.1.1. Seismic forces are computed from the tributary weight of the roof and the walls oriented perpendicular to the direction of analysis; walls parallel to the direction of the seismic forces are typically not considered in the tributary weight, since these walls do not obtain support from the diaphragm in the direction of the seismic force.

<u>N-S direction</u>

Uniform diaphragm loads w_{N1} and w_{N2} are computed as follows (see Figure 6.38):

$$w_{N1} = (0.21 \times 15 \times 185) + \left(0.21 \times 87.5 \times 2 \times \frac{22.67}{2}\right) = 999 \text{ plf}$$

$$w_{N2} = (0.21 \times 15 \times 148) + \left(0.21 \times 87.5 \times 2 \times \frac{22.67}{2}\right) = 883 \text{ plf}$$

Also shown in Figure 6.38 is the shear diagram for the diaphragm.

Figure 6.38
Design Seismic Forces and Shear Forces in the Diaphragm in the N-S Direction, Example 6.10

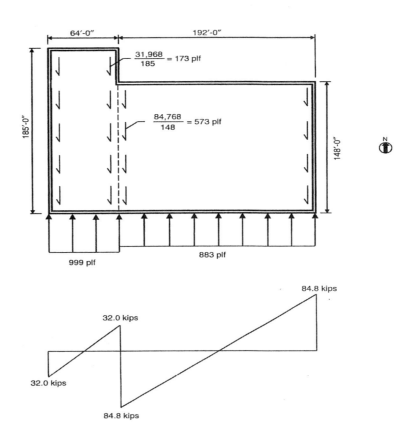

E-W direction

Uniform diaphragm loads w_{E1} and w_{E2} are computed as follows (see Figure 6.39):

$$w_{E1} = (0.21 \times 15 \times 64) + \left(0.21 \times 87.5 \times 2 \times \frac{22.67}{2}\right) = 618 \text{ plf}$$

$$w_{E2} = (0.21 \times 15 \times 256) + \left(0.21 \times 87.5 \times 2 \times \frac{22.67}{2}\right) = 1,223 \text{ plf}$$

Also shown in Figure 6.39 is the shear diagram for the diaphragm.

Figure 6.39
Design Seismic Forces and Shear Forces in the Diaphragm in the E-W Direction, Example 6.10

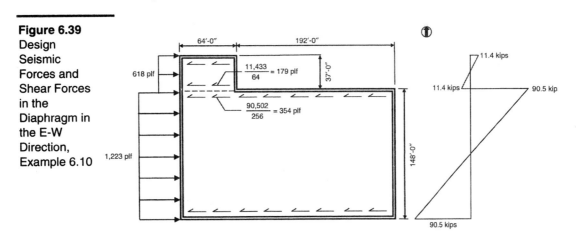

4. Determine connection forces between the diaphragm and the shear walls.

   Since the building has a Type 2 horizontal irregularity and is assigned to SDC D, the diaphragm connection design forces must be increased by 25 percent in accordance with 12.3.3.4. Thus,

   $F_{px} = 1.25 \times 0.21 w_{px} = 0.26 w_{px}$

   This force increase applies to the row of diaphragm nailing that transfers the above diaphragm shears directly to the shear walls (and to the collectors) and to the bolts between the ledger beams and the shear walls.

Part 2: Determine design seismic forces on the steel collector beam in the N-S direction

From the diaphragm shear diagram in Figure 6.38, the maximum collector load is equal to [32.0 × (148/185)] + 84.8 = 110.4 kips tension or compression.

The uniform axial load can be approximated by dividing the maximum load by the length of the collector: 110,400/148 = 746 plf.

The uniform axial load can be used to determine the axial force in any of the beams at any point along their length. For example, at the midspan of the northernmost collector beam, the axial force is equal to 746 × (148 − 37/2) / 1,000 = 97 kips tension or compression. In accordance with 12.3.3.4, this force must be increased by 25 percent unless the collector is designed for the load combinations with overstrength factor of 12.4.3.2.

The collector beams are subsequently designed for the combined effects of gravity and axial loads due to the design seismic forces in accordance with 12.10.2.1.

Part 3: Determine out-of-plane design seismic forces on the precast concrete wall panels

1. Solid wall panels

   According to 12.11.1, structural walls shall be designed for a force normal to the surface equal to $0.4 S_{DS} I_e$ times the weight of the wall. The minimum normal force is 10 percent of the weight of the wall.

   For a solid precast concrete wall panel:

   Weight $W_p = (7/12) \times 0.15 \times 22.67 = 2.0$ kips/ft

   $0.1 \times 2.0 = 0.20$ kips/ft

   $0.4 \times 1.05 \times 1.0 \times 2.0 = 0.84$ kips/ft (governs)

   Distributed load = 0.84/22.67 = 0.04 kips/ft/ft width of wall

   This uniformly distributed load is applied normal to the wall in either direction.

   Anchorage of the precast walls to the flexible diaphragms must develop the out-of-plane force given by Equation 12.11-1:

   $F_p = 0.4 S_{DS} k_a I_e W_p = 0.4 \times 1.05 \times 2 \times 1.0 \times 2.0 = 1.7$ kips/ft $> 0.2 k_a I_e W_p = 0.8$ kips/ft

   where maximum $k_a = [1 + (192/100)] = 2.92 > 2$, use 2.

   Note that the 25 percent increase in the design force for diaphragm connections is not applied to out-of-plane wall anchorage force to the diaphragm (12.3.3.4).

2. Wall panels with openings

   A typical wall panel on the east and west faces of the building is shown in Figure 6.40.

Figure 6.40
Typical Precast Wall Panel on East and West Faces, Example 6.10

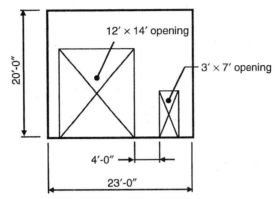

In lieu of a more rigorous analysis, the pier width between the two openings is commonly defined as a design strip. The total weight used in determining the out-of-plane design seismic force is taken as the weight of the design strip plus the weight of the wall tributary to the design strip above each adjacent opening (see Figure 6.41).

Figure 6.41
Design Strip and Tributary Weights, Example 6.10

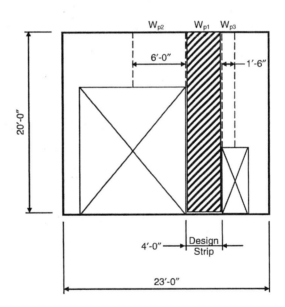

$$W_{p1} = \frac{7}{12} \times 150 \times 4 = 350 \text{ plf}$$

$$W_{p2} = \frac{7}{12} \times 150 \times 6 = 525 \text{ plf}$$

$$W_{p3} = \frac{7}{12} \times 150 \times 1.5 = 131 \text{ plf}$$

The out-of-plane seismic forces are determined by 12.11.1:

$F_{p1} = 0.4 S_{DS} I_e W_{p1} = 0.4 \times 1.05 \times 1.0 \times 350 = 147.0$ plf from 0 to 20 feet

$F_{p2} = 0.4 S_{DS} I_e W_{p2} = 0.4 \times 1.05 \times 1.0 \times 525 = 220.5$ plf from 14 to 20 feet

$F_{p3} = 0.4 S_{DS} I_e W_{p3} = 0.4 \times 1.05 \times 1.0 \times 131 = 55.0$ plf from 7 to 20 feet

The wall is designed for the combination of axial load from the gravity forces and bending and shear from the out-of-plane seismic forces.

6.7.11 Example 6.11 – Retail Building, Simplified Design Method

For the one-story retail building illustrated in Figure 6.42, determine the seismic base shear using the simplified alternative structural design criteria of 12.14.

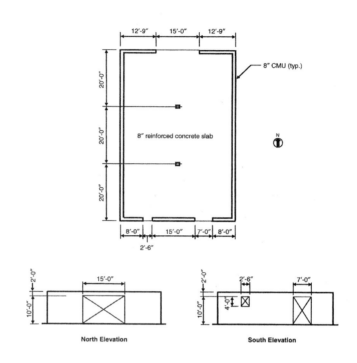

Figure 6.42 Plan and Elevations of One-story Retail Building, Example 6.11

DESIGN DATA

Location:	Seattle, WA (Latitude: 47.60°, Longitude: −122.33°)
Soil classification:	Site Class C
Occupancy:	Business occupancy
Structural system:	Bearing wall system with special reinforced masonry shear walls

SOLUTION

Step 1: Determine the seismic ground motion values from Flowchart 6.3.

Using the USGS "DesignMaps" Web Application, $S_S = 1.37$ and $S_1 = 0.53$.

Using Tables 11.4-1 and 11.4-2, the soil-modified accelerations are $S_{MS} = 1.37$ and $S_{M1} = 0.69$.

Design accelerations: $S_{DS} = \frac{2}{3} \times 1.37 = 0.91$ and $S_{D1} = \frac{2}{3} \times 0.69 = 0.46$

Step 2: Determine if the simplified method of 12.14 can be used for this building.

The simplified method is permitted to be used if the following 12 limitations are met:

1. The structure shall qualify for Risk Category I or II in accordance with Table 1.5-1.

From Table 1.5-1, the Risk Category is II. **O.K.**

2. The Site Class shall not be E or F.

   The Site Class is C in accordance with the design data. **O.K.**

3. The structure shall not exceed three stories in height.

   The structure is one story. **O.K.**

4. The SFRS shall be either a bearing wall system or building frame system in accordance with Table 12.14-1.

   The SFRS is a bearing wall system. **O.K.**

5. The structure has at least two lines of lateral resistance in each of the two major axis directions.

   Masonry shear walls are provided along two lines at the perimeter in both directions. **O.K.**

6. At least one line of resistance shall be provided on each side of the center of mass in each direction.

   The center of mass is approximately located at the geometric center of the building and walls are provided on all four sides at the perimeter. **O.K.**

7. For structures with flexible diaphragms, overhangs beyond the outside line of shear walls or braced frames shall satisfy: $a \leq d/5$.

   The diaphragm in this building is rigid, so this limitation is not applicable.

8. For buildings with diaphragms that are not flexible, the distance between the center of rigidity and the center of mass parallel to each major axis shall not exceed 15 percent of the greatest width of the diaphragm parallel to that axis.

   Assume that the center of mass is at the geometric center of this building (note: the exact location of the center of mass should be computed where it is anticipated to be offset from the geometric center of the building). This limitation is satisfied with respect to the east-west direction, since the center of rigidity and center of mass are on the same line due to the symmetrical layout of the walls on the east and west elevations.

   The center of rigidity must be located in the north-south direction since the north and south walls are not identical. By inspection, the center of rigidity is located closer to the south wall since the stiffness of that wall is greater than the stiffness of the north wall.

   To locate the center of rigidity, the stiffnesses of the north and south walls must be determined. Assuming that the piers are fixed at the top and bottom ends, the stiffnesses (or rigidities) of the walls and piers can be determined by the following:

   Total displacement of pier or wall i: $\delta_i = \delta_{fi} + \delta_{vi}$

   $$\delta_{fi} = \text{displacement due to bending} = \frac{\left(\dfrac{h_i}{\ell_i}\right)^3}{Et}$$

   $$\delta_{vi} = \text{displacement due to shear} = \frac{3\left(\dfrac{h_i}{\ell_i}\right)}{Et}$$

   where h_i = height of pier or wall

ℓ_i = length of pier or wall

t = thickness of pier or wall

E = modulus of elasticity of pier or wall

Stiffness of pier or wall $k_i = 1/\delta_i$

In lieu of a more rigorous analysis, the stiffness of a wall with openings is determined as follows: first, the deflection of the wall is obtained as though it were a solid wall with no openings. Next, the deflection of a solid strip of wall that contains the openings is subtracted from the total deflection. Finally, the deflection of each pier surrounded by the openings is added back.

Table 6.23 contains a summary of the stiffness calculations for the north wall. Similar calculations for the south wall are given in Table 6.24. The pier designations are provided in Figure 6.43.

Figure 6.43 Pier Designations for Stiffness Calculations, Example 6.11

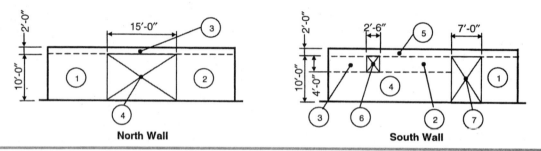

Table 6.23 Stiffness Calculations for North Wall

Pier/Wall	h_i (ft)	ℓ_i (ft)	$\delta_{fi}Et = (h_i/\ell_i)^3$	$\delta_{vi}Et = 3(h_i/\ell_i)$	$\delta_i Et$	k_i/Et
1 + 2 + 3 + 4	12	40.5	0.026	0.889	0.915	---
1 + 2 + 4	10	40.5	−0.015	−0.741	−0.756	---
1	10	12.75	0.483	2.353	---	0.353
2	10	12.75	0.483	2.353	---	0.353
1 + 2	---	---	---	---	1.416	0.706
					1.575 →	0.635

Table 6.24 Stiffness Calculations for South Wall

Pier/Wall	h_i (ft)	ℓ_i (ft)	$\delta_{fi}Et = (h_i/\ell_i)^3$	$\delta_{vi}Et = 3(h_i/\ell_i)$	$\delta_i Et$	k_i/Et	$\delta_i Et$
1 + 2 + 3 + 4 + 5 + 6 + 7	12	40.5	0.026	0.889	0.915	---	---
1 + 2 + 3 + 4 + 6 + 7	10	40.5	−0.015	−0.741	−0.756	---	---
2 + 3 + 4 + 6	10	25.5	0.060	1.177	---	---	1.237
2 + 3 + 6	4	25.5	−0.004	−0.471	---	---	−0.475
2	4	15.0	0.019	0.800	---	1.221	---
3	4	8.0	0.125	1.500	---	0.615	---
2 + 3	---	---	---	---	---	1.836	0.545
2 + 3 + 4	---	---	---	---	---	0.765	1.307
1	10	8.0	1.953	3.750	---	0.175	---
1 + 2 + 3 + 4	---	---	---	---	1.064	0.940	---
						1.223 →	0.818

North wall stiffness = $0.635Et$

South wall stiffness = $0.818Et$

East and west wall stiffness = $1.65Et$

The location of the center of rigidity in the north-south direction can be determined from the following equation:

$$\bar{y}_r = \frac{\sum k_i y_i}{\sum k_i}$$

where y_i is the distance from a reference point to wall i.

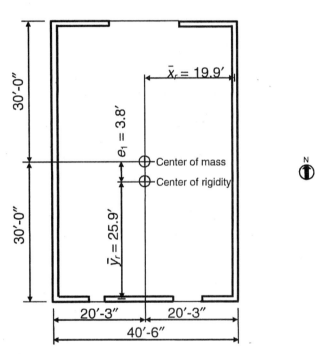

Figure 6.44
Locations of Center of Mass and Center of Rigidity, Example 6.11

Using the centerline of the south wall as the reference line (see Figure 6.44),

$$\bar{y}_r = \frac{0.635Et\left(60 - \frac{7.625}{12}\right)}{0.635Et + 0.818Et} = 25.9 \text{ feet}$$

$$e_1 = \left(30 - \frac{7.625}{2 \times 12}\right) - 25.9 = 3.8 \text{ feet}$$

$0.15 \times 60 = 9.0$ feet > 3.8 feet **O.K.**

In addition, Equations 12.14-2A and 12.14-2B must be satisfied:

$$\sum_{i=1}^{m} k_{2i} d_{1i}^2 + \sum_{j=1}^{n} k_{2j} d_{2j}^2 \geq 2.5\left(0.05 + \frac{e_1}{b_1}\right) b_1^2 \sum_{i=1}^{m} k_{1i}$$

$$\sum_{i=1}^{m} k_{1i} d_{1i}^2 + \sum_{j=1}^{n} k_{2j} d_{2j}^2 \geq 2.5\left(0.05 + \frac{e_2}{b_2}\right) b_2^2 \sum_{j=1}^{n} k_{2j}$$

where the notation is defined in Figure 12.14-1 and 12.14.1.1.

$$\sum_{i=1}^{m} k_{1i}d_{1i}^2 + \sum_{j=1}^{n} k_{2j}d_{2j}^2 = (0.635Et \times 33.5^2) + (0.818Et \times 25.9^2) + (2 \times 1.65Et \times 19.9^2)$$

$$= 2{,}568.2Et$$

$$2.5\left(0.05 + \frac{e_1}{b_1}\right)b_1^2 \sum_{i=1}^{m} k_{1i} = 2.5\left(0.05 + \frac{3.8}{60}\right) \times 60^2 \times (0.635Et + 0.818Et)$$

$$= 1{,}482.1Et < 2{,}568.2Et \quad \textbf{O.K.}$$

$$2.5\left(0.05 + \frac{e_2}{b_2}\right)b_2^2 \sum_{j=1}^{n} k_{2j} = 2.5\left(0.05 + \frac{0}{40.5}\right) \times 40.5^2 \times (2 \times 1.65Et)$$

$$= 676.6Et < 2{,}568.2Et \quad \textbf{O.K.}$$

Thus, all conditions of the eighth limitation are satisfied.

Note that Equations 12.14-2A and 12.14-2B need not be checked where the following three conditions are met:

a. The arrangement of walls is symmetric about each major axis.
b. The distance between the two most separated wall lines is at least 90 percent of the structure dimension perpendicular to that axis direction.
c. The stiffness along each of the lines of resistance considered in Item 2 above is at least 33 percent of the total stiffness in that direction.

In this example, only the second and third conditions are met.

9. Lines of resistance of the lateral-force-resisting system shall be oriented at angles of no more than 15 degrees from alignment with the major orthogonal axes of the building.

   The shear walls in both directions are parallel to the major axes. **O.K.**

10. The simplified design procedure shall be used for each major orthogonal horizontal axis direction of the building.

    The simplified design procedure is used in both directions (see Step 3). **O.K.**

11. System irregularities caused by in-plane or out-of-plane offsets of lateral-force-resisting elements shall not be permitted.

    This building does not have any irregularities. **O.K.**

12. The lateral load resistance of any story shall not be less than 80 percent of the story above.

    Since this is a one-story building, this limitation is not applicable. **O.K.**

Since all 12 limitations of 12.14.1.1 are satisfied, the simplified procedure may be used.

Step 3: Determine the seismic base shear from Flowchart 6.9.

1. Determine S_S, S_{DS} and the SDC from Flowchart 6.4.

   From Step 1, $S_S = 1.37$ and $S_{DS} = 0.91$.

According to 11.6, the SDC is permitted to be determined from Table 11.6-1 alone where the simplified design procedure is used.

For $S_{DS} > 0.50$ and Risk Category II, the SDC is D.

2. Determine the response modification factor, R, from Table 12.14-1.

   For SDC D, a bearing wall system with special reinforced masonry shear walls is required (system A7). For this system, $R = 5$.

3. Determine the effective seismic weight, W, in accordance with 12.14.8.1.

   Conservatively assume that the masonry walls are fully grouted and neglect any wall openings. Also assume a 10 psf superimposed dead load on the roof.

   Weight of masonry walls tributary to roof diaphragm

   $= 0.081 \times \dfrac{12}{2} \times 2(60.0 + 40.5) = 98$ kips

   Weight of roof slab $= \dfrac{8}{12} \times 0.150 \times 60 \times 40.5 = 243$ kips

   Superimposed dead load $= 0.010 \times 60 \times 40.5 = 24$ kips

   $W = 98 + 243 + 24 + = 365$ kips

4. Determine base shear, V, by the equation in 12.14.8.1.

   $$V = \dfrac{FS_{DS}}{R}W = \dfrac{1 \times 0.91 \times 365}{5} = 66 \text{ kips}$$

   where $F = 1$ for a one-story building (see 12.14.8.1).

   Since this is a one-story building, story shear $V_x = V$.

5. Distribute story shear to the shear walls.

   Since the building has a rigid diaphragm, the design story shear is distributed to the shear walls based on the relative stiffness of the walls, including the effects from the torsional moment, M_t, resulting from eccentricity between the locations of the center of mass and the center of rigidity. Note that the simplified procedure does not require accidental torsion (12.14.8.3.2.1).

   For lateral forces in the N-S direction, there is no torsional moment since there is no eccentricity between the center of mass and center of rigidity in that direction. The east and west walls have the same stiffness, so each wall must resist $66/2 = 33$ kips.

   For lateral forces in the E-W direction, the torsional moment is equal to $66 \times 3.8 = 251$ ft-kips.

   The total lateral force to be resisted by the north and south shear walls can be determined from the following equation:

   $$V_{1i} = \dfrac{k_{1i}}{\sum k_{1i}} V_x + \dfrac{d_{1i}k_{1i}}{\sum_{i=1}^{m} k_{1i}d_{1i}^2 + \sum_{j=1}^{n} k_{2j}d_{2j}^2} M_t$$

For the north shear wall:

$$V_{11} = \frac{0.635Et}{0.635Et + 0.818Et} \times 66 + \frac{33.5 \times 0.635Et}{2,568.2Et} \times 251$$

$$= (0.437 \times 66) + (0.0083 \times 251)$$

$$= 28.8 + 2.1 = 30.9 \text{ kips}$$

For the south shear wall:

$$V_{12} = \frac{0.818Et}{0.635Et + 0.818Et} \times 66 - \frac{25.9 \times 0.818Et}{2,568.2Et} \times 251$$

$$= (0.563 \times 66) - (0.0082 \times 251)$$

$$= 37.2 - 2.1 = 35.1 \text{ kips}$$

The east and west shear walls are subjected to a shear force due to the torsional moment for lateral forces in the east-west direction, but that force is less than the 33-kip force that is required for lateral forces in the N-S direction.

6.7.12 Example 6.12 – Nonbuilding Structure

Determine the seismic base shear for the nonbuilding illustrated in Figure 6.45 using (1) 2L4 × 4 × 1/2 braces and (2) 2L4 × 4 × 1/4 braces, given the design data below.

Figure 6.45 Plan and Elevation of Nonbuilding Structure, Example 6.12

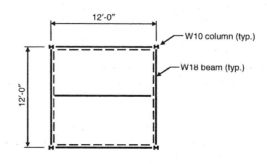

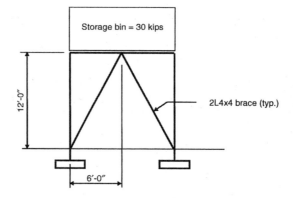

DESIGN DATA

Location:	Phoenix, AZ (Latitude: 33.42°, Longitude: −112.05°)
Soil classification:	Site Class D
Structural system:	Ordinary steel concentrically braced frame

SOLUTION

Part 1: Determine seismic base shear using 2L4 × 4 × 1/2 braces

Determine the seismic base shear from Flowchart 6.11.

This nonbuilding structure is similar to buildings, and the appropriate design requirements from Chapter 15 are used to determine the seismic base shear.

1. Determine S_{DS}, S_{D1} and the SDC from Flowchart 6.4.

    Using the USGS "DesignMaps" Web Application, $S_S = 0.17$ and $S_1 = 0.06$.

    Using Tables 11.4-1 and 11.4-2, the soil-modified accelerations are $S_{MS} = 0.27$ and $S_{M1} = 0.14$.

    Design accelerations: $S_{DS} = \frac{2}{3} \times 0.27 = 0.18$ and $S_{D1} = \frac{2}{3} \times 0.14 = 0.09$

    From IBC Table 1604.5, the Risk Category is I, assuming that the contents of the storage bin are not hazardous and that the structure represents a low hazard to human life in the event of failure.

    From Table 11.6-1, for $0.167 < S_{DS} < 0.33$, the SDC is B.

    From Table 11.6-2, for $0.067 < S_{D1} < 0.133$, the SDC is B.

    Therefore, the SDC is B for this nonbuilding structure.

2. Determine the importance factor, I_e, in accordance with 15.4.1.1.

    Based on Risk Category I, the importance factor, I_e, is equal to 1.0 from Table 1.5-2.

3. Determine the period, T, in accordance with 15.4.4.

    In lieu of a more rigorous analysis, Equation 15.4-6 is used to determine the period, T (note: the approximate fundamental period equations in 12.8.2.1 are not permitted to be used to determine the period of a nonbuilding structure (15.4.4)):

$$T = 2\pi \sqrt{\frac{\sum_{i=1}^{n} w_i \delta_i^2}{g \sum_{i=1}^{n} f_i \delta_i}}$$

where δ_i are the elastic deflections due to the forces, f_i, which represent any lateral force distribution in accordance with the principles of structural mechanics.

For this one-story nonbuilding structure, this equation reduces to

$$T = 2\pi \sqrt{\frac{w}{gk}}$$

where k is the lateral stiffness of the structure.

The stiffness can be obtained by applying a unit horizontal load to the top of the frame. This load does not produce any forces in the columns. Assuming that the elastic shortening of the beams is negligible, only the braces in a given direction contribute to the stiffness of the frame.

From statics, the force in one of the four braces due to a horizontal load of 1 applied to the top of the frame is equal to 0.5592. The horizontal deflection, δ, due to this unit load can be obtained from the following equation from the virtual work method:

$$\delta = \sum u^2 L / AE$$

where u = force in a brace due to the virtual (unit) load = 0.5592

L = length of a brace = $\sqrt{6^2 + 12^2}$ = 13.4 feet = 161 inches

A = area of a 2L4 × 4 × 1/2 brace = 7.49 square inches

E = modulus of elasticity = 29,000 ksi

Thus,

$$\delta = \frac{4 \times 0.5592^2 \times 161}{7.49 \times 29,000} = 0.0009 \text{ inch}$$

The stiffness $k = \frac{1}{\delta} = 1,079$ kips/in.

Therefore, the period, T, is

$$T = 2\pi \sqrt{\frac{30}{386 \times 1,079}} = 0.05 \text{ second}$$

Since the weight of the steel framing is negligible, it is not included in the period calculation.

4. Determine the base shear, V.

Since the period is less than 0.06 second, use Equation 15.4-5 to determine V:

$V = 0.30 S_{DS} W I_e = 0.30 \times 0.18 \times 30 \times 1.0 = 1.6$ kips

Part 2: Determine seismic base shear using 2L4 × 4 × 1/4 braces

The calculations are similar to those in Part 1, except the stiffness and the period of the structure are different due to the use of lighter braces.

$$\delta = \frac{4 \times 0.5592^2 \times 161}{3.87 \times 29,000} = 0.0018 \text{ inch}$$

Stiffness $k = \frac{1}{\delta} = 556$ kips/in.

$$T = 2\pi \sqrt{\frac{30}{386 \times 556}} = 0.07 \text{ second} > 0.06 \text{ second}$$

Therefore, the base shear, V, can be determined by the equivalent lateral force procedure (15.1.3).

Determine seismic response coefficient, C_S.

The value of C_S from Equation 12.8-2 is:

$$C_S = \frac{S_{DS}}{R/I_e} = \frac{0.18}{1.5/1.0} = 0.12$$

where the seismic response coefficient $R = 1.5$ from Table 15.4-1 for an ordinary steel concentrically braced frame with unlimited height, which is permitted to be designed by AISC 360, *Specification for Structural Steel Buildings* (that is, without any special seismic detailing).

Also, C_S must not be less than the larger of $0.044 S_{DS} I_e = 0.008$ and 0.01 (governs) (Equation 12.8-5).

Thus, the value of C_S from Eqaution 12.8-2 governs.

$V = C_S W = 0.12 \times 30 = 3.6$ kips

6.8 References

6.1. Building Seismic Safety Council (BSSC). 2009. *NEHRP Recommended Seismic Provisions for New Buildings and Other Structures*, FEMA P-750. Federal Emergency Management Agency, Washington, DC.

6.2. Building Seismic Safety Council (BSSC). 2010. *Earthquake-Resistant Design Concepts – An Introduction to the NEHRP Recommended Seismic Provisions for New Buildings and Other Structures*, FEMA P-749. Federal Emergency Management Agency, Washington, DC.

6.3. Chopra, Anil K. 2001. *Dynamics of Structures – Theory and Application to Earthquake Engineering*, 2nd Ed. Prentice Hall, NJ.

6.4. American Society of Civil Engineers (ASCE). 1998. *Seismic Analysis of Safety-Related Nuclear Structures*, ASCE 4-8. Reston, VA.

6.5. National Council of Structural Engineering Associations (NCSEA). 2009. *Guide to the Design of Diaphragms, Chords and Collectors Based on the 2006 IBC and ASCE/SEI 7-05*. International Code Council, Washington, DC.

6.6. Mays, Timothy W. 2010. *Guide to the Design of Out-of-Plane Wall Anchorage Based on the 2006/2009 IBC and ASCE/SEI 7-05*. International Code Council, Washington, DC.

6.9 Problems

6.1. A typical floor plan of a five-story residential building is shown in Figure 6.46. Determine the SDC of the structure for the cities identified in Table 6.25 assuming Site Class D.

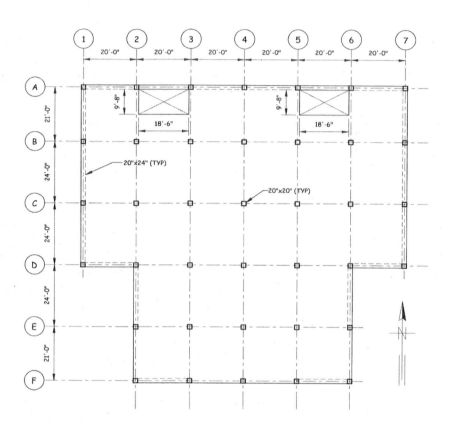

Figure 6.46
Typical Plan of Five-story Residential Building, Problem 6.1

Table 6.25 Locations for Problem 6.1

City	Latitude	Longitude
Berkeley, CA	37.87	−122.28
Boston, MA	42.36	−71.06
Chicago, IL	41.88	−87.63
Denver, CO	39.74	−104.99
Houston, TX	29.76	−95.37
New York, NY	40.72	−74.00

6.2. Given the information in Problem 6.1, determine V and F_x in both directions using the equivalent lateral force procedure for the building located in Berkeley, CA. The first story height is 12 feet and typical floor heights are 10 feet. Assume a 9-inch-thick reinforced concrete slab at all levels and that all concrete is normal weight (density = 150 pcf). Also assume a superimposed dead load of 30 psf on all floors and 10 psf on the roof. A glass curtain wall system is used on all faces of the building, which weighs 8 psf. The SFRS consists of moment-resisting frames in both directions.

6.3. Repeat Problem 6.2 for the building located in New York, NY.

6.4. Given the information in Problems 6.1 and 6.2, determine the diaphragm forces for the building located in New York, NY.

6.5. A typical floor in a 20-story structural steel-framed office building is shown in Figure 6.47. Determine V and F_x in both directions using the equivalent lateral force procedure given the following design data:

- $S_S = 0.46$, $S_1 = 0.20$
- Site Class B
- Dead load per floor (includes superimposed dead load and cladding) = 1,000 kips
- Dead load on roof (includes mechanical equipment and cladding) = 900 kips
- All story heights are 13 feet
- SFRS: steel concentrically braced frames as indicated in Figure 6.47

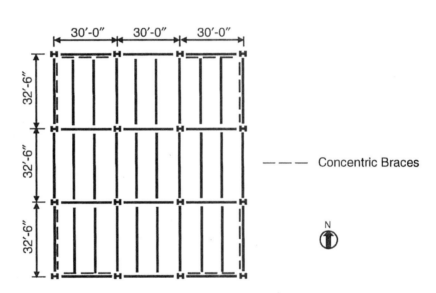

Figure 6.47
Typical Plan of Twenty-story Office Building, Problem 6.5

6.6. The roof of a one-story commercial building is shown in Figure 6.48. Determine if the simplified alternative method of 12.14 can be used to determine the seismic forces given the following design data:

- $S_S = 0.28$, $S_1 = 0.10$
- Site Class C
- All walls are 10-inches thick and are made of the same concrete
- Roof structure consists of open-web joists and metal deck

Determine the seismic base shear assuming the simplified alternative method can be used. For analysis purposes, assume the walls are fixed at the top and bottom.

Figure 6.48 Roof Plan of One-story Commercial Building, Problem 6.7

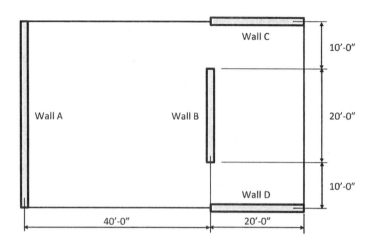

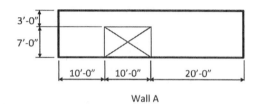

Wall A

6.7. Given the information in Problem 6.6, determine if the simplified alternative method of 12.14 can be used to determine the seismic forces when the roof deck consists of reinforced concrete framing. Determine the seismic base shear assuming the simplified alternative method can be used.

6.8. A 15-kip cooling tower that is braced below its center of mass is supported on the roof of an office building. Determine the seismic design force using $S_{DS} = 1.1$.

6.9. A glass curtain wall panel spans 12 feet between floor levels in an office building that utilizes special steel moment-resisting frames. It has been determined that the maximum deflections at the top and bottom of one of the floors are $\delta_{xAe} = 1.2$ inches and $\delta_{yAe} = 0.80$ inch. These deflections occur 48 feet and 36 feet above the base of the structure, respectively. Determine the seismic relative displacement.

CHAPTER 7
Flood Loads

7.1 Introduction

7.1.1 Overview of Flooding and Flood Loads

In general, *flooding* is the overflow of excess water from a body of water (river, stream, lake, ocean, etc.) onto adjoining land. Depending on local topography, one or more bodies of water can contribute to flooding at a particular site. Storms and tsunamis usually generate the most significant flood hazards. Information on tsunami-generated flood hazards can be found in Appendix M of the IBC.

IBC 202 defines flooding as a general or temporary condition of partial or complete inundation of normally dry land from:

1. The overflow of inland or tidal waters; or

2. The unusual or rapid accumulation or runoff of surface waters from any source.

Along coastlines and the shorelines of large lakes, flooding is caused by wind-driven surges and waves that push water onshore, while along streams and rivers, flooding results from the accumulation of rainfall runoff that drains from upland watersheds.

In order to correctly characterize the potential hazards caused by floodwaters, the following parameters must be investigated at any site regardless of the source:

- Origin of flooding
- Frequency of flooding
- Depth of floodwaters
- Velocity of floodwaters
- Direction of floodwaters
- Duration of flooding
- Effects due to waves
- Effects due to erosion and scour
- Effects due to flood-borne debris

It is evident from the aforementioned parameters that a variety of load types can be exerted on buildings due to flooding (Section 7.4.6 of this publication contains comprehensive information on how to determine flood loads). The effects of these loads can be intensified by erosion and scour, which can lower the ground surface around foundations, causing loss of bearing capacity and resistance to uplift due to lateral loads. Since it is not always possible to quantify each of the above parameters for a particular site, conservative estimates need to be made in order to define the corresponding flood loads.

7.1.2 Overview of Code Requirements

All structures and portions of structures located in flood hazard areas must be designed and constructed to resist the effects of flood hazards and flood loads (IBC 1612.1). As noted above, flood hazards may include erosion and scour whereas flood loads include flotation, lateral hydrostatic pressures (due to static or slow moving water), hydrodynamic pressures (due to fast moving water), wave impact and debris impact.

In cases where a building or structure is located in more than one flood zone or is partially located in a flood zone, the entire building or structure must be designed and constructed according to the requirements of the more restrictive flood zone.

The following sections address the hazards and loads that need to be considered for structures located in flood hazard areas.

7.2 Flood Hazard Areas

The first step in the design for flood loads is to determine if a building or structure is located in a flood hazard area or not. By definition, a *flood hazard area* is the greater of the following two areas (IBC 202):

1. The area within a floodplain subject to a 1-percent or greater chance of flooding in any year; or

2. The area designated as a flood hazard area on a community's flood hazard map, or otherwise legally designated.

The first of these two areas is typically acquired from *Flood Insurance Rate Maps* (FIRMs), which are prepared by the Federal Emergency Management Agency (FEMA) through the National Flood Insurance Program (NFIP). The NFIP is a voluntary program whose goal is to reduce the loss of life and the damage caused by floods, to help victims recover from floods and to promote an equitable distribution of costs among those who are protected by flood insurance and the general public. Conducting flood hazard studies and providing FIRMs and Flood Insurance Studies (FISs) for participating communities are major activities undertaken by the NFIP. Included in FISs are the FIRM, the Flood Boundary and Floodway Map (FBFM), the base flood elevation (BFE) and supporting technical data.

A FIRM is the official map of a community on which FEMA has delineated both the special flood hazard areas and the risk premium zones that are applicable to the community. In general, FIRMs show flood hazard areas along bodies of water where there is a risk of flooding by a *base flood*, that is, a flood having a 1-percent chance of being equaled or exceeded in any given year. A sample FIRM is shown in Figure 7.1.

Figure 7.1
Sample Flood Insurance Rate Map (Source: FEMA)

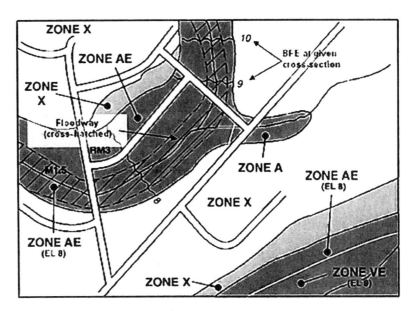

A FIRM that is produced in a digital format is designated a *Digital Flood Insurance Rate Map* (DFIRM). Figure 7.2 contains a sample DFIRM. FIRMs and DFIRMs for specific areas can be obtained from the FEMA Map Service Center website (http://msc.fema.gov/).

Figure 7.2
Sample Digital Flood Insurance Rate Map (Source: FEMA)

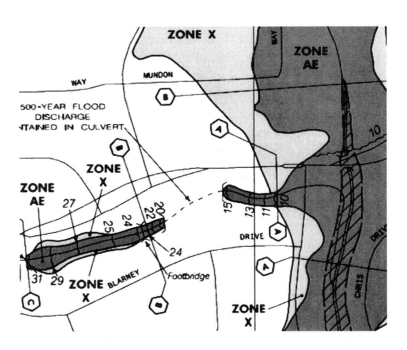

In regards to the return period for floods, the term "100-year flood" is a misnomer. Contrary to popular belief, it is not the flood that will occur once every 100 years, but the flood elevation that has a 1-percent chance of being equaled or exceeded in any given year. The "100-year flood" could occur more than once in a relatively short period of time. The flood elevation that has a 1-percent chance of being equaled or exceeded in any given year is the standard used by most government agencies and the NFIP for floodplain management and to determine the need for flood insurance.

In addition to showing the extent of flood hazards, the FIRMs and DFIRMs also show *base flood elevations* (BFEs) and *floodways*. The BFE is the height to which flood waters will rise during passage or occurrence of the base flood relative to the datum that is used on the flood hazard map. Statistical methods and computer methods that take into account the shape and nature of the floodplain (ground contours and the presence of any buildings, bridges and culverts) were used by FEMA to obtain the BFEs along rivers and streams. In such cases, the BFEs are provided next to river cross-sections on the flood hazard maps (see Figure 7.2). Along coastal areas, BFEs include wave heights and are established considering historical storm and wind patterns. The BFEs for coastal and lake areas are shown in parentheses immediately below the flood zone on the flood hazard maps (see Figures 7.1 and 7.2).

Floodways are channels of a river, creek or other watercourse and adjacent land areas that must be reserved in order to discharge the base flood waters without cumulatively increasing the water surface elevation by more than a designated height. As such, floodways must be kept clear of encroachments (such as fill or buildings) so that the base flood can be discharged without increasing the water surface elevations by more than the designated height given in floodway data tables in FISs.

Some local jurisdictions develop and subsequently adopt flood hazard maps that are more extensive than FEMA maps. In such cases, flood design and construction requirements must be satisfied in the areas delineated by the more extensive maps. Thus, a *design flood* is a flood associated with the greater of the area of a base flood based on a 1-percent or greater chance of flooding in any year or the area legally designated as a flood hazard area by a community.

7.3 Flood Hazard Zones

Flood hazard zones are areas designated on the flood hazard maps that indicate the magnitude and severity of the flood hazards. Such zones are contained in FIRMs.

Table 7.1 contains general descriptions of flood hazard zones and their designations. Comprehensive definitions for all of these zones can be found on the FEMA Map Service Center website (http://msc.fema.gov/).

Table 7.1 FEMA Flood Hazard Zones (Flood Insurance Zones)

Zone	Description
Moderate- to Low-risk Areas	
X*	These zones identify areas outside of the flood hazard area. • Shaded Zone X identifies areas that have a 0.2-percent probability of being equaled or exceeded during any given year. • Unshaded Zone X identifies areas where the annual exceedance probability of flooding is less than 0.2 percent.
High-risk Areas	
A, AE, A1-30, A99, AR, AO and AH	These zones identify areas of flood hazard that are not within the Coastal High Hazard Area; that is, these areas are subject to flooding from riverine (noncoastal) sources.
High-risk Coastal Areas	
V, VE and V1-V30	These zones identify the Coastal High Hazard Area, which extends from offshore to the inland limit of a primary frontal dune along an open coast and any other portion of the flood hazard zone that is subject to high-velocity wave action from storms or seismic sources and to the effects of severe erosion and scour. Such zones are generally based on wave heights (3 feet or greater) or wave runup depths (3 feet or greater).

* Zone B on older FIRMs corresponds to shaded Zone X on more recent FIRMs. Zone C on older FIRMs corresponds to unshaded Zone X on more recent FIRMs.

A designation of "X" is given to those areas with low to moderate risk of flooding. Areas of minimal flood hazard (areas above the 500-year flood level) are designated by the unshaded Zone X. A shaded Zone X is an area of moderate flood hazard (areas between the limits of the 100-year and 500-year floods). On older FIRMs, Zone B corresponds to shaded Zone X and Zone C corresponds to unshaded Zone X.

Special Flood Hazard Areas (high-risk areas) begin with the letter "A" or "V." A Zones are those areas within inland or coastal floodplains where high-velocity wave action is not expected during the base flood. In contrast, V Zones, which are also designated as *Coastal High Hazard Areas,* are those areas within a coastal floodplain where high-velocity wave action from storms or seismic sources can occur during the base flood event.

A *High-risk Flood Hazard Area* is an area where one or more of the following hazards may occur: alluvial fan flooding, flash floods, mudslides, ice jams, high velocity flow, coastal wave heights greater than or equal to 1.5 feet or erosion.

The concept of a *Coastal A Zone* is introduced in Chapter 5 of ASCE/SEI 7 and in ASCE 24-05 (Reference 7.1) to facilitate application of load combinations in Chapter 2 of ASCE/SEI 7. IBC 1612.4 references Chapter 5 of ASCE/SEI 7 and ASCE 24-05 for the design and construction of buildings and structures located in flood hazard areas. The requirements in these documents are covered in the next section of this publication. The NFIP regulations do not differentiate between Coastal and Noncoastal A Zones.

In general, a Coastal A Zone is an area located within a flood hazard area that is landward of a V Zone or landward of an open coast without mapped V Zones (such as the shorelines of the Great Lakes). Wave forces and erosion potential should be taken into consideration when designing a structure for the effects of flood loads in such zones.

To be classified as a Coastal A Zone, the principal source of flooding must be from astronomical tides, storm surges, seiches or tsunamis and not from riverine flooding. Additionally, stillwater flood depths must be greater than or equal to 1.9 feet, and breaking wave heights must be greater than or equal to 1.5 feet during the base flood conditions (see Section 4.1.1 of ASCE 24-05). Stillwater depth is the vertical distance between the ground and the stillwater elevation, which is the elevation that the surface of water would assume in the absence of waves. The stillwater elevation is referenced to the North American Vertical Datum (NAVD), the National Geodetic Vertical Datum (NGVD) or other datum, and it is documented in FIRMs (see Section 7.4.2 of this publication for more information).

The principal sources of flooding in Noncoastal A Zones are runoff from rainfall, snowmelt or a combination of both.

Hazard zones can also be established for areas vulnerable to flooding or inundation by tsunamis. A *tsunami hazard zone map* is defined in Appendix M of the IBC as a map adopted by the community that designates the extent of inundation by a design event tsunami. The map is to be based on that which is developed and provided to the community by either the local state agency or the National Oceanic and Atmospheric Administration (NOAA) under the National Tsunami Hazard Mitigation Program. More information on this program can be found on the following website: http://nthmp.tsunami.gov/. Appendix M also provides provisions for construction within a tsunami hazard zone.

It is recommended to check with the local building official for the most current information on flood hazard areas prior to designing a structure in a flood-prone area.

7.4 Design and Construction

7.4.1 General

According to IBC 1612.4, the design and construction of buildings and structures located in flood hazard areas shall be in accordance with Chapter 5 of ASCE/SEI 7 and ASCE/SEI 24-05 (Reference 7.1). Section 1.6 of ASCE/SEI 24 requires that design flood loads and their combination with other loads be determined by ASCE/SEI 7 (ASCE/SEI 24 references ASCE/SEI 7-02, which was the current edition of the standard at that time).

The provisions of ASCE/SEI 24 are intended to meet or exceed the requirements of the NFIP. Figures 1-1 and 1-2 in ASCE/SEI 24 illustrate the application of the standard and the application of the sections in the standard, respectively. *CodeMaster – Flood Resistant Design* contains a step-by-step procedure on flood-resistant design (Reference 7.2).

The provisions contained in IBC Appendix G, Flood-resistant Construction, are intended to fulfill the floodplain management and administrative requirements of NFIP that are not included in the IBC. IBC appendix chapters are not mandatory unless they are specifically referenced in the adopting ordinance of the jurisdiction.

Other provisions related to construction in flood hazard areas worth noting are found in IBC Chapter 18. IBC 1804.4 prohibits grading in flood hazard areas unless the specific requirements given in the section are met. IBC 1805.1.2.1 requires raised floor buildings in flood hazard areas to have the finished grade elevation under the floor (such as at a crawlspace) to be equal to or higher than outside finished grade on at least one side. The exception permits under-floor spaces in Group R-3 residential buildings to comply with Reference 7.3. This bulletin provides guidance on crawlspace construction and gives the minimum NFIP requirements for crawlspaces constructed in Special Flood Hazard Areas.

7.4.2 Design Flood Elevation

The *design flood elevation* (DFE) is used in the determination of flood loads and is defined as the elevation of the design flood including wave height (see Figure 7.3). For communities that have adopted minimum NFIP requirements, the DFE is identical to the BFE. The DFE exceeds the BFE in communities that have adopted requirements that exceed minimum NFIP requirements. *Freeboard* is defined as the additional depth of water above the BFE that has been approved by a local jurisdiction. This is essentially a factor of safety to account for uncertainties in the determination of flood elevations. As such, it provides an increased level of flood protection, which could reduce flood insurance premiums for structures located in flood hazard areas.

Figure 7.3 Flood Parameters

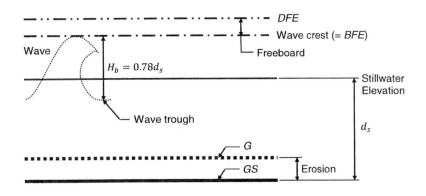

BFE = Base Flood Elevation

DFE = Design Flood Elevation

d_s = Design stillwater flood depth

G = Ground elevation

GS = Lowest eroded ground elevation adjacent to structure

H_b = Breaking wave height

The DFE is also used in determining the minimum elevation of the top of the lowest floor of a building or structure. In general, the IBC and ASCE/SEI 24 require that the lowest floor be elevated to or above the DFE. Table 2-1 in ASCE/SEI 24 contains the minimum elevation of the lowest floor as a function of the structure category, which is defined in Table 1-1, for flood hazard areas other than Coastal High Hazard Areas and Coastal A Zones. Similar information is provided in Table 4-1 of ASCE/SEI 24 for structures located in Coastal High Hazard Areas and Coastal A Zones. Table 7.2 contains a summary of the lowest floor reference based on flood hazard zones.

Table 7.2 Lowest Floor Reference

Zone	Reference	Code Section
A	Top of the floor (walking surface)	IBC 1612.5 ASCE/SEI 24 Section 2.3
V Coastal A	Bottom of the lowest horizontal structural member supporting the lowest floor	IBC 1612.5 ASCE/SEI 24 Section 4.4

7.4.3 Stillwater Flood Depth

The *stillwater flood depth*, d_s, is also used in calculating flood loads. This is defined as the vertical distance between the eroded ground elevation and the *stillwater elevation*. Unless local jurisdictions have adopted a more severe design flood, the stillwater elevation in riverine and lake areas is equal to the BFE published in the FIS and shown on the FIRM.

In coastal areas, the stillwater elevation is the average water level including waves and is published in the FIS (such values are not shown on FEMA FIRMs). Note that the stillwater elevation must be referenced to the same datum that was used in establishing the BFE and the DFE (see Figure 7.3). Guidance on estimating erosion in coastal areas can be found in FEMA'S

Coastal Construction Manual (Reference 7.4). Also, any freeboard that has been added to the 100-year flood should not be used to increase d_s since the load factors in ASCE/SEI 7 were developed for the 100-year flood load.

7.4.4 Breaking Wave Height

Design breaking wave height, H_b, is an important design parameter in the calculation of flood loads at coastal sites. Wave heights at such sites are to be calculated as the heights of depth-limited breaking waves (that is, wave heights that are limited by the depth of the water), which are equivalent to 78 percent of the stillwater depth (see Figure 7.3). Seventy percent of the breaking wave height lies above the stillwater elevation.

7.4.5 Design Flood Velocity

Design flood velocity, V, is used in determining the flood loads in hydrodynamic situations. Estimating such velocities is subject to considerable uncertainty, especially in coastal flood hazard areas. Generally, flood velocity is estimated from historical data, data in the FIS, flood model results or simplified methods. Suitable design values can sometimes be obtained from the local jurisdiction. Flood velocities should be estimated by assuming floodwaters can approach a site in any direction.

The following equations can be used to estimate design flood velocities in coastal areas (Reference 7.4 and ASCE/SEI C5.4.3):

$$V = \frac{d_s}{t} \tag{7.1}$$

$$V = (gd_s)^{0.5} \tag{7.2}$$

Equations 7.1 and 7.2 provide lower and upper bound estimates of the design flood velocity, respectively. In these equations, $t = 1$ second and g is the acceleration due to gravity (32.2 ft/sec^2). The lower bound velocity should be used if the site is

- distant from the flood source
- located in Zone A
- on flat or gently sloping terrain, or
- unaffected by other buildings or obstructions

The upper bound velocity is appropriate if the site is

- near the flood source
- in Zone V
- in Zone AO adjacent to Zone V
- in Zone A subject to velocity flow and wave action
- on steeply sloping terrain, or
- adjacent to other large buildings or obstructions that will confine or redirect floodwaters and increase local flood velocities

Figure 7.4 shows the lower and upper bound velocities, V, as a function of the stillwater flood depth, d_s.

Figure 7.4
Velocity versus Design Stillwater Flood Depth

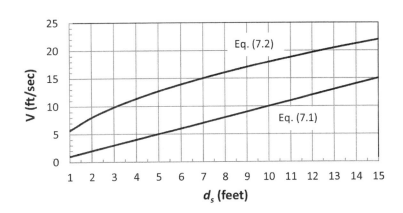

The preceding discussion on the determination of flood velocity is applicable to nontsunami situations. For estimating flood velocities during tsunamis, results from tsunami inundation or evacuation studies should be consulted.

7.4.6 Flood Loads

Overview

Floodwaters can create loads or pressures on surfaces of buildings and structures that are of two basic types: hydrostatic and hydrodynamic. Wave loads are generally considered a special type of hydrodynamic load. Objects that are transported by moving floodwaters can strike a structure causing an impact load.

The following loads are referenced in ASCE/SEI 5.4:

- Hydrostatic loads (ASCE/SEI 5.4.2)
- Hydrodynamic loads (ASCE/SEI 5.4.3)
- Wave loads (ASCE/SEI 5.4.4)
- Impact loads (ASCE/SEI 5.4.5)

Determination of these loads is based on the design flood, which is defined in Section 7.2 of this publication as the greater of the base flood (a flood having a 1-percent chance of being equaled or exceeded in any given year) and the flood adopted by the local jurisdiction.

Loads on walls that are required by ASCE/SEI 24 to break away (i.e., breakaway walls) are given in ASCE/SEI 5.3.3. The minimum design load must be the largest of the following loads: (1) wind load in accordance with ASCE/SEI Chapter 26, (2) seismic load in accordance with ASCE/SEI Chapter 12 or (3) 10 psf. The maximum permitted collapse load is 20 psf, unless the design meets the conditions of ASCE/SEI 5.3.3.

Hydrostatic Loads

Hydrostatic loads occur when stagnant or slowly moving water (velocity less than 5 feet per second; see ASCE/SEI C5.4.2) comes into contact with a building or building component. The water can be above or below the ground surface.

Hydrostatic loads are commonly subdivided into lateral loads, vertical downward loads and vertical upward loads (uplift or buoyancy). The hydrostatic pressure at any point on the surface of a structure or component is equal in all directions and acts perpendicular to the surface.

Lateral hydrostatic pressure is equal to zero at the surface of the water and increases linearly to $\gamma_s d_s$ at the stillwater depth, d_s, where γ_s is the unit weight of water, which is equal to 62.4 pounds per cubic foot for fresh water and 64.0 pounds per cubic foot for saltwater (see Figure 7.5). The total force, F_{sta}, on the width, w, of a vertical element is equal to $\gamma_s(d_s)^2 w/2$ and acts at the point that is two-thirds below the stillwater surface of the water. The magnitude of this type of load is generally not large enough to cause displacement of a building or building element unless there is a substantial difference in water elevation on opposite sides of the building or element.

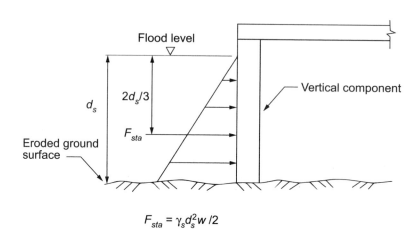

Figure 7.5 Lateral Hydrostatic Force

$$F_{sta} = \gamma_s d_s^2 w / 2$$

In communities that participate in the NFIP, it is required that buildings in V Zones be elevated above the BFE on an open foundation; thus, lateral hydrostatic loads are not applicable. Similarly, it is required that the foundation walls of buildings in A Zones be equipped with openings that allow flood water to enter and exit during design flood conditions; in such cases, the internal and external lateral hydrostatic pressure will equalize resulting in a lateral hydrostatic load that is equal to zero.

The vertical (buoyant) force, F_{bouy}, on a building or object is equal to the unit weight of water, γ_s, times the volume of floodwater displaced by a submerged object. This force must be resisted by the weight of the building or object and any anchorage forces that resist flotation. Since a flood can occur at any time, the live loads in buildings and the contents of objects (for example, the contents of underground storage tanks) should not be included to resist uplift forces. Buoyant forces on a building can be of concern where the actual stillwater flood depth exceeds the design stillwater flood depth. Such forces are also of concern for tanks, swimming pools and underground structures when the soil below them becomes saturated.

Figure 7.6 illustrates the uplift forces on the continuous wall foundation of a building. Buoyant forces are drastically reduced for structures supported on piles or piers (that is, open foundations).

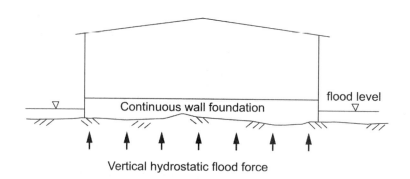

Figure 7.6 Vertical Hydrostatic Force

Hydrodynamic Loads

Hydrodynamic loads are caused by water moving at a moderate to high velocity above the ground level. Similar to wind loads, the loads produced by moving water include an impact load on the upstream face of a building, drag forces along the sides and a negative force (suction) on the downstream face (see Figure 7.7).

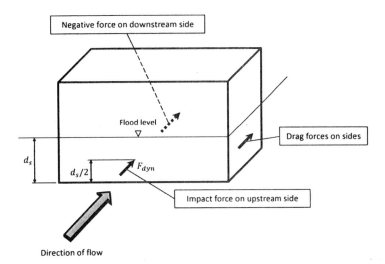

Figure 7.7
Hydrodynamic Loads on a Structure

According to ASCE/SEI 5.4.3, the dynamic effects of moving water are to be determined by a detailed analysis utilizing basic concepts of fluid mechanics. For any flow velocity, the hydrodynamic load, F_{dyn}, can be determined by the following equation (see Section 8.5.9 of Reference 7.4):

$$F_{dyn} = \frac{1}{2}a\rho V^2 A \qquad (7.3)$$

In this equation, a is the coefficient of drag (also referred to as the shape factor), ρ = the mass density of water = γ_w/g, A = surface area of a building or structure normal to the water flow = wd_s and w is the width of the building or structure (note that for completely immersed objects, $A = wh$ where h is the height of the object). The force, F_{dyn}, is assumed to act at the stillwater mid-depth (that is, halfway between the eroded ground surface and the stillwater elevation; see Figure 7.7).

The shape and roughness of the object that is exposed to the moving floodwaters as well as the flow condition are factors that must be considered when selecting the correct value of a. According to ASCE/SEI C5.4.3, drag coefficients for elements that are common in buildings and structures (for example, round and rectangular piles and columns) range from about 1.0 to 2.0. Recommended values of a are 2.0 for square or rectangular piles and 1.2 for round piles (see Section 8.5.9 of Reference 7.4). Because of the uncertainty of actual flood conditions at a particular site, ASCE/SEI 5.4.3 stipulates a minimum value of 1.25.

For other than square, rectangular or round piles, a can be determined based on one of the following ratios:

- For objects completely immersed in water:

  Ratio of width of object (w) to height of object (h)

- For objects not completely immersed in water:

  Ratio of width of object (w) to stillwater flood depth of water (d_s)

Additional information on the proper application of drag coefficients can be obtained from fluid mechanics publications and from Table 8-2 of the *Coastal Construction Manual* (Reference 7.4).

For a water velocity less than or equal to 10 feet per second, the exception in ASCE/SEI 5.4.3 permits the dynamic effects of moving water to be converted into an equivalent hydrostatic load. This is accomplished by increasing the DFE by an equivalent surcharge depth, d_h, which is determined by ASCE/SEI Equation 5.4-1:

$$d_h = \frac{aV^2}{2g} \qquad (7.4)$$

This equivalent surcharge depth is added to the DFE design depth. The resultant equivalent hydrostatic pressure is applied to and uniformly distributed across the vertical projected area of the building or structure that is perpendicular to the direction of flow.

Wave Loads

Wave loads result from water waves propagating over the surface of the water and striking a building or other object. Such loads can be separated into the following four categories:

- **Nonbreaking waves.** The effects of nonbreaking waves can usually be determined using the procedures in ASCE/SEI 5.4.2 for hydrostatic loads on walls and in ASCE/SEI 5.4.3 for hydrodynamic loads on piles.

- **Breaking waves.** These loads are caused by waves breaking on any portion of a building or structure. Although these loads are of short duration, they generally produce the largest magnitude of all of the different types of wave loads.

- **Broken waves.** The loads caused by broken waves are similar to hydrodynamic loads caused by flowing or surging water.

- **Uplift.** Uplift effects are caused by wave run-up striking any portion of a building or structure, deflection or peaking against the underside of surfaces.

Since the load from breaking waves is the highest, this load is used as the design wave load where applicable.

Wave loads must be included in the design of buildings and other structures located in both V Zones (wave heights equal to or greater than 3 feet) and A Zones (wave heights less than 3 feet). Since present NFIP mapping procedures do not designate V Zones in all areas where wave heights greater than 3 feet can occur during base flood conditions, it is recommended that historical flood damages be investigated near a site to determine whether or not wave forces can be significant.

ASCE/SEI 5.4.4 permits three methods to determine wave loads: (1) analytical procedures contained in ASCE/SEI 5.4.4, (2) advanced numerical modeling procedures and (3) laboratory test procedures (physical modeling). The analytical procedures of ASCE/SEI 5.4.4 for breaking wave loads are discussed next.

Breaking Wave Loads on Vertical Pilings and Columns

The net force, F_D, resulting from a breaking wave acting on a rigid vertical pile or column is determined by ASCE/SEI Equation 5.4-4 (see ASCE/SEI 5.4.4.1):

$$F_D = 0.5\gamma_w C_D D H_b^2 \qquad (7.5)$$

In this equation,

C_D = drag coefficient for breaking waves

$\qquad$ = 1.75 for round piles or columns

$\qquad$ = 2.25 for square or rectangular piles or columns

D = pile or column diameter for circular sections

$\qquad$ = 1.4 times the width of the pile or column for rectangular or square sections

H_b = breaking wave height (see Figure 7.3 and ASCE/SEI Equations 5.4-2 and 5.4-3)

$\qquad$ = $0.78d_s = 0.51(BFE - G)$

This load is assumed to act on the pile or column at the stillwater elevation (see Figure 7.8).

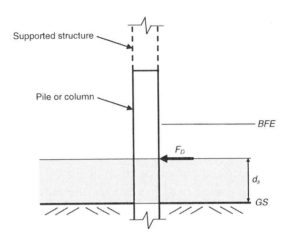

Figure 7.8
Breaking Wave Load on a Vertical Pile or Column

Breaking Wave Loads on Vertical Walls

Two cases are considered in ASCE/SEI 5.4.4.2 for breaking wave loads on vertical walls. In the first case, a wave breaks against a vertical wall of an enclosed dry space (that is, the space behind the vertical wall is dry, which means that no water is present to balance the static component of the wave load acting on the outside of the wall). ASCE/SEI Equations 5.4-5 and 5.4-6 give the maximum pressure, P_{max}, and net force, F_t, respectively, resulting from waves that are normally incident to the wall (that is, the direction of the wave approach is perpendicular to the face of the wall):

$$P_{max} = C_p \gamma_w d_s + 1.2 \gamma_w d_s \tag{7.6}$$

$$F_t = 1.1 C_p \gamma_w d_s^2 + 2.4 \gamma_w d_s^2 \tag{7.7}$$

It is assumed in these equations that the vertical wall causes a reflected (or standing) wave to form against the seaward side of the wall and that the crest of the wave reaches a height of $1.2d_s$ above the stillwater elevation.

The first and second terms in Equation 7.6 are the dynamic wave pressure and the static (hydrostatic) wave pressure, respectively. Both of these pressure distributions are illustrated in ASCE/SEI Figure 5.4-1 (see Figure 7.9). It is assumed that F_t acts at the design stillwater elevation.

Figure 7.9
Breaking Wave Load on a Vertical Wall—Space behind Vertical Wall is Dry

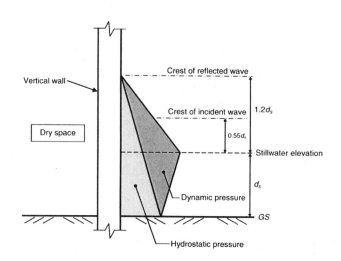

The quantity C_p is the dynamic pressure coefficient, which is given in ASCE/SEI Table 5.4-1 as a function of the risk category (see IBC Table 1604.5 and ASCE/SEI Table 1.5-1 for definitions of risk categories). Risk Category II buildings are assigned a value of C_p corresponding to a 1-percent probability of exceedance, which is consistent with the methods that are used by FEMA in mapping coastal flood hazard areas and in establishing minimum floor elevations. Note that the probability of exceedance is not an annual probability but that associated with a distribution of breaking wave pressures measured during laboratory wave tank tests (see Reference 7.5 for more information). Risk Category I, III and IV buildings are assigned a value of C_p corresponding to 50, 0.2 and 0.1 percent probabilities of exceedance, respectively.

In the second case, a wave breaks against a vertical wall where the stillwater level on both sides of the wall are equal (this can occur, for example, where a wave breaks against a wall equipped with openings, such as flood vents, that allow flood waters to equalize on both sides). The maximum combined wave pressure is computed by Equation 7.6, and the net breaking wave force, F_t, is determined by ASCE/SEI Equation 5.4-7:

$$F_t = 1.1 C_p \gamma_w d_s^2 + 1.9 \gamma_w d_s^2 \tag{7.8}$$

ASCE/SEI Figure 5.4-2 illustrates the pressure distributions in this case (see Figure 7.10).

Figure 7.10
Breaking Wave Load on a Vertical Wall—Stillwater Level Equal on Both Sides of Wall

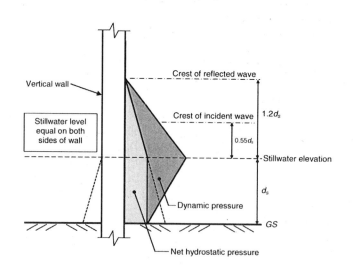

Breaking Wave Loads on Nonvertical Walls

The loads, F_t, determined by Equations 7.7 and 7.8 are to be modified in cases where a wall is not vertical. ASCE/SEI Equation 5.4-8 can be used to determine the horizontal component of a breaking wave load, F_{nv}, in such cases:

$$F_{nv} = F_t \sin^2 \alpha \qquad (7.9)$$

The angle α is the vertical angle between the nonvertical surface of a wall and the horizontal.

Breaking Wave Loads from Obliquely Incident Waves

As noted previously, it is usually assumed in coastal areas that the direction of wave approach is perpendicular to the shoreline. ASCE/SEI Equation 5.4-9 provides a method for reducing breaking wave loads on vertical surfaces that are not parallel to the shoreline:

$$F_{oi} = F_t \sin^2 \alpha \qquad (7.10)$$

In this equation, F_{oi} is the horizontal component of the obliquely incident breaking wave load and α is the horizontal angle between the direction of the wave approach and the vertical surface (see Figure 7.11).

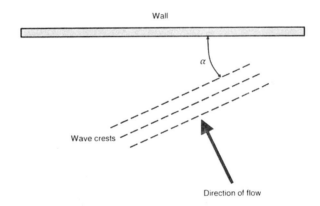

Figure 7.11
Breaking Wave Loads from Obliquely Incident Waves

Impact Loads

Impact loads occur where objects carried by moving water strike a building or structure. The magnitude of these loads is very difficult to predict; however, reasonable allowances can be made for them considering a number of different inherent uncertainties.

Flood Proofing Regulations (Reference 7.6) divides impact loads into three categories: (1) normal impact loads, (2) special impact loads and (3) extreme impact loads. Normal impact loads result from the isolated impacts of commonly encountered objects. The size, shape and weight of waterborne debris may vary according to region; thus, that which is common in one area may not be common in another.

Special impact loads result from large objects such as broken up ice floats and accumulations of waterborne debris. The loads are caused by these objects either striking or resting against the building or structure.

Extreme impact loads result from very large objects (such as boats, barges or parts of collapsed buildings) striking a building or structure. Design for such extreme loads is usually not practical in most cases unless the probability that such an impact load during the design flood is high.

Since most impact loads generated by waterborne debris is of short duration, a dynamic analysis of the building or structure may be appropriate. In cases where the natural period of the structure is much greater than 0.03 second, the impact load can be treated as a static load applied to the structure.

A rational method for calculating normal impact loads on buildings is given in ASCE/SEI C5.4.5. ASCE/SEI Equation C5-3 can be used to determine the impact force, F:

$$F = \frac{\pi W V_b C_I C_O C_D C_B R_{\max}}{2g(\Delta t)} \tag{7.11}$$

In this equation, W is the debris impact weight, which is to be selected considering local or regional conditions. ASCE/SEI C5.4.5 gives the following recommendations:

- **Riverine floodplains.** Trees and logs are the predominate type of debris with weights typically between 1,000 and 2,000 pounds. In riverine areas subject to floating ice, a typical debris weight range is 1,000 to 4,000 pounds.

- **Arid or semiarid regions.** In these regions, wood debris weight is usually less than 1,000 pounds.

- **Alluvial fan areas.** Stones and boulders usually present a much greater debris hazard than wood debris and a typical debris weight should be chosen based on the regional conditions.

- **Coastal areas.** In the Pacific Northwest, large trees and logs suggest a debris weight of 4,000 pounds. In coastal areas where piers and large pilings are prevalent, debris weights may range from 1,000 to 2,000 pounds. Where piers and large pilings are not present, debris will likely be due to failed decks, steps and building components and the corresponding weight is usually less than 500 pounds.

A realistic weight for trees, logs and other large wood-type debris, which are the most common forms of damaging debris, is 1,000 pounds. This is also a reasonable weight for small ice floes, boulders and some man-made objects.

The debris velocity, V_b, depends on the nature of the debris and the velocity of the floodwaters. Smaller pieces of debris usually travel at the velocity of the floodwaters; such items typically do not cause damage to buildings or structures. Large debris, which can cause damage, will likely travel at a velocity less than that of the floodwaters, since it can drag at the bottom or can be slowed by prior collisions with nearby objects. For calculating debris loads, the velocity of the debris should be taken equal to the velocity of the floodwaters.

The importance coefficient, C_I, is tabulated in ASCE/SEI Table C5-1 as a function of risk category. These coefficients are based on a probability distribution of impact loads obtained from laboratory tests and, as expected, the magnitude of this coefficient is larger for more important structures.

The orientation coefficient, C_O, is used to reduce the load determined by ASCE/SEI Equation C5-3 for other than head-on impact loads. Based on measurements taken during laboratory tests, an orientation factor of 0.8 has been adopted.

The depth coefficient, C_D, is used to account for reduced debris velocity due to debris dragging along the bottom in shallow water. The values given in ASCE/SEI Table C5-2 and Figure C5-1 are based on typical diameters of logs and trees and on anticipated diameters of root masses from drifting trees that are likely to be encountered in a flood hazard zone.

The reduction of debris velocity due to screening and sheltering provided by trees or other structures within about 300 feet upstream from a building or structure is accounted for by the blockage coefficient, C_B. Values of C_B are given in ASCE/SEI Table C5-3 and Figure C5-2.

The maximum response ratio, $R_{\max}$, modifies the impact load based on the natural period of the structure and the impact duration, Δt (that is, the time it takes to reduce the velocity of the object to zero). The recommended value of Δt is 0.03 second. Stiff or rigid buildings with natural periods that are similar to the impact duration will see an amplification of the impact load while more flexible buildings with natural periods greater than approximately four times the impact duration will see a reduction of the impact load. It is evident from the tabulated values of R_{max} in ASCE/SEI Table C5-4 that the critical period for buildings or structures subjected to impact flood loads is approximately 0.11 second for $\Delta t = 0.03$ second: buildings with a natural period greater than 0.11 second will see a reduction in the impact load while those with a natural period less than 0.11 second will see an increase in the impact load. ASCE/SEI C5.4.5 contains recommendations on the natural period to use for buildings subject to flood impact loads.

The $\pi/2$ factor in Equation 7.11 is a result of the half-sine form of the impulse load that is used in the derivation of this equation.

ASCE/SEI Equation C5-4, which is the standard drag force equation from fluid mechanics, can be used to determine special impact loads:

$$F = \frac{C_D \rho A V^2}{2} \quad (7.12)$$

In this equation, C_D is the drag coefficient, which is equal to 1.0; ρ is the density of water; A is the projected area of the debris accumulation, which is approximated by the depth of the accumulation times the width of the accumulation that is perpendicular to the direction of the flow; and V is the velocity of the flow upstream of the debris accumulation.

It is assumed that objects are at or near the water surface level when they strike a building. Thus, the loads determined by Equations 7.11 and 7.12 are usually assumed to act at the stillwater flood level; in general, these loads should be applied horizontally at the most critical location at or below the DFE.

7.4.7 Flood Load Combinations

Overview

While Chapter 5 of ASCE/SEI 7 provides the methods to determine the various types of loads due to floodwaters, it does not specify how these individual flood loads get combined into the flood load, F_a, that is to be used in the appropriate load combinations specified in IBC 1605.2.1 and ASCE/SEI 2.3.3. The following discussion provides insight on how F_a should be determined for the building or structure as a whole and for individual structural members.

It is evident from the information presented in the previous sections that the presence of an individual flood load depends on the flood zone in which the building or structure is sited as well as the building or structure type. Thus, the proper determination of F_a depends on these attributes as well.

Flood Load Combinations in A Zones

In an A Zone where the source of the floodwater is from a river or lake, it is reasonable to assume for the analysis of the entire (global) foundation with n piles or columns that one of the piles or columns will be subjected to the impact load, F, and the remaining piles or columns will be subjected to the hydrodynamic load, F_{dyn}; thus, the total flood load is $F_a = F + (n-1)F_{dyn}$.

The flood load, F_a, on an individual pile or column that supports the structure is equal to the sum of the hydrodynamic load, F_{dyn}, and the impact load, F ($F_a = F_{dyn} + F$). These same individual flood loads are used to determine F_a on a solid foundation wall with flood openings that conform to the requirements of ASCE/SEI 24 Section 2.6.1.

In the case of a solid foundation wall without flood openings that meets the dry-floodproofed requirements of ASCE/SEI Section 6.2, F_a is the sum of the hydrostatic load, F_{sta}, the hydrodynamic load, F_{dyn} and the impact load, F ($F_a = F_{sta} + F_{dyn} + F$).

Breaking wave loads need not be considered in A Zones except at sites that are subject to coastal waves loads. Determination of F_a for such sites is the same as that for Coastal A Zones and V Zones where the breaking wave load is equal to F_D for piles and columns or F_t, F_{nv} or F_{oi}, as applicable, for walls.

Flood Load Combinations in Coastal A Zones and V Zones

In a Coastal A Zone or a V Zone, hydrodynamic loads, breaking wave loads and impact loads must be considered.

For the analysis of the entire (global) foundation with n piles or columns, it is reasonable to assume that one of the piles or columns will be subjected to the impact load, F, and the remaining piles or columns will be subjected to the hydrodynamic load, F_{dyn}; thus, the total flood load is $F_a = F + (n-1)F_{dyn}$.

It is usually unrealistic to assume that impact loads occur on all piles or columns at the same time as breaking wave loads. As such, for the case of an individual pile or column in the front row (that is, in the row that experiences the initial effect of the floodwaters), F_a is equal to the larger of F_{dyn} and F_D plus F [F_a = larger of (F_{dyn}, F_D) + F].

As noted previously, solid foundation walls are not permitted by the IBC or ASCE/SEI 24 in Coastal A Zones or V Zones.

7.5 Examples

The following sections contain examples that illustrate the flood load provisions of IBC 1612, Chapter 5 in ASCE/SEI 7 and ASCE/SEI 24-05.

7.5.1 Example 7.1 – Residential Building Located in a Noncoastal A Zone

The plan dimensions of a residential building are 40 by 50 feet. Determine the design flood loads on the perimeter reinforced concrete foundation wall depicted in Figure 7.12 given the design data below and assuming that (1) the source of the floodwaters is a river and that (2) the dry-floodproofing requirements of Section 6.2 of ASCE/SEI 24 are met.

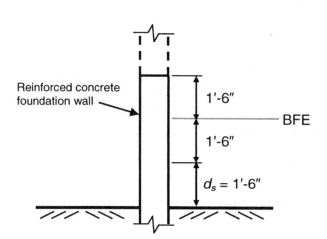

Figure 7.12
Reinforced Concrete Foundation Wall, Example 7.1

Design Data

Location:	Noncoastal A Zone
Design stillwater elevation, d_s:	1 ft-6 in.
Base flood elevation (BFE):	3 ft-0 in.

Solution

In this example, it is assumed that the location jurisdiction has not adopted a more severe flood than that obtained from the FIRM so that the BFE and the DFE are equal.

This residential building is classified as a Risk Category II building in accordance with IBC Table 1604.5. The elevation of the top of the lowest floor relative to the BFE must be greater than or equal to $3 + 1 = 4$ feet to satisfy the requirements of Section 2.3 of ASCE/SEI 24 for Risk Category II buildings located in Noncoastal A Zones (see Table 2-1 of ASCE/SEI 24; the definition of a Category II building in Table 1-1 of ASCE/SEI 24 is equivalent to that of a Risk Category II building in IBC Table 1604.5).

Since it is assumed that the dry-floodproofing requirements of Section 6.2 of ASCE/SEI 24 are met, no flood openings in the wall will be provided; thus, a hydrostatic load will act on the wall. Therefore, the applicable flood loads in this example are hydrostatic, hydrodynamic, and impact.

Step 1: Determine water velocity, V.

Since the building is located in a Noncoastal A Zone, it is appropriate to use the lower bound average water velocity, which is given by ASCE/SEI Equation C5-1 (see Equation 7.1):

$V = d_s/t = 1.5/1.0 = 1.5$ ft/sec

Step 2: Determine lateral hydrostatic load, F_{sta}.

Assuming fresh water (see Figure 7.5),

$$F_{sta} = \frac{\gamma_w d_s^2}{2} = \frac{62.4 \times 1.5^2}{2} = 70 \text{ lb/linear foot of foundation wall}$$

This load acts at 1 ft-0 in. below the stillwater surface of the water (or, equivalently, 6 inches above the ground surface).

Step 3: Determine hydrodynamic load, F_{dyn}.

The hydrodynamic load, F_{dyn}, is determined by Equation 7.3:

$$F_{dyn} = \frac{1}{2}a\rho V^2 A$$

In lieu of a more detailed analysis, the drag coefficient, a, is determined from Table 8-2 of the *Coastal Construction Manual* (Reference 7.4). Since the building will not be completely immersed in water, a is determined by the ratio of longest plan dimension of the building to d_s: $50/1.5 = 33.3$. For a ratio of 33.3, $a = 1.5$ from Table 8-2.

Thus,

$$F_{dyn} = \frac{1}{2}a\rho V^2 A = \frac{1}{2} \times 1.5 \times \frac{62.4}{32.2} \times 1.5^2 \times 1.5 = 5.0 \text{ lb/linear foot of foundation wall}$$

This load acts at 9 inches below the stillwater surface of the water (or, equivalently, 9 inches above the ground surface).

Step 4: Determine impact load, F.

Both normal and special impact loads are determined.

1. Normal impact loads

    ASCE/SEI Equation C5-3 is used to determine normal impact loads:

    $$F = \frac{\pi W V_b C_I C_O C_D C_B R_{max}}{2g(\Delta t)}$$

    As noted in Section 7.4.6 of this publication, guidance on establishing the debris weight, W, is given in ASCE/SEI C5.4.5. It is assumed in this example that $W = 1{,}000$ lb.

    It is reasonable to assume that the velocity of the object, V_b, is equal to the velocity of the water, V. Thus, from Step 1 of this example, $V_b = 1.5$ ft/sec.

    The importance coefficient, C_I, is obtained from ASCE/SEI Table C5-1. For a Risk Category II building, $C_I = 1.0$.

    The orientation coefficient $C_O = 0.8$. This coefficient accounts for impacts that are oblique to the structure.

    The depth coefficient, C_D, is obtained from ASCE/SEI Table C5-2 or, equivalently, from Figure C5-1. For a stillwater depth of 1.5 feet in an A Zone, $C_D = 0.125$ from linear interpolation.

    The blockage coefficient, C_B, is obtained from ASCE/SEI Table C5-3 or, equivalently, from Figure C5-2. Assuming that there is no upstream screening and that the flow path is wider than 30 feet, $C_B = 1.0$.

    The maximum response ratio for impulsive load, R_{max}, is determined from ASCE/SEI Table C5-4. Using the recommended duration of the debris impact load, Δt, of 0.03 second (see ASCE/SEI C5.4.5) and assuming that the natural period of the building is 0.2 second, the ratio of the impact duration to the natural period of the building is $0.03/0.2 = 0.15$. From ASCE/SEI Table C5-4, $R_{max} = 0.6$ from linear interpolation.

    Therefore,

    $$F = \frac{\pi \times 1{,}000 \times 1.5 \times 1.0 \times 0.8 \times 0.125 \times 1.0 \times 0.6}{2 \times 32.2 \times 0.03} = 146 \text{ lb}$$

    This load acts at the stillwater flood elevation and can be distributed over an appropriate width of the foundation wall.

2. Special impact loads

    ASCE/SEI Equation C5-4 is used to determine special impact loads:

    $$F = \frac{C_D \rho A V^2}{2}$$

Using a drag coefficient $C_D = 1$ and assuming a projected area of debris accumulation $A = 1.5 \times 50 = 75$ square feet, the special impact force, F, is

$$F = \frac{1 \times \left(\frac{62.4}{32.2}\right) \times 75 \times 1.5^2}{2} = 164 \text{ lb}$$

This load acts at the stillwater flood elevation and is uniformly distributed over the width of the foundation wall.

Step 5: Determine flood load, F_a.

For the global system, the flood load, F_a, in this case is equal to the summation of the hydrostatic, hydrodynamic and impact loads. These loads act at different locations on the foundation walls (see Figures 7.5, 7.7 and 7.8).

In the design of an individual wall, the hydrostatic, hydrodynamic and impact flood loads make up F_a and are applied to the wall at the locations identified in Figures 7.5, 7.7 and 7.8; these loads are combined with other loads in accordance with the modified strength design load combinations including flood loads in ASCE/SEI 2.3.3(2) or the modified allowable stress design load combinations including flood loads in ASCE/SEI 2.4.2(2).

The design and construction of the foundation, including the foundation walls, must satisfy the requirements of Section 1.5.3 of ASCE/SEI 24.

7.5.2 Example 7.2 – Residential Building Located in an A Zone

For the residential building and design data given in Example 7.1, determine the design flood loads on the perimeter reinforced concrete foundation wall depicted in Figure 7.12, assuming that the source of the floodwaters is from coastal waters and that the dry-floodproofing requirements of Section 6.2 of ASCE/SEI 24 are not met.

SOLUTION

As in Example 7.1, it is assumed that the location jurisdiction has not adopted a more severe flood than that obtained from the FIRM; thus, the BFE and the DFE are equal.

This residential building is classified as a Risk Category II building in accordance with IBC Table 1604.5. The elevation of the top of the lowest floor relative to the BFE must be greater than or equal to $3 + 1 = 4$ feet to satisfy the requirements of Section 2.3 of ASCE/SEI 24 for Risk Category II buildings located in an A Zone (see Table 2-1 of ASCE/SEI 24; the definition of a Category II building in Table 1-1 of ASCE/SEI 24 is equivalent to that of a Risk Category II building in IBC Table 1604.5).

Since it is assumed that the dry-floodproofing requirements of Section 6.2 of ASCE/SEI 24 are not met, nonengineered openings or engineered openings (such as flood vents) must be installed in the foundation wall (see Section 2.6.1 of ASCE/SEI 24). These openings, which must satisfy the requirements of ASCE/SEI 24 Sections 2.6.2.1 and 2.6.2.2, respectively, allow floodwaters to pass through the wall and eliminate lateral hydrostatic loads, F_{sta}, on the wall. The total net area of nonengineered openings must be at least one square inch for each square foot of enclosed area, which in this example is equal to $40 \times 50 = 2,000$ square feet. Therefore, the total net area of openings is equal to 2,000 square inches. Additional details on the location and size of the openings are provided in Section 2.6.2.1 of ASCE/SEI 24. If the engineered opening requirements of Section 2.6.2.2 of ASCE/SEI 24 are satisfied and engineering calculations are performed, the 2,000-square-inch net opening area can be reduced.

Since the wall is subject to coastal waters and openings are provided in the wall to allow passage of floodwaters, the applicable flood loads in this example are hydrodynamic, breaking wave and impact.

Step 1: Determine water velocity, V.

Since the building is located in an A Zone, it is appropriate to use the lower bound average water velocity, which is given by ASCE/SEI Equation C5-1 (see Equation 7.1):

$V = d_s/t = 1.5/1.0 = 1.5$ ft/sec

Step 2: Determine hydrodynamic load, F_{dyn}.

The hydrodynamic load, F_{dyn}, is determined by Equation 7.3:

$$F_{dyn} = \frac{1}{2} a \rho V^2 A$$

In lieu of a more detailed analysis, the drag coefficient, a, is determined from Table 8-2 of the *Coastal Construction Manual* (Reference 7.4). Since the building will not be completely immersed in water, a is determined by the ratio of longest plan dimension of the building to d_s: 50/1.5 = 33.3. For a ratio of 33.3, $a = 1.5$ from Table 8-2.

Thus,

$$F_{dyn} = \frac{1}{2} a \rho V^2 A = \frac{1}{2} \times 1.5 \times \frac{62.4}{32.2} \times 1.5^2 \times 1.5 = 5.0 \text{ lb/linear foot of foundation wall}$$

This load acts at 1 ft-0 in. below the stillwater surface of the water (or, equivalently, 6 inches above the ground surface).

Step 3: Determine breaking wave load, F_t.

Since openings are provided in the wall to allow passage of floodwaters, the breaking wave load, F_t, is determined by ASCE/SEI Equation 5.4-7, which is applicable where free water exists behind the wall (see Equation 7.8):

$$F_t = 1.1 C_p \gamma_w d_s^2 + 1.9 \gamma_w d_s^2$$

For a Risk Category II building, the dynamic pressure coefficient $C_p = 2.8$ from ASCE/SEI Table 5.4-1. Thus,

$$F_t = (1.1 \times 2.8 \times 62.4 \times 1.5^2) + (1.9 \times 62.4 \times 1.5^2)$$

$$= 432 + 267 = 699 \text{ lb/linear foot of foundation wall}$$

This load acts at the stillwater elevation, which is 1 ft-6 in. above the ground surface.

Step 4: Determine impact load, F.

Both the normal and special impact loads are determined exactly the same as that in Example 7.1:

Normal impact load: $F = 146$ lb

Special impact load: $F = 164$ lb

Step 5: Determine flood load, F_a.

For the global system, the flood load, F_a, in this case is equal to the greater of F_{dyn} and F_t plus F. These loads act at different locations on the foundation walls (see Figures 7.7, 7.8 and 7.10).

In the design of an individual wall, the hydrodynamic, breaking wave and impact flood loads make up F_a and are applied to the wall at the locations identified in Figures 7.7, 7.8 and 7.10; these loads are combined with other loads in accordance with the modified strength design load combinations including flood loads in ASCE/SEI 2.3.3(2) or the modified allowable stress design load combinations including flood loads in ASCE/SEI 2.4.2(2).

The design and construction of the foundation, including the foundation walls, must satisfy the requirements of Section 1.5.3 of ASCE/SEI 24.

7.5.3 Example 7.3 – Residential Building Located in a Coastal A Zone

For the residential building described in Example 7.1, determine the design flood loads on the reinforced concrete columns depicted in Figure 7.13 given the design data below. Assume the building is supported on a total of 16 columns.

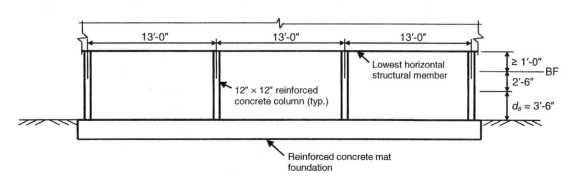

Figure 7.13 Partial Elevation of Residential Building, Example 7.3

Design Data

Location:	Coastal A Zone
Design stillwater elevation, d_s:	3 ft-6 in.
Base flood elevation (BFE):	6 ft-0 in.

Solution

In this example, it is assumed that the location jurisdiction has not adopted a more severe flood than that obtained from the FIRM so that the BFE and the DFE are equal.

The provided data satisfy the criteria of Coastal A Zones given in Section 4.1.1 of ASCE/SEI 24: stillwater depth = 3.5 feet > 1.9 feet and wave height = $0.78d_s$ = 2.7 feet > 1.5 feet.

This residential building is classified as a Risk Category II building in accordance with IBC Table 1604.5. The elevation of the bottom of the lowest supporting horizontal structural member of the lowest floor relative to the BFE must be greater than or equal to 6 + 1 = 7 feet to satisfy the requirements of Section 4.4 of ASCE/SEI 24 for Risk Category II buildings located in Coastal A Zones (see Table 4-1 of ASCE/SEI 24; the definition of a Category II building in Table 1-1 of ASCE/SEI 24 is equivalent to that of a Risk Category II building in IBC Table 1604.5).

The applicable flood loads are hydrodynamic, breaking wave and impact.

Step 1: Determine water velocity, V.

Since the building is located in a Coastal A Zone, it is appropriate to use the upper bound average water velocity, which is given by ASCE/SEI Equation C5-2:

$$V = (gd_s)^{0.5} = (32.2 \times 3.5)^{0.5} = 10.6 \text{ ft/sec}$$

Step 2: Determine hydrodynamic load, F_{dyn}.

Since the water velocity exceeds 10 ft/sec, it is not permitted to use an equivalent hydrostatic load to determine the hydrodynamic load (ASCE/SEI 5.4.3).

The hydrodynamic load, F_{dyn}, is determined by Equation 7.3:

$$F_{dyn} = \frac{1}{2} a \rho V^2 A$$

Based on the recommendations in ASCE/SEI C5.4.3 and Section 8.5.9 of the *Coastal Construction Manual* (Reference 7.4), the drag coefficient, a, is taken as 2.0 for rectangular columns.

Assuming salt water, the hydrodynamic load is

$$F_{dyn} = \frac{1}{2} \times 2.0 \times \left(\frac{64.0}{32.2}\right) \times 10.6^2 \times 3.5 \times \left(\frac{12}{12}\right) = 782 \text{ lb}$$

This load acts at 1 ft-9 in. below the stillwater surface of the water.

Step 3: Determine breaking wave load, F_D.

The breaking wave load is determined by ASCE/SEI Equation 5.4-4, which is applicable for vertical pilings and columns:

$$F_D = 0.5 \gamma_w C_D D H_b^2$$

According to ASCE/SEI 5.4.4.1, the drag coefficient, C_D, is equal to 2.25 for square columns.

The breaking wave height, H_b, is determined by ASCE/SEI Equation 5.4-2:

$$H_b = 0.78 d_s = 0.78 \times 3.5 = 2.7 \text{ feet}$$

Therefore, the breaking wave load on one of the columns is

$$F_D = 0.5 \times 64.0 \times 2.25 \times \left(1.4 \times \frac{12}{12}\right) \times 2.7^2 = 735 \text{ lb}$$

This load acts at the stillwater elevation, which is 3 ft-6 in. above the ground surface.

Step 4: Determine impact load, F.

Both normal and special impact loads are determined.

1. Normal impact loads

    ASCE/SEI Equation C5-3 is used to determine normal impact loads:

    $$F = \frac{\pi W V_b C_I C_O C_D C_B R_{max}}{2g(\Delta t)}$$

    As noted in Section 7.4.6 of this publication, guidance on establishing the debris weight, W, is given in ASCE/SEI C5.4.5. It is assumed in this example that $W = 1,000$ lb.

    It is reasonable to assume that the velocity of the object, V_b, is equal to the velocity of the water, V. Thus, from Step 1 of this example, $V_b = 10.6$ ft/sec.

The importance coefficient, C_I, is obtained from Table C5-1. For a Risk Category II building, $C_I = 1.0$.

The orientation coefficient $C_O = 0.8$. This coefficient accounts for impacts that are oblique to the structure.

The depth coefficient, C_D, is obtained from ASCE/SEI Table C5-2 or, equivalently, from Figure C5-1. For a stillwater depth of 3.5 feet in a Coastal A Zone, $C_D = 0.63$ from linear interpolation.

The blockage coefficient, C_B, is obtained from ASCE/SEI Table C5-3 or, equivalently, from Figure C5-2. Assuming that there is no upstream screening and that the flow path is wider than 30 feet, $C_B = 1.0$.

The maximum response ratio for impulsive load, R_{max}, is determined from ASCE/SEI Table C5-4. Using the recommended duration of the debris impact load, Δt, of 0.03 second (see ASCE/SEI C5.4.5) and assuming that the natural period of the building is 0.2 second, the ratio of the impact duration to the natural period of the building is $0.03/0.2 = 0.15$. From Table C5-4, $R_{max} = 0.6$ from linear interpolation.

Therefore,

$$F = \frac{\pi \times 1,000 \times 10.6 \times 1.0 \times 0.8 \times 0.63 \times 1.0 \times 0.6}{2 \times 32.2 \times 0.03} = 5,212 \text{ lb}$$

This load acts at the stillwater flood elevation.

2. Special impact loads

ASCE/SEI Equation C5-4 is used to determine special impact loads:

$$F = \frac{C_D \rho A V^2}{2}$$

Using a drag coefficient $C_D = 1$ and assuming a projected area of debris accumulation $A = 1 \times 3.5 = 3.5$ square feet, the impact force, F, on one column is

$$F = \frac{1 \times \left(\frac{64.0}{32.2}\right) \times 3.5 \times 10.6^2}{2} = 391 \text{ lb}$$

This load acts at the stillwater flood elevation.

Step 5: Determine flood load, F_a.

For the analysis of the entire (global) foundation with 16 columns, it is reasonable to assume that one of the piles or columns will be subjected to the impact load, F, and the remaining piles or columns will be subjected to the hydrodynamic load, F_{dyn}; thus, the total flood load is $F_a = 5,212 + (15 \times 782) = 16,942$ lb.

For the case of an individual column in the front row, F_a is equal to the larger of F_{dyn} and F_D plus F: $F_a = 782 + 5,212 = 5,994$ lb.

The design and construction of the mat foundation must satisfy the requirements of Section 4.5 of ASCE/SEI 24. The top of the mat foundation must be located below the eroded ground elevation and must extend to a depth sufficient to provide the support to prevent flotation, collapse, or permanent lateral movement under the design load combinations (see Section 1.5.3 of ASCE/SEI 24).

The design and construction of the reinforced concrete columns must satisfy the requirements of ACI 318-11 (Section 4.5.7.3 of ASCE/SEI 24 references ACI 318-02).

7.5.4 Example 7.4 – Residential Building Located in a V Zone

For the residential building described in Examples 7.1 and 7.3, determine the design flood loads on 8-inch-diameter reinforced concrete piles given the design data below. The partial elevation of the building is similar to that shown in Figure 7.13. Assume the building is supported on a total of 16 piles.

DESIGN DATA	
Location:	V Zone
Design stillwater elevation, d_s:	4 ft-6 in.
Base flood elevation (BFE):	7 ft-8 in.

SOLUTION

In this example, it is assumed that the location jurisdiction has not adopted a more severe flood than that obtained from the FIRM so that the BFE and the DFE are equal.

The provided data satisfy the criteria of V Zones given in Section 4.1.1 of ASCE/SEI 24: stillwater depth = 4.5 feet > 3.8 feet and wave height = $0.78d_s$ = 3.5 feet > 3.0 feet.

This residential building is classified as a Risk Category II building in accordance with IBC Table 1604.5. The elevation of the bottom of the lowest supporting horizontal structural member of the lowest floor relative to the BFE must be greater than or equal to 7.67 + 1 = 8.67 feet to satisfy the requirements of Section 4.4 of ASCE/SEI 24 for Risk Category II buildings located in V Zones (see Table 4-1 of ASCE/SEI 24; the definition of a Category II building in Table 1-1 of ASCE/SEI 24 is equivalent to that of a Risk Category II building in IBC Table 1604.5).

The applicable flood loads are hydrodynamic, breaking wave, and impact.

Step 1: Determine water velocity, V.

Since the building is located in a V Zone, it is appropriate to use the upper bound average water velocity, which is given by ASCE/SEI Equation C5-2:

$$V = (gd_s)^{0.5} = (32.2 \times 4.5)^{0.5} = 12.0 \text{ ft/sec}$$

Step 2: Determine hydrodynamic load, F_{dyn}.

Since the water velocity exceeds 10 ft/sec, it is not permitted to use an equivalent hydrostatic load to determine the hydrodynamic load (ASCE/SEI 5.4.3).

The hydrodynamic load, F_{dyn}, is determined by Equation 7.3:

$$F_{dyn} = \frac{1}{2} a \rho V^2 A$$

Based on the recommendations in ASCE/SEI C5.4.3 and Section 8.5.9 of the *Coastal Construction Manual* (Reference 7.4), the drag coefficient, a, is taken as 1.2 for round piles.

Assuming salt water, the hydrodynamic load is

$$F_{dyn} = \frac{1}{2} \times 1.2 \times \left(\frac{64.0}{32.2}\right) \times 12.0^2 \times 4.5 \times \left(\frac{8}{12}\right) = 515 \text{ lb}$$

This load acts at 2 ft-3 in. below the stillwater surface of the water.

Step 3: Determine breaking wave load, F_D.

The breaking wave load is determined by ASCE/SEI Equation 5.4-4, which is applicable for vertical pilings and columns:

$$F_D = 0.5\gamma_w C_D D H_b^2$$

According to ASCE/SEI 5.4.4.1, the drag coefficient, C_D, is equal to 1.75 for round piles.

The breaking wave height, H_b, is determined by ASCE/SEI Equation 5.4-2:

$$H_b = 0.78 d_s = 0.78 \times 4.5 = 3.5 \text{ feet}$$

Therefore, the breaking wave load on one of the piles is

$$F_D = 0.5 \times 64.0 \times 1.75 \times \left(\frac{8}{12}\right) \times 3.5^2 = 457 \text{ lb}$$

This load acts at the stillwater elevation, which is 4 ft-6 in. above the ground surface.

Step 4: Determine impact load, F.

Both normal and special impact loads are determined.

1. Normal impact loads

    ASCE/SEI Equation C5-3 is used to determine normal impact loads:

    $$F = \frac{\pi W V_b C_I C_O C_D C_B R_{max}}{2g(\Delta t)}$$

    As noted in Section 7.4.6 of this publication, guidance on establishing the debris weight, W, is given in ASCE/SEI C5.4.5. It is assumed in this example that $W = 1{,}000$ lb.

    It is reasonable to assume that the velocity of the object, V_b, is equal to the velocity of the water, V. Thus, from Step 1 of this example, $V_b = 12.0$ ft/sec.

    The importance coefficient, C_I, is obtained from Table C5-1. For a Risk Category II building, $C_I = 1.0$.

    The orientation coefficient $C_O = 0.8$. This coefficient accounts for impacts that are oblique to the structure.

    The depth coefficient, C_D, is obtained from ASCE/SEI Table C5-2 or, equivalently, from Figure C5-1. For a V Zone, $C_D = 1.0$.

    The blockage coefficient, C_B, is obtained from ASCE/SEI Table C5-3 or, equivalently, from Figure C5-2. Assuming that there is no upstream screening and that the flow path is wider than 30 feet, $C_B = 1.0$.

    The maximum response ratio for impulsive load, R_{max}, is determined from ASCE/SEI Table C5-4. Using the recommended duration of the debris impact load, Δt, of 0.03 second (see ASCE/SEI C5.4.5) and assuming that the natural period of the building is 0.2 second, the ratio of the impact duration to the natural period of the building is $0.03/0.2 = 0.15$. From Table C5-4, $R_{max} = 0.6$ from linear interpolation.

    Therefore,

    $$F = \frac{\pi \times 1{,}000 \times 12.0 \times 1.0 \times 0.8 \times 1.0 \times 1.0 \times 0.6}{2 \times 32.2 \times 0.03} = 9{,}366 \text{ lb}$$

    This load acts at the stillwater flood elevation.

2. Special impact loads

    ASCE/SEI Equation C5-4 is used to determine special impact loads:

    $$F = \frac{C_D \rho A V^2}{2}$$

    Using a drag coefficient $C_D = 1.0$ and assuming a projected area of debris accumulation $A = 0.67 \times 4.5 = 3.0$ square feet, the impact force, F, on one pile is

    $$F = \frac{1 \times \left(\frac{64.0}{32.2}\right) \times 3.0 \times 12.0^2}{2} = 429 \text{ lb}$$

    This load acts at the stillwater flood level.

Step 5: Determine flood load, F_a.

For the analysis of the entire (global) foundation with 16 piles, it is reasonable to assume that one of the piles or columns will be subjected to the impact load, F, and the remaining piles or columns will be subjected to the hydrodynamic load, F_{dyn}; thus, the total flood load is $F_a = 9{,}366 + (15 \times 515) = 17{,}091$ lb.

For the case of an individual pile in the front row, F_a is equal to the larger of F_{dyn} and F_D plus F: $F_a = 515 + 9{,}366 = 9{,}881$ lb.

The design and construction of the foundation must satisfy the requirements of Section 4.5 of ASCE/SEI 24. Requirements for pile foundations are given in Section 4.5.5.

The design and construction of the reinforced concrete piles must satisfy the requirements of ACI 318-11 (Section 4.5.5.8 of ASCE/SEI 24 references ACI 318-02). Additional design provisions are given in Section 4.5.6 of ASCE/SEI 24.

7.6 References

7.1. American Society of Civil Engineers (ASCE). 2005. *Flood Resistant Design and Construction*, ASCE 24-05. Reston, VA.

7.2. Structures & Code Institute. 2011. *CodeMaster – Flood Resistant Design*. S.K. Ghosh Associates, Palatine, IL.

7.3. Federal Emergency Management Agency. 2001. *Crawlspace Construction for Buildings Located in Special Flood Hazard Areas*, FEMA/FIA-TB-11. Washington, DC.

7.4. Federal Emergency Management Agency. 2011. *Coastal Construction Manual*, FEMA P-55, Washington, DC.

7.5. Walton, T. L., Jr., Ahrens, J. P., Truitt, C. L., and Dean, R. G. 1989. *Criteria for Evaluating Coastal Flood Protection Structures,* Technical Report CERC 89-15. U.S. Army Corps of Engineers, Waterways Experiment Station.

7.6. U.S. Army Corps of Engineers. 1995. *Flood Proofing Regulations,* EP 1165-2-314. Office of the Chief of Engineers.

7.7 Problems

7.1. A 15-foot tall, vertical reinforced concrete wall is located in a V Zone that supports the base of a Risk Category III structure. Determine the design flood loads on the wall assuming the following:

- Design stillwater elevation, d_s = 6 ft-0 in.
- Base flood elevation (BFE) = 10 ft-0 in.

The local jurisdiction has adopted a freeboard of 2 feet above the BFE.

7.2. For the reinforced concrete wall in Problem 7.1, determine the design flood loads on the wall assuming that the wall is nonvertical where the longitudinal axis of the wall is 5 degrees from vertical.

7.3. For the reinforced concrete wall in Problem 7.1, determine the design flood loads on the wall assuming that the wall is vertical and that the waves are obliquely incident to the wall where the horizontal angle between the direction of the waves and the vertical surface is 30 degrees.

7.4. A 125 by 200 foot commercial building is supported by 81 circular reinforced concrete piles that have a diameter of 12 inches each. The site has been categorized as Zone AE with a 100-year stillwater elevation of 10.5 feet NGVD. The lowest ground elevation at the site is 6.0 feet NGVD and a geotechnical report has indicated that 8 inches of erosion is expected to occur during the design flood. The local jurisdiction has not mandated any additional freeboard. Determine the design flood loads on an individual pile.

7.5. For the design data given in Problem 7.4, determine the design floods on an individual pile assuming the building is designated an emergency shelter.

CHAPTER 8
Load Paths

8.1 Introduction

In order to properly ensure the strength and stability of a building as a whole and all of its members, it is important to understand the paths that loads take through a structure. A continuous load path is required from the top of a building or structure to the foundation and into the ground. Structural systems must be capable of providing a clearly defined and uninterrupted load path for the effects due to both gravity and lateral loads.

Section 6.6 of *Design and Construction Guidance for Community Safe Rooms* (Reference 8.1) provides the following analogy on a continuous load path:

> A continuous load path can be thought of as a "chain" running through a building. The "links" of the chain are structural members, connections between members, and any fasteners used in the connections (e.g., nails, screws, bolts, welds, or reinforcing steel). To be effective, each "link" in the continuous load path must be strong enough to transfer loads without permanently deforming or breaking. Because all applied loads (e.g., gravity, dead, live, uplift, lateral) must be transferred into the ground, the load path must continue unbroken from the uppermost building element through the foundation and into the ground.

A continuous load path is essential in order for a building or structure to perform as intended when subjected to the code-prescribed loads; this helps ensure the life safety of its occupants. The general requirements for providing a complete load path that is capable of transferring all loads from their point of origin to the resisting elements are found in IBC 1604.4 and ASCE/SEI 1.3.5. ASCE/SEI 12.1.3 contains specific requirements on continuous load path and interconnection for building structures subject to seismic loads.

In general, the path that loads take through a structure includes the structural members, the connections between the members, and the interface between the foundation and the soil. Each element in the path must be designed and detailed to resist the applicable combinations of gravity and lateral load effects. The level of detailing that is needed to ensure a complete load path depends to some extent on the material and type of structural system involved. For example, the load path in wood frame structures consists of many individual elements; as such, considerable detailing is often required to guarantee that lateral forces are transferred from the out-of-plane walls, through the diaphragm and collectors, and into the supporting shear walls and foundation. Additionally, the structure as a whole must be capable of resisting the effects of overturning and sliding.

The following sections cover typical load paths for both gravity and lateral loads. It is important to keep in mind that even though every building or structure must contain a gravity and lateral-force-resisting system, the two systems are usually interconnected; that is, some or even all of the structural members in a structure can be part of both systems.

8.2 Load Paths for Gravity Loads

The path that gravity loads travel through a conventional building or structure (that is, a building or structure consisting of floor and roof slabs, beams, columns, walls and foundations) is usually straightforward and depends on the type of structural members that are present in the structure. In conventional building construction, a majority of the gravity loads (superimposed dead loads, live loads, rain loads and snow loads) is initially supported by a floor or roof deck (diaphragm), which is usually a concrete slab, a steel metal deck, wood sheathing or any combination thereof. The loads are then transferred to the structural members that support the floor or roof deck. Individual structural members are designed to support their own self-weight and other applicable gravity loads, and all of these loads are transferred to their supporting members. Ultimately, the gravity loads are transmitted to the soil supporting the foundation system.

Crane loads and elevator loads are examples of gravity loads that are not transmitted first to a floor or roof deck. In such cases, the loads are initially transferred to beams (see Figure 3.7 of this publication for an illustration of a typical beam that supports a crane). Like conventional structures, the gravity loads are eventually transferred to the foundations and the adjoining soil.

Figures 8.1 through 8.4 illustrate the general gravity load paths in structures containing conventional framing. These figures are not meant to be comprehensive; their purpose is to exemplify the common paths for gravity loads in such structures.

Shown in Figure 8.1 is a generic structure consisting of a roof or floor slab that is supported by walls and spread footings. Gravity loads are initially supported by the floor or roof slab and are then transferred to the walls, which in turn transfer them to the footings. The ground beneath the footings ultimately supports these loads and the footings must be properly sized based on the properties of the soil.

Figure 8.1 Gravity Load Path—Conventional Structure with Slab, Walls and Footings

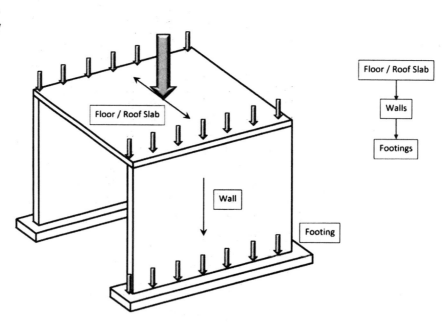

The structure depicted in Figure 8.2 is similar to the one in Figure 8.1 except that the floor or roof slab is supported directly on beams. The gravity loads from the slab are transferred to the beams, which subsequently transfer them to the walls. Like the structure in Figure 8.1, the walls transfer the loads to the footings and into the ground.

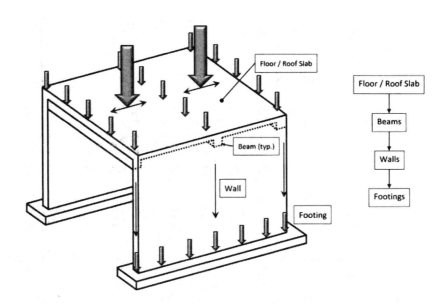

Figure 8.2 Gravity Load Path—Conventional Structure with Slab, Beams, Walls and Footings

Beams are often supported directly by columns in conventional structures. This situation is illustrated in Figure 8.3. The beams are supported by walls on one end and by individual columns at the other end.

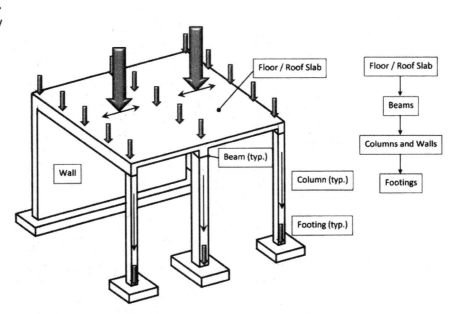

Figure 8.3 Gravity Load Path—Conventional Structure with Slab, Beams, Columns, Walls and Footings

Another common type of conventional framing utilizes a slab, beams, girders and columns (see Figure 8.4; the footings below two of the columns are not shown for clarity). The gravity loads from the slab are transferred to the beams, which are supported at both ends by girders. The girders in turn transfer the loads to the columns (or walls), which are supported by footings.

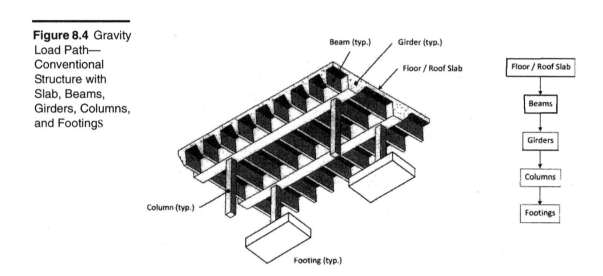

Figure 8.4 Gravity Load Path—Conventional Structure with Slab, Beams, Girders, Columns, and Footings

Although the structures in Figures 8.1 through 8.4 consist of only one story and one bay, the same load paths described above are applicable to multistory and multibay structures. In a multistory building, the gravity loads accumulate over the height of the structure from the top to the bottom. The walls and/or columns in the lowest level must be designed to support the sum of the applicable gravity loads from the roof through the first level.

In certain situations, there may be a need to discontinue one or more columns in a conventionally-framed structure. One common example of this occurs in office buildings where a column-free lobby is desired. In such cases, a transfer beam (or girder) is used to transfer the gravity loads from the column(s) above to one or more columns below (see Figure 8.5). The transfer beam and all of the connections to the columns must be designed to support the gravity loads from above.

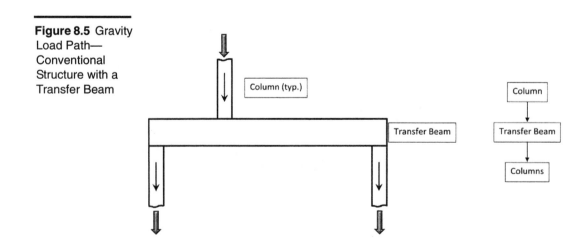

Figure 8.5 Gravity Load Path—Conventional Structure with a Transfer Beam

As noted previously, the connections between the structural members in any structural system must be adequately designed and detailed to ensure a complete load path. For example, the interfaces between the slab and the walls and the walls and the footings in the structure depicted in Figure 8.1 must be designed and detailed for the applicable gravity load effects obtained from the structural analysis at the respective locations. The same is true for all of the other connections in the systems shown in the other figures.

In nonconventional buildings or structures, the gravity load paths may not be as apparent as those described above for conventional ones. For example, in a cable-stayed roof, the gravity loads are transferred from the roof structure to cables, which are connected to masts (see Figure 8.6). The masts subsequently transmit the loads to columns, walls or trusses, for example, which are supported by a foundation system. Tension and shell structures are other types of nonconventional systems where the paths for gravity loads may not be as obvious as those in more conventional framing systems.

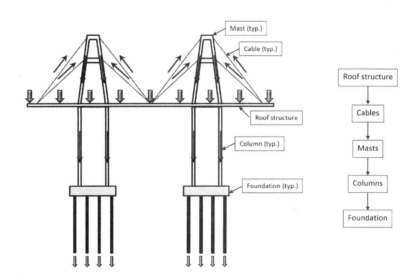

Figure 8.6 Gravity Load Path—Nonconventional Structure with Roof Structure, Cables, Masts, Columns and Foundations

8.3 Load Paths for Lateral Loads

8.3.1 Overview

A continuous load path for lateral loads must be provided for any building or structure. Like in the case of gravity loads, the load path for lateral loads begins at the point of application of the loads. The lateral loads travel through the structural members (and connections) of the lateral-force-resisting system (LFRS) and eventually find their way into the soil surrounding the foundations.

The types of lateral loads that are commonly encountered in the design of building structures are caused by wind, earthquakes and soil pressure. As discussed in Chapter 5 of this publication, the effects of wind must be considered in the design of any above-ground structure; the effects from earthquakes and soil pressure are considered only where applicable. The focus of this section is on load paths for wind and earthquakes.

The paths that lateral loads travel through a building or structure depend on the type of LFRS that has been employed. The LFRS that resists wind loads is referred to as the main windforce-resisting system (MWFRS) as defined in ASCE/SEI 26.2, and that which resists seismic loads is denoted the seismic-force-resisting system (SFRS) as defined in ASCE/SEI 11.2. In either case, the LFRS is the assemblage of structural elements that has been specifically designed to resist the effects from wind or seismic loads. Descriptions of common LFRSs are given below; included in the descriptions are the elements that resist the vertical and lateral loads.

- **Bearing wall system.** A bearing wall system is a structural system where bearing walls support all or a major portion of the vertical loads. Some or all of the walls also provide resistance to the lateral loads (see Figure 8.7).

Figure 8.7
Bearing Wall System

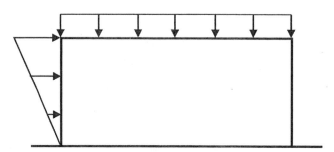

- **Building frame system.** In a building frame system, an essentially complete space frame provides support for the vertical loads. Shear walls or braced frames provide resistance to the lateral loads (see Figure 8.8, which illustrates this system with shear walls).

Figure 8.8
Building Frame System

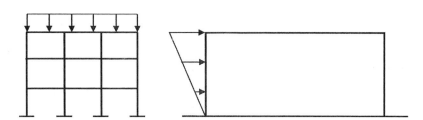

- **Moment-resisting frame system.** A moment-resisting frame system is a structural system with an essentially complete space frame that supports the vertical loads. Lateral loads are resisted primarily by flexural action of the frame members through the joints. The entire space frame or selected portions of the frame may be designated as the LFRS (see Figure 8.9).

Figure 8.9
Moment-resisting Frame System

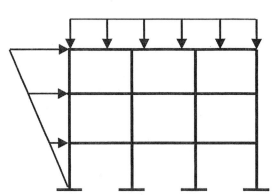

- **Dual system.** In a dual system, an essentially complete space frame provides support for the vertical loads. Moment-resisting frames and shear walls or braced frames provide resistance to lateral loads in accordance with their rigidities. Dual systems are also referred to as shear wall-frame interactive systems (see Figure 8.10, which illustrates this system with shear walls).

Figure 8.10 Dual Frame System

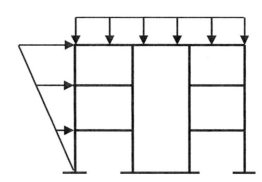

- **Cantilever column system.** In this system, the vertical loads and the lateral loads are resisted entirely by columns acting as cantilevers from their base. This system is usually used in one-story buildings or in the top story of a multistory building.

Information on the specific load paths for wind and seismic loads are covered in the next sections.

8.3.2 Load Paths for Wind Loads

Overview

Figure 8.11 illustrates the continuous path for wind loads through a one-story building with bearing walls. The load path begins at the windward wall: this wall receives the wind pressures and transfers the resulting loads to the diaphragm, which laterally supports the wall at the roof. The wind loads are then transferred from the diaphragm to the walls that are parallel to the direction of wind. These walls, in turn, transfer the wind loads to the spread footing foundation, which is supported by the soil beneath the footing.

Figure 8.11 Load Path for Wind Loads in a One-story Bearing Wall Building

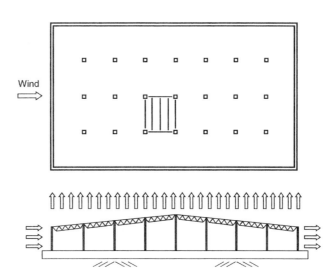

The same sequence of events would occur for a bearing wall building with more than one story (see Figure 8.12) or if moment frames or any other type of LFRS were used instead of or in conjunction with the bearing walls.

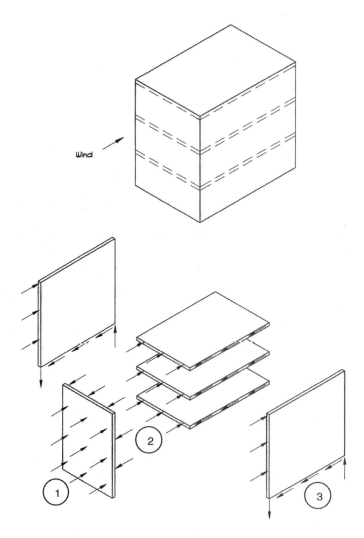

Figure 8.12 Load Path for Wind Loads in a Multistory Bearing Wall Building

1: Wind pressure on windward wall.
2: Diaphragms receive wind loads from the windward wall and distribute them to the walls parallel to the wind pressure.
3: Walls parallel to the wind pressure receive the wind loads from the diaphragms and transfer them to the foundations.

The windward wall (and/or windows and doors) that initially receives the wind pressure is generally perpendicular to the direction of wind. This wall is classified as components and cladding (C&C) for wind in this direction, and the wind pressure on it is determined differently than that on the MWFRS, as shown in Chapter 5 of this publication. For wind in the perpendicular direction, this wall would be part of the MWFRS of the building.

The vertical upward loads on the roof due to wind shown in Figure 8.11 are resisted by the walls and columns and must be properly combined with the applicable gravity load effects that occur on these members. If the wind load effects are large enough, a net upward load could occur on the vertical members and the foundations.

A similar scenario occurs for a wood-frame building (see Figure 8.13). Wind pressure on the windward wall is transferred to the roof diaphragm, which in turn, transfers it to the shear walls parallel to the direction of wind via boundary elements and collectors. Wind loads are then transferred from the shear walls to the reinforced concrete foundation walls via fasteners. Reinforcement in the foundation walls transmits the loads to the concrete footings, which then transfers them to the surrounding soil.

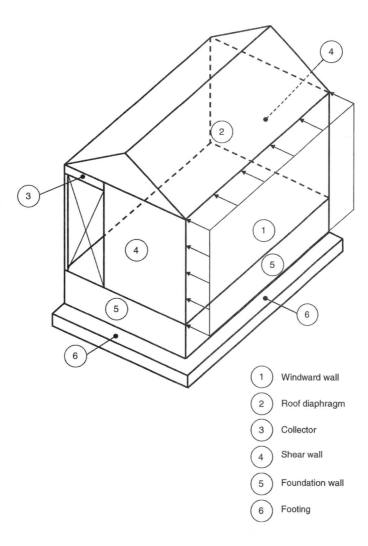

Figure 8.13
Load Path for Wind Loads in a Wood Building

1. Windward wall
2. Roof diaphragm
3. Collector
4. Shear wall
5. Foundation wall
6. Footing

The following sections cover how wind loads propagate through diaphragms and collectors, both of which are essential elements in the lateral load path for any building or structure.

Diaphragms

Overview

Diaphragms are defined in IBC 202 as a horizontal or sloped system that transmits lateral forces to the elements of the LFRS. Horizontal bracing systems are included in this definition. In addition to providing global stability against the effects from lateral loads, diaphragms are commonly the initial point in the load path for gravity loads (see Section 8.2 of this publication). Overall behavior of diaphragms that are subjected to wind loads is discussed below. Additional information can be found in *Guide to the Design of Diaphragms, Chords and Collectors Based on the* 2006 *IBC and ASCE/SEI* 7-05 (Reference 8.2).

Typical diaphragms in building structures include wood sheathing, corrugated metal deck, concrete fill over corrugated metal deck, concrete topping slab over precast concrete planks and cast-in-place concrete slab.

Diaphragm Behavior

As discussed above, wind loads from the windward wall are transferred to diaphragms, which essentially act as beams that span between the walls that are oriented parallel to the direction of wind. Figure 8.14 illustrates wind load propagation at the roof level of the one-story bearing wall building in Figure 8.11 for wind loads in the direction perpendicular to that shown in Figure 8.11.

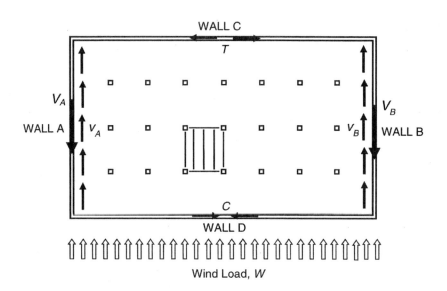

Figure 8.14 Wind Load Propagation from a Diaphragm to the MWFRS

Wall D is the windward wall, which receives the wind load, W. The roof diaphragm, which behaves essentially as a deep flexural member that spans between Walls A and B that are oriented parallel to the direction of the wind, laterally supports Wall D (and Wall C) and receives the wind load from this wall along its windward edge. The web element of the beam transfers the wind loads (shear forces v_A and v_B) along the length of the diaphragm to the supports at Walls A and B. The sum of the forces V_A and V_B at Walls A and B, respectively, is equal to the total wind load, W, at this level.

Note that the roof diaphragm in this case must also resist the negative wind pressure on the roof and the out-of-plane lateral loads caused by negative wind pressure on the leeward and side walls.

Chords behave as flange elements that are at the edges of the diaphragm perpendicular to the wind loads. These elements must be designed to resist the axial tension and compression forces, T and C, respectively, which result from flexural behavior of the diaphragm (see Figure 8.14).

Openings in a diaphragm can have a dramatic effect on the flow of loads through the diaphragm to the MWFRS. Typical elevator and stair openings in a diaphragm will usually not have a major impact on the overall behavior and load distribution. Larger openings can have a significant influence on overall behavior and a more refined analysis is generally warranted to ensure that a complete load path is achieved around the opening. Regardless of the opening size, secondary chord forces will occur due to local bending of the diaphragm segments on either side of the opening; these chord forces need to be accounted for in the design (see Figure 6.27 of this publication).

Diaphragm Flexibility

The manner in which the wind loads are transferred from the diaphragm to the elements of the MWFRS depends on the stiffness of the diaphragm in relation to the vertical-resisting elements. Diaphragms can usually be idealized as either flexible or rigid. According to ASCE/SEI 27.5.4, diaphragms that are constructed of wood panels can be classified as flexible while those constructed of untopped metal deck, concrete-filled metal deck and concrete slabs can be idealized as rigid provided the span-to-depth ratio of the diaphragm is less than or equal to 2.0 for consideration of wind loading. It is possible for a diaphragm to be idealized differently depending on the direction of analysis.

Although the provisions in ASCE/SEI 12.3.1.3 are applicable to seismic loads, they may be used to determine diaphragm flexibility for the case of wind loads as well (see Section 6.3.3 of this publication for more information). Requirements in that section are based on the maximum diaphragm deflection and the average drift of the vertical elements of the MWFRS. If the MWFRS consists of a relatively flexible system, such as a moment frame, the diaphragm may behave as a rigid element, whereas if the MWFRS consists of stiffer elements, such as concrete or masonry walls, the diaphragm may behave as a flexible element. In particular, a diaphragm may be considered flexible where the maximum in-plane deflection of the diaphragm under lateral load is more than two times the average drift of the adjoining vertical elements of the MWFRS (see ASCE/SEI Figure 12.3-1).

It is commonly assumed that flexible diaphragms transfer the wind loads to the elements of the MWFRS by a lateral tributary area basis without any participation of the walls that are perpendicular to the direction of analysis. Thus, assuming the building in Figures 8.11 and 8.14 has a flexible diaphragm, Walls A and B would each carry one half of the total wind load, W, at the roof level.

In the case of rigid diaphragms, wind loads are transferred based on the relative stiffness of the members in the MWFRS considering all such elements at that level in both directions: the stiffer the element, the greater the load it must resist. Torsional effects must also be considered in rigid diaphragms. Torsion occurs where the center of rigidity of the elements of the MWFRS and the centroid of the wind pressure are at different locations. Elements of the MWFRS that are perpendicular to the wind load also resist the effects from torsion. In any situation, a finite element model of the system can be utilized to determine the loads in the diaphragm and in the elements of the MWFRS.

Load Paths for Wind

To ensure a continuous load path, the connections between the diaphragm and the elements of the MWFRS must be properly designed and detailed. The actual connection details depend on the types of materials that are used to construct these systems.

Illustrated in Figure 8.15 is the load path for the building illustrated in Figures 8.11 and 8.14, and it assumes that the diaphragm is a corrugated metal deck and that the walls of the MWFRS are reinforced concrete masonry units. The load path from the diaphragm to the foundation is as follows:

1. The shear force from the diaphragm (v_A or v_B) is transferred to a continuous angle via a welded connection between these elements.

2. The angle transfers the load to the top chord of the joists, which is welded to the base plate that transfers the horizontal and vertical loads to the bond beam through the headed shear studs.

3. Vertical reinforcement in the wall transfers the loads through the load-bearing masonry wall, which is the MWFRS in this direction.

4. At the base of the wall, the vertical reinforcement is spliced to reinforcement in the reinforced concrete footing that transfers the loads to the soil.

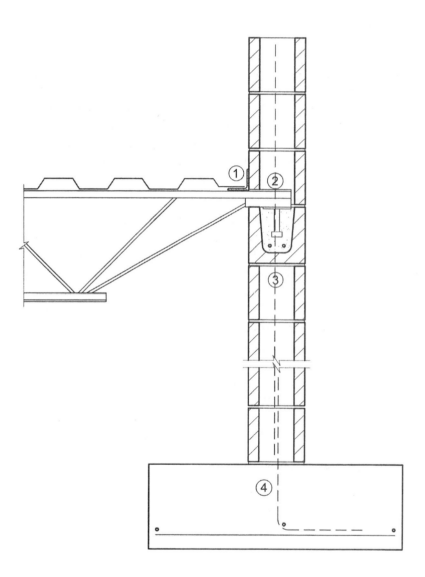

Figure 8.15
Lateral Load Path from Roof to Foundation in a One-story Bearing Wall Building—MWFRS

The connections must be designed for the appropriate load effect combinations to ensure a continuous load path through the system. Also, the individual deck sheets must be adequately fastened together at the seams; the design of the seam fasteners is important to ensure that the shear in the diaphragm is transferred as intended. Additional information on the design of steel deck diaphragms can be found in Reference 8.3 or from steel deck manufacturers. A similar load path would be applicable if reinforced concrete walls were utilized instead of reinforced masonry walls.

Figure 8.16 shows a typical section through Wall D (opposite hand Wall C) of the one-story bearing wall building in Figure 8.14. The load path from the diaphragm to the masonry wall is as follows:

1. The chord forces that develop along the edges of the diaphragm are transferred to a continuous angle via a welded connection between these elements.

2. The angle transfers the load to the masonry bond beam by a connector that is spaced along the length of the wall.

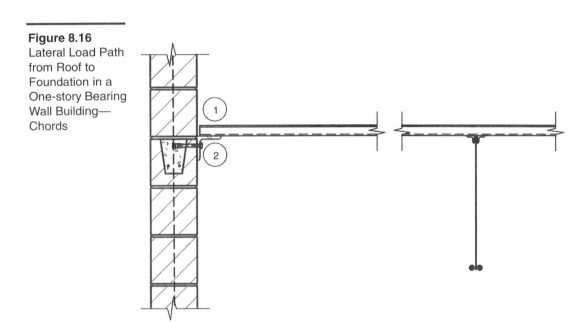

Figure 8.16
Lateral Load Path from Roof to Foundation in a One-story Bearing Wall Building—Chords

Once the load is transferred into the wall, it follows the same path into the foundation as described in Items 3 and 4 above except that in this case, the wall is essentially not load-bearing and is not part of the MWFRS for wind in that direction.

The continuous angle must be designed to resist the combined loads from gravity and the axial tension or compression load from wind. Similarly, the connectors must be designed for the combined effects due to gravity loads, the shear force from the diaphragm due to wind and the out-of-plane tension force due to negative wind pressure on the wall. It is important to note that for wind in the perpendicular direction, this assemblage must resist the diaphragm shear forces that are transferred into Walls C and D, which are now part of the MWFRS. In general, all members and connections must be designed to resist the critical combined effects produced by wind in either direction.

The load paths described above are essentially the same for bearing wall buildings with diaphragms constructed of concrete-filled metal deck, concrete topping slab over precast concrete planks and cast-in-place concrete slab. In all of these cases, dowel bars that are regularly spaced along the length of the masonry or concrete walls are used to transfer the shear forces from the diaphragm into the wall. Figure 8.17 shows the dowel bars for a cast-in-place reinforced concrete system consisting of a slab and wall.

Figure 8.17
Lateral Load Path at the Interface between a Reinforced Concrete Diaphragm and Wall

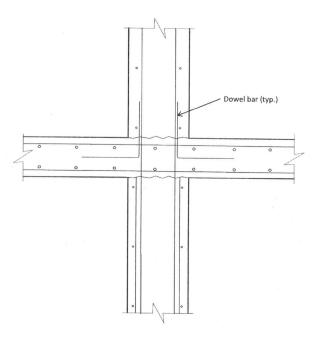

The wind load paths described above for bearing wall systems are fundamentally the same for building frame and dual systems with walls or braces. Regardless of the system or the number of stories in the building, the path begins at the windward wall, continues through the diaphragm to the members of the MWFRS and to the foundation, and ends in the soil surrounding the foundation. All of the members and connections along this path must be properly designed for the most critical combined load effects due to gravity and wind loads.

For moment-resisting frames, the diaphragm must be adequately connected to the beams and columns that form the MWFRS. In a cast-in-place reinforced concrete system, continuity is achieved by providing dowel bars between the diaphragm (reinforced concrete slab) and the beams that support it. Reinforcement that connects the beams to the columns and the columns to the foundations is provided, which completes the load path. Since cast-in-place concrete systems are inherently redundant, a continuous load path for wind loads is usually achieved rather easily.

In the case of a concrete-filled composite steel deck supported by structural steel beams, the concrete diaphragm is usually connected to the steel beams via steel headed stud connectors that are spaced along the length of the beams. These stud connectors, which provide the composite action between the steel beams and the concrete in the metal deck, must be designed for the combined shear flow effects due to gravity and wind loads. The commentary of *Specifications for Structural Steel Buildings* (Reference 8.4) contains information on how to design such connectors for these combined effects.

The load path in wood construction begins at the windward wall, as in all other types of construction, but follows a much more complex path to the soil surrounding the foundation. In general, wind loads are transferred from the exterior out-of-plane walls to diaphragms, which subsequently transfer them to shear walls or frames and then to foundations. The connections between the various elements in the load path typically consist of nails, staples, bolts, straps and other types of steel hardware. Many of these connections are designed and detailed to transmit the effects from lateral loads only (that is, they are not designed to support gravity loads).

Load transfer in a wood diaphragm begins at the sheathing, which is fastened to the top surface of a wood-framed floor or roof usually by nails or staples. The diaphragm itself consists of the sheathing, boundary members and intermediate framing that act as stiffeners. The sheathing provides the required shear capacity to transmit the wind loads to the members of the MWFRS, which typically consist of wood shear walls or frame members. Shear capacity of the diaphragm depends on the panel thickness of the sheathing, size of framing members, the size and spacing of the fasteners that attach the edges of the sheathing to the framing members below, the sheathing layout pattern and the presence of wood blocking below the edges of the sheathing.

Figure 8.18 illustrates the case of wood sheathing supported by floor or roof joists where the short sides of the individual standard 4 by 8 foot wood structural panels are staggered between adjacent rows (this layout results in larger overall shear capacity than one where continuous panel joints are provided in both directions). The fastening pattern shown in the figure corresponds to an unblocked diaphragm, that is, a diaphragm where two of the panel edges are not supported by framing.

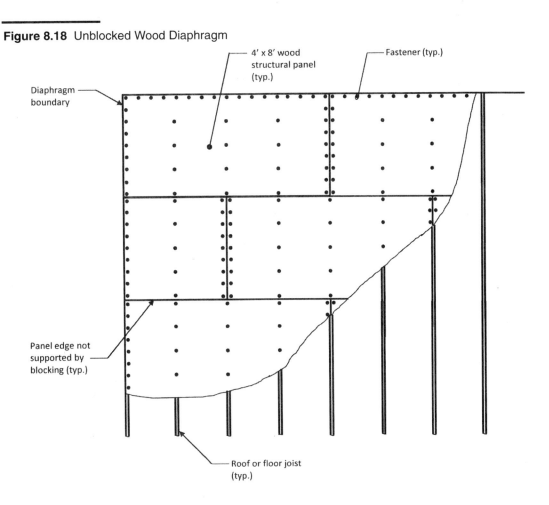

Figure 8.18 Unblocked Wood Diaphragm

Figure 8.19 depicts a blocked diaphragm where all four edges of a panel are supported by framing. In such cases, wood blocking is provided between the joists at the edges of the continuous panel joints and the sheathing is fastened to the blocking. Since a blocked diaphragm contains more fasteners and all panel edges are edge nailed, the overall shear capacity is greater than a similar unblocked diaphragm. Connection requirements can be found in IBC Chapter 23, which also references the 2008 edition of *Special Design Provisions for Wind and Seismic with Commentary* (Reference 8.5).

522 Chapter 8

Figure 8.19 Blocked Wood Diaphragm

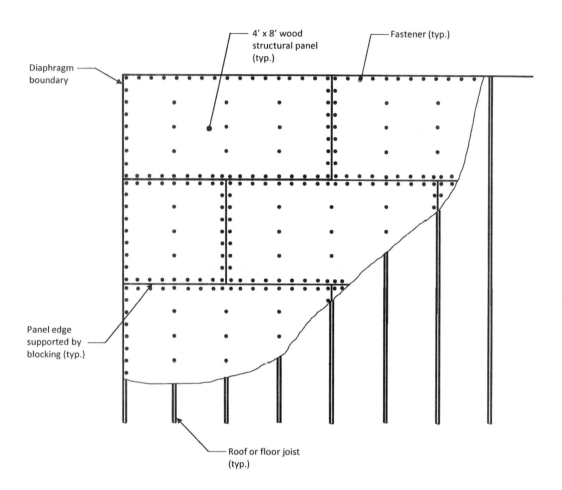

The sheathing fasteners that are along the edges of the diaphragm boundary transfer the wind loads directly to shear walls or frames or to collectors (see Figure 8.13 and the next section for a discussion of load transfer through collectors). Generally, the boundary edge framing member (edge beam, rim joist or blocking) that is attached to the sheathing from below must also be attached to the top plate of the shear wall or, similarly, to a wood nailer that is attached to the top of a beam that is part of a frame.

Illustrated in Figure 8.20 is a typical connection of a rim joist (or blocking) to the top plate of a shear wall. The wind load path in the plane of the walls is as follows:

1. The wind load is transferred from the sheathing to the rim joist via a fastener (the fasteners shown in the figure are nails).

2. The rim joist transfers the wind load to the clip angle.

3. The clip angle transfers the wind load to the top plate of the shear wall below, which is attached to the wall sheathing.

Figure 8.20 Connection of Wood Floor Diaphragm to Wood Shear Wall

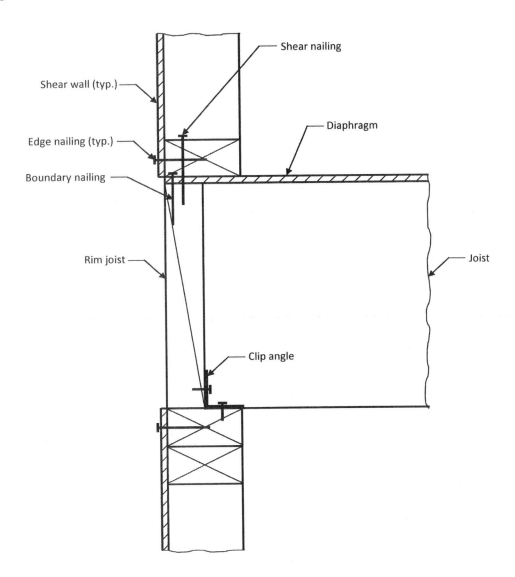

At the base of the shear wall, the sheathing is attached to a foundation plate, which is anchored into a concrete footing. This completes the wind load path. Additional information and details on the overall load path are given below.

Transfer of the chord forces at the ends of the diaphragm must also be considered. In a typical wood-frame building with stud walls, the top plates are usually designed as the chord member (see Figure 8.21). Unless the dimensions of a building are very small, the top plate members are not continuous and, thus, must be spliced together. At least two plates are provided so that the splice in one plate can be staggered with respect to the splice in the other plate(s). In this way, a continuous chord is provided with at least one member being effective at any given point along the chord. Nails can be used to connect the members when chord forces are relatively small. Screws, bolts or steel straps are required for relatively large chord forces; in such cases, more than two plates may be necessary.

Figure 8.21 Transfer of Chord Forces in a Wood-frame Building

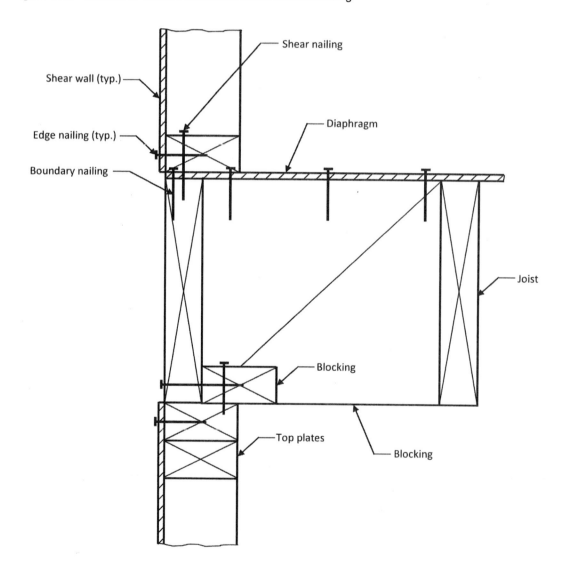

Figure 8.22 shows a typical detail where masonry walls are used in conjunction with horizontal wood framing (a similar detail would be applicable to a cast-in-place, reinforced concrete wall). In such cases, it is usually assumed that the chord forces are resisted by the horizontal reinforcement in the wall at or near the level of the diaphragm. Although it is possible to design the wood ledger to function as the chord member, it is preferable to use the horizontal reinforcement in the wall since the wall is much stiffer than the ledger. It should be noted that this detail shows in-plane shear transfer and not out-of-plane anchorage of the masonry wall to the diaphragm to resist seismic forces (see Section 8.3.3 for more information on load paths for seismic forces).

Figure 8.22 Transfer of Chord Forces in a Wood-frame Building with Masonry Walls

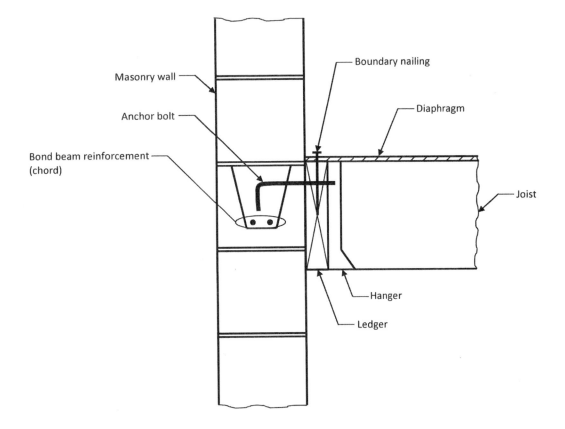

Note: Out-of-plane wall anchorage connections not shown

The complete lateral load path for a two-story wood-frame building is depicted in Figure 8.23, which is given in *Seismic Detailing Examples for Engineered Light-frame Timber Construction* (Reference 8.6). (Note: this lateral load path is applicable to wind loads even though it was originally created for seismic loads.) The load path begins at the roof and floor diaphragms and ends at the soil around the concrete footing. It is evident from the figure that the continuous load path is made of numerous members and connections. Each element and connection in the load path must be designed for the appropriate gravity and wind load combinations.

Figure 8.23 Continuous Lateral Load Path in a Wood-frame Building (from Reference 8.9; used with permission)

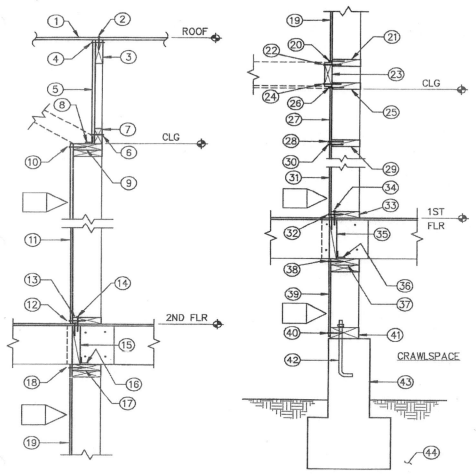

2.2 Load Path Example - for the sections on the following page:

Top of section
1. Roof panel
2. Nail (wood structural panel (panel) boundary nailing) BN
3. Top chord of truss
4. Nail (panel edge nailing) EN
5. Wood structural panel
6. Nail (panel edge nailing) EN
7. Bottom chord of truss
8. FA connector *
9. Upper top plate
10. Nail (panel edge nailing) EN
11. Wood structural panel
12. Nail (panel edge nailing) EN
13. Plate
14. Nail (shear nailing) SN
15. Blocking
16. FA connector *
17. Upper top plate
18. Nail (panel edge nailing) EN
19. Wood structural panel
20. Nail (panel edge nailing) EN
21. Blocking
22. Nail (ledger shear nailing) SN
23. Ledger or blocking
24. Nail (ledger shear nailing) SN
25. Blocking
26. Nail (panel edge nailing) EN
27. Wood structural panel
28. Nail (panel edge nailing) EN
29. Wood structural panel blocking
30. Nail (panel edge nailing) EN
31. Wood structural panel
32. Nail (panel edge nailing) EN
33. Plate
34. Nail (shear nailing) SN
35. Blocking
36. FA connector *
37. Upper top plate
38. Nail (panel edge nailing) EN
39. Wood structural panel
40. Nail (panel edge nailing) EN
41. Plate (foundation plate)
42. Anchor bolt
43. Footing
44. Earth
    Bottom of section (point of application)

* FA connector consists of a bent sheet metal framing connector with nails attaching it to the two the adjacent elements. Toe-nails many be used in lieu of the FA connector when loads are less than 150 pounds per linear foot (1996 SEAOC *Blue Book, Section 802.1.1*).

Collectors

Overview

A *collector* is defined in IBC 202 as a horizontal diaphragm element that is parallel and in line with the applied force that collects and transfers diaphragm shear forces to the vertical elements of the LFRS and/or distributes forces within the diaphragm.

Collectors are used in situations where, in the case of wind loads, the MWFRS does not extend the full length of the building in the direction of analysis (see Figure 8.24 for the case of collector beams with shear walls). As noted in the definition above, the main purpose of a collector is to collect the shear force from the diaphragm over the length where there are no elements of the MWFRS and to transfer this force to the MWFRS.

Figure 8.24
Collector Beams and Shear Walls

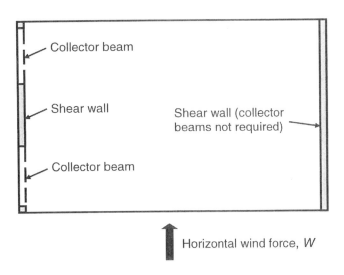

Load Paths for Wind

The load paths that were described previously for wind loads can be modified accordingly in cases where collectors are present. In general, the shear force from the diaphragm must be transferred to the collectors, which in turn, transfer them to the vertical elements of the MWFRS. The general paths before and after this link are the same as outlined above.

Collector forces are obtained from equilibrium. Figure 8.25 illustrates the forces along the line of the wall and collector beams in Figure 8.24. Assuming that the calculated shear force in the wall due to wind is equal to V_W, the unit shear force per length in the diaphragm is equal to $V_W/(\ell_1 + \ell_2 + \ell_3)$ and the unit shear force per length in the wall is equal to V_W/ℓ_2. The net shear forces per unit length are determined by subtracting the unit shear force in the diaphragm from that in the wall. The force at any point along the length of a collector is equal to the net shear force in that segment times the length to that point. For example, at point B the force in the collector is equal to $v_{AB,net}\ell_1$, and at point C it is $v_{CD,net}\ell_3$ or, equivalently, $v_{BC,net}\ell_2 - v_{AB,net}\ell_1$.

Figure 8.25 Unit Shear Forces, Net Shear Forces and Collector Force Diagrams

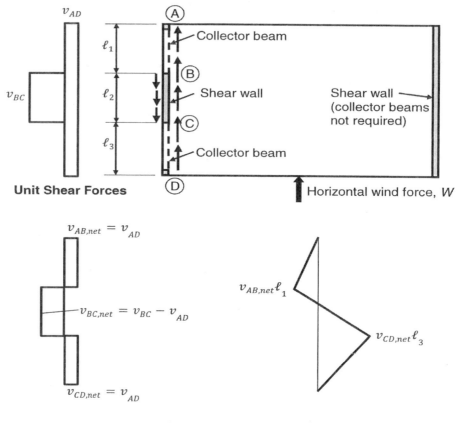

The collector beams must be designed for the appropriate gravity and wind load combinations where the axial force in the collector due to wind load effects is tension or compression. Connections between the diaphragm and the collector beams and between the collector beams and the elements of the MWFRS must also be properly designed and detailed to ensure a continuous load path between these members.

In cast-in-place concrete systems that utilize shear walls, reinforced concrete beams or a portion of the slab adjacent to the walls are typically designated as collectors. The reinforcement in the collectors must be adequately developed into the shear walls to guarantee continuity in this link of the wind load path.

In other types of construction, it is common to use structural steel members as collectors. Illustrated in Figure 8.26 is a roof system that is made up of a wood diaphragm supported on open-web joists, which in turn, are supported by steel columns and precast concrete walls. The precast walls are the MWFRS, and steel collector beams are utilized at the locations shown in the figure to transfer the loads into the walls.

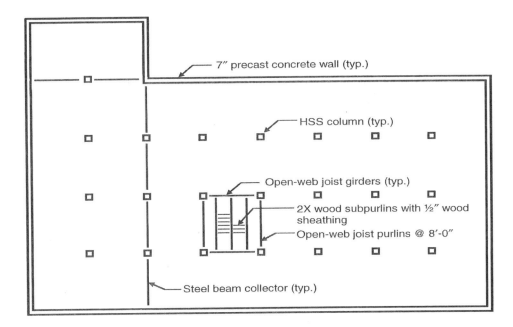

Figure 8.26 Structural Steel Collector Beams

Steel beams are also used as collectors in composite construction, that is, floor or roof systems that utilize a concrete-filled composite metal deck supported by structural steel beams. Commentary Section I7 of *Specifications for Structural Steel Buildings* (Reference 8.4) provides guidelines on the design of collectors and their connections in such cases. Figure C-17.1 in that reference illustrates the shear flow due to gravity and lateral loads at collector beams.

In wood-frame construction, the collector is usually the double top plate of the stud wall. In cases where the top plates are not continuous over openings, a beam or header is often used as the collector over the opening. Where reinforced concrete or masonry walls are present, a wood ledger that is bolted to the wall typically takes on the role of the collector (see Figure 8.27). At interior shear walls, the collector is most often a wood beam that is located in the plane of the floor or roof.

Figure 8.27 Wood Ledger Collector

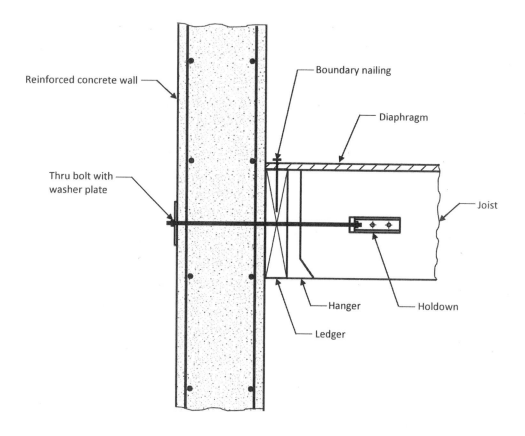

Illustrated in Figure 8.28 is a wood shear wall with openings on both sides. In this case, the shear forces are transferred from the diaphragm to the blocking between the roof joists and then into the collectors (double top plate on one side of the shear wall and the beam on the other side) and shear wall. These forces are then transferred from the shear wall into the reinforced concrete footing through anchor bolts and from the footing into the surrounding soil. All of the connections and members in this load path must be properly designed for the applicable combination of gravity and lateral forces. In lieu of a load path with blocking, one with a continuous rim joist acting as a collector can be used (see Figure 8.20).

Figure 8.28 Collectors in Wood-frame Construction

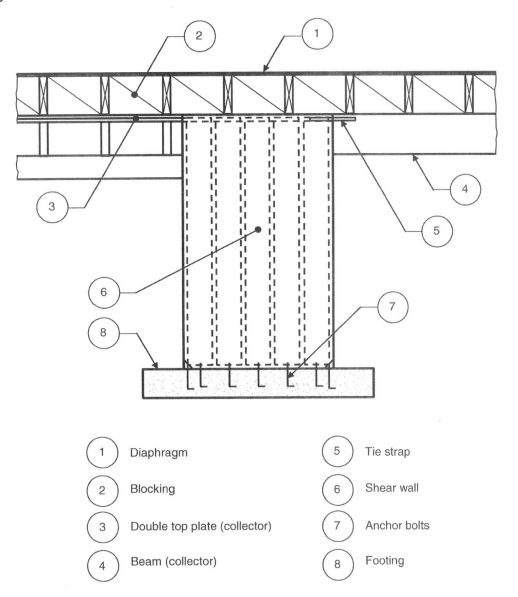

(1) Diaphragm
(2) Blocking
(3) Double top plate (collector)
(4) Beam (collector)
(5) Tie strap
(6) Shear wall
(7) Anchor bolts
(8) Footing

In-depth discussions on load paths for a variety of situations can be found in *The Analysis of Irregular Shaped Structures: Diaphragms and Shear Walls* (Reference 8.7).

8.3.3 Load Paths for Earthquake Loads

Overview

Unlike wind loads, the magnitude of which are proportional to the surface area of a building, earthquake loads are generated by the inertia of the building mass that counteracts ground motion. Although the loads are generated differently, the propagation of the loads through the building and into the ground is assumed to be essentially the same under particular conditions.

It is shown in Chapter 6 of this publication that for certain types of buildings the effects of complex seismic ground motion can be adequately represented by applying a set of static forces over the height of the building. These seismic loads, which are assumed to act at the center of mass at each level in the building or structure, are distributed to the elements of the seismic-force-resisting system in essentially the same way wind loads are distributed by considering the relative stiffness of the diaphragm.

Specific requirements are given in ASCE/SEI 12.1.3 for a continuous load path and interconnection of members in a structure subjected to seismic loads. In particular, all smaller portions of a structure must be tied to the remainder of the structure with elements that have a design strength capable of transmitting the greater of $0.133S_{DS}$ times the weight of the smaller portion or 5 percent of the portion's weight. The purpose of this requirement is to help ensure that a minimum amount of continuity is provided in a structure. In structures without redundant components to resist the seismic effects, the members in the load path must be designed to resist the seismic effects so that the integrity of the structure is preserved. In structures that have a higher degree of redundancy, one or more of the elements in the load path may fail, but other members are available to resist the seismic effects without loss of overall integrity.

A similar requirement that is applicable to connections to supporting elements is given in ASCE/SEI 12.1.4: every connection between each beam, girder or truss to its supporting elements, including diaphragms, must have a minimum design strength of 5 percent of the dead load plus live load reaction on that member. This includes the connections to the foundations.

Additional general information on seismic load paths can be found in *Briefing Paper 1—Building Safety and Earthquakes, Part D* (Reference 8.8).

Diaphragms and Collectors

Diaphragms and collectors that are subjected to seismic loads perform essentially the same roles as those that are subjected to wind loads (see Section 8.3.2).

Chapter 6 of this publication contains information on how to calculate seismic diaphragm design forces at roof and floor levels (see ASCE/SEI 12.10). Special design and detailing requirements must be satisfied for collector elements and their connections in buildings assigned to SDC C and above; similar load combinations are not applicable in the case of wind loads.

In the case of cast-in-place concrete construction, reinforcing bars are designed and detailed to transfer the seismic forces from the diaphragms and collectors into the vertical members of the seismic-force-resisting system (such as walls and columns). Such reinforcing bars must be adequately developed to ensure that load transfer occurs during the design earthquake. Additional information on cast-in-place concrete diaphragms, chords and collectors can be found in Reference 8.9.

The load paths that were described in the previous section for composite steel deck and concrete-filled diaphragms subjected wind loads are basically the same in the case of seismic loads. See Reference 8.10 for more information on these types of diaphragms.

In the case of wood-frame structures, the load path described in Section 8.3.2 for wind loads are essentially the same for seismic loads. As was discussed previously, the lateral load paths for wood structures commonly entail many more members and connections compared to those for other types of constructions. Additional information on the seismic response of wood-frame construction can be found in References 8.5 through 8.7 and 8.11 through 8.14.

8.4 References

8.1. Federal Emergency Management Agency. 2008. *Design and Construction Guidance for Community Safe Rooms*, FEMA 361. Washington, DC.

8.2. National Council of Structural Engineering Associations (NCSEA). 2009. *Guide to the Design of Diaphragms, Chords and Collectors Based on the 2006 IBC and ASCE/SEI 7-05*. International Code Council, Washington, DC.

8.3. Steel Deck Institute (SDI). 2004. *Diaphragm Design Manual*, 3rd Ed (DDM03). Steel Deck Institute, Fox River Grove, IL.

8.4. American Institute of Steel Construction (AISC). 2010. *Specifications for Structural Steel Buildings* (ANSI/AISC 360-10). American Institute of Steel Construction, Chicago, IL.

8.5. American Forest and Paper Association, Inc. (AF&PA). 2008. *Special Design Provisions for Wind and Seismic with Commentary*. American Forest and Paper Association, Inc., ANSI/AF&PA SDPWS-2008, Washington, DC.

8.6. Structural Engineers Association of California (SEAOC). 1997. *Seismic Detailing Examples for Engineered Light-frame Timber Construction*. SEAOC, Sacramento, CA.

8.7. Malone, R.T. and Rice, R.W. 2012. *The Analysis of Irregular Shaped Structures: Diaphragms and Shear Walls*. McGraw Hill, New York, NY.

8.8. Applied Technology Council and Structural Engineers Association of California (ATC/SEAOC). 1999. *Briefing Paper 1 – Building Safety and Earthquakes, Part D: The Seismic Load Path*, ATC/SEAOC Training Curriculum: The Path to Quality Seismic Design and Construction, ATC-48. Redwood City, CA.

8.9. Moehle, J.P., Hooper, J.D., Kelly, D.J., and Meyer T.R. 2010. "Seismic Design of Cast-in-place Concrete Diaphragms, Chords and Collectors: A Guide for Practicing Engineers," *NEHRP Seismic Design Technical Brief No. 3*, NEHRP Consultants Joint Venture (a partnership of the Applied Technology Council and the Consortium of Universities for Research in earthquake Engineering). National Institute of Standards and Technology, NIST GCR 10-917-4, Gaithersburg, MD.

8.10. Sabelli, R., Sabol, T.A., and Easterling, W.S. 2011. "Seismic Design of Composite Steel Deck and Concrete-filled Diaphragms: A Guide for Practicing Engineers," *NEHRP Seismic Design Technical Brief No. 5*, NEHRP Consultants Joint Venture (a partnership of the Applied Technology Council and the Consortium of Universities for Research in earthquake Engineering). National Institute of Standards and Technology, NIST GCR 11-917-10, Gaithersburg, MD.

8.11. Applied Technology Council and Structural Engineers Association of California (ATC/SEAOC). 1999. *Briefing Paper 3 – Seismic Response of Wood-Frame Construction, Part A: How Earthquakes Affect Wood Buildings*, ATC/SEAOC Training Curriculum: The Path to Quality Seismic Design and Construction, ATC-48. Redwood City, CA.

8.12. Applied Technology Council and Structural Engineers Association of California (ATC/SEAOC). 1999. *Briefing Paper 3 – Seismic Response of Wood-Frame Construction, Part B: The Role of Wood-Framed Diaphragms*, ATC/SEAOC Training Curriculum: The Path to Quality Seismic Design and Construction, ATC-48. Redwood City, CA.

8.13. Applied Technology Council and Structural Engineers Association of California (ATC/SEAOC). 1999. *Briefing Paper 3 – Seismic Response of Wood-Frame Construction, Part C: The Role of Wood-Framed Shear Walls*, ATC/SEAOC Training Curriculum: The Path to Quality Seismic Design and Construction, ATC-48. Redwood City, CA.

8.14. Breyer, Donald E., Fridley, Kenneth J., Cobeen, Kelly E., and Pollack, David G. 2007. *Design of Wood Structures–ASD/LRFD*. McGraw Hill, New York, NY.

The National Voice of Structural Engineering

Providing networking, information, education, and advocacy to practicing structural engineers

Who We Are

The **National Council of Structural Engineers Associations (NCSEA)** is comprised of 43 structural engineering associations throughout the United States. NCSEA serves to advance the practice of structural engineering and, as the autonomous national voice for practicing structural engineers, protect the public's right to safe, sustainable and cost effective buildings, bridges, and other structures. NCSEA generates and responds to code changes, promotes structural engineering certification and separate licensure, and promotes the practice of structural engineering to students and the general public. Members also include structural engineering firms as well as companies who provide structural engineering products and services.

What We Do

NCSEA products & services include:
- Education and networking opportunities;
- Monthly structural engineering webinars by nationally known speakers;
- Over 80 recorded webinars;
- STRUCTURE magazine;
- The NCSEA Annual Conference, which includes targeted, specific educational sessions, a trade show, and networking events;
- The Winter Leadership Forum, which gathers structural engineering executives for two days of high-level education;
- The NCSEA Structural Engineering Emergency Response program;
- The NCSEA Structural Engineering Exam online review course;
- The NCSEA Excellence in Structural Engineering Awards Program.
- ...and so much more!

For more information on NCSEA membership, programs and services, visit www.NCSEA.com, or call 312.236.4600.

National Council of Structural Engineers Associations • 645 North Michigan Avenue, Suite 540
Chicago, Illinois 60611 • 312.649.4600 • www.NCSEA.com

Not Currently an ASCE Member?

Make ASCE Your Professional Home Today — and Advance in Your Career!

Join ASCE at www.asce.org/join.
(Reference code 2013JOIN and receive $25 off of membership.)

For more than 160 years, civil engineers have relied on one place to support their life's work: ASCE. Join today and take full advantage of ALL the money-saving benefits and professional opportunities to grow in your civil engineering career.

▶ Leadership & Management Development
▶ Peer Networking
▶ Life-long Learning
▶ Issues & Advocacy

ASCE
AMERICAN SOCIETY OF CIVIL ENGINEERS